NUTRIENTS IN NATURAL WATERS
Herbert E. Allen and James R. Kramer, Editors

pH AND pION CONTROL IN PROCESS AND WASTE STREAMS
F. G. Shinskey

INTRODUCTION TO INSECT PEST MANAGEMENT
Robert L. Metcalf and William H. Luckman, Editors

OUR ACOUSTIC ENVIRONMENT
Frederick A. White

ENVIRONMENTAL DATA HANDLING
George B. Heaslip

THE MEASUREMENT OF AIRBORNE PARTICLES
Richard D. Cadle

ANALYSIS OF AIR POLLUTANTS
Peter O. Warner

ENVIRONMENTAL INDICES
Herbert Inhaber

URBAN COSTS OF CLIMATE MODIFICATION
Terry A. Ferrar, Editor

CHEMICAL CONTROL OF INSECT BEHAVIOR
THEORY AND APPLICATION
H. H. Shorey and John J. McKelvey, Jr.

MERCURY CONTAMINATION A HUMAN TRAGEDY
Patricia A. D'Itri and Frank M. D'Itri

POLLUTANTS AND HIGH RISK GROUPS
Edward J. Calabrese

SULFUR IN THE ENVIRONMENT, Parts I and II
Jerome O. Nriagu, Editor

COPPER IN THE ENVIRONMENT

Part II: Health Effects

COPPER IN THE ENVIRONMENT

Part II: Health Effects

Edited by

JEROME O. NRIAGU

Canada Centre for Inland Waters
Burlington, Ontario, Canada

A WILEY-INTERSCIENCE PUBLICATION

JOHN WILEY & SONS

New York • Chichester • Brisbane • Toronto

Library of Congress Cataloging in Publication Data

Main entry under title:

Copper in the environment.

(Environmental science and technology)
"A Wiley-Interscience publication."
Includes index.
CONTENTS: pt. 1. Ecological cycling.
1. Copper—Environmental aspects. I. Nriagu, Jerome O.

QH545.C6C66 574.5′2 79-15062
ISBN 0-471-04778-3

Printed in the United States of America

10 9 8 7 6 5 4 3 2 1

SERIES PREFACE

Environmental Science Technology

The Environmental Science and Technology Series of Monographs, Textbooks, and Advances is devoted to the study of the quality of the environment and to the technology of its conservation. Environmental science therefore relates to the chemical, physical, and biological changes in the environment through contamination or modification, to the physical nature and biological behavior of air, water, soil, food, and waste as they are affected by man's agricultural, industrial, and social activities, and to the application of science and technology to the control and improvement of environmental quality.

The deterioration of environmental quality, which began when man first collected into villages and utilized fire, has existed as a serious problem under the ever-increasing impacts of exponentially increasing population and of industrializing society. Environmental contamination of air, water, soil, and food has become a threat to the continued existence of many plant and animal communities of the ecosystem and may ultimately threaten the very survival of the human race.

It seems clear that if we are to preserve for future generations some semblance of the biological order of the world of the past and hope to improve on the deteriorating standards of urban public health, environmental science and technology must quickly come to play a dominant role in designing our social and industrial structure for tomorrow. Scientifically rigorous criteria of environmental quality must be developed. Based in part on these criteria, realistic standards must be established and our technological progress must be tailored to meet them. It is obvious that civilization will continue to require increasing amounts of fuel, transportation, industrial chemicals, fertilizers, pesticides, and countless other products; and that it will continue to produce waste products of all descriptions. What is urgently needed is a total systems approach to modern civilization through which the pooled talents of scientists and engineers, in cooperation with social scientists and the medical profession, can be focused on the development of order and equilibrium in the presently disparate segments of the human environment. Most of the skills and tools that are needed are already in existence. We surely have a right to hope a technology that has created such manifold environmental problems is also capable of solving them. It is our hope that this Series in Environmental Sciences and Technology will not only serve to make this challenge more explicit to the established professionals, but that it also will help to stimulate the student toward the career opportunities in this vital area.

Robert L. Metcalf
James N. Pitts, Jr.
Werner Stumm

PREFACE

The environmental awareness of the past decade or so has engendered some detailed studies of the behavior and effects of trace metals in various ecosystems. It is the intention of this book to provide a comprehensive picture of the current biological, chemical, geological, and clinical research pertaining to copper in the environment. While the focus is on copper, several authors have felt it necessary to relate their topics to the broad context of the trace metals in the environment. The transdisciplinary approach should thus be of interest to a wide spectrum of the scientific community; the broad aspects of copper toxicology covered in the volume should also make it invaluable to public health officials and the lay public.

Copper is a key element in our health and our life-styles. The wealth and power of many great empires of the past can be attributed largely to the possession of copper. Today we are no less dependent upon it, for copper plays a vital part in all branches of our engineering, science, and architecture. Furthermore, copper is one of the trace elements essential to the healthy life of many plants and animals, usually occurring as part of the prosthetic group of oxidizing enzymes. On the other hand, copper can be toxic in larger quantities, especially to the lower members of the human food chain. Obviously, a full coverage of all aspects of copper in the environment would fill a large number of volumes. The topics covered here represent areas of major interest, as well as areas in which new and interesting research is going on. The authors have been selected mostly on the basis of their direct knowledge and their working experience in the particular field.

The chapters have been divided into two parts. Part I includes reports on the sources, distribution, behavior, and flow of copper in the environment. Part II deals primarily with the biological, ecological, and health effects of copper. Some overlapping of material between chapters is inevitable in dealing with different environments in which diverse processes are closely interlinked. No attempt has been made to reconcile differing views expressed in some chapters; such controversial issues seem to reflect the uncertain state of present knowledge in a rapidly developing field.

It is a pleasure to acknowledge the excellent effort and cooperation of our distinguished group of contributors. Appreciation is also extended to Wiley-Interscience for invaluable editorial assistance.

JEROME O. NRIAGU

Burlington, Ontario, Canada
May 1979

CONTENTS
PART II

CONTENTS
PART I

COPPER IN THE ENVIRONMENT

Part II: Health Effects

1

ENVIRONMENTAL AND OCCUPATIONAL EXPOSURE TO COPPER

Steven R. Cohen

Department of Dermatology, Yale University School of Medicine, New Haven, Connecticut

1. INTRODUCTION

The aesthetic and practical qualities of copper have been recognized by artisans and crafts workers throughout the ages. In modern times the industrial applications of copper and its compounds have assumed an extraordinary diversity. The distinctive properties of ductility and conductivity of electricity and heat, as well as the toxicity of certain copper salts for microbial life forms, continue to reinforce the utility of this element in our technological society and to ensure its ubiquitous presence in our environment. This chapter focuses on some of the sources of copper in the environment, potential occupational exposures with their attendant health hazard implications, and the toxicity of copper and its compounds for human beings.

2. SOURCES OF COPPER IN THE ENVIRONMENT

Although an exhaustive catalogue of the uses of copper in home, science, and industry will not be compiled here, a limited overview may introduce the reader to the widespread sources of the element in our environment. More detailed commentary on many of the following subject areas can be found elsewhere in this volume.

Copper has been utilized for its malleability and heat conductivity for thousands of years. It is employed to the present day in a great variety of metal craft and sculptural works. As numerous alloys of copper have been discovered, the spectrum of copper usage has expanded to include the manufacture not only of bronze and brass but also of bell metal, gun metal, German silver, aluminum-bronze, phosphor-bronze, manganese-bronze, silicon-bronze, and beryllium-copper among others (Hunter, 1969).

The metal is employed also in castings, sheets, rods, tubing, wire, gas and water piping, roofing materials, cooking utensils, chemical and pharmaceutical equipment, and coins. As a conductor of electricity, metallic copper is used in all gauges of wire, circuitry, coil and armature windings, high-conductivity tubes, and many other applications (Tabershaw, 1977).

Compounds of copper, such as copper acetoarsenite and cupric arsenite, have been used as pigments (Patty, 1963). The former has been variously called "Imperial Sweinfurth" and Vienna, parrot, or Paris green, while the latter has been termed either Scheele's green or Swedish green. Other copper compounds are used as insecticides (copper fluoroarsenate), fungicides (copper sulfate, copper naphthenate), algicides (copper pentachlorophenate, copper dimethylgloxime), and molluscicides. Copper is used to make bronze paint, the paint for ships' bottoms, percussions caps, analytical reagents, and solvents for cellulose in rayon manufacture and in certain electroplating processes.

Among the medicinal applications of the element is the recent utilization of copper-containing intrauterine contraceptive devices for birth control (Gonzalez-Angulo and Aznar-Ramos, 1976; Rubenstein, 1976). Copper is also a component in certain types of dental cement (Reid, 1968) and in dental materials used for periodontal work (Trachtenberg, 1972).

Although this listing is not all inclusive, it underscores the extensive environmental interface between copper and human beings.

3. OCCUPATIONAL EXPOSURE TO COPPER

A partial list of occupations in which exposure to copper may occur is presented in Table 1. The only large-scale studies of workers that include quantitative measurements of ambient and personal exposure (i.e., breathing zone) have been performed on U.S. copper smelter employees.

Wagner (1975) has described the conditions of exposure in 14 U.S. copper smelter facilities that were the subject of industrial hygiene surveys conducted by the National Institute for Occupational Safety and Health. Measurements of the airborne concentrations of As, Pb, Zn, Cd, Mo, Cu, and sulfur dioxide revealed that only the last two occurred consistently at relatively high levels. In regard to copper, the industry-wide average for both area and personal samples exceeded 1 mg/m^3 for dust and fumes in several plant locations. The current federal standard for occupational exposure to copper restricts in-plant exposure to 1 mg/m^3 for dust and 0.1 mg/m^3 for fume emissions (Table 2). However, it was concluded that most of the element occurred as a "nonrespirable" dust, and this dust consisted mainly of a relatively inert copper sulfide.

Urinary levels of copper were determined in the smelter worker studies, but Wagner indicates that methods of collection and analysis were not well controlled for potentially confounding factors. These reservations notwithstanding, a mean urinary copper content of 79 $\mu g/l$ determined from 206 individual samples was

Table 1. Potential Occupational Exposures to Copper and Copper Compounds

Asphalt production	Paint manufacture
Battery production	Pigment production
Copper smelting Coppersmith work	Preservative manufacture and handling: hides, railroad ties, rope, wood, etc.
Electroplating	Rayon manufacture
Metal founding, copper and copper alloys	Tanning
Fungicide production and handling	Wallpaper manufacture
Gem coloring	Water treatment
Insecticide production and handling	Welding and soldering

Table 2. 8-Hour Time-Weighted Average for Copper[a]

In dust	1 mg/m^3[b]
In fumes	0.1 mg/m^3[b]

[a] Based on the occupational standards established by the U.S. Occupational Safety and Health Administration (OSHA).
[b] Approximate milligrams of particulate per cubic meter of air.

well above the 9 to 18 μg range that has been suggested as the normal value in human beings for a 24-hr urinary excretion (Butler and Newman, 1956; Scheinberg and Sternleib, 1958).

Nonetheless, evidence of copper toxicity, acute or chronic, was not revealed in the smelter survey despite the excessive ambient, personal, and urinary levels of copper. Perhaps this relates to the absence in these facilities of free metallic copper and copper oxide, which are the chemical forms most often associated with the so-called metal fume fever of copper and other types of metal exposures.

It should be noted that arsenic contaminants in copper ore have been considered a likely cause of the increased risk of pulmonary carcinoma in smelter worker populations (Pinto and Bennett, 1963; Lee and Fraumeni, 1969; Milham and Strong, 1974; Newman et al., 1976; Rencher et al., 1977) and nonoccupationally exposed residents of at least one copper-mining city (Newman et al., 1976).

One might expect to find a large body of literature with respect to occupational copper exposure owing to the widespread applications of this element in science and industry. However, fewer than half a dozen epidemiologic inquiries in the English language address this subject; and in these, as in the U.S. copper smelter studies, the health hazards related to this metal have been found to be of low order. Yet the dearth of critical, long-term investigations of copper toxicity in the workplace should raise the question of unrecognized, occupationally related, copper-induced illness rather than putting the issue to rest. Of considerable importance in this regard is the very recent discovery of serious to fatal pulmonary (Pimental and Marques, 1969; Villar, 1974) and hepatic abnormalities (Pimental and Menezes, 1977) in a group of Portuguese vineyard sprayers exposed to a copper sulfate-containing fungicide for periods of 3 to 45 years. The details of associated pathologic findings are reviewed in the following section, but this dramatic new evidence concerning long-term copper toxicity implies that it would be prudent to reserve conclusions about the health-hazard implications of prolonged exposure until critical longitudinal studies of worker health become available.

The remaining sections that address occupational exposure to copper are mostly in the form of case reports where the environmental conditions have not been described quantitatively. These case reports and a variety of nonoccupational forms of exposure will be used to illustrate the local and systemic aspects of copper toxicity in human beings.

4. COPPER TOXICITY

The acute and chronic manifestations of copper poisoning in human beings and other animals are dependent on the mode of contact and the milieu in which this contact occurs (see Scheinberg, this volume). Metallic copper has been associated with toxic properties following topical exposure to, or inhalation and ingestion of, the element in its various states. A description of copper dust, fume, and salt intoxication follows.

4.1. Skin, Hair, and Mucous Membranes

An allergic contact dermatitis may result from direct exposure to copper salts or dust; however, from the paucity of published cases, this is probably a rare or unrecognized event. The mechanism by which copper may produce an allergic contact dermatitis is unknown. Saltzer and Wilson (1968) have speculated that copper forms chemical or haptene links with a protein carrier molecule in the skin. It is believed that the antigenicity of this haptene-carrier complex sensitizes a population of thymus-derived lymphocytes, leading to a typical delayed hypersensitivity reaction upon exposure.

The occurrence of an acute urticurial hypersensitivity reaction to a copper-containing dental cement has been reported by Reid (1968). The eruption subsided with the removal of the copper cement; however, no skin testing was performed to assess the precise form of allergic sensitivity. Trachtenberg (1972) described a case of chronic, low-grade gingivitis in an individual with a copper-containing dental prosthesis. The gingivitis resolved completely after the prosthesis was replaced by a device without copper components.

Barkoff (1976) recently detailed the history of a patient with a chronic, recurrent urticarial reaction caused by a copper-containing intrauterine contraceptive device (IUD). The hives that afflicted the 24-year-old woman abated soon after the IUD was removed. Carefully performed cutaneous scratch tests were positive for copper, whereas other components of the IUD gave negative results with similar testing.

The potential for latent toxic reactions to the copper IUD has been implied by several studies. By using electron microscopic and X-ray dispersive analysis, Gonzalez-Angulo and Aznar-Ramos (1976) have demonstrated alterations in the mitochondria of endothelial cells in women wearing the TCU-200 IUD. They postulate that impairment of respiratory mechanisms and energy production at the cellular level may provide an inhospitable environment for the fertilized ovum. The long-range consequences of these ultrastructural alterations have yet to be determined. In another study Rubenstein (1976) observed a low-grade cervicitis in roughly one quarter of the women he studied who were wearing copper IUDs. Papanicolaou smears in these women revealed evidence of a chronic inflammatory cellular reaction.

Another toxic reaction of the integument, particularly hair, to copper is a greenish black discoloration from repeated exposure to copper dust. Many of the standard medical-industrial texts make passing references to this condition but offer no documentation. Parish (1975) has reviewed the subject of occupational skin and hair pigmentation secondary to copper, including references dating to the mid-nineteenth century. He offers ample documentation of the phenomenon related to occupational factors. It is often stated that copper smelters, burnishers, and filers may suffer such ill effects. This author is aware of a case of permanent staining of the scalp in a worker who was chronically exposed to copper dust in a polishing operation. Additionally, Lucas and Ramos (personal communication, 1974) identified a "musical bell tuner" whose job entailed grinding 80% copper bells (to attain an appropriate pitch). The worker developed a striking greenish discoloration of the scalp hair after many years at this job.

In other nonoccupational settings there have been several reports of green hair developing after exposure to copper-contaminated water. Cooper and Goodman (1975) described a minor epidemic of green hair among students at a girls' boarding school. Apparently, this outbreak was related to increased acidification of the city water supply for the purposes of fluoridation. It was determined that excessive quantities of copper were leached from the water pipes, and the girls were exposed to the copper-contaminated water during bathing. Nordlund et al. (1977) relates two similar instances of persons with green-tinted hair which was secondary to frequent shampooing in water that had high concentrations of copper. Swimming pool water containing a copper-based algicide has been implicated as a cause of green hair in two other youngsters (Lampe et al., 1977). Goldsmith and Holmes (1975) have observed several persons with a greenish hue to their hair, and they believe that the common denominator has been the practice of bleaching the hair and later swimming in chlorinated pool water. It is interesting that every nonoccupational instance of green hair, whatever the suspected etiology, occurred in blond or otherwise lightly pigmented individuals. There is no explanation for this predisposition.

There has been one case description of green hair that could not be attributed to an exogenous source of copper. This was seen in a patient with phenylketonuria (Holmes and Goldsmith, 1974).

4.2. Eye

Stokinger (1963) and others (Moeschlin, 1965; American Conference of Governmental Industrial Hygienists, 1971) have described eye contact with copper salts as a cause of conjunctivitis and edema of the eyelids. In severe cases it may also cause turbidity or ulceration of the cornea.

The ophthalmologic consequence of traumatic exposure to copper-containing foreign bodies is usually partial or complete blindness. In most cases this results from mechanical damage to the anterior segment and lens. So-called copper-eye reflects a fast-running cellular integration with liquefaction and retraction of the vitreous structures (Neubauer, 1976).

There are no reports of toxic eye reactions to copper fumes or dust in the literature.

4.3. Respiratory

The most common complaints of workers chronically exposed to excessive concentrations of copper fumes and dust are related to the upper respiratory tract. The symptoms consist of congestion of the nasal mucous membranes and pharynx, and on occasion ulceration with perforation of the nasal septum (White, 1928; American Conference of Governmental Industrial Hygienists, 1971; Hunter, 1969). Quite often there is an associated syndrome of fever and chills which is collectively referred to as "metal fume fever."

Schiötz (1949) and others (White, 1928; Stockinger, 1963; American Conference of Governmental Industrial Hygienists, 1971) have linked copper and other metals to the development of metal fume fever. Presumably, the fumes gain entrance to the body primarily via respiratory routes. In one instance, seven cases of copper fever were observed in a paint factory where copper oxide was being pulverized. This fever has also been reported in workers exposed to copper acetate, in welders exposed to copper fumes (Stockinger, 1963), and in a large number of workers in a Norwegian copper factory (Woldgren and Gorbatow, 1949).

Gleason (1968) reported an interesting study of workers who were involved in lapping (polishing) copper plates. These workers complained of general malaise, not unlike that due to the common cold. They noted sensations of chills or warmth and a "stuffiness of the head." Sampling the ambient air resulted in the finding of copper dust concentrations of 0.75 to 1.20 mg/m^3 in this particular area. Ancillary studies of the entire facility localized the problem to the lapping wheels and prompted management to install a local exhaust ventilation system that ultimately reduced the copper dust concentrations to less than 0.008 mg/m^3. Worker complaints ceased shortly thereafter.

The case described above points out a very important issue in relation to the standards established by federal agencies for "safe" levels of exposure to various chemical and physical agents in the workplace. In this instance the workmen reacted to concentrations of copper dust that were only one tenth of the U.S.

federal standard, considering the 8-hr, time-weighted average in current usage (Table 2). The lesson is that the standards should be considered merely as guide lines that clearly do not obviate the necessity for continued caution and surveillance of medical complaints by the worker.

The traditional notion that metal fume fever is primarily the result of respiratory exposure to the offending element has been questioned by Lyle and coworkers (1973). They describe a typical example of metal fume fever symptomatology, including chills, sweating, headache, exhaustion, and nausea, which developed during and after their patient underwent hemodialysis. A complete cure was effected by the removal of copper-containing parts from the home hemodialysis unit.

The pathogenesis of metal fume fever is unknown. Andrews and Lyons (1957) have shown that divalent copper will bind histamine to bovine serum albumin. The physiologic role of histamine in allergy might therefore lend credence to the postulate that copper binding of histamine to a serum protein provides a means whereby a local exposure (fume inhalation) may produce a systematic syndrome (fume fever) (Stockinger, 1963).

A new disease of the respiratory tract, "vineyard sprayer's lung," has been reported among Portuguese workers handling a fungicide containing a 1 to 2% solution of copper sulfate (Pimentel and Marques, 1969; Villar, 1974). Spraying was carried out 15 to 100 days of the year, and 600 liters of the solution was sprayed each day by each worker. These workers developed interstitial pulmonary lesions, including histiocytic granulomas and associated nodular, fibrohyaline scars. The scars contained abundant deposits of copper. Vineyard sprayer's lung has a course that appears highly variable. The pulmonary lesions may regress, remain stationary, or progress toward a diffuse pulmonary fibrosis (Pimentel and Marques, 1969; Villar, 1974). Lung cancer may develop during the course of the disease, as well (Pimentel and Menezes, 1975). The high incidence of adenocarcinoma and particularly of alveolar cell carcinoma has led Pimentel and colleagues to conclude that lung cancer in these vineyard workers is yet another example of an occupationally related cancer.

4.4. Gastrointestinal

The prompt emetic effect of copper limits its oral toxicity and has led some authors (Poskanzer and Bennett, 1974) to conclude that acute poisoning caused by the ingestion of copper salts is rare. In India, however, copper salt poisoning due to suicidal intent accounts for 33% of all reported poison cases in the New Delhi region of the country (Chuttani et al., 1965). Chuttani and co-workers described in detail 53 cases with their toxic manifestations.

Initially, all the patients in the Chuttani et al. (1965) series complained of a metallic taste (which is commonly the only manifestation of dust toxicity for the gastrointestinal tract), nausea, burning in the epigastrium, and repeated emetus of greenish material (Table 3). Another common complaint was diarrhea, which usually did not exceed five to six stools per day. Ptyalism, not described by Chuttani et al., has been reported by other investigators (Holtzman et al., 1966; American Conference of Governmental Industrial Hygienists, 1971) to occur early in the course of poisoning. In a small percentage of patients with severe toxicity, melena occurred during the first few days postingestion. Most of these findings were due to the direct irritant effect of copper solutions on the gastrointestinal mucosa, which was corroborated by histologic preparations in some of the patients. Permanent damage did not result in any of these cases, as shown by normal barium meal studies 2 weeks after the copper ingestion.

A considerable number of patients developed clinical jaundice during the second or third day. There were two distinct groups of patients in the Chuttani et al. experience, distinguished on the basis of liver function tests and liver biopsy observations. The first group had a more severe clinical course, marked by intense jaundice and markedly abnormal liver function tests. On biopsy, these patients showed centrilobular necrosis and biliary stasis, believed to be the result of acute copper toxicity upon hepatic cells.

The second group had a much more benign course that was characterized by a mild state of jaundice and slightly abnormal liver function tests. This group had an essentially normal histologic pattern on biopsy of the liver, and all findings in these cases were attributed to the intravascular hemolysis that so often accompanies copper salt intoxication.

The acute toxicity of copper for the gastrointestinal tract (as well as the renal, hematologic, cardiovascular, and neurologic systems) seems to be exclusively limited to the ingestion of elemental copper or a salt form. More than likely,

Table 3. Clinical Features of Copper Sulfate Poisoning[a]

Clinical Feature	Number of Cases	Percentage of Cases
Metallic taste, nausea, vomiting, and burning in epigastrium	48	100
Diarrhea	14	29.1
Jaundice	11	23.0
Hemoglobinuria/hematuria	14	29.1
Anuria	13	27.0
Oliguria	5	10.3
Hypotension	4	8.3
Coma	4	8.3
Melena	1	2.1

[a] From Chuttani et al. (1965).

however, a sufficiently high concentration of copper dust would cause similar, albeit less devastating, effects. In the United States the reports of gastrointestinal copper intoxication have been mostly epidemic in nature, directly related to the ingestion of beverages containing a high concentration of copper (Wyllie, 1957; Semple et al., 1960; LeVan and Perry, 1961; Hamel et al., 1977).

Until 3 years ago, gastrointestinal toxicity from long-term exposure to copper was unknown. However, studying vineyard sprayer's lung disease in workers exposed to a copper sulfate-containing algicide, Pimentel and Menezes (1977) have documented extensive liver damage among a group of these men. The pathologic findings included focal and diffuse swelling and proliferation of Kuppfer cells; histiocytic and sarcoid-like granulomatosis; fibrosis of variable degree in the perisinusoidal, portal, and subcapsular areas, accompanied by an atypical proliferation of the sinusoidal lining cells; micronodular cirrhosis; angiosarcoma of the liver; and idiopathic portal hypertension. Histochemical techniques were used to demonstrate abundant deposits of copper within the hepatic lesions of these patients. Especially significant was the finding of yet other copper-containing lesions in the nasal mucosa, kidney, spleen, and lymph nodes. The pathologic implications of the latter types of organ system involvement await further study. Again, it is worthwhile to reemphasize the discovery of still another occupational form of cancer related to prolonged copper exposure. It is most interesting that the hepatic response to copper sulfate as a carcinogen seems to be identical to that reported for vinyl chloride, namely, the rarely encountered angiosarcoma (Creech and Johnson, 1974; Block, 1974; Lee and Harry, 1974; Thomas et al., 1975).

4.5. Hematologic

The insidious events following chronic elemental copper or copper salt ingestion lead to acute and often massive hemolytic crisis. This aspect of copper toxicity is observed in human beings and other animals but is poorly understood. Holtzman et al. (1966), in his one case report, did an in-depth hemolytic evaluation and could find no evidence of a direct or indirect Coombs reaction. He demonstrated a normal G-6-PD level and failed to show the presence of a primary hemoglobinopathy. In sheep the buildup of copper in the liver may take place over a period of weeks or months without any warning before the development of a fatal hemolytic crisis in the brief course of 48 to 72 hr (Todd and Simpson, 1961). At any rate, the hemolytic component of copper intoxication is perhaps the most serious complication, and it is probably responsible for the majority of pathologic reactions that occur in other organ systems.

4.6. Renal

Ingestion accounts for the only reports of renal toxicity to copper. Hemoglobinuria and hematuria were observed in close to one third of the Chuttani et al. (1965) cases during the second and third days postingestion. A rising blood urea nitrogen (BUN) and oliguria or annuria were also very commonly found in the first 24 to 48 hr after the attempted suicide. These findings may be attributed to intravascular hemolysis and the subsequent development of acute tubular necrosis. Microscopic examination of the urine and kidney biopsies in these patients demonstrated the presence of hemoglobin and cellular casts in the urine and congestion of glomeruli, necrosis, and denudation of tubular cells in the histologic specimens. It is possible, however, that the renal damage was due to a direct toxic effect of the copper. This conclusion may be drawn from the animal studies of Vogel (1960), whose work has demonstrated that copper can cause proximal tubular necrosis in mice. It is important to note that hypotension did not precede the development of renal toxicity in the Chuttani et al. series, and this finding maintains a measure of plausability for both the indirect and the direct toxicity hypotheses of pathogenesis.

4.7. Cardiovascular

Hypotension was seen in a small number of the Chuttani et al. patients and has also been described by others after an ingestion (Holtzman et al., 1966; Moeschlin, 1965). This finding is undoubtedly a poor prognostic sign. The pathogenesis of the phenomenon is not well understood, but animal studies by Chatterjee (1963) reportedly imply a direct cardiotoxic action on the myocardium by copper.

4.8. Neurologic

Approximately 8% of the Chuttani et al. (1965) patients developed coma on the third day after ingestion, and two of the four with this finding expired. In these cases, coma was attributed to uremia. It seems unlikely that copper sulfate had a direct toxic effect on the central nervous system, as the patients who did not develop uremia remained conscious throughout their course. Furthermore, Chuttani et al. pointed out that autopsy material showed no evidence of extrapyramidal lesions, suggesting that acute copper sulfate poisoning does not injure the basal ganglia, as observed in such chronic conditions as Wilson's disease (a congenital deficiency of the serum protein ceruloplasmin which results in high levels of copper in the blood).

5. DIAGNOSIS OF COPPER INTOXICATION

The diagnosis of copper toxicity relies heavily upon a careful history and physical examination of the patient, with special reference to determining the type of environmental exposure. An acute exposure to copper fumes or dust is unlikely to produce serious consequences, but in the case of copper salts the outcome may be disastrous. In the latter instance it is probable that there will be a strong index of suspicion concerning the etiologic agent. The history may reveal suicidal intent or accidental exposure to a copper-containing solution in the home or workplace. If the patient is obtunded, aspiration of greenish gastric contents may provide a clue to the etiologic agent. Analysis of the blood for heavy metals is a routine toxicologic screening procedure which is invaluable in documenting the magnitude of the systemic exposure. But what of the chronically exposed individual?

There has been only one attempt to establish guidelines for evaluating the patient or worker with possible chronic exposure to copper (Cohen, 1974). Again, the importance of the history and the physical examination is emphasized as the initial approach to diagnosis. The patient may bring a suggestive pattern of complaints to the attention of medical personnel, such as nasal irritation, metallic taste, and episodes of fever and shaking chills, to name some of the usual symptoms. Of course these symptoms invariably relate to exposure in an occupational setting, and the most important step in the detection of a potential occupational hazard is to visit the workplace. It is a usual and expected industrial hygiene practice to measure ambient levels of suspected toxins; in this case the concentration of copper fumes, dust, or salt mist, among others, would be determined.

If the source of the exposure is not clear but copper intoxication is suspected, the analysis of 24-hr urinary copper levels provides a simple and reliable screening procedure. However, care must be taken to obtain a specimen that is collected and tested in a copper-free system. To rule out exogenous urinary contamination a serum copper level should be performed in all cases where the urine level is increased.

The diagnosis of an allergic contact dermatitis, when copper is the suspected agent, may be confirmed by applying a patch test on the skin of the patient. The patch is impregnated with a 5% solution of copper sulfate (Fisher, 1973). A positive reaction includes the presence of erythema and vesiculation at the test site. It should be noted, however, that the one case of allergic contact dermatitis to copper reported by Saltzer and Wilson (1968) resulted in a positive patch test with a copper sulfate solution as low as 1.25%.

6. TREATMENT OF COPPER TOXICITY

The first and often curative measure in the treatment of copper intoxication is the removal of the offending agent. In cases where systemic toxicity from copper salts presents an immediate and continuing danger to the patient, Poskanzer and Bennett (1974) have suggested prompt gastric lavage with a 1% solution of potassium ferrocyanide, fluid replacement, and control of pain and diarrhea with opiates. The stomach may be alternatively lavaged with milk and egg white emulsion, which will convert the copper into a caseinate and albuminate.

For chronic states of copper intoxication where it is imperative to lower copper levels in the blood, copper chelators such as D-penicillamine and British anti-lewisite (BAL) have been successfully employed (Moeschlin, 1965; Scheinberg and Sternlieb, 1960). As indicated, corticosteroids may be efficacious in more severe cases with jaundice, hemoglobinuria, and hematuria (Gupta et al., 1962).

Vogel (1960) has sounded a note of concern regarding the use of chelators that increase renal copper clearance. His work demonstrates a primary renal toxic effect from copper exposure in rats; however, in the case reported by Holtzman et al. (1966) renal function improved as clearance increased.

Allergic contact dermatitis to copper has been treated successfully with hydrocortisone-containing topical preparations and removal of the offending agent (Saltzer and Wilson, 1968).

7. SUMMARY

The element copper is ubiquitous in nature and has widespread uses in numerous household, scientific, and industrial settings. This review was designed to illustrate various aspects of environmental copper exposure, particularly occupational, with emphasis on potential health hazards and their attendant toxicity for human beings.

Under ordinary circumstances copper is a benign agent, but when human beings and other animals are exposed to excessive concentrations of the metal in any of its forms there may be mild to serious or even life-threatening consequences. The use of the environment or, more specifically, the workplace as a paradigm for toxicity studies is reflected here in the afflictions of the vineyard sprayers. Hence, in the late twentieth century, copper may be added to the growing list of carcinogens. Indeed, the words of Paracelsus seem a fitting note of caution with respect to copper in the environment: "Poison is in everything, and no thing is without poison. Only the dosage makes it either a poison or a remedy."

REFERENCES

American Conference of Governmental Industrial Hygienists (1971). "Copper (as Cu)." In *Documentation of Threshold Limit Values for Substances in the Workroom Air,* 3rd ed., Cincinnati, Ohio, pp. 59–60.

Andrews, A. L. and Lyons, T. D. (1957). "Binding of Histamine and Antihistamine to Bovine Serum Albumin by Mediation with Cu(II)," *Science,* **126,** 561–562.

Barkoff, J. R. (1976). "Urticuria Secondary to a Copper Intrauterine Device," *Int. J. Dermatol.,* **15,** 594–595.

Block, J. B. (1974). "Angiosarcoma of the Liver Following Vinyl Chloride Exposure," *J. Am. Med. Assoc.,* **229,** 53–54.

Butler, E. J. and Newman, G. E. (1956). "The Urinary Excretion of Copper and Its Concentration in the Blood of Normal Human Adults," *J. Clin. Pathol.,* **9,** 157–161.

Chatterjee, P. K. (1963). "An Experimental Study of the Pathogenesis of Copper Sulfate Poisoning." Cited by H. K. Chuttani et al. (1965).

Chuttani, H. K., Gupta, P. S., Gutate, B., and Gupta, D. N. (1965). "Acute Copper Sulfate Poisoning," *Am. J. Med.,* **39,** 849–854.

Cohen, S. R. (1974). "A Review of the Health Hazards from Copper Exposure," *J. Occup. Med.,* **16,** 621–624.

Cooper, R. and Goodman, J. (1975). "Green Hair," *N. Engl. J. Med.,* **292,** 483–484.

Creech, J. L. and Johnson, M. N. (1974). "Angiosarcoma of Liver in the Manufacture of Polyvinyl Chloride," *J. Occup. Med.,* **16,** 150–151.

Fisher, A. A. (1973). *Contact Dermatitis,* 2nd ed. Lea and Febiger, Philadelphia, p. 373.

Gleason, R. P. (1968). "Exposure to Copper Dust," *Am. Ind. Hyg. Assoc. J.,* **29,** 461–462.

Goldsmith, L. A. and Holmes, L. B. (1975). "Green Hair," *N. Engl. J. Med.,* **292,** 484.

Gonzalez-Angulo, A. and Aznár-Ramos, R. (1976). "Ultrastructural Studies on the Endometrium of Women Wearing TCU-200 Intrauterine Device by Means of Transmission and Scanning EM and X-Ray Dispersive Analysis," *Am. J. Obstet. Gynecol.,* **125,** 170–179.

Gupta, P. S., Bhargava, S. P., and Sharma, M. L. (1962). "Acute Copper Sulfate Poisoning with Special Reference to Its Management with Corticosteroid Therapy," *J. Assoc. Phys. India,* **10,** 267.

Hamel, A. J., Drawbaugh, R., McBean, A. M., Watson, W. N., and Withrell, L. E. (1977). "Outbreak of Acute Gastroenteritis Due to Copper Poisoning—Vermont," *Morb. Mortal. Wkly. Rep.,* **26,** 218, 223.

Holmes, L. B. and Goldsmith, L. A. (1974). "The Man with Green Hair," *N. Engl. J. Med.,* **291,** 1037.

Holtzman, N. A., Elliot, D. A., and Heller, R. H. (1966). "Copper Intoxication: Report of a Case with Observations on Ceruloplasmin," *N. Engl. J. Med.,* **275,** 347–352.

Hunter, D. (1969). *The Diseases of Occupations,* 4th ed. Little Brown, Boston, pp. 163, 421.

Lampe, R. M., Henderson, A. L., and Hansen, G. H. (1977). "Green Hair," *J. Am. Med. Assoc.,* **237,** 2092–2094.

Lee, A. M. and Fraumeni, J. F. (1969). "Arsenic and Respiratory Cancer in Man: An Occupational Health Study," *J. Natl. Cancer Inst.*, **42,** 1045–1052.

Lee, F. I. and Harry, D. S. (1974). "Angiosarcoma of the Liver in a Vinyl-Chloride Worker," *Lancet,* **1,** 1316–1318.

LeVan, J. H. and Perry, E. I. (1961). "Copper Poisoning on Shipboard," *Public Health Rep.,* **76,** 334.

Lyle, W. H., Payton, J. C., and Hui, M. (1973). "Hemodialysis and Copper Fever," *Lancet,* **1,** 1324–1325.

Milham, S. and Strong, T. (1974). "Human Arsenic Exposure in Relation to a Copper Smelter," *Environ. Res.,* **7,** 176–182.

Moeschlin, S. (1965). "Metals, Copper, Copper Sulfate." In *Poisoning, Diagnosis and Treatment,* 4th ed. (translated by Jenifer Bickel), Grune and Stratton, New York, pp. 124–125.

Neubauer, H. (1976). "Vitreous Metallosis," *Adv. Ophthalmol.,* **33,** 202–204.

Newman, J. A., Archer, V. E., Saccomanno, G., Kuscher, M., Auerbach, O., Grondahl, R. D., and Wilson, J. C. (1976). "Histologic Types of Bronchogenic Carcinoma among Members of Copper-Mining and Smelting Communities." In U. Saffioti and J. K. Wagoner, eds., *Ann. N.Y. Acad. Sci.,* **271,** 260–268.

Nordlund, J. J., Hartley, C., and Fister, J. (1977). "On the Cause of Green Hair," *Arch. Dermatol.,* **113,** 1700.

Parish, L. C. (1975). "Green Hair." *N. Engl. J. Med.,* **292,** 483.

Patty, F. A. (1963). "Arsenic, Phosphorus, Selenium, Sulfur, and Tellurium." In D. W. Fasset and D. D. Irish, eds., *Industrial Hygiene and Toxicology,* Vol. II, 2nd ed., Wiley-Interscience, New York, p. 874.

Pimentel, J. C. and Marques, F. (1969). "Vineyard Sprayer's Lung: A New Occupational Disease," *Thorax,* **24,** 678–688.

Pimentel, J. C. and Menezes, A. P. (1975). "Lung Cancer and Copper Inhalation in Vineyard Sprayers." In *Proceedings of the Fifth Congress of the European Society of Pathology,* Vienna, p. 140.

Pimentel, J. C. and Menezes, A. P. (1977). "Liver Disease in Vineyard Sprayers," *Gastroenterology,* **72,** 275–283.

Pinto, S. S. and Bennett, B. M. (1963). "Effect of Arsenic Trioxide Exposure on Mortality," *Arch. Environ. Health,* **7,** 583–591.

Poskanzer, D. C. and Bennett, I. L. (1974). "Heavy Metals: Copper." In M. M. Wintrobe, G. N. Thorn, R. D. Adams, et al., eds., *Harrison's Principles of Internal Medicine.* McGraw-Hill, New York, pp. 667–671.

Reid, D. J. (1968). "Allergic Reaction to Copper Cement," *Br. Dent. J.,* **124,** 92–94.

Rencher, A. C., Carter, M. W., and McKee, D. W. (1977). "A Retrospective Epidemiological Study of Mortality at a Large Western Copper Smelter," *J. Occup. Med.,* **19,** 754–758.

Rubenstein, E. (1976). "Is the Copper-Releasing Intrauterine Contraceptive Device Able to Induce Unphysiologic Cell Differentiation in the Uterine Cervix?" *Am. J. Obstet. Gynecol.,* **125,** 277.

Saltzer, E. I. and Wilson, J. W. (1968). "Allergic Contact Dermatitis Due to Copper," *Arch. Dermatol.*, **98,** 375–376.

Scheinberg, I. H. and Sternleib, I. (1958). "Copper Metabolism," *Pharm. Rev.*, **12,** 355–381.

Scheinberg, I. H. and Sternleib, I. (1960). "Penicillamine as the Basis of Therapy in Wilson's Disease." In M. J. Seven and L. A. Johnson, Eds., *Metal Binding in Medicine.* J. B. Lippincott, Philadelphia, p. 275.

Schiötz, E. H. (1949). "Metal Fever Produced by Copper Dust." In *Proceedings of the 9th International Congress on Industrial Medicine,* London, p. 798.

Semple, A. B., Parry, W. H., and Phillips, D. E. (1960). "Acute Copper Poisoning Outbreak Traced to Contaminated Water from a Corroded Geyser," *Lancet,* **2,** 700–701.

Stokinger, H. E. (1963). "The Metals (Excluding Lead)—Copper." In F. A. Patty, Ed., *Industrial Hygiene and Toxicology,* Vol. II, 2nd ed., Wiley-Interscience, New York, pp. 1033–1037.

Tabershaw, I. (1977). "Chemical Hazards. Copper." In *Occupational Diseases: A Guide to Their Recognition.* U.S. Public Health Serv. Publ. (NIOSH) 77-181, U.S. Government Printing Office, Washington, D.C.

Thomas, L. B., Popper, H., Beck, P. D., et al. (1975). "Vinyl-Chloride-Induced Liver Disease. From Idiopathic Portal Hypertension (Banti's Syndrome) to Angiosarcomas," *N. Engl. J. Med.*, **292,** 17–22.

Todd, J. R. and Simpson, R. H. (1961). "Methemoglobin in Chronic Copper Poisoning of Sheep," *Nature,* **191,** 89–90.

Trachtenberg, D. I. (1972). "Allergic Response to Copper—Its Possible Gingival Implications," *J. Periodontol.*, **43,** 705–707.

Villar, T. G. (1974). "Vineyard Sprayer's Lung: Clinical Aspects," *Am. Rev. Respir. Dis.*, **110,** 545–555.

Vogel, F. S. (1970). "Nephrotoxic Properties of Copper under Experimental Conditions in Mice with Special Reference to the Pathogenesis of Renal Alterations in Wilson's Disease," *Am. J. Pathol.*, **36,** 699–711.

Wagner, W. L. (1975). *Environmental Conditions in U.S. Copper Smelters.* U.S. Public Health Serv. Publ. 1975-657-603/5546, U.S. Government Printing Office, Washington, D.C., 35 pp.

White, R. P. (1928). *The Dermatergoses or Occupational Afflictions of the Skin,* 3rd ed. H. K. Lewis, London, p. 194.

Wyllie, J. (1957). "Copper Poisoning at a Cocktail Party," *Am. J. Public Health,* **47,** 617.

2

HUMAN HEALTH EFFECTS OF COPPER

I. Herbert Scheinberg

Albert Einstein College of Medicine, Bronx, New York

1. INTRODUCTION

Copper is essential to human life and yet can be toxic at high concentrations. Deficiency of the element is very unusual, probably because of an ill-understood genetic control mechanism that appears to facilitate absorption of copper at times of dietary scarcity. Copper toxicosis, which can be fatal, rarely if ever occurs in human beings (see, however, Cohen, this volume), except when there is an inherited defect in another homeostatic process that facilitates excretion in the face of a dietary excess of the element. This extraordinarily effective defense against copper toxicosis, despite the inherent toxicity of this (and every other) heavy metal, contrasts sharply with the apparent lack of any mechanisms mitigating the toxicity of the nonessential heavy metals—lead and mercury, for example.

Copper thus appears to exemplify the fact that only when a metal has been selected as essential by the evolutionary process do specific, genetically transmitted mechanisms develop that buffer the organism against environmentally induced deficiency and toxicosis. No such mechanisms appear to have evolved to protect human beings against the toxicosis of nonessential metals.

2. COPPER DEFICIENCY IN HUMAN BEINGS

A consideration of dietary supply, metabolic requirements, and loss of copper show why deficiency of this element is rare in human beings. The dietary content of almost any Western diet is between 2 and 5 mg Cu/day, and it is difficult to construct a diet with less than 1 mg Cu. The total body copper of an adult is about 100 to 150 mg. Less than about 50 μg Cu is lost in the daily urine, and the fecal content of the metal is accounted for by the part of the dietary copper that is not absorbed and by copper excreted via bile. Loss of copper with blood amounts to less than 0.1 mg/dl.

It is interesting to compare copper with iron in these respects. A normal Western diet contains somewhat more iron—perhaps 10 to 15 mg—than copper, but this has to supply a body content of 2 to 5 g in the face of loss, with blood, that amounts to 50 mg/dl. Since, despite a roughly equivalent supply, but much greater requirements and loss, deficiency of iron is unusual (except in the face of chronic, or severe acute, blood loss), it would seem a priori that deficiency of copper is likely to be even more unusual. That is the case.

The distribution of copper in human tissues and organs is shown in Table 1. The newborn contains considerably more hepatic copper than the older child or the adult but only about one quarter of the plasma copper. Except for these changes with age the tissue concentrations of copper and its distribution are remarkably constant throughout life.

Virtually all of the copper in human beings exists as an integral part of one or another specific copper protein. The copper proteins that have been isolated or are known to exist in human beings are tabulated in Table 2. In these 16 distinctive copper-protein complexes, the chemical character of the element is made uniquely specific by the molecular environment of the protein in which it is bound (see Weser et al., this volume). Equally important, the potential toxicity of ionic copper is thereby avoided, since there are no free copper ions to modify or inactivate the body's other proteins.

The physiologic roles of only four of the copper proteins listed in Table 2 are known. Cytochrome oxidase is an essential link in the electron transfer system. Tyrosinase—probably consisting of several copper proteins—catalyzes the oxidation of tyrosine to melanin in several steps. Superoxide dismutase, a copper protein found in many actively metabolizing cells, catalyzes the dismutation of the superoxide ion to hydrogen peroxide and water. Monoamine oxidase (or lysyl oxidase) converts the ε-amino group of lysine to an aldehyde group, a reaction required for the cross-linking of both collagen and elastin that is essential to the integrity of a number of connective tissues. The physiologic functions of the other copper proteins listed in Table 2 are unknown.

Table 1. Copper Content of Human Tissues and Body Fluids[a]

Tissue	Mean Content (μg/g dry weight)	
	Normal	Wilson's Disease
Adrenal	7.4	17.64
Aorta	6.7	—
Bone	4.2	—
Brain	23.9	—
Caudate nucleus	—	212.0
Cerebellum	—	261.0
Frontal lobe cortex	—	118.0
Globus pallidus	—	254.5
Putamen	—	313.5
Cornea	—	92.9
Erythrocytes (per 100 ml packed red blood cells)	89.1	—
Hair	23.1	—
Heart	16.5	12.7
Kidney	14.9	96.2
Leukocytes (per 10^9 cells)	0.9	—
Liver	25.5	584.2
Lung	9.5	15.5
Muscle	5.4	25.9
Nails	18.1	—
Ovary	8.1	5.2
Pancreas	7.4	4.2
Placenta	13.5	—
Prostate	6.5	—
Skin	2.0	5.2
Spleen	6.8	5.6
Stomach and intestines	12.6	22.9
Thymus	6.7	—
Thyroid	6.1	—
Uterus	8.4	—
Aqueous humor	12.4	—
Bile (common duct)	1050	173
Cerebrospinal fluid	27.8	—
Gastric juice	28.1	—
Pancreatic juice	28.4	—
Plasma, Wilson's disease	50.0	—
Saliva	31.7	—
Serum		
Female	120.0	—
Male	109.0	—
Newborn	36.0	—
Sweat		
Female	148.0	—
Male	55.0	—
Synovial fluid	21.0	—
Urine (24 hr)	18.0	—

[a] Modified from Sternlieb and Scheinberg (1977).

Table 2. Mammalian Copper Proteins[a]

Protein	Isolated from: Species	Organ or Tissue
Albocuprein I	Human	Brain
Albocuprein II	Human	Brain
Ceruloplasmin	Numerous, including human	Plasma
Copper-chelatin (L-6-D)	Human, rat	Liver
Cytochrome *c* oxidase	Numerous	Heat, liver, etc.
3,4-Dihydroxyphenylethylamine β-hydroxylase	Cattle	Adrenals
Dopamine β-hydroxylase	Cattle	Adrenals
Ferroxidase II	Human	Serum
Hepatomitochondrocuprein	Human, cattle	Liver
Lysyl oxidase	Chicken	Cartilage
Mitochondrial monoamine oxidase	Human, rat, cattle	Liver, brain
Pink copper protein	Human	Erythrocytes
Plasma/serum monoamine oxidase	Human, rabbit, pig	Plasma/serum
Superoxide dismutase (cytocuprein)		
Cerebrocuprein	Human	Brain
Erythrocuprein	Human	Erythrocytes
Hemocuprein	Human	Blood
Hepatocuprein	Human	Liver
Tryptophan-2,3-dioxygenase	Rat	Liver
Tyrosinase	Human	Skin, eye

[a] Modified from Sternlieb and Scheinberg (1977).

Human deficiency of copper is manifested by insufficient activity of one or another of these copper proteins. Acquired deficiency, though rare, is seen in certain newborn infants who have diarrhea and who are fed only milk, a poor source of copper. Kwashiorkor is also complicated by copper deficiency, though this is generally overshadowed by the lack of many other essential components of the diet. Prolonged parenteral hyperalimentation has also caused copper deficiency, as have instances of very severe malabsorption (Sternlieb and Janowitz, 1964).

To some extent, it is possible to deplete the normal individual of copper by prolonged therapy with the chelating agent D-penicillamine: the copper-penicillamine complex is readily excreted in the urine. Yet, although many patients with rheumatoid arthritis and cystinuria have been treated with D-penicillamine, very few, if any, have required dietary supplementation with copper. Urinary excretion of copper is generally increased with severe proteinuria; and although this may result in abnormally low concentrations of plasma ceruloplasmin, there appears to be no need for copper supplementation.

An X-linked disorder of copper transport, Menkes' disease, results in copper deficiency that is always fatal by 3 to 5 years of age (Danks et al., 1972, 1973). The afflicted boys are mentally retarded, perhaps as a result of deficiency of cytochrome oxidase, and possess hair of an unusual kinky or steely structure, possibly because of an insufficient activity of monoamine or lysyl oxidase. This disease constitutes evidence that a normal genetic mechanism exists that prevents deficiency of copper by actively facilitating its absorption and transport across various membranes. Neither oral nor parenteral treatment with copper has significantly altered the course of the disease or prolonged life.

3. COPPER TOXICOSIS IN HUMAN BEINGS

3.1. Acquired

Acquired toxicosis due to copper is rare and usually is not serious. It is seen in acute form most dramatically in suicidal attempts where several grams of a copper salt are ingested. Or it may be the consequence of the application of similar amounts of copper salts to burned skin. Nausea and vomiting occur first, on oral ingestion, and are soon followed by deepening coma, oliguria, hepatic necrosis, and death (Chuttani et al., 1965). Treatment with chelating agents, such as D-penicillamine, or hemo- or peritoneal dialysis is seldom effective.

Copper can be mildly toxic, producing nothing more than nausea, vomiting, diarrhea, and malaise, when acidic food or beverage is allowed to remain in contact with metallic copper for prolonged periods. Such toxicity has been seen when carbonated water remains in contact with a copper check valve, or when whiskey sours have been mixed in copper cocktail shakers.

Acquired chronic copper toxicosis in human beings is even more unusual than the acute form. Because the liver probably normally excretes the small amount of copper required to maintain the normal zero copper balance, several forms of chronic liver disease are associated with excessive amounts of hepatic copper. This condition occurs most strikingly in biliary atresia (Sternlieb et al., 1966) and primary biliary cirrhosis (Fleming et al., 1974), where concentrations of copper can be seen which approach the indubitably toxic levels characteristic of Wilson's disease (see Section 3.2). Hemodialysis, generally accomplished through the use of copper-containing semipermeable membranes, and sometimes with solutions of low pH passing through copper tubing, has also caused the deposition of excessive amounts of hepatic copper, when carried on for months (Klein et al., 1972). It is even possible that copper, known to be absorbed from copper-containing intrauterine contraceptive devices (certainly not a physiological route), may accumulate in the course of years of use of these devices (see Cohen, this volume; Sorenson, this volume).

In none of these three instances, however, is there conclusive evidence that excess copper is significantly toxic, although preliminary clinical data suggest that the pathologic process in primary biliary cirrhosis may be ameliorated if D-penicillamine is used to promote excretion of the metal (Deering et al., 1977).

Under any other circumstances than those discussed, it seems almost impossible to overwhelm the body's defense against copper toxicosis by oral ingestion of the element. Chilean copper miners eat their lunch, for as long as 20 years, in an atmosphere cloudy with 2% copper ore dust. Analyses of the copper contents of liver and blood, in a small number of these miners, have been normal. Vineyard workers who spray grapes with copper arsenate have shown some toxic effects, but these are not clearly related to copper.

3.2. Hereditary

Hereditary copper toxicosis is known as Wilson's disease. This is an autosomal recessive disorder, first described in one of its clinical forms in 1912. The experimental evidence currently available suggests that patients with this illness have inherited a defect in the hepatic lysosomal excretion of copper, and that the copper so retained is probably derived from ceruloplasmin.

Ceruloplasmin, like all plasma proteins, with the clear exception of albumin, is a glycoprotein and is removed from the circulation when two or more of its terminal sialic acid residues are removed (Frieden, this volume). Removal of the sialyl residues exposes terminal galactosyl residues, and these are rapidly bound to a hepatic receptor in the liver with immediate removal of the protein from the circulation (see Figure 1).

Very rapidly, after its binding to the hepatic receptor, asialoceruloplasmin is transported into lysosomes of the hepatocytes, where its copper and galactose are removed from the protein, as shown in Figure 2. The copper freed from this presumed pathway of ceruloplasmin catabolism is apparently excreted into the bile. The only direct evidence in support of this hypothesis derives from an experiment in which radiocopper was administered to a woman (W.D.) with Wilson's disease and to another woman (C.) without Wilson's disease, 18 hr before cholecystectomy (Sternlieb et al., 1973). At operation, samples of liver and bile were collected and analyzed for total copper and radiocopper. The results (Tables 3 and 4) show as expected that the total concentrations of copper in all the subcellular fractions of the liver of W.D. are higher than those in C. Unexpectedly (and a more interesting finding) the specific activity of copper in the bile of W.D. not only is much lower than the specific activity in the bile of C., but also is virtually identical to the specific activity in the lysosomal fraction of W.D. Indeed, the specific activity of no other subcellular fraction of her liver is similar to that of copper in the bile.

Considered together, the studies on the catabolism of asialoceruloplasmin

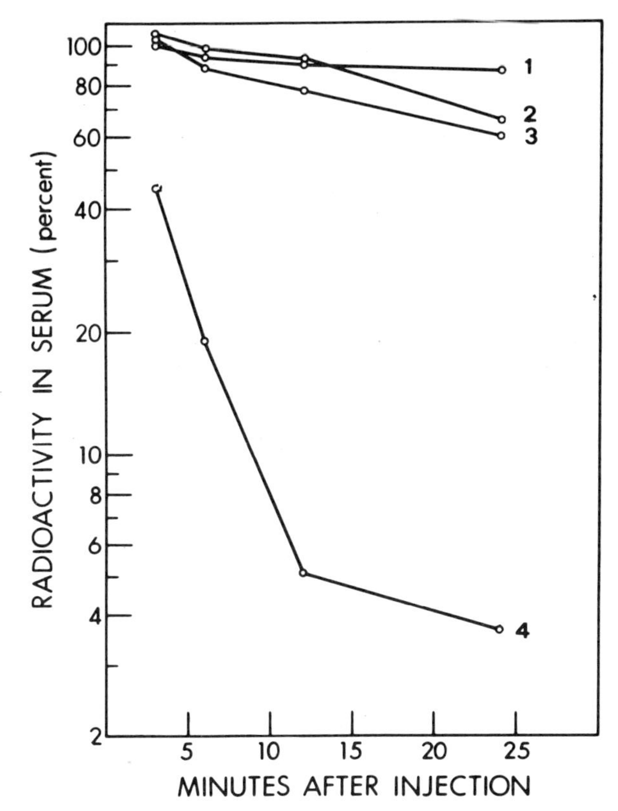

Figure 1. Disappearance from the serum of ^{64}Cu-labeled native and modified rabbit ceruloplasmin injected intravenously. Each point is the average value obtained from two animals. 1, Ceruloplasmin; 2, oxidized ceruloplasmin; 3, asialogalactoceruloplasmin; 4, asialoceruloplasmin. From Morell et al. (1968).

and the clinical data suggest that the normal individual may excrete copper derived from the catabolism of ceruloplasmin into bile and that there is a defect in this mechanism in the patient with Wilson's disease. Table 5 gives calculations based on this hypothesis and shows that the amount of copper retained as a result of defective excretion of ceruloplasmin copper is of the same order as that retained in patients with Wilson's disease.

There is one other item of evidence that the catabolism of ceruloplasmin is followed by biliary excretion of at least part of its copper. In experiments with rabbits, 5% of the radiocopper contained in intravenously administered ceruloplasmin was found both in the bile and in the intestines of the animals (Aisen et al., 1964). This suggests that, except for the dietary copper that is not absorbed, fecal copper is derived from ceruloplasmin.

Whether or not Wilson's disease is produced by the mechanism just discussed, the natural history of this ailment can be completely explained by the effects

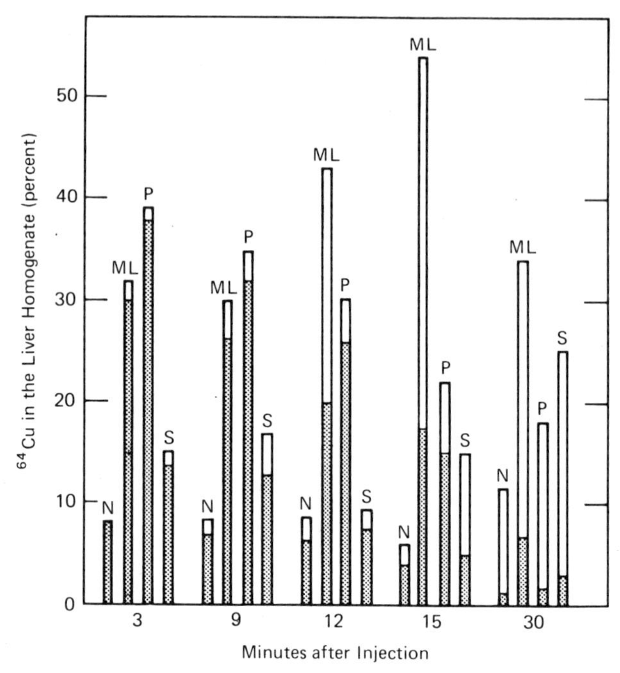

Figure 2. Intracellular distribution of ^{64}Cu in fractions of rat liver as a function of time after injection of 300 μg of human [^{64}Cu]asialoceruloplasmin (ASCPN). The height of each bar represents ^{64}Cu in the fraction; the shaded portion of the bar represents ^{64}Cu recovered in immunoprecipitable ASCPN. Values are the means for two to five rats. At a given time, the sum of ^{64}Cu recovered in each fraction was 89 to 96% of the radioactivity. N, nuclear fraction; ML, crude mitochondrial and lysosomal fraction; P, microsomal fraction; S, cytosol. Of the acid phosphatase activity present in the homogenate, the mitochondrial-lysosomal fraction contained 45.1% ± 4.3 with *p*-nitrophenyl phosphate as substrate and 35.6% ± 3.1 with 1-glycerophosphate as substrate. From Gregoriadis et al. (1970).

of copper toxicosis. At any stage of their lives, more than 5 times as much copper is found in the livers of untreated patients with Wilson's disease as the upper limit of normal persons. In a few patients, the concentration of hepatic copper can reach 50 times the normal level! From 95 to 97% of the patients show less than 20 mg ceruloplasmin/dl, the lower limit of the normal range (Scheinberg and Gitlin, 1952; Sternlieb and Scheinberg, 1968).

Copper accumulates in the liver as the infant grows. Although evidence of histologic damage can be seen in early infancy, clinical illness is never observed before the age of 5 years. The pathogenic sequence generally involves the appearance of an excess of fat in hepatocytes, followed by necrosis that can be accompanied by acute or chronic hepatitis. Ultimately the necrotic liver cells are replaced by fibrous tissue with scarring and mixed micro- and macronodular cirrhosis. At this stage there may be either remarkably few symptoms or severe,

Table 3. Distribution of Copper in Bile and Liver[a]

Source	Concentration (μg/g or μg/ml) Control	Wilson's Disease
Common duct bile	10.5	1.73
Gall bladder bile	27.5	—
Liver homogenate	7.1	62.2
Nuclear fraction	0.41	6.8
Mitochondrial fraction	0.60	9.4
Lysosomal fraction	0.37	15.9
Microsomal fraction	0.68	3.3
Cytosol	5.78	21.1

[a] From Sternlieb et al. (1973).

not infrequently fatal, parenchymal or circulatory disorders of the liver (Sternlieb, 1972; Sternlieb and Scheinberg, 1974).

Throughout the years during which hepatic copper is accumulating, there is probably a release of some of the element into the plasma. Whether asymptomatic or overtly ill, patients with Wilson's disease who have not been treated uniformly manifest elevated plasma levels of free, non-ceruloplasmin-bound copper. In addition, although this seems paradoxical in view of the continued accumulation of copper, the urinary excretion of the element may reach 300 to 500 μg/day, which is almost 10 times the output of copper in a normal individual's urine. Both of these characteristic findings undoubtedly reflect the diffusion of copper from the hepatic excesses into plasma, and thence into urine.

When there is acute necrosis of a large number of copper-laden hepatocytes, a more sudden release of copper from the liver into the plasma may occur. This is probably the cause of the acute hemolytic anemia that is seen in about 10% of patients with Wilson's disease, often as their first clinical manifestation of the disorder (Deiss et al., 1970). The copper, release of which from dying liver

Table 4. Specific Activity of ^{64}Cu in Bile and Liver[a]

Source	Specific Activity Control	Wilson's Disease
Common duct bile	4.13	0.087
Liver homogenate	4.42	1.610
Nuclear fraction	1.46	0.265
Mitochondrial fraction	1.53	0.160
Lysosomal fraction	2.27	0.082
Microsomal fraction	1.95	0.637
Cytosol	4.64	3.950

[a] From Sternlieb et al. (1973).

Table 5. Hypothetical Pathogenesis of Wilson's Disease

	Normal "Subject," 17 Years Old	"Patient" with Wilson's Disease
Plasma ceruloplasmin (mg/dl)	30	8.0
Total body ceruloplasmin (mg)	990	275
Fractional daily turnover of ceruloplasmin	0.2	0.2
Daily ceruloplasmin catabolism (mg)	200	55
Daily copper freed from ceruloplasmin (μg)	600	165
Yearly copper freed from ceruloplasmin (mg)	220	60
Assumed yearly accumulation (mg) of copper	(0–10)	(60)
Total accumulation (mg) of body copper in 10 years	—	600

cells is analogous to the release of malarial parasites from dying erythrocytes, is almost certainly the cause of this hemolysis since much greater concentrations of free copper in the plasma are observed at the height of the hemolytic crisis than are seen in patients without this complication.

In most patients a more gradual diffusion of free plasma copper into many tissues and organs of the body produces more insidious, disastrous effects. Copper toxicosis to the central nervous system causes neurologic disorders that are peculiarly limited to the motor systems and may mimic multiple sclerosis, Parkinsonism, chorea, and/or dystonia. The resulting incapacity may be severe. Anarthria, dysphagia, and the other manifestations of dystonic type, or resting and intention tremors, complete the picture of total disability in the most advanced stages of the disease (Wilson, 1912; Denny-Brown, 1964).

Despite Wilson's original description, copper accumulates not only in the lenticular nuclei but also in all portions of the brain. It is not surprising, therefore, that in addition to neurologic disorders psychiatric disturbances of almost every variety are also seen. Schizophrenia, manic-depressive psychosis, severe neurotic disturbances, and bizarre behavioral changes all may occur as part of Wilson's disease (Scheinberg et al., 1968; Goldstein et al., 1968).

Diffusion of the free plasma copper into the kidney produces observable disturbances of glomerular as well as proximal and distal tubular function (Bearn, 1972). Hematuria, proteinuria, and excretion of all of the components of the urine seen in Fanconi's syndrome can be noted. Nevertheless, significant clinical renal insufficiency rarely occurs as part of Wilson's disease.

An important effect of the diffusion of copper into body tissues is dramatically visible as Kayser-Fleischer rings. These are deposits of copper in Descemet's membrane of the cornea, generally detectable in diffuse room light in blue-eyed individuals, but almost always requiring examination with a slit-lamp in brown-eyed subjects. These rings are always present when Wilson's disease has produced symptomatic disturbances in the central nervous system. Until recently, it was thought that they were not seen in any other disorder but Wilson's disease. However, it is now clear that in certain other disorders of the liver, of the types mentioned above where there is accumulation of large excesses of hepatic copper in association with prolonged cholestasis, structures indistinguishable from Kayser-Fleischer rings may be seen (Fleming et al., 1975).

Copper also affects the bones and probably causes the ill-defined arthralgias and X-ray manifestations that are seen in a moderate percentage of patients with Wilson's disease.

As is apparent from Table 1, excessive concentrations of copper are present in a number of tissues, but definitive correlation of these with pathologic effects other than those described has not been recorded. Clinical manifestations of Wilson's disease may appear initially to be endocrinologic. Primary or secondary amenorrhea or irregularity of periods is probably related to hepatic rather than ovarian dysfunction.

Even in women whose menstrual cycles are not demonstrably disordered, the incidence of miscarriage, or spontaneous abortion, is high if Wilson's disease has not been treated. This may be related to an excess of free plasma copper ions diffusing into uterine secretions and producing the same interference with implantation of the fertilized ovum or blastula that is attributed to the action of copper-containing intrauterine contraceptive devices (see Sorenson, this volume).

Specific treatment of Wilson's disease is accomplished by removal of the excess deposits of copper. Not only does such treatment produce dramatic therapeutic and prophylactic results, but also, through these, it constitutes proof of the etiologic role of copper in this disease.

Chelation therapy was first proposed by Cumings in 1948, following the demonstration of the large excess of tissue and organ copper in patients with Wilson's disease. British anti-lewisite (BAL) was soon used by Denny-Brown; it produced modest improvement in patients' symptomatology and even some fading of Kayser-Fleischer rings (Cumings, 1951; Denny-Brown and Porter, 1951).

In 1957 Walshe introduced D-penicillamine as an oral form of chelation therapy, and this is now the treatment of choice for patients with this disorder (Scheinberg and Sternlieb, 1960; Sternlieb and Scheinberg, 1964; Cartwright, 1978; Walshe, 1973; Sorenson, this volume). With continual and life-long administration of penicillamine, in a dose of about 1 g daily, there is dramatic clinical improvement. Asymptomatic patients, diagnosed by biochemical criteria, most specifically deficiency of plasma ceruloplasmin and excess of hepatic copper, can be kept free of all clinical manifestations of the disorder by treatment with penicillamine and appear to have as long lives as individuals without Wilson's disease (Sternlieb and Scheinberg, 1968).

Specific treatment with D-penicillamine should be instituted no matter how desperately ill the patient appears to be when the diagnosis of Wilson's disease is first made. Total disability due to neurologic dysfunction and severe psychiatric disturbances may be relieved completely or improved dramatically. If the disturbance is primarily hepatic, and very severe, therapeutic results are not so easily obtained, and the patient may die no matter how much copper is removed. Patients with the most severe forms of neurologic disease may benefit in only limited fashion, if at all. The effects of copper toxicosis on speech are the most difficult to dispel completely.

It is very important, therefore, to diagnose patients in the asymptomatic state. Although screening programs of large populations have been attempted, asymptomatic subjects are generally discovered only when they are examined because a blood relative has been found to have Wilson's disease. Asymptomatic relatives of patients with Wilson's disease, or *any* individual in whom there are found to be less than 20 mg ceruloplasmin/dl and more than 250 μg Cu/g dry liver, must be considered to have Wilson's disease. They should be treated with D-penicillamine as intensively and for the duration of their lives in exactly the manner as patients who have symptomatic forms of the illness.

Validation of the efficacy of prophylactic treatment of asymptomatic patients with Wilson's disease now rests on over 100 patients who have been so treated, some for as long as 20 years. Overt clinical manifestations of the disorder are seen only in individuals who clearly have not adhered to therapy. Either patients may refuse to take penicillamine, and admit it, or various biochemical and clinical laboratory tests may indicate that they are not taking their medication despite their protestations. Prime among these biochemical indices of nonadherence to the therapeutic regimen is a concentration of free, or nonceruloplasmin, copper in plasma greater than about 20 μg/dl. Well treated patients almost uniformly exhibit no more than 10 to 15 μg free Cu/dl.

Penicillamine may produce hypersensitivity reactions of an allergic nature (which can almost always be circumvented) or, after months or years of therapy, toxic effects on the kidneys, lungs, skin, bone marrow, joints, pericardial or pleural membranes, neuromuscular junctions, or thymus that are more difficult to obviate. In particular, the appearance of a nephrotic syndrome or, worse, Goodpasture's syndrome (Sternlieb et al., 1975) requires at least temporary, if not permanent, cessation of therapy with penicillamine.

Since Wilson's disease is, if untreated, progressive and uniformly fatal, it is encouraging that triethylenetetramine (TETA) has recently been introduced by Walshe and appears to be able to induce sufficient cupriuria to effect continued therapy in patients who are irreversibly intolerant to penicillamine (Walshe, 1970, 1973).

4. SUMMARY

Copper is essential to human life but is also potentially lethally toxic. Two distinct and specific genetic regulatory mechanisms exist in normal human beings. One appears to facilitate the transmembranous transfer of copper. Absence of this mechanism results in a hereditary form of copper deficiency which is fatal despite attempts at either oral or parenteral treatment with copper. A second mechanism, transmitted as an autosomal recessive, seems to be essential to the excretion of ceruloplasmin copper from hepatic lysosomes into bile, and may be responsible for the maintenance of the normal zero balance of copper. In patients in whom this second type of genetically transmitted mechanism is absent, Wilson's disease, or human chronic copper toxicosis, ultimately develops and is fatal if untreated.

Toxicosis due to copper is avoided in genetically normal individuals both by the action of this second mechanism and also by virtue of the fact that the essentiality of copper acts only through specific copper proteins. Such a metal-protein complex reduces the concentration of free copper ions throughout the body and thus avoids the toxicity that is as inherent in copper ions as it is in ions of such nonessential metals as lead and mercury.

Acquired deficiency of copper can be effectively treated by administration of the element. Acquired or inherited toxicosis of this metal can be treated by use of one of three available chelating agents, all of which promote the excretion of copper. Sorenson discusses the therapeutic uses of copper in detail in Chapter 5.

REFERENCES

Aisen, P., Morell, A. G., Alpert, S., and Sternlieb, I. (1964). "Biliary Excretion of Caeruloplasmin Copper," *Nature,* **203,** 873–874.

Bauman, L. K. (1960). "The Copper Content in Tissues of Patients with Hepatolenticular Degeneration," *Zh. Nevropatol. Psikhiatr.,* **60,** 1141–1145 (in Russian).

Bearn, A. G. (1972). "Wilson's Disease." In J. B. Stanbury, J. G. Wyngaarden, and D. S. Fredrickson, Eds., *The Metabolic Basis of Inherited Disease,* 3rd ed. McGraw-Hill, New York, pp. 1033–1050.

Bickel, H., Neale, F. C. and Hall, G. (1957). "A Clinical and Biochemical Study of Hepatolenticular Degeneration (Wilson's Disease)," *Am. J. Med.,* **26,** 527–558.

Cartwright, G. E. (1978). "The Diagnosis of Treatable Wilson's Disease," *N. Engl. J. Med.,* **298,** 1347–1350.

Cartwright, G. E. et al. (1954). "Studies on Copper Metabolism. XIII: Hepatolenticular Degeneration," *J. Clin. Invest.,* **33,** 1487–1501.

Chuttani, H. K. et al. (1965). "Acute Copper Sulfate Poisoning," *Am. J. Med.,* **39,** 849–854.

Cumings, J. N. (1951). "The Effects of B.A.L. in Hepatolenticular Degeneration," *Brain,* **74,** 10–22.

Cumings, J. N. (1959). "Copper. Hepatolenticular Degeneration." In *Heavy Metals and the Brain,* Blackwell, Oxford, pp. 3–71.

Danks, D. M. et al. (1972). "Menkes' Kinky-Hair Syndrome," *Lancet,* **1,** 1100–1102.

Danks, D. M. et al. (1973). "Menkes' Kinky Hair Disease: Further Definition of the Defect in Copper Transport," *Science,* **179,** 1140–1142.

Deering, T. B. et al. (1977). "Effect of D-Penicillamine on Copper Retention in Patients with Primary Biliary Cirrhosis," *Gastroenterology,* **72,** 1208–1212.

Deiss, A., Lee, G. R., and Cartwright, G. E. (1970). "Hemolytic Anemia in Wilson's Disease," *Ann. Intern. Med.,* **73,** 413–418.

Denny-Brown, D. (1964). "Hepatolenticular Degeneration (Wilson's Disease): Two Different Components," *N. Engl. J. Med.,* **270,** 1149–1156.

Denny-Brown, D. and Porter, H. (1951). "The effect of BAL (2,3-Dimercaptopropanol) on Hepatolenticular Degeneration (Wilson's Disease)," *N. Engl. J. Med.,* **245,** 197–925.

Fleming, C. R. et al. (1974). "Copper and Primary Biliary Cirrhosis," *Gastroenterology,* **67,** 1182–1187.

Fleming, C. R. et al. (1975). "Pigmented Corneal Rings in a Patient with Primary Biliary Cirrhosis," *Gastroenterology,* **69,** 220–225.

Goldstein, N. P. et al. (1968). "Psychiatric Aspects of Wilson's Disease (Hepatolenticular Degeneration): Results of Psychometric Tests during Long-Term Therapy," *Am. J. Psychiatr.,* **124,** 1555–1561.

Gregoriadis, G., Morell, A. G., Sternlieb, I., and Scheinberg, I. H. (1970). "Catabolism of Desialylated Ceruloplasmin in the Liver," *J. Biol. Chem.,* **245,** 5833–5837.

Klein, W. J., Jr. et al. (1972). "Acute Copper Intoxication," *Arch. Intern. Med.,* **129,** 578–582.

Leu, M. L. et al. (1971). "Muscle Copper, Zinc, and Manganese Levels in Wilson's Disease: Studies with the Use of Neutron-Activation Analysis," *J. Lab. Clin. Med.,* **76,** 432–438.

Morell, A. G. et al. (1968). "Physical and Chemical Studies on Ceruloplasmin. V: Metabolic Studies on Sialic Acid-Free Ceruloplasmin *in vivo*," *J. Biol. Chem.,* **243,** 155–159.

Scheinberg, I. H. and Gitlin, D. (1952). "Deficiency of Ceruloplasmin in Patients with Hepatolenticular Degeneration (Wilson's Disease)," *Science,* **116,** 484–485.

Scheinberg, I. H. and Sternlieb, I. (1960). "The Long-Term Management of Hepatolenticular Degeneration (Wilson's Disease)," *Am. J. Med.,* **29,** 316–333.

Scheinberg, I. H., Sternlieb, I., and Richman, J. (1968). "Psychiatric Manifestations in Patients with Wilson's Disease." In D. Bergsma, I. H. Scheinberg, and I. Sternlieb, Eds., *Wilson's Disease—Birth Defects Original Article Series.* National Foundation March of Dimes, New York, pp. 85–87.

Sternlieb, I. (1972). "Evolution of the Hepatic Lesion in Wilson's Disease (Hepatolenticular Degeneration)," *Prog. Liver Dis.,* **4,** 511–525.

Sternlieb, I. and Janowitz, J. D. (1964). "Absorption of Copper in Malabsorption Syndromes," *J. Clin. Invest.*, **43,** 1049–1055.

Sternlieb, I. and Scheinberg, I. H. (1968). "Prevention of Wilson's Disease in Asymptomatic Patients," *N. Engl. J. Med.,* **278,** 352–359.

Sternlieb, I. and Scheinberg, I. H. (1974). "Wilson's Disease." In F. Schaffner, S. Sherlock, and C. M. Leevy, Eds., *The Liver and Its Diseases.* Intercontinental Medical Book Corp., New York, pp. 328–336.

Sternlieb, I. and Scheinberg, I. H. (1977). "Human Copper Metabolism." In *Medical and Biologic Effects of Environmental Pollutants*—Copper. National Academy of Sciences, Washington, D.C., pp. 29–54.

Sternlieb, I., Harris, R. C., and Scheinberg, I. H. (1966). "Le Cuivre dans la Cirrhose Biliaire de l'Enfant," *Rev. Int. Hepatol.,* **16,** 1105–1110.

Sternlieb, I. et al. (1973). "Lysosomal Defect of Hepatic Copper Excretion in Wilson's Disease (Hepatolenticular Degeneration)," *Gastroenterology,* **64,** 99–105.

Sternlieb, I., Bennett, B., and Scheinberg, I. H. (1975). "D-Penicillamine Induced Goodpasture's Syndrome in Wilson's Disease," *Ann Intern. Med.,* **82,** 673–676.

Tu, J.-B., Blackwell, R. Q., and Hou, T.-Y. (1963). "Tissue Copper Levels in Chinese Patients with Wilson's Disease," *Neurology,* **13,** 155–159.

Walshe, J. M. (1956). "Penicillamine, a New Oral Therapy for Wilson's Disease," *Am. J. Med.,* **21,** 487–495.

Walshe, J. M. (1970). "Triethylene Tetramine" (letter), *Lancet,* **2,** 154.

Walshe, J. M. (1973). "Copper Chelation in Patients with Wilson's Disease," *Quart. J. Med.,* **42,** 441–452.

Wisniewski, H. M. et al. (1967). "Quantitative Topography of Copper in Wilson's Disease and in Porto-systemic Encephalopathy," *Neuropatol. Pol.,* **5,** 92–103.

3

TERATOGENIC EFFECTS OF COPPER

Lucille S. Hurley

Carl L. Keen

Department of Nutrition, University of California, Davis, California

1. INTRODUCTION

The essentiality of copper in the diet of mammals was first reported in 1928 by Hart and his associates. A few years later, Neal et al. (1931) suggested that copper deficiency could occur in livestock under normal grazing conditions. This idea was soon verified by Sjollema (1933), who described a disease of plants and animals in parts of Holland that could be controlled by administration of copper. An important role of copper in the perinatal development of mammals was demonstrated by the work of Bennetts and Chapman (1937) and Bennetts and Beck (1942), who showed that enzootic ataxia of lambs, a disease affecting the developing fetus, could be prevented by administration of copper to the ewe during pregnancy.

This chapter presents a survey of the teratogenic effects of both copper deficiency and copper excess during development. Although teratology has classically been concerned with structural abnormalities apparent at birth, the term has more recently been expanded to include the study of functional abnormalities as well as those occurring in the early postnatal period (Wilson, 1972). In keeping with this definition of teratology, we shall discuss abnormalities arising in the early postnatal as well as in the prenatal period. Reviews on the teratogenic effects of metals other than copper in mammalian embryos have recently been prepared by Ferm (1972), Hurley (1977), and Williams (1977).

2. TERATOGENIC CONSEQUENCES OF COPPER DEFICIENCY

2.1. Neurological Abnormalities

Lambs

Knowledge of the importance of copper in prenatal development arose from studies of the disease "enzootic ataxia," which occurred in newborn lambs in certain geographical areas of western Australia. The disorder was characterized by spastic paralysis, especially of the hind limbs, severe incoordination of movement, blindness in some cases, and anemia. The nervous disorder was initially classified as a demyelinating encephalopathy, similar to Schilder's encephalitis in man. The brains of affected animals were smaller than normal, with collapsed cerebral hemispheres and shallow convolutions. Throughout the brain there was a marked paucity of normal myelin. The abnormalities in the brain and nervous system were apparently responsible for the ataxia that these animals exhibited. Lambs with enzootic ataxia usually died within a short time after birth or were born dead (Bennetts, 1932; Innes, 1936; Innes and Shearer, 1940). In the case of twins, the disease affected both members of the pair, but the time of onset of the ataxia could vary (Stewart, 1932).

The identification of copper as an etiologic factor in enzootic ataxia was made by Bennetts and Chapman (1937), who found that the disease was accompanied by subnormal levels of copper in the liver and blood of ataxic lambs and in the blood and milk of their mothers. Additionally, correction of these low levels by copper supplementation to the lambs was found to arrest but not reverse the demyelination process. Furthermore, the disease could be prevented by supplementing the ewe with copper during early pregnancy, indicating that an insufficient amount of the element during prenatal development brought about the syndrome (Bennetts and Beck, 1942; Allcroft et al., 1959).

The fundamental neural lesions of enzootic ataxia have not yet been firmly agreed upon. Barlow (1963) did not consider gross cerebral lesions to be an essential part of the pathology. The basic abnormality, according to this investigator, is a degeneration of the nerve fibers in well-defined tracts of the spinal cord, with chromatolysis and necrosis of neurons in the brain stem and spinal cord. He argued that enzootic ataxia as found in western Australia is the same disease as "swayback," occurring in Britain (Barlow, 1960). Although many investigators have supported this point of view (Mills and Williams, 1962; Howell et al., 1969), some have argued against classifying the two diseases together (Symonds, 1975). In this chapter swayback and enzootic ataxia are considered pathologically identical.

The extent of neural degeneration in ataxic lambs that occurs *in utero* and its time of onset are not well established. Differences in the quantity of myelin are not demonstrable prenatally (O'Dell et al., 1976), but neural degeneration has been found as early as 90 days of gestation (Smith et al., 1977). Since myelination in this species begins at approximately 70 to 80 days of fetal age in the ventral column of the spinal cord (Barlow, 1969), the lesions reported by Smith et al. (1977) were apparent only 10 days after the initiation of myelination, and at a time when cytochrome oxidase activity increases rapidly in the normal brain (Barlow, 1969). The development of lesions during this time period is in keeping with the report of Allcroft et al. (1959) that copper supplementation to the ewe was efficacious until mid-gestation. Myelin aplasia, rather than excessive myelin degeneration, has also been suggested as a mechanism for the development of neural lesions observed in ataxic lambs, since Howell and coworkers (1969) reported pathological changes in neurons and spinal cords in which products of degenerating myelin were not detected.

Ultrastructurally, neurons from copper-deficient, ataxic lambs were found to have a progressive lack of Nissel substance, abnormalities of Golgi apparatus, increased neurofibrils, and abnormal neuronal and astroglial processes. The distal peripheral cytoplasm of the neuron appeared to undergo necrosis as the cell body lost the ability to support it. Degeneration of the axon was then followed by destruction of the myelin sheath (Cancilla and Barlow, 1966a, 1966b, 1968; Howell et al., 1969; Fell et al., 1965).

The major biochemical lesion in the brain of an affected animal appears to be a reduction in the copper-containing terminal respiratory enzyme, cytochrome oxidase, due to low levels of brain copper (Howell and Davison, 1959). Mills and Williams (1962) have demonstrated that the activity of cytochrome oxidase is specifically reduced in the large motor neurons of the red nucleus of the brain, an area whose degeneration is associated with the onset of ataxia. These investigators consider that a low level of cytochrome oxidase activity may lead to tissue anoxia, since, in an *in vitro* system, the capacity of mitochondria from copper-deficient animals to oxidize substrates was severely reduced (Gallagher et al., 1956; King et al., 1977). It is not clear, however, whether tissue anoxia due to low cytochrome oxidase levels occurs *in vivo* as well. Gallagher and Reeve (1971) have reported that there is a causal relationship between low cytochrome oxidase activity and phospholipid production in rat liver mitochondria from copper-deficient animals. Apparently, the low activity of cytochrome oxidase is not sufficient for the production of the ATP needed for phospholipid synthesis. If a similar effect occurs also in the brain, the amyelination seen there could result from a lack of the phospholipids needed for myelin synthesis.

Conditioned Deficiency. Copper deficiency in the pregnant ewe may also be induced by conditions other than a low dietary intake of the element. A copper-deficient state can be precipitated by high intakes of dietary sulfate and molybdate which apparently make dietary copper unavailable (Kirchgessner et al., this volume). In newborn lambs of ewes receiving such rations, there was low copper concentration in the brain and a deficiency of cytochrome oxidase in the motor neurons. Ataxia similar to that seen in enzootic ataxia was frequently found (Mills and Fell, 1960; Suttle and Field, 1968).

Other dietary intakes that apparently result in conditioned copper deficiency in pregnant ewes have also been reported. In Iceland, Palsson and Grimsson (1953) found that the offspring of ewes fed on seaweed showed the symptoms and the morphological characteristics of enzootic ataxia. The lambs had demyelination of the cerebrum and other abnormalities of the brain. Although the dietary copper intake was adequate, the pregnant ewes showed low levels of the element in blood, and in affected lambs the liver copper content was subnormal. The disorder could be largely prevented by giving a copper supplement to the pregnant ewes. Roberts et al. (1966a, 1966b) have reported additional cases of teratogenicity due to hypocupremia of unknown cause, perhaps related to inability of the dam to utilize dietary copper. More recently, Ghergariu (1978) has proposed that high levels of cadmium and lead in forage can induce hypocuprosis even when the concentration of copper appears to be sufficient.

Experimental Animals

Neonatal ataxia, similar to that seen in lambs affected with enzootic ataxia, has been reported in newborn, copper-deficient goats (Owen et al., 1965), swine (Cancilla and Barlow, 1970; Wilkie, 1959), guinea pigs (Everson et al., 1967), and rats (Carlton and Kelly, 1969; Zimmerman et al., 1976).

Guinea Pigs. Abnormalities of the nervous system similar to those seen in lambs with enzootic ataxia were found in the newborn offspring of female guinea pigs fed a copper-deficient diet during growth and pregnancy. These offspring, which had subnormal levels of liver copper, showed a high incidence of ataxia and gross abnormalities of the brain at birth. The brains of the copper-deficient pups were pale and translucent, with many small hemorrhagic areas. Cerebellar folia were often absent or abnormal in appearance. Throughout the brain there was a paucity of myelin, and that which was present was abnormally low in phospholipid concentration. The pups that survived birth usually died within the first month of life of aneurisms of the aortic arch. In addition to the neural abnormalities in deficient pups, immature kidneys were a frequent finding (Everson et al., 1968). Structural abnormalities are discussed in Section 2.2.

Rats. In female rats, copper deficiency during pregnancy has pronounced deleterious effects on embryonic and fetal development. Diets that are severely deficient in copper usually cause fetal resorption or stillbirths (Hall and Howell, 1969; Dutt and Mills, 1960; Spoerl and Kirchgessner, 1975a, 1975b). If the diets are only moderately deficient in copper, the offspring have a high incidence of numerous abnormalities, and high postnatal mortality (O'Dell et al., 1961; Spoerl and Kirchgessner, 1976). Structural anomalies are discussed in Section 2.2.

In a report by Carlton and Kelly (1969), the newborn pups from copper-deficient dams were found to be hyperirritable, with convulsive seizures following stimulation. Gross neural lesions included focal pale areas in the cerebral cortex and corpus striatum, prominent cerebral edema, and cortical necrosis. Unlike lambs afflicted with enzootic ataxia, the rats did not show nerve fiber degeneration in the brain stem or spinal cord. It is interesting, however, that similar pathology was described by Roberts et al. (1966a, 1966b) in nonneonatal ataxic, copper-deficient sheep.

Copper-deficient rat pups, like offspring of copper-deficient guinea pigs, had low contents of myelin and copper in the brain. However, in contrast to guinea pigs, brain phospholipid concentration and composition were not altered in deficient rat pups (Dipaolo et al., 1974; Zimmerman et al., 1976). Prohaska and Wells (1975) have reported that mitochondria isolated from the brains of copper-deficient rat pups were enlarged and abnormal in shape, and had low levels of cytochrome oxidase activity. Oxidation of succinate and glutamate was also found to be depressed. Peroxidation of brain lipids was not increased despite subnormal activity of the copper metalloenzyme, superoxide dismutase.

2.2. Growth and Structural Abnormalities

Growth Retardation

Among the early observations on the effect of dietary copper deficiency was growth retardation in rats (McHargue, 1926), which could be reversed by the reintroduction of copper to the diet (Hart et al., 1928). This finding has been confirmed by later reports using diets containing more recently recognized essential nutrients which could have been lacking in the earlier studies (Moore et al., 1964; Lei, 1977). Growth retardation has also been reported in copper-deficient lambs (Bennetts, 1932), foals (Bennetts, 1935), calves (Cunningham, 1950; Davis, 1950), puppies (Baxter, 1951), chicks (O'Dell et al., 1961), turkeys (Simpson et al., 1967), guinea pigs (Tsai, 1964), mice (Guggenheim, 1964), rabbits (Hunt and Carlton, 1965), and swine (Lahey et al., 1952; Lillie and Frobish, 1978). Inability to utilize copper properly has been suggested as part of the explanation of the small size of chondrodysplastic Alaska malamutes (Brown et al., 1977).

It should be emphasized that, in most reports dealing with the effects of copper deficiency on growth rates, differences in caloric intake between the control and deficient groups are rarely considered. In one case where a pair-fed group was included, the proportion of growth retardation that was independent of caloric intake was found to be minimal (Buckingham et al., unpublished data). Certain specific abnormalities of the skeletal and cardiovascular systems due to copper deficiency are considered in greater detail in the following sections.

Skeletal Malformations

The earliest observations of skeletal defects attributable to copper deficiency were those of Bennetts (1932, 1935) in lambs affected with enzootic ataxia in western Australia. These animals were found to have poorly developed, light, brittle bones with thin cortices. In animals with enzootic ataxia, rib fractures were common findings. In contrast, skeletal lesions were not found in lambs born of dams supplemented with copper during pregnancy (Bennetts and Beck, 1942). Skeletal lesions similar to those described by Bennetts were subsequently reported to occur in New Zealand in lambs afflicted with peat scours (Cunningham, 1946), as well as in Britain associated with the syndrome of swayback, which can also be prevented by copper supplementation (Butler and Barlow, 1963; Barlow et al., 1960). In all three cases—swayback, enzootic ataxia, and peat scours—the disease is confined to animals born to dams grazing pasture land inadequate in copper, but the type and the frequency of skeletal lesions vary geographically. Although the pathogenesis is still not understood, the recently acquired ability to produce the syndrome in lambs experimentally will probably lead to a clarification of the problem (Lewis et al., 1967; Suttle and Angus, 1976, 1978).

In the lambs examined by Cunningham (1950), fractures of the humerus, in particular, were found in the majority of animals with swayback. The fractures were described as osteoporotic-like, although blood calcium and phosphorus were not abnormal. These findings are similar to those reported in lambs in western Australia (Bennetts and Beck, 1942), but are in contrast to data collected in Scotland, in which skeletal malformations were usually not observed in lambs affected with enzootic ataxia (Barlow et al., 1960). In South Africa a disorder similar to swayback and known as "lamkruis" has been described in which a deformity of the fetlock joint is characteristic (Schulz et al., 1951).

Skeletal abnormalities have also been found in calves raised on copper-deficient soils, although the lack of copper is often not the only causative agent (Cunningham, 1950). In the United States the development of a rachitic type of malformation in some calves was prevented by copper supplementation (Davis, 1950). In these animals, as well as in copper-deficient foals (Bennetts, 1935), bone malformations, including wide epiphyses and beaded ribs, were found.

The most extensive research on the skeletal system of copper-deficient animals has been conducted on dogs and swine. In dogs Baxter (1951) and co-workers (Baxter and Van Wyk, 1953; Baxter et al., 1953) found that most puppies born to females fed copper-deficient diets became lame within 2 to 4 months of age. The leg bones were usually severely deformed, with the foreleg characteristically bowed outward, and the hind legs bowed inward. Wrist joints were often hyperextended, with both wrists and elbows showing bony enlargements. Spontaneous fractures were of frequent occurrence. Radiological examination of the bones demonstrated an evenly distributed reduced density and lower than normal cortical thickness. The marrow cavity was usually normal or slightly hyperplastic. Rarefaction of the trabeculae in the spongiosa was common. Epiphyses and metaphyses were deformed, appearing compressed and broadened. Growth of the bone, judged by length, maturation, and epiphyseal ossification centers, appeared normal. Calcium and phosphorus values and the ratio of calcium phosphate to calcium carbonate were similar in bones from control and deficient animals. Circulating levels of calcium, inorganic phosphorus, alkaline phosphatase, ascorbic acid, potassium, vitamin D, and carbon dioxide and blood pH were not affected by copper deficiency. Baxter (Baxter and Van Wyk, 1953; Baxter et al., 1953) has concluded from the chemical and histological data that the skeletal lesion is associated with an impairment of osteogenesis, with resultant thinning of the cortex and trabeculae of long bones, but with unimpaired growth of epiphyseal cartilage. It is important to note that, although copper supplementation can arrest this syndrome, the damage that occurred before the supplementation is irreversible (Baxter and Van Wyk, 1953).

Skeletal malformations were also reported in copper-deficient swine by Teague and Carpenter (1951) and Lahey et al. (1952). Follis et al. (1955) demonstrated that the skeletal lesions found in pigs were essentially identical to those in copper-deficient puppies, and concluded that copper deficiency resulted in a bone with abnormal osteoblastic activity, but with normal osteoclastic and chondroblastic activity (Follis, 1958).

Skeletal abnormalities have also been reported in copper-deficient chicks (Gallagher, 1957; O'Dell et al., 1961; Savage et al., 1962; Carlton and Henderson, 1964). The lesions differ from those of swine and puppies; in contrast to the hyperplasia found in these animals (Baxter and Van Wyk, 1953; Baxter et al., 1953; Follis, 1955), copper-deficient chicks show severe hypoplasia of the long bones (Carlton and Henderson, 1964). Concentrations of calcium, phosphorus, and magnesium were normal in bone of copper-deficient chicks, but copper concentration was low (Rucker et al., 1969). Catalase activity also was normal, but both amine oxidase and cytochrome oxidase activity were significantly below normal in copper-deficient chick bone. There was also a significantly higher ratio of soluble to insoluble collagen in the bone (Rucker et al., 1969). Because of the low activity of the copper enzyme amine oxidase, Rucker et al. (1975) suggested that the increased fragility of copper-deficient bones results from the low number of cross-links present in the collagenous matrix. This hypothesis was supported by the observation that increasing the cross-links of copper-deficient bone collagen by sodium borhydride and formaldehyde produced a bone strength approaching normal (Rucker et al., 1975).

There is one report of postnatal copper deficiency in a human infant with accompanying skeletal malformations (Cordano et al., 1964), including osteoporosis, metaphyseal abnormalities, and frequent fractures. Copper supplementation was effective in treatment.

Surprisingly, the rat does not appear to develop postnatal skeletal abnormalities due to copper deficiency (Underwood, 1977), although skeletal anomalies from prenatal copper deficiency have been reported (O'Dell et al., 1961). The reason for this difference between the rat and other species studied is unexplained.

Cardiovascular Lesions

Despite the early observation of Bennetts (1932) that cardiac hemorrhages were characteristic of enzootic ataxia, current interest in the effect of copper deficiency on the cardiovascular system stems from the observation of high mortality rates in copper-deficient chickens (O'Dell et al., 1961) and swine (Carnes et al., 1961) and their correlation with severe hemorrhages. Similar hemorrhages occurred in newborn offspring of copper-deficient rats (O'Dell et al., 1961). Since the coagulation time of the newborn copper-deficient rat pups was similar to that of controls, it was hypothesized that the lesion was structural in nature. It is now known that the high mortality rate of copper-deficient rats, chickens, and swine is usually due to internal hemorrhage caused by rupture of the heart, aorta, or other large vessels, with smaller lesions throughout the cardiovascular system (Shields et al., 1962; Coulson and Carnes, 1963). Lesions of a similar nature have been reported in copper-deficient turkey poults (Savage et al., 1962), swine (Carnes et al., 1961), rabbits (Hunt and Carlton, 1965; Nystrom 1965), guinea pigs (Everson et al., 1967), and cattle afflicted with "falling disease" (Bennetts et al., 1948).

Nystrom (1965, 1968) has examined the vascular lesions of copper-deficient rabbits. The walls of the internal and common carotid arteries in deficient fetuses were found to have an endothelium normal in appearance, with sparse, poorly developed elastica. Sections of the middle cerebral artery were low in elastin content, and the elastin that was present did not have the concise fibrillar arrangement seen in the control animals. Collagen concentration and arrangement, on the other hand, were reported to be normal. Morphological observations similar to those of Nystrom were reported by Everson and her colleagues (1967) in guinea pigs deprived of copper *in utero*.

In copper-deficient chick embryos, low levels of elastin in the tunica media were correlated with a small number of elastin-secreting fibroblasts, while the number of mesenchymal cells present was higher than normal (Simpson et al., 1967). These findings could suggest a retardation in cell maturation in the tunica media, with a resulting decrease in elastin synthesis. Support for this view can be found in a recent report concerning aortic lesions in copper-deficient chick embryos from 10 days of incubation to 8 days of age posthatching (Anna et al., 1977). There was a delay in the appearance of elastin, and an overall reduction in the amount of elastin formed in these copper-deficient chicks.

There is considerable evidence, therefore, that copper deficiency during the fetal period produces retarded and diminished formation of elastin. Furthermore, the elastin that is made does not undergo the normal maturation steps. Elastin from copper-deficient animals may contain as little as half the cross-linking amino acids found in normal elastin (Starcher et al., 1964; Hill et al., 1967). The primary biochemical lesion responsible for the reduction in cross-links is thought to be a decrease in lysyl oxidase activity (Rucker and Tinker, 1977). This enzyme, for which copper is a cofactor, catalyzes the oxidation of certain peptidyl lysine and hydroxylysine residues to peptide aldehydes, which initiate the cross-linking mechanism required for connective tissue stability (Rucker and Tinker, 1977). Copper deficiency leads to decreased lysyl oxidase activity, with a concomitant decrease in elastin cross-links, but reintroduction of copper to the diet rapidly restores lysyl oxidase activity to normal (Harris, 1976). However, rapid reversal of the connective tissue damage does not occur.

It has been suggested that the low tensile strength of the vascular walls in copper-deficient animals could be due in part to reduced cross-linking of the collagenous matrix, in addition to the reduced cross-linking in the elastin matrix (Coulson and Carnes, 1963). However, Ganezer et al. (1976) were unable to detect a reduction in the cross-links present in the collagen of copper-deficient swine.

In summary, it is clear that a dietary deficiency of copper is related to certain abnormalities of connective tissue. If the deficiency occurs during the developmental period, retardation of skeletal growth and development and abnormalities of the cardiovascular system result. The pathogenesis of the cardiovascular abnormalities undoubtedly involves the low activity of lysyl oxidase, leading to a reduction of elastin cross-links and therefore a decrease in tensile strength, but asynchrony of cell maturation is also a factor. The mechanisms responsible for the skeletal malformations are unresolved, but may perhaps be related to an effect of copper deficiency on collagen formation.

2.3. Copper Deficiency in Human Infants

Although copper deficiency can arise postnatally by a variety of means (Al-Rashid and Spangler, 1971; Karpel and Pender, 1972; Graham and Cordano, 1976), it is not clear to what extent hypocupremia contributes to teratogenesis in human beings. Morton et al. (1976) have reported a statistically significant correlation between low copper in drinking water and neural tube defects in south Wales. Wald and Hambidge (1977), however, were unable to demonstrate a relationship between maternal serum copper concentration and the incidence of anencephaly.

Two cases have been reported in which abnormal children were born to mothers who had received large doses of D-penicillamine during pregnancy. In the first case, Mjølnerød et al. (1971) described the infant of a woman who had taken 2 g of penicillamine daily throughout pregnancy for the treatment of cystinuria. The child had severe connective tissue abnormalities, including hyperflexibility of the joints, fragility of the veins, varicosities, and lax skin, and died of septicemia at 7 weeks of age. After the report of this case, it was suggested that the malformations might have been due to hypocupremia in the mother induced by the penicillamine treatment (Hurley, 1977). Such an interpretation seems reasonable in view of the copper-chelating properties of this drug and its ability to interfere with connective tissue metabolism in the adult (Harris and Sjoerdsma, 1966; Beer and Lyle, 1966; Rupp and Weser, 1976).

In the second, more recent case (Solomon et al., 1977) an abnormal child was born to a mother who had received penicillamine during her pregnancy for the treatment of seropositive rheumatoid arthritis. The infant was small for dates and had a flattened face with a broad nasal ridge, low-set ears, a short neck, and a general looseness of the facial skin. Body skin was also unusually lax and wrinkled. Bilateral inguinal hernias were present, and the knees and hips had fixed flexion contratures. The child died at 3 days of age of undetermined causes. Unfortunately, trace element analyses of tissues were not presented in either of these reports.

Some investigators have questioned the teratogenicity of penicillamine, since use of the drug during pregnancy for the control of Wilson's disease has not apparently increased the frequency of malformed infants (Corcorcan and Castles, 1977; Scheinberg and Sternlieb, 1975; Walshe, 1977). The apparent absence of serious effects of the drug on the fetuses of patients with Wilson's disease, however, could be due to the extremely high concentration of copper in these patients compared to that of the normal population. A similarly high concentration of copper may not be present in patients treated for cystinuria or rheumatoid arthritis. The amount of penicillamine given may also be a critical factor. Scheinberg and Sternlieb (1975) have suggested that the maximum dose for the treatment of Wilson's disease during pregnancy should be 250 mg/day, but the dosage used in the case reported by Solomon et al. (1977) was 900 mg/day, and in that of Mjølnerød et al. (1971), 2000 mg/day. Lyle (1978) reviewed eight cases of women who had received penicillamine throughout pregnancy (four for cystinuria, four for rheumatoid arthritis) and found that all had had normal infants, but the dosage of penicillamine was not given.

It should also be noted that penicillamine removes other trace elements such as zinc, as well as copper, so that the abnormalities reported from penicillamine cannot be attributed with certainty to copper deficiency. It is clear that teratogenicity due to hypocupremia does occur in the genetic disorder Menkes' disease (see Section 4.2).

3. TERATOGENIC CONSEQUENCES OF EXCESS COPPER

3.1. Animals

It is known that small amounts of metabolic copper from intrauterine loops or wires can prevent mammalian embryogenesis by blocking implantation and blastocyst development (Chang et al., 1970; Zipper et al., 1969), but the teratogenicity of excess copper has not been established. Studies of copper intrauterine devices have failed to find any abnormalities in the offspring of rats, hamsters, rabbits, or guinea pigs exposed to a high intrauterine copper environment (Chang and Tatum, 1973; Moo-Young and Tatum, 1974). In ewes, feeding sublethal doses of copper did not cause any fetal malformations (James et al., 1966). In addition, despite numerous reports of copper toxicosis under natural grazing conditions in pigs, sheep, and cattle, there are no references to abnormal young being born to these animals (Underwood, 1977).

In contrast, Ferm (1974) has reported that intravenously injected copper is teratogenic. Both copper citrate and copper sulfate were found to be potent teratogens when injected on day 8 of pregnancy in the hamster. Fetal resorption, kinked tail, thoracic and supraumbilical ventral hernia, craniorachischisis, microphthalmia, cleft lip, and ectopic cordis were among the abnormalities reported. Hanlon and Ferm (1974) have suggested that the teratogenicity of copper is specific in that similar malformations were not obtained by injection of the divalent cations zinc and cadmium.

3.2. Human Beings

There do not appear to be any reports in the literature of teratogenesis in human beings induced by excess copper. In particular, no reports could be found of abnormalities in the offspring of mothers with untreated Wilson's disease.

Thus, although copper teratogenicity has been demonstrated by intravenous injection, the element has not been shown to be teratogenic when excess amounts are consumed, or when it is present at high concentrations in the uterine environment.

4. COPPER MUTANTS

The importance of copper during perinatal development is demonstrated by several genetic mutants that possess errors of copper metabolism. Interactions occurring between copper and genetic factors and affecting development can be classified into two groups. The first type of interaction involves strain differences which produce differential responses to a dietary deficiency of the element; the second type involves a single mutant gene, whose phenotypic expression may resemble the characteristic signs of a deficiency or toxicity of an element, and whose expression can be reduced or prevented by pre-, post-, or perinatal nutritional manipulation (Hurley, 1976, 1977).

4.1. Animal Models

Representative of the first category of gene-nutrient interactions is Wiener's observation (1966) of a marked influence of the breed of sheep on the incidence of swayback in a given geographical location. The Welsh genotype was associated with the lowest incidence of swayback, and the Blackface genotype with the highest (Wiener and Field, 1969). These differences in the frequency of swayback have been correlated with differences in blood copper concentration, sheep of the Blackface genotype having levels markedly lower than those of the Welsh genotype (Wiener and Field, 1969). Further evidence that the metabolism of copper differs in these two species is the observation that copper toxicity occurs at a lower level of copper intake in the Welsh genotype than in the Blackface genotype (Wiener and MacLeod, 1970).

Recent embryo transfer studies have demonstrated that the increased frequency of swayback in the Blackface genotype is primarily due to an abnormality in maternal copper metabolism. Wiener et al. (1978) have reported that the liver copper concentration of the newborn is determined by the breed of the dam, not of the lamb. It is not known whether placental transfer or maternal plasma copper concentration is abnormal. The early postnatal liver copper concentration of the lamb was also found to be dependent on the breed of the dam. Wiener has suggested that this may be due to a difference in the copper concentration of the dam's milk, since the Welsh dams had a higher concentration than that of the Blackface breed. The copper concentration of milk has been shown to influence liver copper stores in the offspring both in dietary copper deficiency (Kirschgessner and Spoerl, 1975b) and in copper excess (Keen et al., unpublished data).

Examples of the second category of gene-nutrient interactions are certain mutant genes in mice and human beings that produce alterations in copper metabolism. The autosomal recessive gene *crinkled* (*cr*) in mice causes, phenotypically, a smooth coat, thin skin, delayed pigmentation, and early postnatal mortality. Hair bulb development is retarded and abnormal; only one (straight) of the four normal types of hairs is found (Falconer et al., 1952). Thus the phenotypic characteristics are similar to the signs of copper deficiency seen in several species (see preceding sections).

The relationship between the mutant gene *crinkled* and copper metabolism was therefore investigated by supplementing mice during pregnancy and lactation with a high level of dietary copper (500 ppm Cu). The supplementation treatment increased the survival of crinkled mutants from 32 to 59% in the early neonatal period. In these offspring, skin and epidermal thickness, pigmentation, and hair bulb number and development were all substantially improved by supplementation when compared to the crinkled young of dams fed a diet normal in copper concentration (Hurley and Bell, 1975).

Copper supplementation during the perinatal period was also found to reduce the severity of brain abnormalities in the crinkled mutant, including a high level of brain cholesterol esters. In addition, copper supplementation during the perinatal period reduced the high concentration of sulfatides in the brains of the young mutants and the high concentration of cerebrosides in the brains of older mutants. Histological examination demonstrated abnormalities of the myelin sheath, which, coupled with biochemical data, suggested neural degeneration. Behavioral abnormalities were occasionally observed. All of these signs were ameliorated by the higher copper diet (Theriault et al., 1977).

Low levels of liver copper have also been found in the crinkled mutant from 18 days of gestation to 21 days of age. After this time liver copper concentration is similar to that of controls. Brain copper concentrations have not been found to be different between the mutant and its control (Keen and Hurley, 1978). Intestinal transport of copper is depressed in the newborn crinkled mutant compared to normal control mice, although intestinal copper-binding proteins have not been shown to be different (Bronner et al., 1978). Whereas some effects of the crinkled gene appear to be restricted to the perinatal period, others, such as low activity of the copper metalloenzyme superoxide dismutase, persist into adulthood (Keen and Hurley, 1978).

Another mutant gene in mice that is related to copper metabolism is *quaking* (*qk*). Mice homozygous for this gene have intermittent axial body tremors beginning about 10 days after birth and persisting throughout life. Such animals also show abnormalities of myelin and reduced levels of brain cerebrosides and sulfatides (Sidman et al., 1964; Baumann et al., 1968). When mice heterozygous for the mutant gene *quaking* were given a diet containing a high level of copper (250 ppm) during pregnancy and lactation, the frequency of tremors in their mutant offspring (homozygous for *qk*) was greatly reduced. Brain copper concentrations, which were found to be subnormal in the nonsupplemented quaking mutant, were of normal levels in the high copper diet group (Keen and Hurley, 1976).

Mice with the mutant gene *mottled* also have defects of copper metabolism. These X-linked mutants consist of a number of different alleles at the mottled locus. Many of the alleles are lethal for the heterozygous male. In the animals that survive birth, there are severe dilution of hair pigmentation and persistent neurological disturbances, including a mild sustained tremor and general inactivity (Gruneberg, 1969). These mice have reduced levels of noradrenalin, cytochrome *c* oxidase, and superoxide dismutase. Copper concentration of the intestinal wall and kidney tends to be high in the mutant, with subnormal levels occurring in the liver and brain, suggesting an error in copper transport (Hunt, 1974; Prins and Van den Hamer, 1978; Evans and Reis, 1978). Mottled mice have defects in collagen and elastin cross-linking, leading to aortic aneurisms (Rowe et al., 1974), which have been correlated with decreased lysyl oxidase activities (Rowe et al., 1977). Fibroblasts isolated from the mottled mouse bearing the *blotchy* allele have been reported to have normal ceruloplasmin (ferroxidase) levels and to contain what appears to be a nonmetallothionein copper-binding protein in abnormally high concentrations (Starcher et al., 1978). Intraperitoneal injections of copper have been shown to ameliorate some of the effects of the mottled genes (Hunt, 1976), but feeding trials with high copper diets have not been reported. One of the more interesting aspects of the mottled mutants is the similarity of their symptoms to those of the human X-linked genetic disease Menkes' kinky hair syndrome (Section 4.2), for which the mottled mouse has been suggested as a model (Danks, 1975).

In the two cases described above in which dietary supplementation ameliorated copper metabolic errors (*crinkled* and *quaking*), it should be noted that the time of intervention was critical. Prenatal supplementation with copper was of maximum benefit for the crinkled mutant, while postnatal supplementation was most advantageous for the quaking mutant (Dungan et al., unpublished data). Information gained from these and other copper mutants should be helpful in gaining a better understanding of copper metabolism.

4.2. Menkes' Kinky Hair Syndrome in Human Beings

Menkes' kinky hair syndrome, an X-linked genetic disorder, was first described in 1962 (Menkes, 1962). The disease is characterized by progressive degeneration of the brain and spinal cord in infants. Bone, cardiovascular, and enzymatic abnormalities similar to those described for copper-deficient animals are also found. Death usually occurs before 3 years of age. In 1972 Danks et al. (1972a, 1972b) demonstrated an abnormality of copper metabolism in infants with this syndrome. Subsequent work suggested that there is a generalized defect of copper transport (Danks et al., 1973). However, despite increased serum copper and ceruloplasmin levels with copper therapy (Bucknall et al., 1973; Dekaban and Steusing, 1974; Grover and Scrutton, 1975), the long-term prognosis of these patients has not been greatly improved (Garnica et al., 1977). One possible explanation for this failure of copper therapy is that it is started when the disease is already in an advanced state (Bucknall et al., 1973) and that reversal does not occur. The recent observation of abnormal copper metabolism *in utero* in the Menkes fetus (Grover and Henkin, 1976) supports this hypothesis. It is thus possible that, as in the case of the crinkled mutant, prenatal intervention is required for efficacious treatment.

Danks (1975) has proposed that the primary biochemical lesion in Menkes' syndrome and in mottled mice is the presence of an abnormal protein with a high avidity for copper. According to this hypothesis, when small or normal amounts of copper are available to the cell, the majority of it becomes sequestered by the mutant protein so that other copper-dependent processes in the cell have a relative deficiency of the element. If the concentration of copper in the cell is increased until the abnormal protein is saturated, additional copper could be utilized, but the level at which this occurs might be toxic (Danks et al., 1978).

One method by which such an abnormal protein might be bypassed would be to supply the element in a form which could be absorbed and incorporated without passing through the mutant protein. Nitrilotriacetic acid (NTA), which can form a ligand with copper, has been utilized in the treatment of Menkes' syndrome by Henkin and Grover (1978). These investigators have suggested that NTA may bypass the metabolic block of Menkes' disease, although only preliminary results are available. Unpublished data from our laboratory indicate that the NTA ligand of copper is also effective in ameliorating the effects of the mutant gene *crinkled* in mice.

5. CONCLUDING REMARKS

It is obvious that many unanswered questions remain concerning the role of copper, in deficiency or excess, as a teratogen. One of the subject areas receiving attention at present is the dynamics of maternal-fetal utilization of copper (Henkin et al., 1971; Williams and Bremner, 1976; Moss et al., 1974). A better understanding of this subject may clarify the mechanisms of copper teratogenicity. The utilization of copper mutants both as models of human disorders and as tools to elucidate the metabolic pathways of copper is also of great potential value. Although the degree to which copper abnormalities contribute to human teratogenicity is not clear, the relative ease with which other species are affected by copper deficiency warrants continued inspection of the role of this element in human development.

REFERENCES

Al-Rashid, R. A. and Spangler, J. (1971). "Neonatal Copper Deficiency," *N. Engl. J. Med.*, **285,** 841–843.

Allcroft, R., Clegg, F. G., and Uvarov, O. (1959). "Prevention of Swayback in Lambs," *Vet. Rec.*, **71,** 884–889.

Anna, K., Agnes, J., and Jellinek, H. (1977). "Lesions Caused by Copper Depletion in the Chicken: Light and Electron Microscopic Study," *Paroi Arterielle* (*Arterial Wall*), **4**(7), 45–57.

Barlow, R. M. (1963). "Further Observations on Swayback. I: Transitional Pathology," *J. Comp. Pathol.*, **73,** 51–60.

Barlow, R. M. (1969). "The Foetal Sheep: Morphogenesis of the Nervous System and Histochemical Aspects of Myelination," *J. Comp. Neurol.*, **135,** 249–269.

Barlow, R. M., Purver, D., Butler, E. J., and MacIntyre, J. I. (1960). "Swayback in South-East Scotland. II: Clinical, Pathological and Biochemical Aspects," *J. Comp. Pathol.*, **70,** 411–428.

Baumann, N. A., Jacque, C. M., Pollet, S. A., and Harpin, M. L. (1968). "Fatty Acid and Lipid Composition of the Brain of a Myelin Deficient Mutant, the "Quaking" Mouse," *Eur. J. Biochem.*, **5,** 340–344.

Baxter, J. H. (1951). "Bone Disorder in Copper-Deficient Puppies," *Am. J. Physiol.*, **167,** 766.

Baxter, J. H. and Van Wyk, J. J. (1953). "A Bone Disorder Associated with Copper Deficiency. I: Gross Morphological, Roentgenological, and Chemical Observation," *Bull. Johns Hopkins Hosp.*, **93,** 1–23.

Baxter, J. H., Van Wyk, J. J., and Follis, R. H., Jr. (1953). "A Bone Disorder Associated with Copper Deficiency. II: Histological and Chemical Studies on the Bones," *Bull. Johns Hopkins Hosp.*, **93,** 25–39.

Beer, W. E. and Lyle, W. H. (1966). "Penicillamine for the Treatment of Darier's Disease and Other Disorders of Keratin Formation," *Lancet*, **ii,** 1337–1340.

Bennetts, H. W. (1932). "Enzootic Ataxia of Lambs in Western Australia," *Aust. Vet. J.*, **8,** 137–142, 183–184.

Bennetts, H. W. (1935). "The Prevention of So-Called "Rickets" in Lambs and Foals," *J. Dept. Agric. W. Aust.* (Series 2), **12,** 305–310.

Bennetts, H. W. and Beck, A. B. (1942). "Enzootic Ataxia and Copper Deficiency of Sheep in Western Australia," *Aust. Coun. Sci. Ind. Res. Bull.*, **147,** 1–52.

Bennetts, H. W. and Chapman, F. E. (1937). "Copper Deficiency in Sheep in Western Australia: A Preliminary Account of the Aetiology of Enzootic Ataxia of Ewes," *Aust. Vet. J.*, **13,** 138–149.

Bennetts, H. W., Beck, A. B., and Harley, R. (1948). "The Pathogenesis of "Falling Disease": Studies of Copper Deficiency in Cattle," *Aust. Vet. J.*, **24,** 237–244.

Bronner, F., Golub, E. E., Ueng, T. H., and Bossak, C. (1978). "Intestinal Copper Transport and Copper-Binding Protein (CuBP) in Crinkled (*cr/cr*) Mice," *Fed. Proc.*, **3,** 721.

Brown, R. G., Hoag, G. N., Smart, M. E., Boechner, G., and Subdcn, R. E. (1977). "Alaskan Malamute Chondrodysplasia. IV: Concentrations of Zinc, Copper and Iron in Various Tissues," *Growth,* **41,** 215–220.

Buckingham, K. W., Heng-Khoo, C. S., Lefevre, M., Rucker, R. B., and Julian, L. M. (1978). "Morphology of Copper Deficient Chick Lung," (in preparation).

Bucknall, W. D., Haslam, R. H. A., and Holtzman, N. A. (1973). "Kinky Hair Syndrome: Response to Copper Therapy," *Pediatrics,* **52,** 653–657.

Butler, E. J. and Barlow, R. M. (1963). "Factors Influencing the Blood and Plasma Copper Levels of Sheep in Swayback Flocks," *J. Comp. Pathol.,* **73,** 107–118.

Cancilla, P. A. and Barlow, R. M. (1966a). "Structural Changes of the Central Nervous System in Swayback (Enzootic Ataxia) of Lambs. I: Electron Microscopy of the Cerebral Lesion," *Acta Neuropathol.,* **6,** 260–265.

Cancilla, P. A. and Barlow, R. M. (1966b). "Structural Changes of the Central Nervous System in Swayback (Enzootic Ataxia) of Lambs. II: Electron Microscopy of the Lower Motor Neuron," *Acta Neuropathol.,* **6,** 251–259.

Cancilla, P. A. and Barlow, R. M. (1968). "Structural Changes of the Central Nervous System in Swayback (Enzootic Ataxia) of Lambs. IV. Electron Microscopy of the White Matter of the Spinal Cord," *Acta Neuropathol.,* **11,** 294–300.

Cancilla, P. A. and Barlow, R. M. (1970). "Experimental Copper Deficiency in Miniature Swine," *J. Comp. Pathol.,* **80,** 315–319.

Carlton, W. W. and Henderson, W. (1964). "Skeletal Lesions in Experimental Copper-Deficiency in Chickens," *Avian Dis.,* **8,** 48–55.

Carlton, W. W. and Kelly, W. A. (1969). "Neural Lesions in the Offspring of Female Rats Fed a Copper Deficient Diet," *J. Nutr.,* **97,** 42–52.

Carnes, W. H., Shields, G. S., Cartwright, G. E., and Wintrobe, M. M. (1961). "Vascular Lesions in Copper Deficient Swine," *Fed. Proc.,* **20,** 118.

Chang, C. C. and Tatum, H. J. (1970). "A Study of the Antifertility Effect of Intrauterine Copper," *Contraception,* **1,** 265–270.

Chang, C. C. and Tatum, H. J. (1973). "Absence of Teratogenicity of Intrauterine Copper Wire in Rats, Hamsters, and Rabbits," *Contraception,* **7,** 413–434.

Chang, C. C., Tatum, H. J., and Kinel, F. A. (1970). "The Effect of Intrauterine Copper and Other Metals on Implantation in Rats and Hamsters," *Fertil. Steril.,* **21,** 274–278.

Corcoran, R. and Castles, W. J. B. (1977). "Penicillamine Therapy and Teratogenesis," *Br. Med. J.,* **1,** 838.

Cordano, A., Baertl, J. M., and Graham, G. G. (1964). "Copper Deficiency in Infancy," *Pediatrics,* **34,** 324–336.

Coulson, W. F. and Carnes, W. H. (1963). "Cardiovascular Studies on Copper-Deficient Swine. V: The Histogenesis of the Coronary Artery Lesions," *Am. J. Pathol.,* **43,** 945–954.

Cunningham, I. J. (1946). "Copper Deficiency in Cattle and Sheep on Peat Land," *N. Z. J. Sci. Technol.,* **27A,** 381.

Cunningham, I. J. (1950). "Copper and Molybdenum in Relation to Diseases of Cattle and Sheep in New Zealand." In W. D. McElroy and B. Glass, Eds., *A Symposium on Copper Metabolism.* Johns Hopkins Press, Baltimore, Md., pp. 246–273.

Danks, D. M. (1975). "Steely Hair, Mottled Mice and Copper Metabolism," *N. Engl. J. Med.,* **293,** 1147–1148.

Danks, D. M., Campbell, P. E., Gillespie, J. M., Walker-Smith, J., Blomfield, J., and Turner, B. (1972a). "Menkes Kinky Hair Syndrome," *Lancet,* **i,** 1100–1102.

Danks, D. M., Campbell, P. E., Stevens, B. J., Mayne, V., and Cartwright, E. (1972b). "Menkes Kinky Hair Syndrome: An Inherited Defect in Copper Absorption with Widespread Effects," *Pediatrics,* **50,** 188–201.

Danks, D. M., Cartwright, E., Stevens, B. J., and Townley, R. R. W. (1973). "Menkes Kinky Hair Disease: Further Definition of the Defect in Copper Transport," *Science,* **179,** 1140–1142.

Danks, D. M., Camakaris, J., and Stevens, B. J. (1978). "The Cellular Defect in Menkes' Syndrome and in Mottled Mice." In M. Kirchgessner, Ed., *Trace Element Metabolism in Man and Animals*—3. Technischen Universität München, Freising-Weihenstephan, Germany, pp. 401–404.

Davis, G. K. (1950). "The Influence of Copper on the Metabolism of Phosphorus and Molybdenum." In W. D. McElroy and B. Glass, Eds., *A Symposium on Copper Metabolism.* Johns Hopkins Press, Baltimore, Md., pp. 216–229.

Dekaban, A. S. and Steusing, J. K. (1974). "Menkes Kinky Hair Disease Treated with Subcutaneous Copper Sulphate," *Lancet,* **ii,** 1523.

Dipaolo, D. V., Kanfer, J. N., and Newberne, P. M. (1974). "Copper Deficiency and the Central Nervous System. Myelination in the Rat: Morphological and Biochemical Studies," *J. Neuropathol. Exp. Neurol.,* **33,** 226–236.

Dungan, D. D., Keen, C. L., Theriault, L. L., and Hurley, L. S. "Effect of Copper Supplementation Prenatally versus Postnatally on Mutant Characteristics of Crinkled and Quaking Mice" (in preparation).

Dutt, B. and Mills, C. F. (1960). "Reproductive Failure in Rats Due to Copper Deficiency," *J. Comp. Pathol.,* **70,** 120–125.

Evans, G. W. and Reis, B. L. (1978). "Impaired Copper Homeostasis in Neonatal Male and Adult Female Brindled (Mo^{br}) Mice," *J. Nutr.,* **108,** 554–560.

Everson, G. J., Tsai, C. H., and Wang, T. (1967). "Copper Deficiency in the Guinea Pig," *J. Nutr.,* **93,** 533–540.

Everson, G. J., Shrader, R. E., and Wang, T. (1968). "Chemical and Morphological Changes in the Brains of Copper-Deficiency Guinea Pigs," *J. Nutr.,* **96,** 115–125.

Falconer, D. S., Fraser, A. S., and King, J. W. B. (1952). "The Genetics and Development of "Crinkled," a New Mutant in the House Mouse," *J. Genet.,* **50,** 324–344.

Fell, B. F., Mills, C. F., and Boyne, R. (1965). "Cytochrome Oxidase Deficiency in the Motor Neurons of Copper-Deficient Lambs: A Histochemical Study," *Res. Vet. Sci.,* **6,** 170–177.

Ferm, V. H. (1972). "The Teratogenic Effects of Metals on Mammalian Embryos." Chapter 2 in D. H. M. Wollam, Ed., *Advances in Teratology.* Logos Press: London, pp. 51–75.

Ferm, V. H. and Hanlon, D. P. (1974). "Toxicity of Copper Salts in Hamster Embryonic Development," *Biol. Reprod.*, **11,** 97–101.

Follis, R. (1958). *Deficiency Disease.* Thomas, Springfield, Ill., pp. 56–64.

Follis, R. H., Jr., Bush, J. A., Cartwright, G. E., and Wintrobe, M. M. (1955). "Studies on Copper Metabolism. XVIII: Skeletal Changes Associated with Copper Deficiency in Swine," *Bull. Johns Hopkins Hosp.*, **97,** 405–414.

Gallagher, C. H. (1957). "The Pathology and Biochemistry of Copper Deficiency," *Aust. Vet. J.*, **33,** 311–317.

Gallagher, C. H. and Reeve, V. E. (1971). "Copper Deficiency in the Rat. Effect on Synthesis of Phospholipids," *Aust. J. Exp. Biol. Med. Sci.*, **49,** 21–31.

Gallagher, C. H., Judah, J. D., and Rees, K. R. (1956). "The Biochemistry of Copper Deficiency and Enzymological Disturbances, Blood Chemistry and Excretion of Amino Acids," *Proc. R. Soc.*, **B145,** 134–150.

Ganezer, K. S., Hart, M. L., and Carnes, W. H. (1976). "Tensile Properties of Tendon in Copper-Deficient Swine," *Proc. Soc. Exp. Biol. Med.*, **153,** 396–399.

Garnica, A. D., Frias, J. L., and Rennert, O. M. (1977). "Menkes' Kinky Hair Syndrome: Is It a Treatable Disorder?" *Clin. Gen.*, **11,** 154–161.

Ghergariu, S. (1978). "Some Factors Affecting the Incidence of Swayback in Lambs. In M. Kirchgessner, Ed., *Trace Element Metabolism in Man and Animals*—3. Technischen Universität München, Freising-Weihenstephan, Germany, p. 500.

Graham, G. G. and Cordano, A. (1976). "Copper Deficiency in Human Subjects." In A. S. Prasad, Ed., *Trace Elements in Human Health and Disease,* Vol. I: *Zinc and Copper,* pp. 263–272.

Greunberg, H. (1969). "Threshold Phenomena versus Cell Heredity in the Manifestation of Sex-Linked Genes in Mammals," *J. Embryol. Exp. Morphol.*, **22,** 145–179.

Grover, W. D. and Henkin, R. I. (1976). "Trichopoliodystrophy (TPD): A Fetal Disorder of Copper Metabolism," *Ped. Res.*, **10,** 448.

Grover, W. D. and Scrutton, M. D. (1975). "Copper Infusion Therapy in Trichopoliodystrophy," *J. Pediat.*, **86,** 216–220.

Guggenheim, K. (1964). "The Role of Zinc, Copper and Calcium in the Etiology of the "Meat Anemia," *Blood,* **23,** 786–794.

Guggenheim, K., Tal, E., and Zor, U. (1964). "The Effect of Copper on the Mineralization of Bones of Mice Fed on a Meat Diet," *Br. J. Nutr.*, **18,** 529–535.

Hall, G. A. and Howell, J. McC. (1969). "The Effect of Copper Deficiency on Reproduction in the Female Rat," *Br. J. Nutr.*, **23,** 41–45.

Hanlon, D. P. and Ferm, V. H. (1974). "Possible Mechanisms of Metal Ion Induced Teratogenesis," *Teratology,* **9,** A-19.

Harris, E. D. (1976). "Copper-Induced Activation of Aortic Lysyl Oxidase *in vivo*," *Proc. Natl. Acad. Sci.*, **73**(2), 371–374.

Harris, E. D. and Sjoerdsma, A. (1966). "Collagen Profile in Various Clinical Conditions," *Lancet,* **ii,** 707–711.

Hart, E. B., Steenbock, H., Waddell, J., and Elvehjem, C. A. (1928). "Iron in Nutrition. VII: Copper as a Supplement to Iron for Hemoglobin Building in the Rat," *J. Biol. Chem.*, **77,** 797–812.

Henkin, R. I. and Grover, W. D. (1978). "Trichopoliodystrophy (TPD): New Aspects of Pathology and Treatment." In M. Kirchgessner, Ed., *Trace Element Metabolism in Man and Animals*—3. Technischen Universität München, Friesing-Weihenstephan, Germany, pp. 405–408.

Henkin, R. I., Marshall, J. R., and Meret, S. (1971). "Maternal-Fetal Metabolism of Copper and Zinc at Term," *Am. J. Obstet. Gynecol.,* **110**(1), 131–134.

Hill, C. H., Starcher, B., and Kim, C. (1967). "Role of Copper in the Formation of Elastin," *Fed. Proc.,* **26,** 129–133.

Howell, J. McC. and Davison, A. N. (1959). "The Copper Content and Cytochrome Oxidase Activity of Tissue from Normal and Swayback Lambs," *Biochem. J.,* **72,** 365–368.

Howell, J. McC., Davison, A. N., and Oxberry, J. (1969). "Observations on the Lesions in White Matter of the Spinal Cord of Swayback Sheep," *Acta Neuropathol.,* **12,** 33–41.

Hunt, C. E. and Carlton, W. W. (1965). "Cardiovascular Lesions Associated with Experimental Copper Deficiency in the Rabbit," *J. Nutr.,* **87,** 385–393.

Hunt, D. M. (1974). "Primary Defect in Copper Transport Underlies Mottled Mutants in the Mouse," *Nature,* **249,** 852–854.

Hunt, D. M. (1976). "A Study of Copper Treatment and Tissue Copper Levels in the Murine Congenital Copper Deficiency, Mottled," *Life Sci.,* **19,** 1913–1920.

Hurley, L. S. (1976). "Trace Elements and Teratogenesis," *Med. Clin. N. America,* **60,** 771–778.

Hurley, L. S. (1977). "Nutritional Deficiencies and Excess. In J. G. Wilson and F. C. Fraser, Eds., *Handbook of Teratology*. Plenum, New York, pp. 361–408.

Hurley, L. S. and Bell, L. T. (1975). "Amelioration by Copper Supplementation of Mutant Gene Effects in the Crinkled Mouse," *Proc. Soc. Exp. Biol. Med.,* **149,** 830–834.

Innes, J. R. M. (1936). "Swayback"—a Demyelinating Disease of Lambs with Affinities to Schilder's Encephalitis in Man," *Vet. Rec.,* **48,** 1539–1549.

Innes, J. R. M. and Shearer, G. D. (1940). "Swayback"—a Demyelinating Disease of Lambs with Affinities to Schilder's Encephalitis in Man," *J. Comp. Pathol. Ther.,* **53,** 1–41.

James, L. S., Lazor, V. A., and Binns, W. (1966). "Effects of Sublethal Doses of Certain Minerals on Pregnant Ewes and Fetal Development," *Am. J. Vet. Res.,* **27,** 132–135.

Karpel, J. T. and Pender, V. T. (1972). "Copper Deficiency in Long Term Parenteral Nutrition," *J. Pediat.,* **80,** 32–36.

Keen, C. L. and Hurley, L. S. (1976). "Copper Supplementation in Quaking Mutant Mice: Reduced Tremors and Increased Brain Copper," *Science,* **193,** 244–246.

Keen, C. L. and Hurley, L. S. (1978). "Developmental Abnormality of Copper Metabolism in the Mouse Mutant Crinkled," *Fed. Proc.,* **37,** 332.

Keen, C. L., Lönnerdal, B., Sloan, M. V., and Hurley, L. S. "Effect of Dietary Zinc, Copper, and Iron Chelates of Nitrilotriacetic Acid (NTA) on Trace Metal Concentrations in Rat Milk, Maternal, and Pup Tissues" (in preparation).

King, R. A., Osborne-White, W. S., and Smith, R. M. (1977). "Does Cytochrome Oxidase Become Rate-Limiting in Severe Copper Deficiency?" *Proc. Aust. Biochem. Soc.,* **10,** 31.

Kirchgessner, V. M. and Spoerl, R. (1975a). "Zum Trachtigkeitsanabolismus an Kupfer in Abhangigkeit von der Cu-versorgung," *Z. Tierphysiol., Tierernaehr. Futtermittelkd.,* **36,** 75–86.

Kirchgessner, M. and Spoerl, R. (1975b). "Zur Cu-konzentration von Rattenmilch bei Unterschiedlichen Cu-gehalten der Diaten," *Arch. Tierernaehr.,* **25,** 505–512.

Lahey, M. E., Gubber, G. J., Chase, M. S., Cartwright, G. E., and Wintrobe, M. M. (1952). "Studies on Copper Metabolism. II: Hematologic Manifestations of Copper Deficiency in Swines," *Blood,* **7,** 1053–1074.

Lei, K. Y. (1977). "Cholesterol Metabolism in Copper-Deficient Rats," *Nutr. Rep. Int.,* **15**(5), 597–605.

Lewis, G., Terlecki, S., and Allcroft, R. (1967). "The Occurrence of Swayback in Lambs of Ewes Fed a Semi-purified Diet of Low Copper Content," *Vet. Rec.,* **81,** 415–416.

Lillie, R. J. and Frobish, L. T. (1978). "Effect of Copper and Iron Supplements on Performance and Hematology of Confined Sows and Their Progeny through Four Reproductive Cycles," *J. Anim. Sci.,* **46,** 678–685.

Lyle, W. H. (1978). "Penicillamine in Pregnancy," *Lancet,* **i,** 606–607.

McHargue, J. S. (1926). "Further Evidence That Small Quantities of Copper, Manganese and Zinc Are Factors in the Metabolism of Animals," *Am. J. Physiol.,* **77,** 245–255.

Menkes, J. H., Alter, M., Steigleder, G. K., Weakley, D. R., and Sung, J. H. (1962). "A Sex-linked Recessive Disorder with Retardation of Growth, Peculiar Hair and Focal Cerebellar Degeneration," *Pediatrics,* **29,** 764–779.

Mills, C. F. and Fell, B. F. (1960). "Demyelination of Lambs Born of Ewes Maintained on High Intakes of Sulphate and Molybdate," *Nature,* **185,** 20–22.

Mills, C. F. and Williams, R. B. (1962). "Copper Concentration and Cytochrome-Oxidase and Ribonuclease Activities in the Brains of Copper Deficient Lambs," *Biochem. J.,* **85,** 629–632.

Mjølnerød, O. K., Dommerud, S. A., Rasmussen, K., and Gjeruldsen, S. T. (1971). "Congenital Connective-Tissue Defect Probably due to D-Penicillamine Treatment in Pregnancy," *Lancet,* **i,** 673–675.

Moore, T., Constable, B. J., Day, K. C., Impey, S. G., and Symonds, K. R. (1964). "Copper Deficiency in Rats Fed on Raw Meat," *Br. J. Nutr.,* **18,** 135–146.

Moo-Young, A. J. and Tatum, H. J. (1974). "Copper Levels in Maternal and Fetal Tissues of Rabbits Bearing Intrauterine Copper Wires," *Contraception,* **9,** 487–496.

Morton, M. S., Elwood, P. C., and Abernethy, M. (1976). "Trace Elements in Water and Congenital Malformations of the Central Nervous System in South Wales," *Br. J. Prev. Soc. Med.,* **30,** 36–39.

Moss, B. R., Madsen, F., Hansard, S. L., and Gamble, C. T. (1974). "Maternal-Fetal Utilization of Copper by Sheep," *J. Anim. Sci.,* **38**(2), 475–479.

Neal, W. M., Becker, R. B., and Shealy, A. L. (1931). "A Natural Copper Deficiency in Cattle Rations," *Science,* **74,** 418–419.

Nystrom, S. H. M. (1965). "Cytological Aspects of the Pathogenesis of Intercranial Aneurysms." In W. S. Fields and A. L. Sahs, Eds., *Intercranial Aneurysms and Subarachnoid Hemorrhage.* Thomas, Springfield, Ill., pp. 40–69.

Nystrom, S. H. M. (1968). "Studies on the Effect of Experimental Hypocupremia on the Rabbit Circulatory System, with Special Reference to the Vessels of the Neck and Base of the Brain," *Acta Neuropathol.,* **10,** 209–217.

O'Dell, B. L., Hardwick, B. C., and Reynolds, G. (1961). "Mineral Deficiencies of Milk and Congenital Malformations in Rats," *J. Nutr.,* **73,** 151–157.

O'Dell, B. L., Smith, R. M., and King, R. A. (1976). "Effect of Copper Status on Brain Neurotransmitter Metabolism in the Lamb," *J. Neurochem.,* **26,** 451–455.

Owen, E. C., Proudfoot, R., Robertson, J. M., Barlow, R. M., Butler, E. J., and Smith, B. S. W. (1965). "Pathological and Biochemical Studies on an Outbreak of Swayback in Goats," *J. Comp. Pathol.,* **75,** 241–251.

Palsson, P. A. and Grimsson, H. (1953). "Demyelination in Lambs from Ewes Which Feed on Seaweeds," *Proc. Soc. Exp. Biol. Med.,* **83,** 518–520.

Prins, H. W. and Van den Hamer, C. J. A. (1978). "Cu-Content of the Liver of Young Mice in Relation to a Defect in the Cu-Metabolism." In M. Kirchgessner, Ed., *Trace Element Metabolism in Man and Animals*—3. Technischen Universität München, Freising-Weihenstephan, Germany, pp. 397–400.

Prohaska, J. R. and Wells, W. W. (1975). "Copper Deficiency in the Developing Rat Brain: Evidence for Abnormal Mitochondria," *J. Neurochem.,* **25,** 221–228.

Roberts, H. E., Williams, B. M., and Harvard, A. (1966a). "Cerebral Edema in Lambs Associated with Hypocurosis, and Its Relationship to Swayback. I: Biochemical Findings," *J. Comp. Pathol.,* **76,** 279–284.

Roberts, H. E., Williams, B. M., and Harvard, A. (1966b). "Cerebral Edema in Lambs Associated with Hypocuprosis, and Its Relationship to Swayback. II: Histopathological Findings," *J. Comp. Pathol.,* **76,** 285–290.

Rowe, D. W., McGoodwin, E. B., Martin, G. R., Sussman, M. D., Grahn, D., Faris, B., and Franzblau, C. (1974). "A Sex-Linked Defect in the Cross-Linking of Collagen and Elastin Associated with the Mottled Locus in Mice," *J. Exp. Med.,* **139**(1), 180–192.

Rowe, D. W., McGoodwin, E. B., Martin, G. R., and Grahn, D. (1977). "Decrease in Lysyl Oxidase Activity in the Aneurysm-Prone, Mottled Mouse," *J. B. C.* **252,** 939–942.

Rucker, R. B. and Tinker, D. (1977). "Structure and Metabolism of Arterial Elastin," *Int. Rev. Exp. Pathol.,* **17,** 1–47.

Rucker, R. B., Parker, H. E., and Rogler, J. C. (1969). "Effect of Copper Deficiency on Chick Bone Collagen and Selected Bone Enzymes," *J. Nutr.,* **98,** 57–63.

Rucker, R. B., Riggins, R. S., Laughlin, R., Chan, M. M., Chen, M., and Tom, K. (1975). "Effects of Nutritional Copper Deficiency on the Biomechanical Properties of Bone and Arterial Elastin Metabolism in the Chick," *J. Nutr.,* **105,** 1062–1070.

Rupp, H. and Weser, U. (1976). "Reactions of D-Penicillamine with Copper in Wilson's Disease," *Biochem. Biophys. Res. Commun.,* **72,** 223–229.

Savage, J. E., Ross, D. A., Reynolds, G., and O'Dell, B. L. (1962). "Copper Deficiency and Beta-Aminopropionitride Toxicity in Turkeys," *Fed. Proc.,* **21,** 311.

Scheinberg, H. I. and Sternlieb, I. (1975). "Pregnancy in Penicillamine Treated Patients with Wilson's Disease," *N. Engl. J. Med.,* **293,** 1300–1302.

Schulz, K. C. A., van der Merwe, P. K., van Rensburg, P. J. J., and Swart, J. S. (1951). "Studies in Demyelinating Diseases of Sheep Associated with Copper Deficiency," *Onderstepoort J. Vet Res.,* **25,** 35–77.

Shield, G. S., Coulson, W. F., Kimball, D. A., Carnes, W. H., Cartwright, G. E., and Wintrobe, M. M. (1962). "Studies on Copper Metabolism. XXXII: Cardiovascular Lesions in Copper Deficient Swine," *Am. J. Pathol.,* **41,** 603–622.

Sidman, R. L., Dickie, M. M., and Appel, S. H. (1964). "Mutant Mice (Quaking and Jimpy) with Deficient Myelination in the Central Nervous System," *Science,* **144,** 309–311.

Simpson, C. F., Jones, J. E., and Harms, R. H. (1967). "Ultrastructure of Aortic Tissue in Copper Deficient and Control Chick Embryos," *J. Nutr.,* **91,** 283–291.

Sjollema, B. (1933). "Kupfermangel als Ursacge von Krankheiten bei Pflazen und Iteren," *Biochem. Z.,* **267,** 151–156.

Smith, R. M., Fraser, F. J., and Russel, G. R. (1977). "Enzootic Ataxia in Lambs: Appearance of Lesions in the Spinal Cord during Foetal Development," *J. Comp. Pathol.,* **87,** 119–128.

Solomon, L., Abrams, G., Dinner, M., and Berman, L. (1977). "Neonatal Abnormalities Associated with D-Penicillamine Treatment during Pregnancy," *N. Engl. J. Med.,* **196,** 54–55.

Spoerl, V. R. and Kirchgessner, M. (1975a). "Veranderungen des Cu-status und der Coeruloplasimaktivitat von Mutterlichen und saugenden Ratten bei gestaffelter Cu-verorgung," *Z. Tierphysiol., Tierernaehr. Futtermittelkd.,* **35,** 113–127.

Spoerl, V. R. and Kirchgessner, M. (1975b). "Zur Auswirkung eines Cu-mangels an Ratten wahrend der Aufzucht und graviditat auf die Reproduktion," *Z. Tierphysiol., Tierernaehr. Futtermittelkd.,* **35,** 321–328.

Spoerl, R. and Kirchgessner, M. (1976). "Cu-mangelschaden bei reproduzierenden Ratten," *Zb. Vet. Med.,* **A23,** 131–138.

Starcher, B. C., Hill, C. H., and Matrone, G. (1964). "Importance of Dietary Copper in the Formation of Aortic Elastin," *J. Nutr.,* **82,** 318–322.

Starcher, B. C., Madaras, J. A., Perry, R. F., and Hill, C. H. (1978). "Abnormal Cellular Copper Metabolism in the Blotchy Mouse," *Fed. Proc.,* **3,** 325.

Stewart, W. L. (1932). "Swingback (Ataxia) in Lambs," *Vet. J.,* **88,** 133–137.

Suttle, N. F. and Angus, K. W. (1976). "Experimental Copper Deficiency in the Calf," *J. Comp. Pathol.,* **86,** 595–608.

Suttle, N. R. and Angus, K. W. (1978). "Effects of Experimental Copper Deficiency on the Skeleton of the Calf," *J. Comp. Pathol.,* **88,** 137–148.

Suttle, N. F. and Field, A. C. (1968). "Effect of Intake of Copper, Molybdenum and Sulphate on Copper Metabolism in Sheep. II: Copper Status of the Newborn Lamb," *J. Comp. Pathol.,* **78,** 363–370.

Symonds, C. P. (1975). "Multiple Sclerosis and the Swayback Story," *Lancet,* **i,** 155–156.

Teague, H. S. and Carpenter, L. E. (1951). "The Demonstration of a Copper Deficiency in Young Growing Pigs," *J. Nutr.,* **43,** 389–399.

Theriault, L. L., Dungan, D. D., Simons, S., Keen, C. L., and Hurley, L. S. (1977). "Lipid and Myelin Abnormalities of Brain in the Crinkled Mouse," *Proc. Soc. Exp. Biol. Med.,* **155,** 549–553.

Tsai, M. C., Everson, G. J., and Shrader, R. (1964). "Copper Deficiency in the Guinea Pig," *Fed. Proc.,* **23,** 133.

Underwood, E. J. (1977). Chapter 3 in *Trace Elements in Human and Animal Nutrition.* Academic Press, New York, pp. 56–108.

Wald, N. and Hambidge, M. (1977). "Maternal Serum-Copper Concentration and Neural-Tube Defects," *Lancet,* **ii,** 560.

Walshe, J. M. (1977). "Pregnancy in Wilson's Disease," *Quart. J. Med.,* New Series, **XLVI,** 73–83.

Wiener, G. (1966). "Genetic and Other Factors in the Occurrence of Swayback in Sheep," *J. Comp. Pathol.,* **76,** 435–447.

Wiener, G. and Field, A. C. (1969). "Copper Concentrations in the Liver and Blood of Sheep of Different Breeds in Relationship to Swayback History," *J. Comp. Pathol.,* **79,** 7–14.

Wiener, G. and MacLeod, N. S. M. (1970). "Breed, Body Weight, and Age as Factors in the Mortality Rate of Sheep Following Copper Injection," *Vet. Rec.,* **86,** 740–743.

Wiener, G., Wilmut, I., and Field, A. C. (1978). "Maternal and Lamb Breed Interactions in the Concentration of Copper in Tissues and Plasma of Sheep." In M. Kirchgessner, Ed., *Trace Element Metabolism in Man and Animals*—3. Technischen Universität München, Freising-Weihenstephan, Germany, pp. 469–472.

Wilkie, W. J. (1959). "Mineral Deficiencies in Pigs," *Aust. Vet. J.,* **35,** 209–216.

Williams, R. B. (1977). "Trace Elements and Congenital Abnormalities," *Proc. Nutr. Soc.,* **36,** 25–32.

Williams, R. B. and Bremner, I. (1976). "Copper and Zinc Deposition in the Foetal Lamb," *Proc. Nutr. Soc.,* **35,** 86A.

Wilson, J. G. (1972). "Environmental Effects on Developmental Teratology." In N. A. Assali, Ed., *Pathophysiology of Gestation.* Academic Press, New York, pp. 269–320.

Zimmerman, A. W., Mattieu, J-M., Quarles, R. H., Brady, R. O., and Hsu, J. M. (1976). "Hypomyelination of Copper-Deficient Rats," *Arch. Neurol.,* **33,** 111–119.

Zipper, J., Medel, M., and Prager, R. (1969). "Suppression of Fertility by Intrauterine Copper and Zinc in Rabbits: A New Approach to Intrauterine Contraception," *Am. J. Obstet. Gynecol.,* **105,** 529–534.

4

BIOCHEMICAL AND PATHOLOGICAL EFFECTS OF COPPER DEFICIENCY

Clifford H. Gallagher

Department of Veterinary Pathology, University of Sydney, Sydney, New South Wales, Australia

1. INTRODUCTION

Copper was identified as a constituent of biological tissues early in the nineteenth century. Meissner (1817) detected the element in plants, Devergie and Hervy (1838) in vertebrates, and Harless (1847) in invertebrates. A function of copper in the metabolism of invertebrates was suggested by Fredericq (1878), who isolated the copper protein hemocyanin. However, the metabolic importance of the element in plants and higher animals was not suspected until the 1920s, when diseases due to copper deficiency began to be recognized. As with other essential nutrients, research into diseases caused by copper deficiency has elucidated much of the role of copper in metabolism.

2. COPPER DEFICIENCY IN VERTEBRATES

Involvement of copper in the metabolism of higher animals was suggested by McHargue (1926), who found that small amounts of the element improved the rate of growth of rats fed a synthetic diet. More definite evidence came from the work of Hart et al. (1925, 1928), who showed that rabbits and rats on milk diets developed anemia that did not respond to iron therapy unless accompanied by copper. The requirement of copper in addition to iron for treatment of anemia in animals fed a milk diet was confirmed by McHargue et al. (1928), Waddell et al. (1929), and Underhill et al. (1933).

The importance of copper in the metabolism of vertebrates was soon emphasized by recognition of the natural occurrence of disease due to copper deficiency in cattle, sheep, and goats in many parts of the world (McElroy and Glass, 1950; Underwood, 1956). Subsequently disease syndromes due to copper deficiency have been identified in many species, including human beings (Underwood, 1971; Danks et al., 1972).

The common syndrome of copper deficiency in young animals is characterized by slow growth, achromotrichia, coat changes, loss of weight, emaciation, generalized edema, and death, in chronological order (Gallagher, 1964). A typical illustration of the disease is given in Figure 1, which shows the failure of growth and the loss of hair pigment in a copper-deficient rat as compared with a litter-mate control of the same sex.

Other overt clinical and pathological signs of copper deficiency vary with the species and tend to occur in a particular species or group of species.

Anemia is usually, but not always, present and is not as early a sign of deficiency as retarded growth or achromotrichia. In severe deficiency anemia occurs in all species because of the requirement of copper for iron metabolism, as shown below, but the type of anemia varies with the species. For example, pigs (Lahey et al., 1952), rats (Smith and Medlicott, 1944), and rabbits (Smith et al., 1944) develop a microcytic, hypochromic anemia; dogs (Van Wyk et al., 1953), an

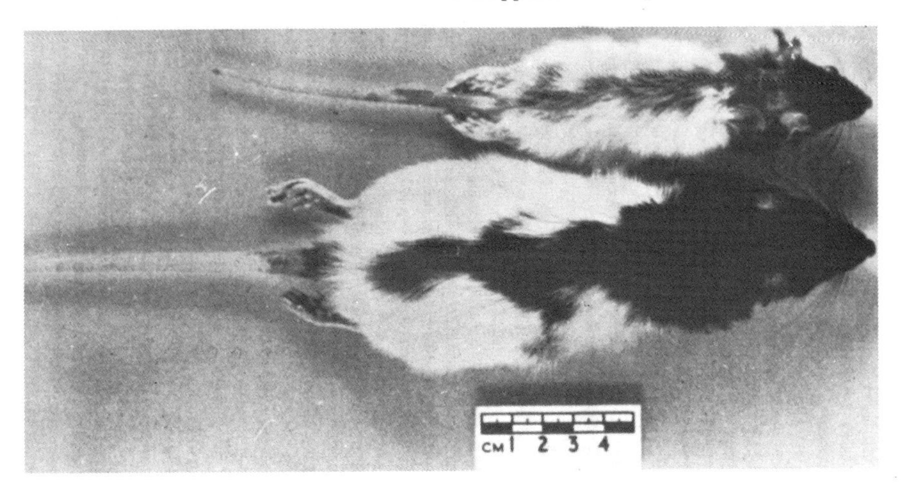

Figure 1. A copper-deficient and a control male rat from the same litter.

anemia characterized by relatively normal red cell indices, but reduction in numbers; cattle (Bennetts and Hall, 1939; Bennetts et al., 1941), a macrocytic, hypochromic anemia; and sheep (Bennetts and Beck, 1942; Marston and Lee, 1948; Marston et al., 1948), a macrocytic, hypochromic anemia in adults, but a microcytic, hypochromic anemia in lambs.

Skeletal deformities and fractures occur in many species in copper deficiency because of decreased tensile strength of bone. Such skeletal abnormalities have been reported in dogs (Baxter et al., 1953), pigs (Teague and Carpenter, 1951; Lahey et al., 1952; Furstman and Rothman, 1972), sheep (Cunningham, 1950; Suttle et al., 1972), cattle (Cunningham, 1950; Davis, 1950; Irwin et al., 1974; Smith et al., 1975), chickens (Gallagher, 1955), and children (Menkes et al., 1962).

Lesions of hypomyelination of the central nervous system (CNS), leading to locomotor disturbances of gait or posture, occur naturally in lambs born of copper-deficient ewes (Bennetts and Chapman, 1937; Bull et al., 1938; Innes, 1939). Nervous disorders have also been reported to accompany copper deficiency in goats, pigs, guinea pigs, and rats (Underwood, 1971). Lesions in the CNS can be produced under experimental conditions in the rat (Di Paolo et al., 1974; Zimmerman et al., 1976) and the guinea pig (Everson et al., 1967) and are a characteristic feature of Menkes' disease, which produces copper deficiency in children (Menkes et al., 1962; Danks et al., 1972).

Some species deficient in copper are prone to die suddenly. Cattle on copper-deficient land may die quickly from myocardial failure due to widespread lesions of myocardial necrosis, atrophy, and replacement fibrosis. Such deaths are most common when the cattle are forced to move, and have led to the farmer's designation of "falling disease" for copper deficiency in this species (Bennetts and Hall, 1939; Bennetts et al., 1941, 1942, 1948). Sudden death due to copper deficiency may also occur in chickens (O'Dell et al., 1961), pigs (Shields et al., 1962) and turkeys (Graham, 1977), because of rupture of one of the major blood vessels, usually the aorta.

An overt expression of copper deficiency in some species is abnormal development of hair or wool in addition to loss of pigment. This is particularly evident in the wool of sheep, which assumes an appearance described by the term "steely wool" because of the influence of copper deficiency on keratinization (Marston, 1948). Changes in the hair of children with Menkes' disease has led to the designation "Menkes' kinky hair syndrome" (Menkes et al., 1962). The frank alopecia or absence of hair and dermatosis seen in copper-deficient rabbits (Smith and Ellis, 1947) and cats (Davis, 1950) are probably extreme manifestations of the same fundamental disturbance.

Other signs of copper deficiency have been reported but are of less significance to the general syndrome or are of doubtful validity (Gallagher, 1964).

3. BIOCHEMISTRY

Early studies on copper deficiency were mainly concerned with the relationship of copper to hematopoiesis (Elvehjem, 1935; Schultze, 1940; Cartwright, 1947; Marston, 1948). Little attention was devoted to the effect of copper deficiency on enzyme systems, although decreased activities of cytochrome oxidase (Cohen and Elvehjem, 1934; Schultze, 1939, 1941) and catalase (Schultze and Kuiken, 1941) were reported. However, these disturbances were observed in rats fed a diet that was suboptimal in other nutritional factors, and consequently they could have been due to a number of causes. Other workers reported that catalase activity was not impaired by copper deficiency in rats (Lahey et al., 1952) and was actually increased by copper deficiency in mice (Adams, 1953).

A comprehensive study of the effects of an uncomplicated deficiency of copper in the rat on enzyme systems and on synthetic processes was reported by Gallagher (1955) and Gallagher et al. (1956a, 1956b). It was found that a moderate to advanced depletion of copper did not affect the level or activity of the following: liver-slice respiration, anaerobic glycolysis, citric acid cycle, fatty acid oxidation, amino acid oxidation, oxidative phosphorylation, oxidation of glucose and pyruvate by brain mitochondria, catalase, NADH-cytochrome *c* reductase, transmethylase, choline oxidase, isocitric dehydrogenase, succinic dehydrogenase, malic dehydrogenase, glutamic dehydrogenase, urinary amino acids, or plasma concentrations of protein, magnesium, calcium, sodium, potassium, and inorganic phosphate. At this stage of deficiency liver copper was reduced to about 3 μg/g dry weight compared with control values of about 15 μg/g. The activity of cytochrome oxidase was much reduced, and succinoxidase activity also declined.

The loss of cytochrome oxidase activity was severe and progressive from a very early stage of the deficiency and accounted for the fall in succinoxidase activity, as this system has a much higher Q_{O_2} than other respiratory dehydrogenase systems and therefore a greater requirement for cytochrome oxidase. Heme *a*, the prosthetic group of cytochrome oxidase, was observed to be almost absent from copper-deficient tissues, and it was suggested that this was the limiting component of the cytochrome oxidase system in copper deficiency. This suggestion was strongly supported by Lemberg et al. (1962), who found that the depletion of heme *a* in heart muscle of copper-deficient pigs corresponded quantitatively to the loss of cytochrome oxidase activity.

Liver mitochondria from moderately deficient rats were found to be very susceptible to "aging," with a rapid loss of selective semipermeability on standing in the cold compared with control mitochondria similarly treated.

Extreme copper deficiency caused a grave loss of the capacity of mitochondria to oxidize any substrate; this was probably due largely to the negligible activity of cytochrome oxidase at this stage of deficiency but possibly also to an accelerated and exaggerated "aging" effect with the loss of soluble respiratory cofactors, like the NAD coenzymes, through increased mitochondrial permeability. Extreme deficiency also decreased NADH-cytochrome *c* reductase activity and increased the activity of isocitric dehydrogenase. The increased activity of isocitric dehydrogenase may simply reflect the release from control exerted by tissue copper levels on this sulfydryl-dependent enzyme as copper is depleted. The same phenomenon has been observed in copper-deficient tomato leaves (Nason, 1952).

The most significant enzymological disturbance observed by Gallagher et al. (1956a) was the early, severe, and progressive loss of activity of the terminal enzyme of respiration, cytochrome oxidase. A considerable decrease in the activity of this enzyme occurred before the deficient animals showed slower growth than normal. As early as 4 weeks on the deficient diet, liver cytochrome oxidase activity had fallen by about 76%, while the concentration of copper in the liver, about 8.5 μg/g dry weight, was still approximately half that of control rats. The loss of cytochrome oxidase activity was progressive with copper depletion until there was an immeasurably small activity in the liver during the final stages of the disease. Recovery of cytochrome oxidase activity quickly follows the administration of copper to depleted rats.

An interesting tissue specificity was shown in the depletion of cytochrome oxidase activity in copper-deficient rats. Brain, particularly, and kidney lose cytochrome oxidase activity much more slowly than do bone marrow, heart muscle, and, especially, liver. However, the loss of cytochrome oxidase activity in all tissues is great and progressive with copper depletion.

The influence of copper deficiency on synthesis in the rat was also studied by Gallagher et al. (1956b), who found that the synthesis of phospholipids both *in vitro* and *in vivo* was depressed considerably early in the course of copper deficiency. The synthesis of protoheme was depressed to a degree paralleling the level of anemia. Inconstant depression of the rate of incorporation of [^{14}C] glycine into total plasma and liver protein and into plasma albumin and globulin fractions was also found.

Moderate to severe copper deficiency did not change the ability of rat enzyme systems to synthesize RNA or long-chain fatty acids at normal rates (Gallagher, 1955; Gallagher et al., 1956b).

The two major biochemical disturbances in the copper-deficient rat found in the above studies were the early, progressive, and profound loss of cytochrome oxidase activity and the early depression of phospholipid synthesis. Subsequent research by Gallagher and Reeve (1971a, 1971b, 1971c, 1976) has been directed toward determining the relationship, if any, between these two biochemical aberrations.

The failure of phospholipid synthesis was traced by Gallagher et al. (1956b) to the esterification of fatty acyl-CoA and L-glycerol-3-phosphate to form phosphatidic acids. As this is not an oxidative reaction, it was difficult to envisage the loss of cytochrome oxidase activity as a direct causal factor in depressing the synthesis of phospholipid. However, it was possible that reduced potential for oxidative metabolism influenced the production, or the availability at the reaction site, of the esterifying enzyme, acyl-CoA: L-glycerol-3-phosphate acyltransferase.

Gallagher and Reeve (1971a) found that copper-deficiency does not alter the level of acyl-CoA: L-glycerol-3-phosphate acyltransferase in either liver microsomes or mitochondria, nor does it affect the activities of the other key enzymes of palmitoyl-CoA metabolism, palmitoyl-CoA hydrolase in liver microsomes or mitochondria, and palmitoyl-CoA: carnitine palmitoyltransferase in liver mitochondria. The unexpected finding that copper deficiency depressed phospholipid synthesis by liver mitochondria but not by liver microsomes led to identification of the biochemical lesion. The effect of copper depletion on mitochondrial phospholipid synthesis was found to be eliminated by the addition of 1 m*M* ATP to the incubation medium. It was further found that the influence of copper deficiency on mitochondria could be mimicked by 0.1 to 1 m*M* cyanide, which inhibits cytochrome oxidase and depresses phospholipid synthesis in a manner which is reversed by 1 m*M* ATP.

Further research by Gallagher and Reeve (1971b) into the effect of copper deficiency on adenine nucleotide metabolism elucidated the problem. It was found that copper deficiency does not influence the synthesis of ATP by liver mitochondria, nor does it alter the concentration of ATP in whole liver homogenates or in isolated mitochondria, microsomes, and supernatant. However, copper deficiency considerably reduces the affinity of mitochondria to bind adenine nucleotides on membrane preparations. This defect leads to failure of ATP transport across the inner membrane (Winkler and Lehninger, 1968), resulting in abnormally high accumulation of ATP within mitochondria incubated *in vitro*. As a result, exogenous ATP is required for optimal rates of phospholipid synthesis by the key enzymes concerned, acid: CoA ligase and acyl-CoA: L-glycerol-3-phosphate acyltransferase, which are both located on the outer mitochondrial membrane (Norum et al., 1966; Bygrave, 1969; Zborowski and Wojtczak, 1969). The effect of copper deficiency on adenine nucleotide metabolism and consequently on phospholipid synthesis is analogous to that of atractyloside, the glycoside that inhibits the binding of ATP to the inner mitochondrial membrane and hence its translocation across the membrane, imposing a requirement for exogenous ATP for optimal phospholipid synthesis (Winkler et al., 1968; Winkler and Lehninger, 1968; Bygrave, 1969; Zborowski and Wojtczak, 1969).

Inhibition of the synthesis of phospholipid, a major component of myelin (Johnson et al., 1950), by cyanide, the cytochrome oxidase inhibitor, in a manner analogous to copper deficiency lent support to the possibility of a causal relationship between cytochrome oxidase activity and phospholipid synthesis. Like copper deficiency in some species, administration of the respiratory inhibitors potassium cyanide, sodium azide, carbon monoxide, and nitrous oxide (Weston Hurst, 1944; Lumsden, 1950) leads to loss of myelin, the predominantly phospholipid complex, from the central nervous system. The study reported by Gallagher and Reeve (1976) was designed to investigate possible links between copper status, cytochrome oxidase, adenine nucleotide binding, and phospholipid synthesis in liver mitochondria. The following three possibilities were considered:

1. The cytochrome oxidase complex itself is the adenine nucleotide binding site or translocator. Some evidence against this view was published by Gallagher et al. (1975), who found that chronic cyanide poisoning of rats did not reduce cytochrome oxidase activity but decreased the affinity of mitochondrial membranes to bind ADP by 24%.
2. Impaired cytochrome oxidase activity affects adenine nucleotide binding and/or translocation indirectly, perhaps by altering the ionic charge on membrane sites. Another possibility is that decreased cytochrome oxidase activity increases the proportion of unsaturated fatty acids in the mitochondrial membrane and thus decreases the affinity to bind adenine nucleotides, as with oleate treatment (Wojtczak and Zaluska, 1969). Copper deficiency was shown by Gallagher and Reeve (1971c) to increase significantly the proportion of polyunsaturated fatty acids in liver mitochondria.
3. Adenine nucleotide binding and/or translocation depends directly on some form of copper that is not associated with cytochrome oxidase.

The results were against cytochrome oxidase as the binding site and/or translocator of adenine nucleotides. Tenfold increase of the specific activity of cytochrome oxidase did not increase the affinity to bind adenine nucleotides. Aging or dialysis of mitochondria at 0 to 2°C for 18 hr did not reduce cytochrome oxidase activity or copper content but considerably lowered the capacity to take up [^{14}C]ADP. These findings agreed with those of Winkler (1969), who concluded that the distribution of cytochome oxidase as measured polarographically in inner membrane preparations of mitochondria is not related to the ADP-binding sites.

The second possibility, that cytochrome oxidase activity per se influences ADP binding indirectly, either by altering the charge on mitochondrial membranes, since adenine nucleotides and the ADP-binding inhibitors, cyanide, atractyloside, and diethyldithiocarbamate, are anions, or by increasing the degree of unsaturation of mitochondrial fatty acids, as does copper deficiency, was also discounted. Neither fluoride nor azide at concentrations strongly inhibitory to cytochrome oxidase inhibited either ADP binding or phospholipid synthesis. On the other hand, atractyloside strongly depresses ADP binding and phospholipid synthesis by mitochondria but has no influence on cytoehrome oxidase activity.

The third possibility, that copper, in a form other than in association with cytochrome oxidase, is a component of the adenine nucleotide binding site and/or translocator, is strongly suggested by the evidence and warrants further study for these reasons:

1. Increased copper content of mitochondrial membranes increases ADP binding.
2. Copper deficiency depresses tissue copper levels as well as cytochrome oxidase activity.
3. Cyanide and diethyldithiocarbamate, which inhibit cytochrome oxidase, adenine nucleotide binding, and phospholipid synthesis, also form strong complexes with tissue copper. The binding sites on the inner mitochondrial membrane and/or the translocator for adenine nucleotides may well contain copper.

Copper deficiency in the rat lowers significantly the phospholipid content of liver mitochondria (Gallagher and Reeve, 1971c), showing that the influence of copper status on the synthesis of phospholipids by liver mitochondria is not an artifact of *in vitro* incubation conditions. The total quantity of phospholipids in liver or brain is not altered significantly by copper deficiency, nor is the proportional distribution of specific phospholipids in liver or brain homogenates.

Other studies on the biochemistry of copper deficiency have been related to specific aspects of the problem rather than to the general syndrome. These will be discussed in Section 4 in relation to the biochemical bases of the clinical and pathological signs of copper deficiency.

4. BIOCHEMICAL BASIS OF SYNDROME

Copper is an essential component of many enzyme systems (Peisach et al., 1966;

O'Dell, 1976). However, not all of the cuproprotein enzymes have their activities so decreased in copper deficiency as to be metabolically limiting and cause overt clinical or pathological signs of disease. Future research may implicate at least some of these enzymes in the syndrome of copper deficiency, but the present discussion will be limited to what appear to be clear or strongly suggestive relationships between biochemical aberrations and the disease syndrome.

4.1. Retarded Growth, Loss of Weight, Emaciation, and Death

Poor growth, loss of weight, emaciation, and, in general, death are probably related to the earliest and most severe biochemical lesion in copper deficiency, the rapidly progressive loss of cytochrome oxidase activity. Cytochrome oxidase has a large functional reserve, especially in the liver, and clinical signs do not become apparent until 70 to 80% of its activity has been lost. Even at this stage the oxidation by liver mitochondria *in vitro* of most respiratory substrates proceeds at normal rates (Gallagher et al., 1956a). Evidence of limiting cytochrome oxidase activity is shown, however, in depression of succinoxidase activity. The succinoxidase system has a much higher Q_{O_2} than any other mitochondrial dehydrogenase system and thus requires more cytochrome oxidase. The high metabolic demands of young, growing animals are progressively restricted by diminishing cytochrome oxidase activity with copper depletion, producing the clinical signs of retarded growth and cessation of growth far below the normal species weight. Continued depletion of copper and loss of cytochrome oxidase activity restrict oxidative metabolism below the body maintenance level, leading to loss of weight, emaciation, and death. Rats killed in advanced copper deficiency have immeasurably low cytochrome oxidase activity, and their liver mitochondria are unable to oxidize any substrate at a measurable rate (Gallagher et al., 1956a).

Loss of cytochrome oxidase activity is not the only cause of death in copper deficiency. In some species death may occur precipitately. Copper-deficient cattle may die suddenly from heart failure (Bennetts and Hall, 1939; Bennetts et al., 1941, 1942, 1948), and poultry, pigs, and turkeys from rupture of a major blood vessel (O'Dell et al., 1961; Shields et al., 1962; Graham, 1977). Another specific cause of death in young lambs is hypomyelination of the central nervous system (Bull et al., 1938; Bennetts and Chapman, 1937; Innes, 1939). These conditions will be discussed below.

4.2. Edema

Generalized edema occurs during the terminal stages of copper deficiency when oxidative metabolism in the tissues is depressed almost to extinction by the loss of cytochrome oxidase activity. Edema is clearly not a specific effect of copper deficiency but simply reflects the loss of control of fluid balance consequent to the breakdown of cellular metabolism. The likely contributory factors to the movement of fluid out of the circulatory system into the tissue spaces and body cavities are myocardial, hepatic, and endothelial failure from severe depression or inactivation of cellular respiration, leading to increased venous hydrostatic pressure, depressed albumin synthesis and hence low plasma protein, and increased vascular permeability, respectively. Anemia present at this stage of deficiency is also likely to contribute to the development of edema. Alteration of vascular permeability and edema may be a consequence of abnormalities of vessel walls from defective synthesis of elastin and collagen.

4.3. Achromotrichia

Polyphenol oxidase (tyrosinase), the enzyme catalyzing the formation of melanin, is a cuproprotein. There is no reason to doubt that the loss of hair, wool, and feather color in copper deficiency is due to decreased activity of this enzyme.

4.4. Anemia

A high rate of metabolism in bone marrow is essential for normal erythropoiesis. As the cytochrome oxidase activity of bone marrow is severely depressed early in copper deficiency (Schultze, 1941), it is probable that anemia results largely from impairment of oxidative metabolism.

In the copper-deficient rat, in which anemia is a relatively late sign, it was found that the degree of anemia paralleled the decrease in synthesis of protoheme (Gallagher, 1955; Gallagher et al., 1956b). More recently, Williams et al. (1976) provided further evidence for the dependence of intracellular iron metabolism and the synthesis of protoheme on cytochrome oxidase activity by studying heme synthesis in copper-deficient cells. It was found that iron uptake from transferrin by copper-deficient reticulocytes was 52% of normal and the rate of heme synthesis was 33% of normal. Hepatic mitochondria from copper-deficient animals were deficient in cytochrome oxidase activity and failed to synthesize heme from ferric iron and protoporphyrin at the normal rate. The rate of heme synthesis correlated with the cytochrome oxidase activity. As heme synthesis from ferric iron and protoporphyrin was enhanced by succinate and inhibited by malonate, antimycin A, azide, and cyanide, the authors proposed that an intact electron transport system is required for the reduction of ferric iron to provide a pool of ferrous iron for heme synthesis.

A more specific influence of copper deficiency on the metabolism of iron than that operating through cytochrome oxidase activity may yet be identified. For example, Wagner and Tephly (1975) have suggested a possible role of copper in the regulation of heme synthesis via ferrochelatase. These authors found that ferrochelatase activity could be inhibited by cobalt and manganese to produce a dose-dependent decrease in hepatic cytochrome P-450 content. On the other hand, copper stimulated ferrochelatase activity, and there was an interaction between Cu^{2+} and Fe^{2+} in the system such that the K_m of Fe^{2+} was dependent upon the concentration of Cu^{2+}. Dialysis of hepatic mitochondrial preparations inactivated ferrochelatase in a manner reversible by adding Cu^{2+}. In addition, copper, but not iron, reversed the inhibition of ferrochelatase activity produced by lead or cobalt. These results suggest that cytochrome P-450 content may be reduced during copper deficiency because of decreased ferrochelatase activity and a consequent defect in heme synthesis. Synthesis of other hemoproteins may similarly be influenced by copper deficiency through depression of ferrochelatase activity.

The possible role of copper in the metabolism of iron generally has been the subject of much research and speculation in relation to anemia in copper deficiency. Osaki et al. (1966) proposed the hypothesis that ferroxidase activity is required to convert Fe^{2+} absorbed from the gastrointestinal tract into Fe^{3+} for transport by transferrin, the iron-binding protein of plasma, and that depression of ferroxidase activity in copper deficiency lowers the oxidation of Fe^{2+} below the rate required for normal absorption and transport of iron. The transport protein for copper, ceruloplasmin, is a ferroxidase (Williams et al., 1974; Frieden, this volume) and is considerably lowered in concentration by copper deficiency. Bohnenkamp and Weser (1976) found that the ferroxidase activity of ceruloplasmin was decreased to 15% of control values on the 15th day after rats were placed on a copper-deficient diet. Other ferroxidases are present in plasma, and there is evidence that ferroxidase II also contains copper (Topham et al., 1975). It is likely that interference with iron absorption and transport contributes to the development of anemia in copper deficiency.

An observation of importance in relation to anemia is that erythrocytes have a shorter life span than normal in copper deficiency (Cartwright and Wintrobe, 1964; Lee et al., 1968). The reason for this is not clear but may be related to the toxic effect of the superoxide anion, O_2^-, a highly reactive free radical produced by the oxidation of some substrates by molecular oxygen (e.g., xanthine oxidation by xanthine oxidase). Decomposition and thus detoxification of the superoxide anion are catalyzed by the cuprozinc enzyme superoxide dismutase, which is present in red cells (Keele et al., 1971; McCord and Fridovich, 1969). Williams et al. (1975) and Bohnenkamp and Weser (1976) have shown that erythrocyte superoxide dismutase (erythrocuprein) is reduced in copper deficiency in pigs and rats.

4.5. Skeletal Pathology

Spontaneous fractures of bone are common in copper-deficient animals, and skeletal lesions, consisting basically of osteoporosis with rarefaction of cortices and trabeculae, have been observed in naturally occurring or experimental copper deficiency in cattle (Cunningham, 1950; Davis, 1950), sheep (Cunningham, 1950), pigs (Teague and Carpenter, 1951; Lahey et al., 1952), dogs (Baxter et al., 1953), and chickens (Gallagher, 1955, 1957). Rucker et al. (1969) found that bone from copper-deficient chickens contains a higher proportion of soluble collagen than normal with fewer aldehydes. It was concluded that collagen cross-linking, which is preceded by aldehyde generation on peptide chains, is impaired in copper deficiency. The consequent failure of collagen maturation in the organic matrix of bone accounts for its greater fragility and structural abnormalities. The basic biochemical defect lies in the failure of aldehyde production on peptide residues of soluble collagen by amine oxidases which have been shown to contain copper (Yamada and Yasunobu, 1962; Buffoni and Blaschko, 1964; Buffoni et al., 1968; Mondovi et al., 1967; Hill, 1969; Siegel and Martin, 1970; Rucker and O'Dell, 1971; Rucker and Goettlich-Rieman, 1972). Of particular interest is the connective tissue amine oxidase isolated from bovine aorta by Rucker and O'Dell (1971) and from rabbit aorta by Rucker and Goettlich-Rieman (1972). This enzyme appears to be copper dependent as its activity decreases in copper deficiency and can be partially reactivated by the addition of Cu^{2+} *in vitro* (Bird et al., 1966; Chou et al., 1968).

The failure of cross-linking of lysine residues of polypeptides clearly plays a part in the production of faulty collagen in copper deficiency, as it does in producing defective elastin, as lysyl oxidase is a cuproprotein (Hill, 1969; Siegel and Martin, 1970).

4.6. Cardiovascular Pathology

Blood Vessel Walls

Copper-deficient pigs (Shields et al., 1962), chickens (O'Dell et al., 1961), and turkeys (Graham, 1977) may die suddenly from rupture of a major blood vessel. The underlying pathology was found to be derangement of elastic tissue in the vessel walls, and hemorrhage was often preceded by dissecting aneurysm (O'Dell et al., 1961; Shields et al., 1962). Antecedent lesions were found in both large and small arteries, producing defective elastic laminae and lower tensile strength from depressed maturation of collagen (Bader, 1973).

As with collagen, a clue to the underlying biochemical mechanism of defective formation of elastin is shown in the increased proportion of tropoelastin, the soluble precursor of elastin, in arteries and tendons from copper-deficient animals (Sandberg et al., 1969; Roensch et al., 1972; Whiting et al., 1974; Foster et al.,

1975; Rucker et al., 1975). Tropoelastin has an amino acid content similar to that of elastin except that it contains more lysine and no desmosine. Desmosine and isodesmosine, the cross-links of mature, insoluble elastin, are formed from the peptide-bound lysines of tropoelastin by oxidative deamination to semialdehydes, three of which, with a fourth lysyl side chain, produce desmosine and isodesmosine (Bader, 1973). The oxidative deamination of lysine to semialdehydes is catalyzed by an amine oxidase, which is depressed in activity in copper-deficient chicken aorta (Hill, 1969). This enzyme has been identified as lysyl oxidase, a cuproprotein that oxidizes the lysyl side chains of tropoelastin to produce the aldehyde precursors of desmosine and isodesmosine (Siegel and Martin, 1970). Copper deficiency, like lathyrism, in which lysyl oxidase is inhibited by the toxic principle β-(N-γ-glutamyl)aminopropionitrile (Schilling and Strong, 1954), causes failure of maturation of elastin and thus vascular and tendon pathology by depressed activity of lysyl oxidase.

Myocardium

Copper-deficient cattle are prone to die of sudden heart failure, especially when disturbed or exercised, giving rise to the designation "falling disease" (Bennetts and Hall, 1939; Bennetts et al., 1941, 1942, 1948). Autopsy reveals focal widespread lesions of myocardial necrosis, atrophy, and replacement fibrosis in most cases, indicative of repeated episodes. Myocardial lesions involving necrosis have also been described by Kelly et al. (1974) in young rats whose dams were fed a copper-deficient diet.

Cardiac hypertrophy has been reported in copper-deficient rats and pigs (Hill, 1969).

Heart muscle is a very active tissue and requires a high level of cytochrome oxidase activity to maintain its oxidative metabolism. Cytochrome oxidase is severely depleted from the myocardium in copper deficiency. Consequently, the extra oxidative demand made on the heart by forced exercise, when most deaths of copper-deficient cattle occur, may exceed the available oxidative capacity and result in respiratory failure, leading to focal or generalized myocardial necrosis. Kelly et al. (1974) found that deaths from heart disease occurred suddenly without previous symptoms in one or more rat pups in litters from copper-deficient dams. Sudden deaths could be precipitated by pressure on the thorax. The hearts of the deficient young rats were enlarged and pale and did not show other vascular lesions, so the primary effect of copper deficiency was on the myocardium. All areas of the copper-deficient rat hearts showed histochemical evidence of reduced cytochrome oxidase activity, even in hearts that were normal in appearance.

Although myocardial abnormalities precede apparent changes in vessel walls, at least in the rat, it is likely that depressed and faulty synthesis of elastin and collagen in coronary vessels and cardiac connective tissue will augment cardiac pathology with developing copper deficiency. It is improbable, for example, that replacement fibrosis will be as effective as in normal animals in the healing of focal areas of myocardial necrosis in copper deficiency due to failure of collagen maturation.

4.7. Coat Changes

It appears likely that a mechanism similar to those in collagen and elastin synthesis, involving copper in protein cross-linkages, will be elucidated to explain the depressed development of keratin which leads to abnormal wool and hair in copper deficiency (Underwood, 1971; Danks et al., 1972).

Studies by Marston (1948) on copper-deficient sheep exhibiting the characteristic sign of "steely wool" showed that the rate of oxidation of sulfydryl groups of prekeratin to the disulfide linkages of keratin is greatly retarded, allowing time for internal disorientation of the fibrillae before their structure is fixed by cross-linkage. Consequently, the physical properties of the wool are changed, with a loss of characteristic crimp or waviness, tensile strength, elasticity, and affinity for dyes. Although the exact manner in which copper influences the oxidation of sulfydryl groups in the skin is unknown, Cu^{2+} itself is a powerful catalyst of sulfyldryl oxidation.

4.8. Central Nervous System Pathology

Attention was first drawn to the role of copper deficiency in producing disease of the central nervous system (CNS) by the occurrence of locomotor incoordination in lambs on soil low in copper. The disease was called neonatal or enzootic ataxia, or swayback because it was characterized by an unsteady gait and stance (Bull et al., 1938; Bennetts and Chapman, 1937; Innes, 1939).

The lesions consist of symmetrical hypomyelination of the spinal cord, involving mostly the ventral and dorsolateral aspects of the white matter, and diffuse, symmetrical hypomyelination of the cerebral white matter, extending from small foci in the centrum ovale to gross hypomyelination, degeneration, liquefaction, and porencephalic cavitation involving almost all of the white matter in the hemispheres.

Sheep will breed when deficient in copper, and the fetal copper levels are lower than the dam's (Marston, 1952). Consequently, the lamb can be severely deficient in copper in late gestation and at birth, when myelin is being laid down most rapidly.

Nervous disorders have been reported to result from copper deficiency in sheep, goats, pigs, guinea pigs, and rats (Underwood, 1971) and are a feature of Menkes' syndrome in human neonates with an inherited defect in copper metabolism (Danks et al., 1972).

Faulty myelination of the CNS is more difficult to produce by copper deficiency in species other than sheep but can be achieved experimentally in the neonatal offspring of guinea pigs (Everson et al., 1968) and rats (Di Paolo et al., 1974; Zimmerman et al., 1976) by breeding from successive generations of copper-deficient dams. Di Paolo et al. (1974) found that the hypomyelination of copper-deficient rat neonates showed less cerebrosides, cholesterol, and sulfatides, but no difference in phospholipid content. However, the more recent study by Zimmerman et al. (1976) showed that copper-deficient rat pups had substantially less myelin, a reduction of 56%, than did controls but that the lack of myelin was of a general nature, as the qualitative composition was not different from that in controls.

Inhibitors of cytochrome oxidase, such as KCN, NaN_3, and N_2O, produce hypomyelination of the CNS when large doses are administered repeatedly (Weston Hurst, 1944; Lumsden, 1950). It appears likely that low cytochrome oxidase activity plays a role in hypomyelination of the CNS in copper deficiency. The sheep, the species most prone to hypomyelination in copper deficiency, loses even more of its CNS cytochrome oxidase activity relative to other tissues during copper depletion than does the rat (Howell and Davison, 1959). Involvement of the loss of cytochrome oxidase activity in producing brain lesions is also suggested by the results published by Kelly et al. (1974), who found decreased staining for cytochrome oxidase in copper-deficient rat pups in all areas of the brain, but especially in the cerebral cortex, basal ganglia, and thalamic regions, in agreement with the location of histopathological lesions.

Phospholipid is a major component of myelin (Johnson et al., 1950), and copper deficiency depresses the synthesis of phospholipids by rat liver mitochondria (Gallagher et al., 1956b; Gallagher and Reeve, 1971a), producing a general depletion of phospholipid in the mitochondria *in vivo* (Gallagher and Reeve, 1971c). However, there is no evidence as yet that depressed phospholipid synthesis contributes to hypomyelination of the CNS in copper deficiency. Phospholipid synthesis in the CNS in naturally occurring hypomyelination in copper-deficient lambs has not been studied. Rats made deficient in copper after weaning from normal dams do not show lesions in the CNS, and they synthesize phospholipids in the CNS at normal rates (Gallagher et al., 1956b) to produce normal concentrations of these substances (Gallagher and Reeve, 1971c). Neonatal rats made deficient by the special technique of breeding from copper-deficient dams do show lesions of hypomyelination, but there is some difference of opinion about their phospholipid status. Di Paolo et al. (1974) found lower levels of cerebrosides, cholesterol, and sulfatides but not of phospholipid in the brains of the copper-deficient neonates, but Zimmerman et al. (1976) reported that all components of the myelin complex were depressed. In any case there is no evidence for specific involvement of impaired synthesis of phospholipid in the hypomyelination of copper deficiency.

4.9. Genetic Defects

Menkes' Kinky Hair Syndrome

Menkes et al. (1962) described an X-linked, recessively inherited syndrome in male children characterized by slow growth, progressive cerebral degeneration, convulsions, temperature instability, scorbutic-like bone changes, and peculiar steel-like hair, *pili torti.* Death occurred before 3 years of age. The disease was shown by Danks et al. (1972) to be due to copper deficiency induced by a defect in absorption of copper from the intestine. Lott et al. (1975) found that the absorption block was not absolute as it could be overcome by high doses of copper given orally. Further research by Danks et al. (1973) established that the defect was not in the absorption of copper from the intestinal lumen by the brush border of the mucosal epithelium, but in the transport of copper within these cells or across the membrane on their serosal aspect. Holtzman (1976) suggested that metallothionein (Starcher, 1969) or some other binding protein in intestinal epithelium may have increased affinity for copper in Menkes' syndrome. Clearly, further definition of the basic defect is required.

The pathological lesions described by Danks et al. (1972) for arteries, bone, and hair are typical of nutritional copper deficiency. Unusual features of Menkes' syndrome are the convulsions and progressive cerebral degeneration, shown clinically as mental retardation and pathologically as cystic degeneration and gliosis. These are explicable, in large part at least, by the widespread degenerative change in the cerebral arterial walls, producing tortuosity and variation in the lumina of the vessels from fragmentation of the internal elastic lamina and thickening of the intima.

Quaking or Mottled Mouse Syndrome

A syndrome analogous to Menkes' syndrome occurs in a mutant strain of mice (Hunt, 1974, 1976), characterized by mottling of the coat, tremors, and low tissue copper. Keen and Harley (1976) found that the tremors could be greatly reduced, and the brain copper levels of the offspring brought to normal, by high supplementation of the dams' diet with copper. Hunt (1974) found that, as in Menkes' syndrome, the mottled mice absorb copper from the intestinal lumen at normal rates to give a higher content within the intestinal wall than in controls but fail to release the copper from intestinal cells. The mottled mouse syndrome is an X-linked, heritable characteristic like Menkes' syndrome and offers an excellent experimental model for further study of the mechanisms controlling copper binding, transport, and release in intestinal epithelium.

Recently considerable interest has been focused upon the possibility that the CNS disturbances in copper deficiency induced by inherited defects and, possibly, by dietary deficiency result from alteration of catecholamine metabolism. O'Dell et al. (1976) measured the catecholamine and serotonin concentrations in the anterior brain stems of ataxic copper-deficient lambs with and without the intravenous administration of copper, as compared with nonataxic lambs from the same flock given copper. Copper treatment significantly increased the dopamine concentration in ataxic lambs and raised the norepinephrine toward the higher levels in copper-treated, nonataxic lambs. Serotonin levels were not altered. The results, although on small numbers, suggest that dopamine levels are low in the striated area of copper-deficient brain, as they are in Parkinson's disease in human beings, which produces similar ataxic signs and tremors. The rise in norepinephrine with copper administration is not surprising as dopamine-β-hydroxylase, which catalyzes its synthesis, is a cuproprotein (Friedman and Kaufman, 1965). Low levels of brain norepinephrine have also been reported for copper-deficient neonatal rats by Prohaska and Wells (1974), and Morgan and O'Dell (1977) found low levels of dopamine and norepinephrine in similar animals. Reduced levels of brain noradrenaline were reported by Hunt and Johnson (1972a, 1972b) and Hunt (1977) in mutant mice because of depression of conversion of dopamine by dopamine-β-hydroxylase. The significance of altered catecholamine levels in the brain in relation to the CNS disorders in copper deficiency remains to be elucidated.

4.10. Mitochondrial Ultrastructure

Enlargement of liver mitochondria was reported by Goodman and Dallman (1967) and Dallman and Goodman (1970) to result from a mild degree of copper deficiency in the rat. Subsequently it was shown by Gallagher et al. (1973) that severe copper deficiency produces grossly enlarged and misshapen hepatocyte mitochondria which occupy most of the cytoplasmic space. A similar effect was observed by Suzuki (1969) and Suzuki and Kikkawa (1969) after administration of the copper-chelating compound Cuprizone (biscyclohexanone oxaldihydrazone). Similar but less dramatic changes in mitochondrial morphology have been reported for the intestinal mucosa (Fell et al., 1975) and myocardium (Leigh, 1975) of copper-deficient cattle and for the brain of copper-deficient neonatal rats (Prohaska and Wells, 1975).

Gallagher et al. (1973) postulated that the most likely explanation for the giant, misshapen mitochondria in copper deficiency is a change in their composition, altering the membrane surface properties and permeability. The following mitochondrial membrane components and properties are known to be influenced by copper deficiency: cytochrome oxidase activity is severely and progressively depressed, and its prosthetic group heme *a* depleted (Gallagher et al., 1956a); phospholipid synthesis is impaired (Gallagher et al., 1956b; Gallagher and Reeve, 1971a); the phospholipid content is decreased, and the proportion of polyunsaturated fatty acids increased (Gallagher and Reeve, 1971c); and the affinity of mitochondrial membranes to bind and transport adenine nucleotides is decreased (Gallagher and Reeve, 1971b). Similarly, for riboflavin-deficient mouse livers, Tandler et al. (1968) postulated that the production of giant mitochondria might be due to the change in mitochondrial membrane phospholipids causing a loss of membrane integrity, leading to water uptake and fusion of adjacent mitochondria from the change in surface properties and adhesiveness.

5. CONCLUDING REMARKS

By virtue of its peculiar characteristics copper, like iron, has become part of the fabric of life. The higher the order of evolution, the more intricate and varied has become the role of copper. Deny the tissues the trace quantities of copper required, and metabolism begins to disintegrate at many points. The syndrome of copper deficiency is an overt expression of the interaction of the many aberrations of metabolism induced by depletion of copper from the enzyme systems that are dependent upon it. Some signs of copper deficiency are clearly related chiefly to failure of one particular aspect of copper metabolism, like the vascular degenerative changes produced by inadequate maturation of elastin secondary to depressed activity of the cuproenzyme lysyl oxidase. Other aspects of the syndrome, such as anemia or hypomyelination, cannot be assigned to a single major defect of metabolism, at least at this stage of knowledge. Probably even the most obvious causal relationships between biochemical and pathological lesions in copper deficiency are modified by other failings because of the complex involvement of copper in metabolism. It seems most unlikely that each pathological manifestation of copper deficiency will be shown with advancing knowledge to be attributable to one biochemical defect in a particular pathway of copper metabolism. Further research may, of course, establish the role, if any, of other known cuproenzymes, like uricase, superoxide dismutase, and tryptophan-2,3-dioxygenase, in the syndrome of copper deficiency and may identify cuproenzymes as yet unrecognized.

REFERENCES

Adams, D. H. (1953). "The Effect on Mouse Liver Catalase Activity and Blood-Haemoglobin Level of a Milk Diet Deficient in Iron, Copper and Manganese," *Biochem. J.*, **54,** 328–336.

Bader, L. (1973). "Disorders of Elastic Tissue: a Review," *Pathology,* **5,** 269–289.

Baxter, J. H., Van Wyk, J. J., and Follis, R. H. (1953). "A Bone Disorder Associated with Copper Deficiency. II: Histological and Chemical Studies on the Bones," *Johns Hopkins Hosp. Bull.,* **93,** 25–31.

Bennetts, H. W. and Beck, A. B. (1942). "Enzootic Ataxia and Copper Deficiency of Sheep in Western Australia," *Commonw. Aust. Sci. Ind. Res. Org. Bull.,* **147,** 1–52.

Bennetts, H. W. and Chapman, F. E. (1937). "Copper Deficiency in Sheep in Western Australia: a Preliminary Account of the Etiology of Enzootic Ataxia of Lambs and an Anaemia of Ewes," *Aust. Vet. J.,* **13,** 138–149.

Bennetts, H. W. and Hall, H. T. B. (1939). "Falling disease of Cattle in the South-west of Western Australia," *Aust. Vet. J.,* **15,** 152–159.

Bennetts, H. W., Beck, A. B., Harley, R., and Evans, S. T. (1941). "Falling Disease of Cattle in the South-west of Western Australia. 2: Studies of Copper Deficiency in Cattle," *Aust. Vet. J.,* **17,** 85–93.

Bennetts, H. W., Harley, R., and Evans, S. T. (1942). "Studies on Copper Deficiency of Cattle: the Fatal Termination (Falling Disease)," *Aust. Vet. J.,* **18,** 50–63.

Bennetts, H. W., Beck, A. B., and Harley, R. (1948). "The Pathogenesis of Falling Disease," *Aust. Vet. J.,* **24,** 237–244.

Bird, D. W., Savage, J. E., and O'Dell, B. L. (1966). "Effect of Copper Deficiency and Inhibitors on the Amine Oxidase Activity of Chick Tissues," *Proc. Soc. Exp. Biol. Med.,* **123,** 250–254.

Bohnenkamp, W. and Weser, U. (1976). "Copper Deficiency and Erythrocuprein (2Cu,2Zn-Superoxide Dismutase)," *Biochim. Biophys. Acta,* **444,** 396–406.

Buffoni, F. and Blaschko, H. (1964). "Benzylamine Oxidase and Histamine: Purification and Crystallization of an Enzyme from Pig Plasma," *Proc. R. Soc. London,* **B161,** 153–167.

Buffoni, F., Della Corte, L., and Knowles, P. F. (1968). "The Nature of Copper in Pig Plasma Benzylamine Oxidase," *Biochem. J.,* **106,** 575–576.

Bull, L. B., Marston, H. R., Murnane, D., and Lines, E. W. L. (1938). "Ataxia in Young Lambs," *Commonw. Aust. Sci. Ind. Res. Org. Bull.,* **113,** 23–27.

Bygrave, F. L. (1969). "Studies on the Biosynthesis and Turnover of the Phospholipid Component of the Inner and Outer Membrane of Rat Liver Mitochondria," *J. Biol. Chem.,* **244,** 4768–4772.

Cartwright, G. E. (1947). "Dietary Factors Concerned in Erythropoiesis," *Blood,* **2,** 111–153, 256–298.

Cartwright, G. E. and Wintrobe, M. M. (1964). "The Question of Copper Deficiency in Man," *Am. J. Clin. Nutr.,* **15,** 94–110.

Chou, W. S., Savage, J. E., and O'Dell, B. L. (1968). "Relation of Monoamine Oxidase Activity and Collagen Crosslinking in Copper Deficient and Control Tissues," *Proc. Soc. Exp. Biol. Med.,* **128,** 948–952.

Cohen, E. and Elvehjem, C. A. (1934). "The Relation of Iron and Copper to the Cytochrome Oxidase Content of Animal Tissues," *J. Biol. Chem.,* **107,** 97.

Cunningham, I. J. (1950). In W. D. McElroy and B. Glass, Eds., *Copper Metabolism, a Symposium on Animal, Plant and Soil Relationships.* Johns Hopkins Press, Baltimore, Md., p. 216.

Dallman, P. R. and Goodman, J. R. (1970). "Enlargement of Mitochondrial Compartment in Iron and Copper Deficiency," *Blood,* **35,** 496–505.

Danks, D. M., Campbell, P. E., Stevens, B. J., Mayne, V., and Cartwright, E. (1972). "Menkes' Kinky Hair Syndrome: An Inherited Defect in Copper Absorption with Widespread Effects," *Pediatrics,* **50,** 188–201.

Danks, D. M., Cartwright, E., Stevens, B. J., and Townley, R. R. W. (1973). "Menkes' Kinky Hair Disease: Further Definition of the Defect in Copper Transport," *Science,* **179,** 1140–1142.

Davis, G. K. (1950). In W. D. McElroy and B. Glass, Eds., *Copper Metabolism, a Symposium on Animal, Plant and Soil Relationships.* Johns Hopkins Press, Baltimore, Md., p. 216.

Devergie, A. and Hervy O. (1838). *Ann. Hyg.* (Paris), **20,** 463.

Di Paolo, R. V., Kanfer, J. N., and Newberne, P. M. (1974). "Copper Deficiency and the Central Nervous System. Myelination in the Rat: Morphological and Biochemical Studies," *J. Neuropathol. Exp. Neurol.,* **33,** 226–236.

Elvehjem, C. A. (1935). "The Biological Significance of Copper and Its Relation to Iron Metabolism," *Physiol. Rev.,* **15,** 471–507.

Everson, G. J., Tsai, H. C., and Wang, T. I. (1967). "Copper Deficiency in the Guinea Pig," *J. Nutr.,* **93,** 533–540.

Fell, B. F., Dinsdale, D., and Mills, C. F. (1975). "Changes in the Enterocyte Mitochondria Associated with Deficiency of Copper in Cattle," *Res. Vet. Sci.,* **18,** 274–281.

Foster, J. A., Shapiro, R., Voynow, P., Crombie, G., Faris, B., and Franzblau, C. (1975). "Isolation of Soluble Elastin from Lathyritic Chicks. Comparison to Tropoelastin from Copper-Deficient Pigs," *Biochemistry,* **14,** 5343–5347.

Fredericq, L. (1878). "La digestion des Matieres Albuminoides chez Quelques Invertebres," *Arch. Zool. Exp. Gen.,* **7,** 391–400.

Friedman, S. and Kaufman, S. (1965). "3,4-Dihydroxyphenylethylamine-β-hydroxylase: Physical Properties, Copper Content and Role of Copper in the Catalytic Activity," *J. Biol. Chem.,* **240,** 4763–4773.

Furstman, L. and Rothman, R. (1972). "The Effect of Copper Deficiency on the Mandibular Joint and Alveolar Bone of Pigs," *J. Oral Pathol.,* **1,** 249–255.

Gallagher, C. H. (1955). "Pathological and Enzymological Disturbances in Copper and Allied Deficiencies." Ph.D. thesis, University of London.

Gallagher, C. H. (1957). "The Pathology and Biochemistry of Copper Deficiency," *Aust. Vet. J.,* **33,** 311–317.

Gallagher, C. H. (1964). *Nutritional Factors and Enzymological Disturbances in Animals.* Crosby Lockwood, London.

Gallagher, C. H. and Reeve, Vivienne E. (1971a). "Copper Deficiency in the Rat: Effect on Synthesis of Phospholipids," *Aust. J. Exp. Biol. Med. Sci.,* **49,** 21–31.

Gallagher, C. H. and Reeve, Vivienne E. (1971b). "Copper Deficiency in the Rat: Effect on Adenine Nucleotide Metabolism," *Aust. J. Exp. Biol. Med. Sci.,* **49,** 445–451.

Gallagher, C. H. and Reeve, Vivienne E. (1971c). "Copper Deficiency in the Rat: Effect on Liver and Brain Lipids," *Aust. J. Exp. Biol. Med. Sci.,* **49,** 453–461.

Gallagher, C. H. and Reeve, Vivienne E. (1976). "Interrelationships of Copper, Cytochrome Oxidase and Adenine Nucleotide Binding," *Aust. J. Exp. Biol. Med. Sci.,* **54,** 593–600.

Gallagher, C. H., Judah, J. D., and Rees, K. R. (1956a). "The Biochemistry of Copper

Deficiency. 1: Enzymological Disturbances, Blood Chemistry and Excretion of Amino Acids," *Proc. R. Soc. London,* **B145,** 134–150.

Gallagher, C. H., Judah, J. D., and Rees, K. R. (1956b). "The Biochemistry of Copper Deficiency. 2: Synthetic Processes," *Proc. R. Soc. London,* **B145,** 195–205.

Gallagher, C. H., Reeve, Vivienne, E., and Wright, R. (1973). "Copper Deficiency in the Rat: Effect on the Ultra-structure of Hepatocytes," *Aust. J. Exp. Biol. Med. Sci.,* **51,** 181–189.

Gallagher, C. H., Reeve, Vivienne, E., and Wright, R. (1975). "Copper Deficiency in the Rat: Relationship to Chronic Cyanide Poisoning," *Aust. J. Exp. Biol. Med. Sci.,* **53,** 343–348.

Goodman, J. R. and Dallman, P. R. (1967). "Mitochondrial Morphology and Cytochrome Oxidase in Copper Deficiency in the Rat." In C. J. Arcenaux, Ed., *Proceedings of the 25th Annual Meeting of the Electron Microscopy Society of America.* Claitor's Book Store, Louisiana, p. 164.

Graham, C. L. (1977). "Copper Levels in Livers of Turkeys with Naturally Occurring Aortic Rupture," *Avian Dis.,* **21,** 113–116.

Harless, E. (1847). "Ueber das blaue Blut einiger wirbellosen Thiere und dessen Kupfergehalt," *Arch. Anat. Physiol.,* 148–156.

Hart, E. B., Steenbock, H., Elvehjem, C. A., and Waddell, J. (1925). "Iron in Nutrition. I: Nutritional Anaemia on Whole Milk Diets and the Utilisation of Inorganic Iron in Haemoglobin Building," *J. Biol. Chem.,* **65,** 67.

Hart, E. B., Steenbock, H., Waddell, J., and Elvehjem, C. A. (1928). "Iron in Nutrition. VII: Copper as a Supplement to Iron for Haemoglobin Building in the Rat," *J. Biol. Chem.,* **77,** 797–812.

Hill, C. H. (1969). "A Role of Copper in Elastin Formation," *Nutr. Rev.,* **27,** 99–100.

Holtzman, N. A. (1976). "Menkes' Kinky Hair Syndrome: A Genetic Disease Involving Copper," *Fed. Proc.,* **35,** 2276–2280.

Howell, J. M. and Davison, A. N. (1959). "The Copper Content and Cytochrome Oxidase Activity of Tissues from Normal and Swayback Lambs," *Biochem. J.,* **72,** 365–368.

Hunt, C. E. and Carlton, W. W. (1965). "Cardiovascular Lesions Associated with Experimental Copper Deficiency in the Rabbit," *J. Nutr.,* **87,** 385–393.

Hunt, D. M. (1974). "Primary Defect in Copper Transport Underlies Mottled Mutants in the Mouse," *Nature,* **249,** 852–854.

Hunt, D. M. (1976). "A Study of Copper Treatment and Tissue Copper Levels in the Murine Congenital Copper Deficiency, Mottled," *Life Sci.,* **19,** 1913–1919.

Hunt, D. M. (1977). "Catecholamine Biosynthesis and the Activity of a Number of Copper-Dependent Enzymes in the Copper-Deficient Mottled Mouse Mutants," *Comp. Biochem. Physiol.,* **57C,** 79–83.

Hunt, D. M. and Johnson, D. R. (1972a). "Aromatic Amino Acid Metabolism in Brindles (Mo^{br}) and Viable-Brindled (Mo^{vbr}), Two Alleles at the Mottled Locus in the Mouse," *Biochem. Genet.,* **6,** 31–40.

Hunt, D. M. and Johnson, D. R. (1972b). "An Inherited Deficiency in Noradrenaline Biosynthesis in the Brindled Mouse," *J. Neurochem.,* **18,** 7–16.

Innes, J. R. M. (1939). "Swayback: a Demyelinating Disease of Lambs with Affinities to Schilder's Encephalitis and Its Prevention by Copper," *J. Neurol. Psychiatr.*, **2,** 323–334.

Irwin, M. R., Poulos, P. W., Jr., Smith, B. P., and Fisher, G. L. (1974). "Radiology and Histopathology of Lameness in Young Cattle with Secondary Copper Deficiency," *J. Comp. Pathol.*, **84,** 611–621.

Johnson, A. C., McNabb, A. R., and Rossiter, R. J. (1950). "Chemistry of Wallerian Degeneration," *Arch. Neurol. Psychiatr.*, **64,** 105–121.

Keele, B. B., McCord, J. M., and Fridovich, I. (1971). "Further Characterisation of Bovine Superoxide Dismutase and Its Isolation from Bovine Heart," *J. Biol. Chem.*, **246,** 2875–2880.

Keen, C. L. and Harley, L. S. (1976). "Copper Supplementation in Quaking Mutant Mice: Reduced Tremors and Increased Brain Copper," *Science*, **193,** 244–246.

Kelly, W. A., Kesterson, J. W., and Carlton, W. W. (1974). "Myocardial Lesions in the Offspring of Female Rats Fed a Copper-Deficient Diet," *Exp. Mol. Pathol.*, **20,** 40–56.

Lahey, M. E., Gubler, C. J., Chase, M. S., Cartwright, G. E., and Wintrobe, M. M. (1952). "Studies on Copper Metabolism. II: Hematologic Manifestations of Copper Deficiency in Swine," *Blood*, **7,** 1053–1074.

Lee, G. R., Nacht, S., Lukens, J. N., and Cartwright, G. E. (1968). "Iron Metabolism in Copper Deficient Swine," *J. Clin. Invest.*, **47,** 2058–2069.

Leigh, L. C. (1975). "Changes in the Ultrastructure of Cardiac Muscle in Steers Deprived of Copper," *Res. Vet. Sci.*, **18,** 282–287.

Lemberg, R., Newton, N., and Clarke, L. (1962). "Decrease of Haem *a* Content in the Heart Muscle of Copper-Deficient Swine," *Aust. J. Exp. Biol. Med. Sci.*, **40,** 367–372.

Lott, I. T., DiPaolo, R., Schwartz, D., Janowska, S., and Kanfer, J. N. (1975). "Copper Metabolism in the Steely Hair Syndrome," *N. Engl. J. Med.*, **292,** 197–199.

Lumsden, C. E. (1950). "Cyanide Leucoencephalopathy in Rats and Observations on the Vascular and Ferment Hypotheses of Demyelinating Diseases," *J. Neurol. Psychiatr.*, **13,** 1–15.

McCord, J. M. and Fridovich, I. (1969). "Superoxide Dismutase: An Enzymic Function for Erythrocuprein (Hemocuprein)," *J. Biol. Chem.*, **244,** 6049–6055.

McElroy, W. D. and Glass, B., Eds. (1950). *Copper Metabolism: a Symposium on Animal, Plant and Soil Relationships.* Johns Hopkins Press, Baltimore, Md.

McHargue, J. S. (1926). "Further Evidence That Small Quantities of Copper, Manganese and Zinc Are Factors in the Metabolism of Animals," *Am. J. Physiol.*, **77,** 245–255.

McHargue, J. S., Healy, D. J., and Hill, E. S. (1928), "The Relation of Copper to the Hemoglobin Content of Rat Blood," *J. Biol. Chem.*, **78,** 637.

Marston, H. R. (1948). "Nutritional Factors Involved in Wool Production by Merino Sheep. II: The Influence of Copper Deficiency on the Nature of The Fleece," *Aust. J. Sci. Res., B: Biol. Sci.*, **1,** 362–287.

Marston, H. R. (1952). "Cobalt, Copper and Molybdenum in the Nutrition of Animals and Plants," *Physiol. Rev.*, **32,** 66–121.

Marston, H. R. and Lee, H. J. (1948). "The Effects of Copper Deficiency and of Chronic Overdosage with Copper on Border-Leicester and Merino Sheep," *J. Agric. Sci.*, **38,** 229–240.

Marston, H. R., Lee, H. J., and McDonald, I. W. (1948). "Cobalt and Copper in the Nutrition of Sheep," *J. Agric. Sci.*, **38,** 216–221.

Meissner, W. (1817). "Sur la Presence du Cuivre dans les Cendres des Vegetaux," *Ann. Chim. Phys.*, **4,** 106.

Menkes, J. H., Alter, M., Steigleder, G. K., Weakley, D. R., and Sung, J. H. (1962). "A Sex-Linked Recessive Disorder with Retardation of Growth, Peculiar Hair, and Focal Cerebral and Cerebellar Degeneration," *Pediatrics*, **29,** 764–779.

Mondovi, B., Rotilio, G., Costa, M. T., Finazzi-Agro, A., Chiancone, E., Hansen, R. E., and Beinert, H. (1967). "Diamine Oxidase from Pig Kidney: Improved Purification and Properties," *J. Biol. Chem.*, **242,** 1160–1167.

Morgan, R. F. and O'Dell, B. L. (1977). "Effect of Copper Deficiency on the Concentration of Catecholamines and Related Enzyme Activities in the Rat Brain," *J. Neurochem.*, **28,** 207–213.

Nason, A. (1952). "Metabolism of Micronutrient Elements in Higher Plants. II: Effect of Copper Deficiency on the Isocitric Enzyme in Tomato Leaves," *J. Biol. Chem.*, **198,** 643–653.

Norum, K. R., Farstad, M., and Bremer, J. (1966). "The Submitochondrial Distribution of Acid:CoA Ligase (AMP) and Palmitoyl-CoA:Carnitine Palmitoyltransferase in Rat Liver Mitochondria," *Biochem. Biophys. Res. Commun.*, **24,** 797–804.

O'Dell, B. L. (1976). "Biochemistry of Copper," *Med. Clin. North Am.*, **60,** 687–703.

O'Dell, B. L., Hardwick, B. C., Reynolds, G., and Savage, J. E. (1961). "Connective Tissue Defects in the Chick Resulting from Copper Deficiency," *Proc. Soc. Exp. Biol. Med.*, **108,** 402–405.

O'Dell, B. L., Smith, R. M., and King, R. A. (1976). "Effect of Copper Status on Brain Neurotransmitter Metabolism in the Lamb," *J. Neurochem.*, **26,** 451–455.

Osaki, S., Johnson, D. A., and Frieden, E. (1966). "The Possible Significance of the Ferrous Oxidase Activity of Ceruloplasmin in Normal Human Serum," *J. Biol. Chem.*, **241,** 2746–2751.

Peisach, J., Aisen, P., and Blumberg, W. E. (1966). *The Biochemistry of Copper*. Academic Press, New York.

Prohaska, J. R. and Wells, W. W. (1974). "Copper Deficiency in the Developing Rat Brain: a Possible Model for Menkes' Steely Hair Disease," *J. Neurochem.*, **23,** 91–98.

Prohaska, J. R. and Wells, W. W. (1975). "Copper Deficiency in the Developing Rat Brain: Evidence for Abnormal Mitochondria," *J. Neurochem.*, **25,** 221–228.

Roensch, L. F., Savage, J. E., and O'Dell, B. L. (1972). "Purification and Characterization of Tropoelastin from Copper Deficient Chick Aorta," *Fed. Proc.*, **31,** 480.

Rucker, R. B. and Goettlich-Rieman, W. (1972). "Properties of Rabbit Aorta Amine Oxidase," *Proc. Soc. Exp. Biol. Med.*, **139,** 286–289.

Rucker, R. B. and O'Dell, B. L. (1971). "Connective Tissue Amine Oxidase. I: Purification of Bovine Aorta Amine Oxidase and Its Comparison with Plasma Amine Oxidase," *Biochim. Biophys. Acta,* **235,** 32–43.

Rucker, R. B., Parker, H. E., and Rogler, J. C. (1969). "Effect of Copper Deficiency on Chick Bone Collagen and Selected Bone Enzymes," *J. Nutr.,* **98,** 57–63.

Rucker, R. B., Riggins, R. S., Laughlin, R., Chan, M. M., Chen, M., and Tom, K. (1975). "Effects of Nutritional Copper Deficiency on the Biomechanical Properties of Bone and Arterial Elastin Metabolism in the Chick," *J. Nutr.,* **105**(8), 1062–1070.

Sandberg, L. B., Weissman, N., and Smith, D. W. (1969). "The Purification and Partial Characterization of a Soluble Elastin-like Protein from Copper-Deficient Porcine Aorta," *Biochemistry,* **8,** 2940–2945.

Schilling, E. D. and Strong, F. M. (1955). "Isolation, Structure and Synthesis of a *Lathyrus* factor from *L. odoratus*," *J. Am. Chem. Soc.,* **77,** 2843–2845.

Schultze, M. O. (1939). "The Effect of Deficiencies in Copper and Iron on the Cytochrome Oxidase of Rat Tissues," *J. Biol. Chem.,* **129,** 729.

Schultz, M. O. (1940). "Metallic Elements and Blood Formation," *Physiol. Rev.,* **20,** 37–67.

Schultze, M. O. (1941). "The Relation of Copper to Cytochrome Oxidase and Hematopioetic Activity of the Bone Marrow of Rats," *J. Biol. Chem.,* **138,** 219.

Schultze, M. O. and Kuiken, K. A. (1941). "The Effect of Deficiencies in Copper and Iron on the Catalase Activity of Rat Tissues," *J. Biol. Chem.,* **137,** 727.

Shields, G. S., Coulson, W. F., Kimball, D. A., Carnes, W. H., Cartwright, G. E., and Wintrobe, M. M. (1962). "Studies on Copper Metabolism. XXXII: Cardiovascular Lesions in Copper-Deficient Swine," *Am. J. Pathol.,* **41,** 603–617.

Siegel, R. C. and Martin, G. R. (1970). "Collagen Crosslinking. Enzymatic Synthesis of Lysine Derived Aldehydes and the Production of Crosslinked Components," *J. Biol. Chem.,* **245,** 1653–1658.

Smith, B. P., Fisher, G. L., Poulos, P. W., and Irwin, M. R. (1975). "Abnormal Bone Development and Lameness Associated with Secondary Copper Deficiency in Young Cattle," *J. Am. Vet. Med. Assoc.,* **166,** 682–688.

Smith, G. E. and Ellis, G. H. (1947). "Copper Deficiency in Rabbits. Achromotrichia, Alopecia and Dermatosis," *Arch. Biochem.,* **15,** 81–87.

Smith, S. E. and Medlicott, M. (1944). "The Blood Picture of Iron and Copper Deficiency Anemias in the Rat," *Am. J. Physiol.,* **141,** 354–358.

Smith, S. E., Medlicott, M., and Ellis, G. H. (1944). "The Blood Picture of Iron and Copper Deficiency Anemias in the Rabbit," *Am. J. Physiol.,* **142,** 179–181.

Starcher, B. C. (1969). "Studies on the Mechanism of Copper Absorption in the Chick," *J. Nutr.,* **97,** 321–326.

Suttle, N. F., Angus, K. W., Nisbet, D. I., and Field, A. C. (1972). "Osteoporosis in Copper-Depleted Lambs," *J. Comp. Pathol.,* **82,** 93–97.

Suzuki, K. (1969). "Giant Hepatic Mitochondria: Production in Mice Fed with Cuprizone," *Science,* **163,** 81–82.

Suzuki, K. and Kikkawa, Y. (1969). "Status Spongiosis of CNS and Hepatic Changes Induced by Cuprizone (Biscyclohexanone Oxaldihydrazone), *Am. J. Pathol.,* **54,** 307–325.

Tandler, B., Erlandson, R. A., and Wynder, E. L. (1968). "Riboflavin and Mouse Hepatic Cell Structure and Function. I: Ultrastructural Alterations in Simple Deficiency," *Am. J. Pathol.,* **52,** 69–95.

Teague, H. S. and Carpenter, L. E. (1951). "The Demonstration of a Copper Deficiency in Young Pigs," *J. Nutr.,* **43,** 398–399.

Topham, R. W., Sung, C. S., Morgan, F. G., Prince, W. D., and Jones, S. H. (1975). "Functional Significance of the Copper and Lipid Components of Human Ferroxidase—II, *Arch. Biochem. Biophys.,* **167,** 129–137.

Underhill, F. A., Orten, J. M., Mugrage, E. R., and Lewis, R. C. (1933). "The Effect of the Prolonged Feeding of a Milk-Iron-Copper Diet to Rats," *J. Biol. Chem.,* **99,** 469.

Underwood, E. J. (1956). *Trace Elements in Human and Animal Nutrition.* Academic Press, New York.

Underwood, E. J. (1971). *Trace Elements in Human and Animal Nutrition.* 3rd Ed. Academic Press, New York, pp. 57–115.

Van Wyk, J. J., Baxter, J. H., Akeroyd, J. H., and Motulsky, A. G. (1953). "The Anemia of Copper Deficiency in Dogs Compared with That Produced by Iron Deficiency," *Johns Hopkins Hops. Bull.,* **93,** 41–46.

Waddell, J., Steenbock, H., and Hart, E. B. (1929). "Iron in Nutrition. X: The Specificity of Copper as a Supplement to Iron in the Cure of Nutritional Anaemia," *J. Biol. Chem.,* **84,** 115.

Wagner, G. S. and Tephly, T. R. (1975). "A Possible Role of Copper in the Regulation of Haem Biosynthesis through Ferrochelatase," *Adv. Exp. Med. Biol.,* **58,** 343–354.

Weston Hurst, E. (1944). "A Review of Some Recent Observations on Demyelination," *Brain,* **67,** 103–124.

Whiting, A. H., Sykes, B. C., and Partridge, S. M. (1974). "Isolation of Salt-Soluble Elastin from Ligamentum Nuchae of Copper-Deficient Calf," *Biochem. J.,* **141,** 573–575.

Williams, D. M., Lee, G. R., and Cartwright, G. E. (1974). "Ferroxidase Activity of Rat and Ceruloplasmin," *Am. J. Physiol.,* **227,** 1094–1097.

Williams, D. M., Lynch, R. E., Lee, G. R., and Cartwright, G. E. (1975). "Superoxide Dismutase Activity in Copper-Deficient Swine," *Proc. Soc. Exp. Biol. Med.,* **149,** 534–536.

Williams, D. M., Loukopoulos, D., Lee, G. R., and Cartwright, G. E. (1976). "Role of Copper in Iron Metabolism," *Blood,* **48:** 77–85.

Winkler, H. H. (1969). "Localization of the Atractyloside-Sensitive Nucleotide Binding Sites in Rat Liver Mitochondria," *Biochim. Biophys. Acta,* **189,** 152–161.

Winkler, H. H. and Lehninger, A. L. (1968). "The Atractyloside-Sensitive Nucleotide Binding Site in a Membrane Preparation from Rat Liver Mitochondria," *J. Biol. Chem.,* **243,** 3000–3008.

Winkler, H. H., Bygrave, F. L., and Kehninger, A. L. (1968). "Characteristics of the Atractyloside-Sensitive Adenine Nucleotide Transport System in Rat Liver Mitochondria," *J. Biol. Chem.,* **243,** 20–28.

Wojtczak, L. and Zaluska, H. (1969). "Inhibition by Oleate of Binding and Exchange of ATP by Mitochondrial Membranes," *Biochim. Biophys. Acta,* **189,** 455–456.

Yamada, H. and Yasunobu, K. T. (1962). "Monoamine Oxidase. II: Copper, One of the Prosthetic Groups of Plasma Monoamine Oxidase," *J. Biol. Chem.,* **237,** 3077–3082.

Zborowski, J. and Wojtczak, L. (1969). "Phospholipid Synthesis in Rat Liver Mitochondria," *Biochim. Biophys. Acta,* **187,** 73–84.

Zimmerman, A. W., Matthieu, J. M., Quarles, R. H., Brady, R. O., and Hsu, J. M. (1976). "Hypomyelination in Copper-Deficient Rats. Prenatal and Postnatal Copper Replacement," *Arch. Neurol.,* **33,** 111–119.

5

THERAPEUTIC USES OF COPPER

John R. J. Sorenson

College of Pharmacy, University of Arkansas for Medical Sciences, Little Rock, Arkansas

1. INTRODUCTION

With regard to this subject, I am indebted to Whitehouse and Walker (1978) for calling my attention to a pertinent remark by Louis Pasteur, quoted by Bickel (1972): "Remember always that some ideas that seem dead and buried may at one time or another rise up to life again, more vital than ever before." It is probably true that an idea which is not understood appears to die. Fortunately, a valid idea is revitalized when it is rediscovered and additional information is provided to increase our understanding. It is uncertain, however, how many such rediscoveries are required to allow an idea to live to its fullest.

The therapeutic uses of copper must have been rediscovered many times during the last 5000 years. The following is offered as an interesting historical account from the biogeochemical point of view, which will increase in significance as the rest of this chapter is read.

Copper was found by Neolithic man, who probably had no hesitation about adorning himself with ornaments of it, just as human beings do today. The bright hues of the many copper minerals took the fancy of people on all parts of the earth in ancient times and led to the smelting of huge quantities of copper and bronze. Copper was originally discovered and mined in Cyprus, which is the Greek word for copper, and was considered to be sacred to Aphrodite, the Greek goddess of love and beauty (An Onlooker's Notebook, 1974). Ramses III had vast quantities of copper smelted from malachite [$Cu(OH)_2CuCO_3$], which he offered to the gods of Egypt, according to the Harris Papyrus. Further north, in Wadi Arabah, King Solomon's men mined and smelted great quantities of copper sulfide (CuS) ore.

Copper and its compounds have been used for their remedial effects since the beginning of recorded history. The Egyptians are reported to have used copper as an antiseptic for healing wounds and to sterilize drinking water as early as 3000 B.C. (Osterberg et al., 1975). Whitehouse and Walker (1978) also point out that the Ebers Papyrus (ca. 1550 B.C.) reports the use of copper acetate [$Cu_2(OAc)_4$], Cu sulfate ($CuSO_4$), and pulverized metallic copper for the treatment of trachoma, a granulomatus inflammation of the eye. It is certain that one of the Durotrigan defenders of Maden Castle in Dorset was wearing a spiral bronze ring on one of his toes, perhaps to ward off gout or rheumatism, when Vespasian's men caught up with him in A.D. 43 (An Onlooker's Notebook, 1974). Similar rings have been unearthed from the lake villages of Somerset, which date back to the Iron Age. Modern copper-containing folklore remedies for the treatment of arthritis include the wearing of copper bracelets, the use of copper-containing vinegar in cooking, and the dietary consumption of foods which have high contents of copper, such as shellfish, mushrooms, and nuts (Whitehouse, 1976).

This historical account of the use of copper and its compounds for supposed therapeutic purposes is interesting in the light of modern evidence concerning the use of this element and its compounds in the treatment of deficiencies, fertility control, prophylaxis of disease, and antibiosis, and as antiarthritic, as well as antineoplastic, agents. In spite of these newer rediscoveries, these applications and their mechanisms of action are still only partially understood.

It is hoped that this review will stimulate the reader's interest not only in copper metabolism, but also in all essential metal metabolism and the interrelationships of the various essential metal-dependent processes. This may suggest interesting new approaches to more effective therapy of some diseases, and new treatments for diseases that have resisted successful therapeutic solutions using classical approaches.

Although some of the material presented in this manuscript appeared in two recent reviews (Osterberg et al., 1975; Sorenson, 1978), it seemed worthwhile to include the information contained in them for the readers of this volume.

2. TREATMENT OF COPPER DEFICIENCY IN INFANTS

Anemia in infants with or without infection that is resistant to therapy with iron alone was originally attributed to copper deficiency by Josephs (1931). Two milliliters per kilogram of body weight per day of a 10% solution of ferric ammonium citrate, along with 1 ml of a 0.5% solution of copper sulfate ($CuSO_4 \cdot 5H_2O$) per kilogram of body weight per day, accelerated the rate of hemoglobin (Hb) synthesis, compared to the group given only the iron solution. Subsequently, Sturgeon and Brubaker (1956) reported that milk-fed infants who had hypochromic microcytic anemia, hypoferremia, increased total iron-binding capacity, and hypoproteinemia were copper deficient and that this could be confirmed by hypocupremia. They also pointed out that milk is known to have a very low copper content and that the requirement for this element in growing infants is very high. Symptoms of copper-deficiency anemia included a mild degree of edema of the face and/or limbs, pallor of the skin and oral mucous membranes, upper respiratory infections, heart murmurs, vomiting, diarrhea, and marked irritability. These symptoms were corrected with 0.5 ml of a 10% solution of copper sulfate ($CuSO_4$) given daily along with oral ferric sulfate [$Fe_2(SO_4)_3$], intramuscular (i.m.) iron-dextran, or packed red blood cells. True iron-deficiency anemia was recognized as being associated with higher than normal serum copper concentrations, which distinguished this deficiency from copper deficiency (Sturgeon and Brubaker, 1956). The higher than normal copper concentrations associated with true iron-deficiency anemia are probably the result of a positive feedback mechanism associated with the need for active erythropoiesis.

Graham and Cordano (1969) pointed out that milk-fed infants also develop neutropenia, leukopenia, severe bone demineralization, and lack ossification of growth centers. These symptoms were dramatically reversed by oral supplementation providing 800 μg Cu/kg body weight/day along with 15 to 60 mg Fe/day in a fortified cow's milk formula. It was suggested that neutropenia and leukopenia were early indicators of copper deficiency because of their consistant occurrence and their unequivocal response to copper supplementation. The anemia was recognized as a later manifestation of copper deficiency, as it is in swine copper deficiency (Lee et al., 1968), caused by chronic malabsorption, diarrhea, and/or reduced copper intake associated with exclusive milk feeding.

Griscom et al. (1971) found epiphyseal separations, extensive subperiosteal new bone growth, metaphyseal cupping, many rib fractures, systemic porosis, and enlarged costochondral junctions in very small premature infants who were copper deficient. These infants also had anemia with stomatocytes, hypoproteinemia, and moderate to severe neutropenia.

One female infant had seizures following cardiorespiratory arrest with hypoxia. Apneic spells and a systolic murmur were also seen in this patient, who had severe microcytic, hypochronic anemia. Her bones were normal at 30 days, but at age 77 days she had a grossly underdeveloped skeleton. This patient was also hydrocephalic and had leg edema, perhaps due to cardiac insufficiency. She died of aspiration pneumonitis at 78 days of age. Severe brain pathology and a marked decrease in liver copper, compared to the level in term infants, were discovered at autopsy. Although there was no mention of hair abnormality and the infant was female, the symptoms would seem to support a diagnosis of Menkes'-like syndrome (see Section 3).

The other two reported infants tolerated a solid food diet well, and their symptoms remitted. Although these patients were not treated with copper, it was recognized that the clinical picture was consistent with copper deficiency due to placental insufficiency. Zinc and manganese deficiencies were also considered, however, since these had been pointed out to lead to fetal stunting and malformations in animals (Winick, 1970).

Al-Rashid and Spangler (1971) suggested that deficiency of copper alone probably has never been observed, copper deficiency does occur along with other deficiencies. In infants a disorder characterized by hypocupremia, hypoferremia, neutropenia, hypoproteinemia, and anemia may better be viewed as a multiple deficiency state, as originally suggested by Sturgeon and Brubaker (1956).

Al-Rashid and Spangler (1971) reported a pale 3-month-old male infant with anemia (22% hematocrit), "a central nervous system (CNS) disorder," numerous episodes of sepsis with fever and apnea, erythematous scaly skin lesions over the forehead and behind the ears, and a systolic murmur. The lower extremities were kept in flexion and were painful to motion, and muscle tone was markedly decreased. Blood studies revealed low serum copper and copper oxidase, leukopenia with neutropenia, and elevated platelet and reticulocyte counts. Bone marrow aspirate was also found to be markedly abnormal. Bone X-rays demonstrated widened anterior ribs and a trabecular pattern in both humeri with flared, cupped, and irregular metaphyses, periosteal reaction, and a healing fracture at midshaft of the right femur. This infant's symptoms are also very much like those reported for Menkes' syndrome (see Section 3). Treatment with 50 ml of packed red blood cells (a source of iron) and 10 drops of a 10% $CuSO_4$ solution daily (2.5 mg $CuSO_4$/day) produced a remarkable normalization of symptoms within a month, and the infant was discharged on a maintenance dose of $CuSO_4$.

Seely et al. (1972) suggested that copper deficiency might be the result of feeding premature infants with an iron-fortified formula, but Al-Rashid and Spangler (1972) provided evidence that discounted this possibility. The agranulocytopenia and epiphyseal and periosteal abnormalities observed by Seely et al. (1972) are now known to persist in spite of iron therapy, because of lack of sufficient copper-dependent enzymatic activity required for iron utilization. These abnormalities are promptly corrected after treatment with oral $CuSO_4$ solution.

Karpel and Peden (1972) found typical copper deficiency, characterized by anemia and low serum copper and ceruloplasmin (Cp), in a female infant with postoperative short bowel syndrome who had been maintained for 8.5 months on total parenteral nutrition. This patient responded promptly to intravenous (i.v.) infusion of 22 μg Cu/kg/day. These observations were confirmed by Ashkenazi et al. (1973) in a patient who had undergone i.v. alimentation for only 3 months following neonatal bowel surgery. Unusual diets or parenteral nutrition lasting for as short as 6 weeks was also suggested as a cause of copper deficiency in other infants. Oral therapy with 2.5 mg $CuSO_4$/day reversed the blood and bone changes.

After copper supplementation Karpel and Peden (1972) noted a sudden decrease in serum iron. This observation is especially significant. Since copper is known to be required for the utilization of iron by bone marrow and other tissues, steps should be taken to avoid iron deficiency following copper therapy, that is, both elements should be monitored closely, and iron supplemented also, if required.

Ashkenazi et al. (1973) also commented on an unusual case of neonatal copper deficiency (very low serum copper and Cp levels) in a 6-month-old male infant who reportedly received a milk diet for 3 months because he vomited everything but milk. This child was lethargic, failed to gain weight, and was severely depigmented with pale skin and fair hair ("almost albino"). He had distended veins, psychomotor retardation, hypotonia, and occasional sebhorreic dermatitis. Disinterest in his surroundings and failure to follow light suggested blindness. The child had definite anemia, despite iron deposition in mature red blood cell

(RBC) precursors and other marrow-derived cells. Ulnar, radial, tibial, femoral, and fibular bones were characterized as osteoporotic with distal metaphyseal cupping and blurred margins. Rather than neonatal copper deficiency, this patient might also have been diagnosed as having Menkes' syndrome (see Section 3). However, it is still possible that, under certain circumstances, copper deficiency can produce many of the same clinical effects as are seen in Menkes' syndrome (Camakaris and Danks, personal communication).

A 10% solution of $CuSO_4$, given orally in increasing doses of from 1 to 3 mg/day, produced rapid and dramatic normalization of this patient's clinical and hematologic symptoms. Anemia and other blood dyscrasias were reversed in 18 days. Values for serum iron and iron-binding capacity, which were high-normal before copper therapy, diminished to an incipient iron deficiency. Copper therapy was then discontinued and iron therapy initiated.

The patient became interested in his surroundings and food, and for the first time sight was apparent. He stopped vomiting and gained weight. Simultaneously with clinical and hematologic improvement, serum copper and Cp values returned to normal. After 4 weeks of therapy the bone marrow, Hb, and white blood cells had returned to normal. Bone abnormalities were no longer found. Distension of scalp veins decreased, and the psychomotor and behavioral retardations were no longer apparent. At 1 year of age he was a normal, active, sighted child, although somewhat myopic.

The successful treatment of this patient with copper and iron, as well as the one described by Al-Rashid and Spangler (1971), merit special mention. Both of these patients might have been diagnosed as having Menkes' syndrome (Section 3.1) rather than copper deficiency. In the past, when this syndrome was recognized, it was unsuccessfully treated with copper alone. Perhaps Menkes' syndrome would be better treated with copper and iron in addition to supplementation with the remaining essential metals.

Infants with marasmus of kwashiorkor who were treated with skimmed milk diets developed hypochromic anemia, neutropenia, and osteoporosis, which responded to copper therapy (Alexander, 1974). However, these children also received iron therapy in some form, as well as an improved diet that may have included other essential metals.

Ashkenazi et al. (1973) recommended the prevention of copper-deficiency syndrome by treating all premature infants with 100 to 500 μg Cu/day for several months, as well as infants who required intravenous alimentation, were malnourished, or had a malabsorption syndrome. Perhaps this recommendation should now be extended to include iron and other essential metal supplements. It would seem to be worthwhile to analyze plasma and formed elements of blood for all of the essential metals as a part of routine blood analysis done on newborns as well as hospitalized infants. Perhaps these analyses would provide the physician with an early indication of essential metal deficiencies or increased demands in the disease state.

3. TREATMENT OF MENKES' SYNDROME

3.1. Symptoms

In 1962 Menkes et al. (1962) described a sex-linked recessive syndrome in male infants characterized by (*a*) onset within the first few weeks or months of life; (*b*) lack of awareness or interest in surroundings (lethargy); (*c*) retarded growth, in spite of eating well; (*d*) sparse, lusterless, hypopigmented to ivory-white wiry hair which was twisted about its axis, varied in diameter along the hair shaft, and fractured at regular intervals; (*e*) hypoplasia of the dental enamel; (*f*) marked cerebral (white and gray matter) degeneration with cerebellar cortex atrophy associated with a lack of behavioral or mental development, leading to a vegetative decerebrate state with seizures; and (*g*) resulting in death at 7 months to 3.5 years of age.

For various reasons Menkes' syndrome or disease is also referred to as kinky hair disease (Aguilar et al., 1966), trichopoliodystrophy (French et al., 1972), Menkes' steely hair disease (Danks et al., 1973b), and congenital hypocupremia (Osaka et al., 1977). It was once thought to be rare, but more probably the syndrome has gone undiagnosed since it has recently been estimated to occur in one child out of every 35,000 live births (Danks et al., 1972a). Also, the syndrome is now associated with atypical copper metabolism resembling copper deficiency (Danks et al., 1972a; Menkes, 1972).

In addition to the symptoms originally described by Menkes et al. (1961) many others have been found in association with this syndrome. They include spastic quadriparesis, irritability, anorexia, and noisy respirations (Bray, 1965; Walker-Smith et al., 1973); thick, dry skin, constipation, dehydration, optic disc pallor, vertical eye movement, moderate hypochromic anemia, and death due to respiratory infection (Aguilar et al., 1966); diarrhea, fever, twitching of the left side of the mouth, and profuse drooling (Billings and Degnan, 1971; Volpintesta, 1974); hypothermia (French et al., 1972; Danks et al., 1972; Bucknall et al., 1973; Grover and Scrutton, 1975); frog posture, marked hypotonia, apnea (which may be equivalent to the noisy respirations reported by Bray, 1965), lip smacking, blinking, fever without known cause, moniliasis, cough, rapid respiration, and no response to auditory stimuli (French et al., 1972); characteristic facial features (Danks, 1972) or "cherubism" (Grover et al., in press), labile blood pressure, prolonger neonatal hyperbilirubinemia, inguinal hernia, sparse, short, coarse eyebrows, normal serum iron and iron-binding capacity (Bucknall et al., 1973); vacuolation and deposition of ferritin granules in young bone-marrow-derived cells (Ashkenazi et al., 1973); vomiting, feeding difficulties, and, again, death associated with bronchopneumonia, pyelonephritis, and meningitis (French et al., 1972; Walker-Smith et al., 1973); decreased Hb, hypotroteinemia, and hypoalbuminemia (Volpintesta, 1974; Adams et al., 1974); low hematocrit (Adams et al., 1974; Lott et al., 1975); seborrhea or scaly scalp

(Danks et al., 1972a); extreme susceptibility to infection and death due to septicemia (Danks et al., 1972a); battered-child-like syndrome (Adams et al., 1974); and salaam spasms with EEG hypoarrhythmia (Osaka et al., 1977). Many of these symptoms have been confirmed by subsequent observation, but their variety also suggests a great deal of variation in the severity and expression of this genetic syndrome.

3.2. Heterozygous Females

Although the diagnosis of Menkes' syndrome requires that the infant be male, heterozygous females have also been found to be symptomatic. Wiry or twisted hair shafts and abnormal skin pigmentation have been found in living female siblings (Danks et al., 1972a; Volpintesta, 1974). Two living sisters and the mother of one Menkes' patient had higher than normal serum copper and copper oxidase concentrations. A female sibling of another Menkes' patient died at the age of 7 months after a short illness of unknown etiology complicated by vomiting and aspiration (Grover and Scrutton, 1975).

3.3. Brain and Blood Vessel Pathology

The severe cerebral and cerebellar degenerations observed by Menkes et al. (1962) were confirmed (Aguilar et al., 1966; Wesenberg et al., 1969) and attributed to neuronal degeneration (Bray, 1965). Angiograms revealed unusually tortuous cerebral, carotid, and spinal cord arteries (Wesenberg et al., 1969; Danks et al., 1972b; Osaka et al., 1977), bilateral subdural effusions (Aguilar et al., 1966), and a very slow rate of blood flow (Bucknall et al., 1973), which is consistent with dilated, irregularly thickened medial and intimal elastic laminae (Danks et al., 1972b). Similar arterial changes have been found in the femoral, ilial, visceral, and renal arteries, as well as the abdominal aorta (Wesenberg et al., 1969; Danks et al., 1972b; Singh and Bresnan, 1973). Osaka et al. (1977) found a diffuse avascular area about 2 cm in thickness in both hemispheres which may have been due to chronic subdural effusions.

Brain atrophy and a marked lack of myelination have been correlated with grossly defective brain lipid metabolism (Aguilar et al., 1966; O'Brien and Sampson, 1966; French et al., 1972; Bucknall et al., 1973; Lou et al., 1974; Vagn-Hansen et al., 1973; Mallekaer, 1974) in association with decreased activity of the coperer-dependent enzymes monoamine oxidase (Bucknall et al., 1973) and cytochrome *c* oxidase (French et al., 1972) or an increase in autooxidation of the polyunsaturated brain lipids (O'Brien and Sampson, 1966; French et al., 1972). This decrease in polyunsaturated lipids was found in older patients but not in a younger patient (Lou et al., 1974) who apparently had not progressed to the same state of autooxidative or peroxidative degeneration. The increase in lipid peroxidation can be attributed to the lack of another copper-dependent enzyme, superoxide dismutase (Pederson and Aust, 1973).

Cerebral white and gray matter copper concentrations have been reported to be markedly lower than normal (Walker-Smith et al., 1973) and normal (Reske-Nielsen et al., 1973) in infants who died, but markedly above normal in an aborted fetus suspected of having Menkes' syndrome (Horn, 1976). This seems to be consistent with the notion that these children are born with at least a partially adequate supply of copper but gradually become depleted through malabsorption or inadequate intake. All of the degenerative changes found in the brains of Menkes' syndrome patients are consistent with changes in copper-deficient animals.

Copper deficiency in rats has been shown to cause reduced whole brain weight, cerebellar growth, and myelination (Prohaska and Wells, 1974), accompanied by defective brain lipid synthesis (Gallagher and Reeve, 1971; Di Paolo and Newberne, 1972, 1973), decreased cerebellar cytochrome *c* oxidase (Gallagher et al., 1956; Prohaska and Wells, 1974), and decreased superoxide dismutase activity and norepinephrine concentrations (Prohaska and Wells, 1974). A decrease in norepinephrine associated with copper deficiency is also consistent with neuronal degeneration. Copper-deficient sheep also develop cerebellar and neuronal degeneration in association with a lack of myelination (Adelstein and Vallee, 1962), along with diminished cytochrome *c* oxidase activity (Howell and Davidson, 1959; Mills and Williams, 1962). Guinea pigs fed copper-deficient diets also show marked underdevelopment of brain myelin (Everson et al., 1968).

3.4. Bone Pathology

Aguilar et al. (1966) were the first to report pathologic bone change in association with Menkes' syndrome. They found spur formations on ulnar and femoral bones along with periosteal new bone formations and marked flaring of the costocondral junctions. Wesenberg et al. (1969) confirmed these observations and extended this list of bone abnormalities to include bilateral symetrical metaphyseal spurring of the long bones with normal bone density; diffuse anterior flaring of the ribs; extensive Wormian bone formation in the posterior region of the skull, which was microcephalic; underdeveloped facial bones; diaphyseal periosteal reaction of the femora, humeri, and fibulae; and periosteal thickening of the scapulae and clavicles which becomes more pronounced at age 4 to 6 months. At 1 year of age the femoral, humeral, and fibular bone changes observed earlier were no longer evident. However, there was a mild diffuse osteoporosis with moderately delayed maturation. Anterior flaring of the ribs was persistent. Metaphyseal spurring (Billings and Degnan, 1971) and Wormian skull bones (French et al., 1972) were confirmed, along with the observation that it was common to find healing fractures of the ribs and limb bones. Copper deficiency in animals is known to reduce bone growth and, with increasing age, produces osteoporosis, which accounts for the easily broken bones.

3.5. Ocular Pathology

The ocular pathology noted by Menkes et al. (1962) and Aguilar et al. (1966), but not by Bray (1965), was observed to be much more severe in two cases seen by Seelenfreund et al. (1968). They found microcysts in the pigmented epithelium of the iris, a markedly decreased number of retinal ganglion cells, atrophy of the optic nerve associated with markedly fewer nerve fibers, an increase in glial cells, and a lack of myelin. Billings and Degnan (1971) were the first to find electroretinographic changes consistent with blindness. In other patients the retinal cones were normal but the rods had poor function (Singh and Bresnan, 1973). These observations again indicate the variation in symptoms found in these patients.

3.6. Atypical Copper Metabolism

In 1972 Danks et al. (1972a, 1972b) reported that marked copper deficiency, manifested by low total-serum copper and copper oxidase values as well as extremely low liver copper concentrations, had been found in seven infants with Menkes' syndrome. Since serum copper is derived from liver stores of this element, low serum copper values coupled with low liver copper values represent severe copper depletion. Six of these patients were premature, and it has since been pointed out that the onset of Menkes' syndrome is likely to occur sooner in these infants than in term infants because there is less stored liver copper in prematures (Dekaban et al., 1974). A very low level of liver copper seems to be a constant feature of Menkes' syndrome, and has even been found in a fetus suspected of having this syndrome (Heydorn et al., 1975).

Normally serum copper and Cp levels in full-term infants are lower than adult levels (Gubler et al., 1953; Neumann and Sass-Kortsak, 1967), but these levels are much lower in infants with Menkes' syndrome than in normal term infants (Grover et al., 1978). A recently observed progressive decrease in copper concentration from higher than normal during the first day of life to lower than normal at the end of 1 month (Grover et al., in press) may account for the original observation of Menkes et al. (1962) that the serum Cp was "unremarkable." Higher than normal plasma copper and Cp levels, accompanied by decreased plasma zinc values, suggest that these neonates were physiologically responding to the stress of disease and that copper was being mobilized from the liver to the blood and then to other tissues. Because of already low liver copper stores, this mobilization may rapidly bring about a total depletion of the stored element.

Depletion of liver copper stores may also result from an apparent lack or blockade of intestinal absorption (Danks et al., 1972a, 1972b, 1973b). However, it is unclear as to whether the lack of systemic absorption following oral intake is due to a blockade of absorption or an accumulation of copper by copper-deficient intestinal tissues (see also Section 3.7). The basis for the lack of systemic absorption following oral intake has not been firmly established (Bucknall et al., 1973; Garnica et al., 1973; Danks et al., 1978). Accumulation of copper by copper-deficient intestinal tissues seems to be a possibility since tissues of infants are generally richer in this element per unit of weight than are those of adults (Underwood, 1977) and adult duodenal copper levels (Tipton and Cook, 1963) are nearly sixfold greater than those found in Menkes patients (Danks et al., 1972a, 1972b, 1973b). Accumulation of copper by copper-deficient tissues is consistent with the results obtained in cell culture studies.

Cultured fibroblasts from the skin of Menkes patients and heterozygous females showed intense metachromasia, unlike normal control cells, when primary cultures were grown in medium of low copper content (Danks et al., 1973b; Danks, 1975). This metachromasia was most pronounced with cells from Menkes patients but disappeared when copper was supplied to the culture medium and reappeared when medium with a low copper content was used. The disappearance of this metachromasia may be due in part to an accumulation of copper since cultured-skin-derived fibroblasts from Menkes patients have been shown to have a fivefold higher copper concentration, whereas cultured heterozygote mothers' cells had normal concentrations (Goka et al., 1976). This accumulation of copper in cultured skin fibroblasts has been confirmed in two additional studies (Horn, 1976; Danks et al., 1978).

Determinations of the copper contents of other tissues have revealed normal levels in the cerebrospinal fluid (Grover and Henkin, 1978) but higher than normal concentrations in brain, spleen, and kidney (Heydorn et al., 1975; Horn et al., 1976), pancreas, muscle, lung, skin, and placenta (Horn et al., 1976), and cord blood (Grover et al., in press). The higher than normal levels support the suggestion that peripheral tissue accumulation of absorbed copper may account for depleted liver stores and low blood levels.

An apparent increased tissue need for copper, coupled with increased fecal (Danks et al., 1972b) and urinary (Horn, 1976; Grover and Henkin, 1978) excretion of this element, may be taken as evidence for the rapid turnover of copper in this disease state. The rapid turnover may provide another reason for low liver and blood copper levels.

A state of copper deficiency is consistent with many other symptoms of Menkes' syndrome. Danks et al. (1972a, 1972b) found a normal hair amino acid composition with more than normal free-sulfhydryl groups to be consistent with an absence of copper-dependent disulfide cross-linking. Hair copper levels were found to be normal in one patient at 9 months of age but below normal at 11 months (Singh and Bresnan, 1973), but to be higher than normal in two infants, aged 4 and 6 months (Osaka et al., 1977). Copper deficiency and the lack of lysyl oxidase activity also accounted for fragmentation of the internal elastic laminae and internal proliferation, eventually resulting in occlusion as well as the scorbutic bone changes. Deficiency also explained hair and skin depigmentation caused by decreased tyrosine hydroxylase activity, peroxidation of unsaturated fatty acids due to decreased brain superoxide dismutase activity, and hypotonia due to decreased cytochrome *c* oxidase activity. Since copper may be required for the activity of antibiotics (see Section 6), death due to infection or septicemia may also be viewed as another symptom of copper deficiency.

Studies of serum copper-dependent enzymes have shown normal levels of serum superoxide dismutase (SOD) but no dopamine-β-hydroxylase (DBH) activity (Grover and Henkin, 1978). However, the normal muscle cytochrome *c* oxidase and SOD activities were ascribed to heterogeneity associated with the biochemical abnormalities related to this syndrome (Zelkowitz et al., 1976).

Red blood cells have been reported to contain normal (Danks et al., 1972b), above normal (Grover and Henkin, 1978), and less than but nearly normal (Osaka et al., 1977) copper levels. It was suggested that these levels could account for the absence, in Menkes' syndrome, of anemia or neutropenia, which is usually associated with copper deficiency (Danks et al., 1972b). However, it is not difficult to reconcile the absence of anemia with the presence of low serum copper and Cp levels. Patients with Wilson's disease have low serum copper and Cp levels but are not anemic. It may be better to accept the absence of anemia in some Menkes patients as a variation in copper-deficiency anemia in infants (Al-Rashid and Spangler, 1971) or a variation in the degree of hypocupremic anemia in Menkes' syndrome, since more severe disease is associated with moderate hypochromic anemia (Aguilar et al., 1966), vacuolation and deposition of ferritin granules in young bone-marrow-derived cells (Ashkenazi et al., 1973), and decreased Hb, hypoproteinemia, and hypoalbuminemia (Volpintesta, 1974; Adams et al., 1974), as well as low hematocrit (Adams et al., 1974; Lott et al., 1975) in the presence of normal serum iron and iron-binding capacity (Bucknall et al., 1973). Normal RBC copper has alternatively been interpreted as indicating a preferential access to available copper for RBC superoxide dismutase (Singh and Bresnan, 1973). For a recent review of copper transport and utilization in Menkes' syndrome see Danks (1977).

Nearly identical disturbances in copper metabolism and symptoms of Menkes' syndrome have been found to occur in mottled mutant mice (Hunt, 1974, 1976). These mice mutants have been found to be excellent for the study of defects associated with Menkes' syndrome (Hunt, 1974, 1976; Danks et al., 1978; Prins and Van den Hamer, 1978), and it may be possible to develop new approaches to the treatment of Menkes' syndrome with this model (Danks et al., 1978). Since *in utero* diagnosis has been achieved with amniotic fibroblast culture studies (Goka et al., 1976), early therapy is possible.

3.7. Approaches to Therapy

After the recognition that Menkes' syndrome might be related to copper deficiency, Danks et al., (1972b) treated two patients. One patient received daily oral doses of 1 mg $CuSO_4$ for 8 weeks, and the other received 2.5 mg/day orally for 4 weeks after having been given 80 μg $CuCl_2$ i.v. for 2 days. The patient given 1 mg/day had a small rise in total serum copper and copper oxidase activity, began to eat solid foods, and showed improved growth, but there was no change in his neurologic state. The patient given oral and i.v. therapy had higher Cp levels, which may have accounted for the observed slight neurologic improvement and normalized body temperature. Improvement ceased when serum copper levels returned to the initial low levels and the hypothermia returned. The disappointing lack of neurologic improvement following oral therapy with copper salts caused this approach to be abandoned.

Oral absorption was thought to have been impaired when 75 to 97% of the administered dose was recovered in the feces within 48 hr (Danks et al., 1972b). However, another group of patients given oral or i.v. $CuCl_2$ had longer retention times than normal (Dekaban et al., 1974). Fecal and urinary excretion was less than expected and liver copper was higher than normal, even though oral absorption was only one fourth of the control values. The questions concerning impaired oral absorption, therefore, have not been fully answered (see also Section 3.6).

Oral therapy with higher doses of $CuSO_4$ (300 μg/kg/day: Bucknall et al., 1973, or 1 to 10 mg/day for 6 months: Walker-Smith et al., 1973) also failed to increase plasma copper or copper oxidase levels, but i.v. infusions of 90 μg Cu/kg/day or 136 mg human Cp did raise these values (Bucknall et al., 1973). The half-life for Cp was shorter than normal and interpreted to be due to nutritional deficiency. A smaller quantity of $Cu_2(OAc)_4$ (22 μg Cu/kg/day given i.v. for 10 days) also failed to increase serum copper or Cp values (Garnica et al., 1973). The amount of copper given in this treatment appears to have been too small to bring about an increase in the blood values of this element. This is consistent with the suggestion that normal infant requirements are 200 to 500 μg Cu/day (Alexander et al., 1974). For optimal therapy of this syndrome it seems reasonable to suggest a daily dose that would provide enough copper to achieve the usual plasma increase of 2 to 3 times the normal plasma copper level, observed as a general response to many disease and stress states.

Normal serum copper values were obtained after $Cu_2(OAc)_4$ was given by infusion, 150 μg Cu/kg/day, for 5 days (Grover and Scrutton, 1975). Continued administration of 190 to 220 μg Cu/kg/day once or twice weekly maintained elevated hepatic and serum concentrations. After 3 months of therapy a 6-month-old patient reached the functioning level of a 4-month-old infant, but an older child demonstrated progressive loss of neurologic function and died. Examination of the surviving patient at 8 hr of life had revealed signs of Menkes' syndrome. He had jaundice and received an exchange transfusion which may have provided iron as well as copper and Cp. He did not have cerebral vascular lesions at 6 weeks of age and appeared to develop neurologically up to age 5 months without seizures or characteristic eye abnormalities, although his scalp hair did not grow.

In an attempt to clarify the nature of the impaired intestinal absorption of copper in other patients, copper-loading studies were performed with doses of $CuSO_4$ supplying 520 μg/kg/day or 20 μg Cu/kg per i.v. infusion (Lott et al., 1975) and 1 mg $Cu_2(OAc)_4$ given orally or i.m. (Van den Hamer and Prins, 1978). Increases in serum copper and Cp were found to be greatest when the parenteral treatments were given. Serum copper increased with all four treatments, and Cp values rose in all but the oral $CuSO_4$ treatment. Both oral and i.v. $CuSO_4$ treatments caused a moderate improvement in hypothermia and hair pigmentation, but neither improved neurologic status. The therapeutic results obtained with $Cu_2(OAc)_4$ were not reported.

In an attempt to facilitate the absorption and utilization of copper, a number of different complexes of the element have been used to treat Menkes' syndrome. Giving copper complexes of penicillamine and ethylenediaminetetraacetic acid (EDTA) orally failed to increase the absorption of copper, while i.v. administration of copper-albumin (Danks et al., 1973a) or i.m. doses of Cu-EDTA (Walker-Smith et al., 1973) raised serum copper and Cp levels to normal. However, after 4 to 6 weeks of Cu-EDTA therapy the Cp levels were again low, and treatment on two occasions, 3 months apart, produced no change in neurologic status. The patient died even though the liver copper had been raised to the normal adult level.

Oral therapy with copper-nitrilotriacetic acid (Cu-NTA) has been found to increase serum copper and Cp as well as renal, cerebrospinal fluid (CSF), liver, and urine copper concentrations (Grover and Henkin, 1978). These changes were accompanied by an increase in muscle mitochondrial succinate activity, but DBH activity was unchanged. One child treated with Cu-NTA for 2.5 years is now over 5 years old, is seizure free, and is growing slowly. These data indicate that Cu-NTA can facilitate gastrointestinal absorption of a utilizable form; however, the lack of change in DBH suggests a defect in selective copper binding, storage, and/or utilization, perhaps in the CNS, even though copper levels in blood and other tissues were corrected. Parenteral copper therapy with Cu-NTA brought about normal blood and hepatic copper levels, but the values in cerebral cortex and white matter remained significantly decreased and renal dysfunction was not corrected (Grover et al., in press). Parenteral administration of amino acid-copper complexes supplying 90 μg Cu/kg has been suggested as an alternative therapeutic approach (Osterberg et al., 1975).

The recognition that copper deficiency was associated with Menkes' syndrome was an important step forward in the treatment of this disorder. Results obtained with copper therapy alone have been disappointing, but this also may be viewed as another advance: it is now recognized that other measures must be taken to successfully treat this syndrome. Such measures might include delivery of copper in a complexed form which would allow exchange to yield copper-dependent enzymes even in the presence of altered metal-bonding components in Menkes cells, or delivery of utilizable copper to the fetus to promote normal development of the CNS (Camakaris and Danks, personal communication).

The patient described by Grover and Scrutton (1975), Grover and Henkin (1978), and Grover et al. (in press), who has survived the longest and is now over 5 years old (Grover, personnel communication), represents more than the limited success achieved by others, even though the child still suffers from severe mental retardation. The medical history of this diagnosed Menkes patient is unique in another respect. This child was discovered to be jaundiced during the initial examination 8 hr after birth and was given an exchange transfusion. As a result, he received a source of iron, copper, and Cp in addition to other blood components. Since Menkes' syndrome patients usually do not have anemia or low blood iron levels, iron therapy is not employed.

However, Ashkenazi et al. (1973) noted, in a patient who might have been diagnosed as having Menkes' syndrome but was instead diagnosed as suffering from neonatal copper deficiency, that copper supplementation produced a marked reduction in serum iron. Copper therapy was discontinued, and iron therapy begun. This child had a total remission of the Menkes-like symptoms. A similar remission was reported for another patient with Menkes-like syndrome after treatment with iron (blood transfusion) and copper supplementation (Al-Rashid and Spangler, 1971). Perhaps patients with Menkes' syndrome would be better treated if given iron along with copper, or copper and iron plus supplements of the other essential metals, including zinc, since deficiency of a single essential metal is unlikely.

Unfortunately, there have been too few studies of the other essential metals in blood and tissues of Menkes' syndrome patients. Serum or plasma zinc values 3 times the normal (Singh and Bresnan, 1973), equal to the normal (Walker-Smith et al., 1973; Grover et al., in press), and less than the normal (Grover et al., in press) values reported by Henkin et al. (1973) have been found for these patients. This apparent inconsistency may also be related to the variation of disease state in the patients studied. Normal urinary zinc levels and low amniotic fluid zinc levels have also been reported (Grover et al., in press). The etiology and pathology of this disease might be better understood if more information concerning the other essential metals were available.

4. REGULATION OF FERTILITY

In considering the need for an intrauterine device (IUD) to regulate fertility, Osterberg et al. (1975) pointed out that this method of contraception offers many advantages over oral ones. More efficient contraception, resulting in fewer pregnancies, and greater tolerance, as shown by less pain, bleeding, spontaneous expulsion, and infection, have led to the acceptance of a newer device to which is attached a flattened metallic copper coil. This copper IUD has become one of the most popular contraceptives to date, even in the highly developed countries (Osterberg et al., 1975; Oster and Salgo, 1975).

The copper device may have another advantage over the plain one in that the gonococcicidal action of copper (Fiscina et al., 1973) may prevent the spread of gonorrheal infection to the uterus (Spellacy et al., 1974). This is not surprising since, as will be pointed out in Section 6, copper and its compounds are known to have antibacterial activity.

The copper IUD was developed as a result of the work of Zipper et al. (1969), who demonstrated that metallic copper wire prevented uterine implantation of fertilized ova in rabbits. The mechanism of this antifertility effect is not well established. However, it is likely that copper complexes formed by reaction of the uterine contents with either Cu(I) and/or Cu(II), leached from the metallic copper coil by endometrial fluids, are absorbed and induce a uterine state of pseudopregnancy. This is consistent with the pseudopregnancy produced in rats after intravenous injections of inorganic copper salts (Fevold et al., 1936; Dury and Bradbury, 1941; Hiroi et al., 1969), which can form copper complexes *in vivo*. Mixtures of a large variety of complexing agents and $Cu_2(OAc)_4$, which no doubt resulted in the formation of copper complexes, have been demonstrated to prevent pregnancy when instilled into the oviducts of fertile rabbits (Biological Concepts, Inc., 1973).

The proliferation of endometrial mucosal stroma in contact with copper observed by Zipper et al. (1969) is characteristic of a local estrogenic effect (Oster and Salgo, 1975) and is consistent with the physiologic effects of estrogens. Metallic copper wire is known to increase the amount of estrogen specific binding protein in the rat uterus (Adadevoh and Dada, 1973; Aedo and Zipper, 1973), and it is well known that estrogens increase serum or plasma concentrations of low molecular weight copper complexes and Cp (Markowitz et al., 1955; Clemetson, 1968; Halstead et al., 1968).

Low concentrations (10^{-5} to 10^{-6} M) of copper stimulate the contraction of uterine smooth muscle to a greater degree and for a longer period than the maximum response achieved with oxytocin, the naturally occurring uterine stimulant (Salgo and Oster, 1973). The physiologic mechanism of this smooth muscle contraction may involve copper complex modulation of prostaglandin syntheses. Lee and Lands (1972) reported that addition of $CuSO_4$ or mixtures of $CuSO_4$ and diethyldithiocarbamate, which is known to complex with copper, to seminal vesicle homogenates increased the synthesis of prostaglandin $F_{2\alpha}$ ($PGF_{2\alpha}$) and concomitantly decreased the synthesis of prostaglandin E_2 (PGE_2). These observations were confirmed by Maddox (1973) and Vargaftig et al. (1975), using $CuCl_2$. In addition, Boyle et al. (1976) found that the copper complex of aspirin [$Cu(II)_2(aspirinate)_4$] caused the same modulation of prostaglandin synthesis. All of these findings are consistant with Caton's (1971) results which demonstrated that $PGF_{2\alpha}$ caused pronounced uterine contractions, whereas diethyldithiocarbamate inhibited uterine contractions by decreasing the synthesis of $PGF_{2\alpha}$ (Daniel, 1964; Salgo and Oster, 1973).

Also, since $PGF_{2\alpha}$ has anti-inflammatory (vasoconstrictor) activity, an increase in its synthesis could also account for the decreased inflammatory reaction and reduced uterine bleeding associated with the copper IUD, in comparison to the plain device.

After very careful consideration Oster and Salgo (1975) have postulated that free radicals produced in the corrosive processes leading to the release of copper from the copper IUD are responsible for the prevention of implantation and the contraceptive action of this device. However, this view is in conflict with the observed destruction of superoxide radicals and the termination of free radical processes associated with the superoxide dismutase mimetic activity of amino acid-copper complexes, which are likely to be formed by amino acids and copper in endometrial fluids. Like the copper-dependent enzyme superoxide dismutase (Fridovich, 1975), copper complexes of lysine, glycylhistadine, tryptophan, glycylhistadylleucine, and histadine (Brigelius et al., 1974, 1975; Paschen and Weser, 1975; De Alvare et al., 1976) have been shown to have superoxide dismutase mimetic activity. In addition, other non-amino-acid copper complexes were also found to have superoxide dismutase mimetic activity (De Alvare et al., 1976; Weser et al., 1978). This is of interest since many of the amino acid- and non-amino-acid copper complexes have anti-inflammatory activity (see Section 7.2), which provides an additional explanation of the observed anti-inflammatory effect and reduced uterine bleeding associated with the copper IUD.

In addition to prostaglandin modulation and superoxide mimetic activities, some of these copper complexes have been found to induce lysyl oxidase, another copper-dependent enzyme (Harris, personal communication). Since remodeling of uterine connective tissue may play a role in preventing implantation, this aspect of biochemistry also warrants further attention.

For additional information on this subject, the reader is directed to a recent review by Oster and Salgo (1977).

5. PROPHYLAXIS OF INFECTIONS

The antibacterial and antiviral properties of metallic copper gauze filters and

$CuCl_2$ solutions (100 $\mu g/ml$) are used in hospitals for continuous sterilization of air-conditioned air, nebulizers, and effluent air of respiratory therapy apparatus to prevent pulmonary infections (Deane et al., 1970; Jordon and Nassar, 1971). A 0.1-$\mu g/ml$ concentration of copper added to natural waters is also effective in killing snails that serve as hosts to a human liver fluke which has been estimated to cause 200 million cases of schistosomiasis (Osterberg et al., 1975). Similarily, malaria is prevented with inorganic copper salts because this element has mosquito larvicidal activity.

6. TREATMENT OF INFECTIONS

In the 1920s it was recognized that the incidence of tuberculosis was much less in copper miners than in the general population. This led to the successful use of copper oxide (CuO) in the treatment of tuberculosis before 1940. However, in 1940 several additional copper complexes were reported to have antitubercular activity, and it was suggested that one of them, the copper complex sodium 3-(allylcuprothiouredo)-1-benzoate (Cupralene) (**1**), was superior to gold as

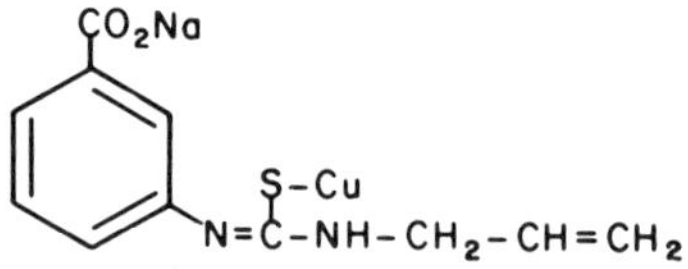

CO_2Na ; S–Cu ; N=C–NH–CH_2–CH=CH_2 **1**

therapy for tuberculosis (Goralewski, 1940; Arnold, 1940; Tuchler and Razenhofer, 1940).

Subsequent research led to the discovery that *p*-aminosalicylic acid (PAS) was an active antitubercular agent, and it was suggested that the antitubercular activity of this drug was due to its Cu(II) complex (Carl and Marquardt, 1949), which was prepared and found to be 10 times as active as PAS itself (Roth et al., 1951). Studies of Cu(II) (*p*-aminosalicylate)$_2$ (**2**) revealed also that it was

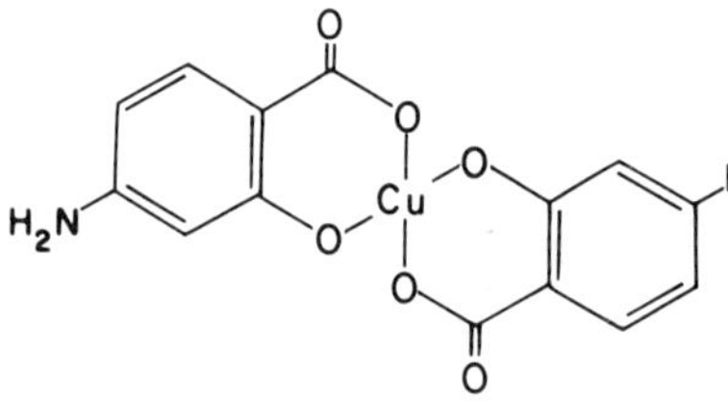

O ; O ; O ; Cu ; H_2N ; O ; O ; NH_2 ; O **2**

30 times more lipid soluble than PAS (Foye, 1955; Foye and Duvall, 1958). The enhanced activity of the copper complex was attributed to this increased lipid solubility, which facilitated penetration of the fatty outer membrane of *Mycobacterium tuberculosis*.

Similarly, additions of copper were found to enhance the *in vitro* activity of isonicotinylhydrazide (INH) (Sorkin et al., 1952). This led to several demonstrations that the copper complex of INH was the active form of this antitubercular agent (Cymerman-Craig et al., 1955, 1956; Reiber and Bemski, 1969). Voyatzakis et al. (1968) found that the copper complex of a substituted hydrazide, *o*-nitrobenzylidenoisonicotinoylhydrazide, was the most effective antitubercular complex of 11 derivatives studied.

However, we appear to have taken a step backward. Copper complexes are known to have antitubercular activity in human beings, and the copper complex of INH is more effective than INH itself. Nevertheless, the drug of choice in the treatment of tuberculosis is INH and not its copper complex (Citron, 1972).

After considering the matter of copper complexation with INH, Krivis and Rabb (1969) suggested that a 1 : 1 cuprous complex was more likely to be formed with INH than a cupric complex. They pointed out that the copper complex formed with INH followed the reduction of Cu(II) to Cu(I), so that 1 : 1 cuprous complexes were formed upon addition of Cu(II) to solutions containing INH. Krivis and Rabb also suggested that the formation of Cu(I) complexes of thiosemicarbazones accounted for the synergistic effect of copper with the antitubercular agent *p*-acetamidobenzaldehyde thiosemicarbazone (thiacetazone).

Copper complexes of thioureas and *N*-substituted thioureas have also been suggested to account for the effectiveness of the parent compounds as antitubercular agents (Buu-Hoi and Xuong, 1953). Alternatively, Ueno (1956) suggested that the parent thioureas acted by removing copper from some *M. tuberculosis* copper-dependent enzyme and that the antitubercular activity was dependent on the relative stabilities of the enzyme complex and the thiourea complex. However, the activities of two series of *N*-substituted (2-pyridyl and 4-pyryl) thioureas (Glasser and Doughty, 1962) could not be correlated with copper complex stability (Johnson, 1966).

In 1944 Albert et al. (1953) found that Cu(II)(8-hydroxyquinoline)$_2$ (**3**) was

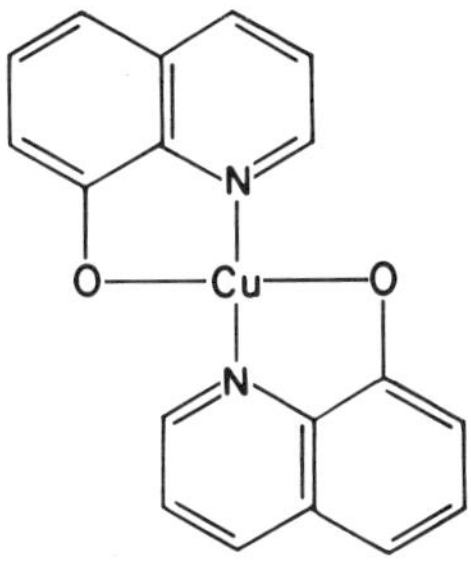

3

much more effective as a bactericidal and fungicidal agent than was 8-hydroxyquinoline (8-HQ). Copper complexes of halo-substituted 8-HQ have been found by Gershon (1974) to be even more effective than Cu(II)(8-HQ)$_2$. In addition, both 8-HQ (Auld et al., 1974) and its copper complex (Levinson et al., 1978, and in press) have been found to have antiviral activity. It was suggested that the parent compound inhibited the RNA-dependent DNA polymerase of chick myeloblastosis virus and its copper complex (**3**) produced the same effect with the RNA-dependent DNA polymerase of the tumerogenic Rous sarcoma virus.

Isatin-β-thiosemicarbazone was originally found to be a potent antivaccinia viral agent in mice, and the activity of this compound was attributed to complex formation (O'Sullivan and Sadler, 1961). Subsequently the derivative *N*-methylisatin-3-thiosemicarbazone (M-IBT) proved to be a very effective antiviral agent in human beings for the prevention of morbidity and mortality due to smallpox epidemics (Schubert, 1966). Levinson et al. (1973) also found that addition of copper to M-IBT enhanced the *in vitro* inactivation of Rous sarcoma virus, as well as other RNA tumor viruses and herpes virus, an effect that was later shown to be due to direct association of the copper complex with nucleic acids of Rous sarcoma virus, rather than to inhibition of the zinc-requiring RNA-dependent DNA polymerase (Mikelens et al., 1976). Subsequently, Levinson et al. (1978, and in press) suggested that this and other RNA tumor viruses were inactivated with the copper complex, but not M-IBT, as a result of bonding to single- and double-stranded DNA and RNA.

The enhancement of antibiotic activity has been extended to include many classes of complexing agents, including the naturally occurring antibiotics (Albert, 1961). Doluizio and Martin (1963a) found that the more active tetracyclines formed 2:1 copper complexes of high stability, while the less active tetracyclines formed 1:1 complexes of lower stability. The antitubercular activity of streptomycin has been attributed to a 3:1 copper complex (Foye et al., 1955) since this complex, containing three copper atoms, provides higher blood levels of this drug for a longer time *in vivo*. It has been pointed out that other well-known antibiotics such as penicillin and bacitracin are also complexing agents (Schubert, 1966).

The medical need for new wide-spectrum antibiotics has led to the discovery of copper myxin (**4**) [Cu(II)(6-methoxy-1-phenazinol-5,10-dioxide)$_2$], which

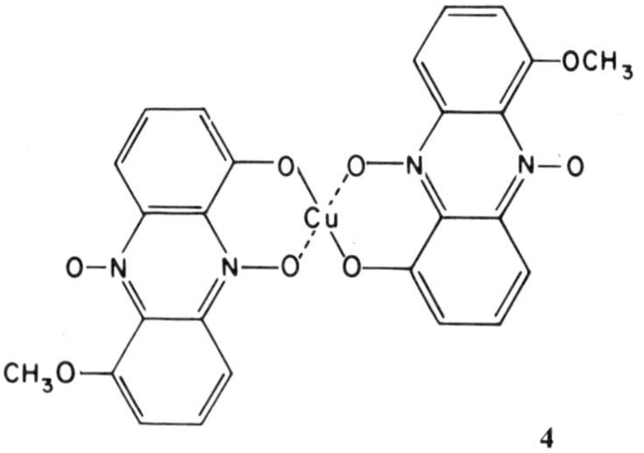

4

is very effective against gram-positive and gram-negative bacteria, fungi, and yeasts (Peterson et al., 1966). Its activity *in vivo* and *in vitro* indicates that it is more effective than penicillin, streptomycin, nitrofurazone, oxytetracyline, and sulfadimethoxine as an antibacterial agent and more effective than tolnaftate, nystatin, and amphotericin B as an antifungal agent. The results obtained with this copper complex raise the question of whether copper complexes of other synthetic or naturally occurring antibiotics might be more potent and effective as wide-spectrum antibiotics than the parent compounds.

Another copper complex, fluoropsin C (**5**) [Cu(II)(*N*-methylthioformohydroxamate)$_2$], has also been found to be very effective against gram-positive and gram-negative bacteria (Shirahata et al., 1970; Itoh et al., 1970). Studies in mice indicate that the $LD_{50}:ED_{50}$ ratio is large and, as a result, safe for drug use. It is also of interest that this complex was found to have potent *in vitro* antitumor activity.

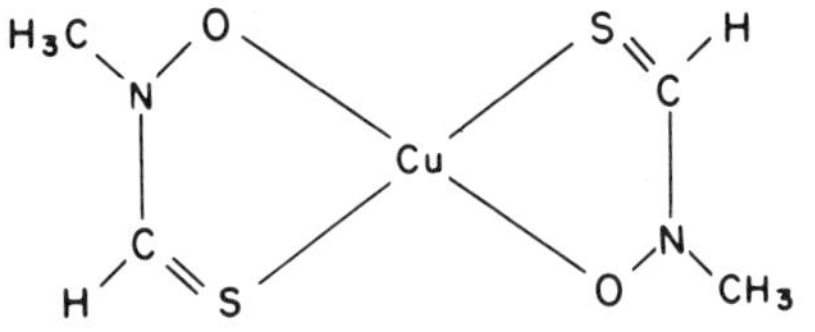

Cuphen [Cu(II)(3,4,7,8-tetramethyl-1,10-phenanthroline)$_2$] (**6**) has been suggested as being very useful for the treatment of moniliasis (*Candida albicans*), pustular *Acne vulgaris*, staphylococcal infections, and vaginitis due to either *C. albicans* or *Trichomonas vaginalis* (Butler et al., 1970).

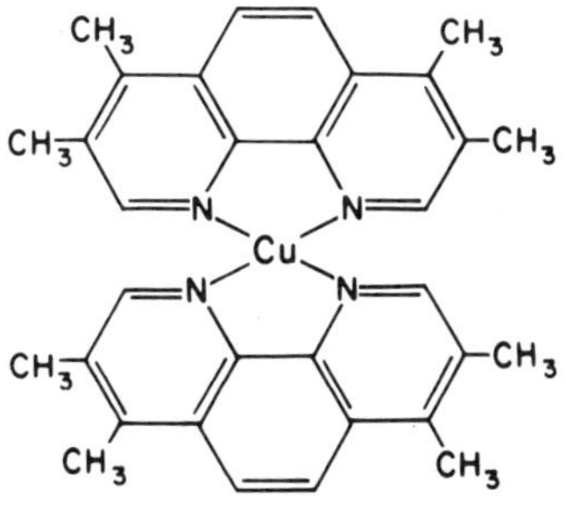

Consistently with these observations, many of the copper complexes reported in Section 7.2 to have anti-inflammatory and antiulcer activities (Sorenson, 1976) were also found to have antiviral activity against influenza A, strain 575 virus; antibacterial activity against *Diplococcus pneumoniae, Escherichia coli,* and *Erwinia* sp.; antiprotozoal activity against *Tetrahymena gellii, Trichomonas vaginalis,* and *Trichomonas foetus;* antihelmintic activity against *Turbatrix aceti;* antifungal activity against *Trichophyton mentagrophytes* and *Candida albicans;* and antialgal activity against *Chlorella vulgaris* in *in vitro* test systems (Sorenson, unpublished observations).

Osterberg et al. (1975) have pointed out that liniments and salves prepared with dilute concentrations of inorganic copper salts are also effective as fungicidal and bactericidal preparations. They are especially useful in treating scalp infections such as dandruff, seborrheic skin disease, and warts since they do not cause contact dermatitis or other allergic reactions. These authors also suggested wearing copper barrettes, chains, or bracelets as an alternative to the use of liniments or salves to control fungal or bacterial skin afflictions. It was thought to be unlikely that absorption of copper from these devices, in the form of amino acid complexes, would be harmful since only small amounts of the element would be absorbed. Support for this suggestion has recently been provided by Walker et al. (1977).

7. TREATMENT OF RHEUMATOID AND OTHER DEGENERATIVE CONNECTIVE TISSUE DISEASES

A historical review of the use of copper complexes in the treatment of rheumatoid and other degenerative connective tissue diseases has been published (Sorenson and Hangarter, 1977). This review provides an account of the therapeutic results obtained with copper complexes used from 1940 to 1971 for the treatment of 1500 patients with acute or chronic rheumatoid arthritis, rheumatic fever, ankylosing spondylitis, staphylococcal spondylitis, gonococcal arthritis, chronic gouty arthritis, chronic polyarticular synovitis, coxitis, disseminated spondylitis, arthritis with psoriasis, Reiter's syndrome, lupus erythematosus, sarcoidosis, arthrosis deformans, erythema nodosum, sciatica (with and without lumbar involvement), cervical spine-shoulder syndrome, lumbar spine syndrome, or osteoarthritis. Even though control studies of these clinical applications were not done, the results revealed that copper complexes are useful antiarthritic drugs. Treatment with these copper complexes produced no recognized copper-induced toxicity, and there was no gastrointestinal irritation, which is a common toxicity associated with the use of modern antiarthritic drugs.

After the reports (Goralewski, 1940; Arnold, 1940; Tuchler and Razenhofer, 1940) that Cupralene (1) was superior to gold therapy for the treatment of tuberculosis, Fenz (1941) used this drug to treat rheumatoid arthritis (RA). Modest improvement was obtained with injections of low doses of Cupralene, but exceptional results were claimed with some patients given larger doses 3 times per week. Treatment was completed after 10 to 14 injections, corresponding to 0.54 to 1.24 g Cu. "Excellent results" were achieved in 21% of the cases, and 41% received "important benefits," compared with 31% who "improved moderately" and 7% who experienced no effect (Fenz, 1941, 1951). Unfortunately, the terms "excellent," "important," and "improved moderately" were not fully defined as to changes in objective or subjective symptoms of the disease.

Fenz also reported that therapy with Cupralene brought about remission of the anemia associated with RA. This was consistent with the observations of others who found that anemic girls working in a copper mine were soon relieved of their anemia and what is now known to be a copper-dependent iron mobilization process required for hemoglobin synthesis (Hart et al., 1928; Lee et al., 1968, 1976; also see Frieden, this volume). Fenz also reported that toxic side effects such as those that occur relatively often with gold therapy were not observed with copper therapy.

Forestier (1944) also recognized that gold therapy was associated with a high risk of toxic manifestations and in his search for drugs of lesser toxicity began to use Cupralene in 1942. In 1944 he published preliminary results, which were patterned after his 20 years of clinical experience with the use of gold compounds. His first group of patients consisted of 33 RA, 4 ankylosing spondylitic, and 6 osteoarthritic patients. Of these, 45% or 14 were reported to have had "very good" results from the Cupralene therapy, 20% or 6 had "favorable" results, and 35% or 11 had "moderate improvement"—results that were consistent with those obtained by Fenz (1941, 1951), although these clinical classifications were also undefined. Subsequently, Forestier and Certonciny (1946) found that increasing the dose to 47.5 to 95 mg Cu per i.v. injection and increasing the duration of therapy, as well as the total dose per series (380 to 950 mg Cu), increased the percentage of patients experiencing "very good" and "favorable" results.

Separation of these patients into groups by disease category illustrated that not all were equally benefited from Cupralene. More patients (38%) with RA had "very good" responses if the duration of their disease was less than 1 year. There were no patients in the "moderate" effect group and only 1 (8%) in the "no effect" group whose disease was of shorter duration. When the disease was of more than 1 year's duration, only 1 patient (4%) obtained "very good" results, while 8 (35%) had "good" results, 9 (39%) had "moderate" results, and 5 (22%) experienced no beneficial effect. All patients with chronic polyarticular synovitis were favorably affected, but none of the responses was rated as "very good." No beneficial effect was observed in the disease categories of ankylosing spondylitis, acute articular rheumatism, or monoarthritis. In contrast to these, a "good" response was observed with 1 patient who had staphylococcal spondylitis. In summary, these data demonstrated that 72% or 56 individuals received some benefit with Cupralene therapy, while 28% or 14 patients experienced no beneficial effects.

In a continuing search for more effective therapy, Forestier et al. (1948) reported their results with a new copper complex, bis[8-hydroxyquinoline di(diethylammonium sulfonate)] Cu(II) (Dicuprene or Cuprimyl) (**7**). Although

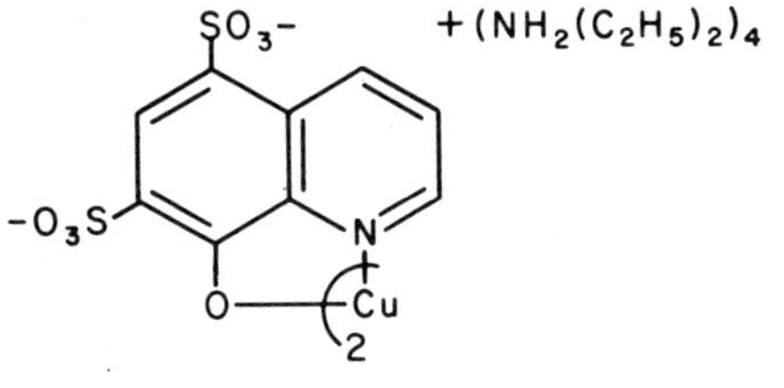

SO_3^- $+(NH_2(C_2H_5)_2)_4$

^-O_3S N Cu O 2

7

this compound had been found by Michez and Ortegal (1945), as later cited by Forestier et al. (1955), to be less effective orally, Forestier et al. (1948) considered Dicuprene to be somewhat superior to Cupralene because it was less irritating and could be given by both i.v. and i.m. routes of administration.

Dicuprene (6.5% Cu) mixed with novacaine was given i.m. or i.v. 2 or 3 times a week (32.5 to 65 mg Cu), with a total dose of 390 to 780 mg Cu per series. As in gold therapy, this copper compound was given in series of injections, with rest periods between series. These intervals were not to exceed 1 month between the first two series, and 2 to 3 months between following series. Any one series of injections was not to exceed 15 to 20 doses, and treatment was to be continued for several series to ensure that all signs of active arthritis disappeared and the erythrocyte sedimentation rate (ESR) returned to normal.

The results with Dicuprene therapy were only "good" when it was used to treat RA that had been active for less than a year, but these effects were obtained even if the patient had become resistant to gold therapy. From the results obtained in this study it appeared that patients who had RA for longer than a year benefited the most from Dicuprene. Initial patient therapy with Dicuprene showed that 3 (23%), 7 (54%), 2 (15%), and 1 (8%) obtained "very good," "good," "moderate," and "no" effects, respectively. However, some patients treated with Dicuprene after toxic or intolerance reactions to gold therapy were also benefited. Two (10%), 9 (50%), and 2 (10%), respectively, progressed to the first three stages of relief. Unlike Cupralene, Dicuprene Therapy produced "very good" results in 2 (6%) patients with ankylosing spondylitis and 1 patient with disseminated spondylitis. In addition, of 4 patients with gonococcal arthritis 2 achieved "good" effects, with 1 each in the two lesser effects categories. Treatment of 3 patients with chronic polyarticular gout produced "very good" results in 2 and "good" results in the other. Of 4 patients with chronic polyarticular synovitis 2 experienced "very good" results, and the others "good" or "moderate" benefit. Two patients with monoarthritis obtained "very good" or "good" results. The least affected category was coxitis; of the 2 patients treated only 1 obtained "moderate" relief. In 1946 Christin, in a personal communication to Forestier, reported that patients with both RA and psoriasis received simultaneous amelioration of both articular and skin lesions when treated with Dicuprene.

In 1949 and 1950 Forestier and others (Forestier and Certonciny, 1949; Forestier et al., 1950) presented summary comparisons of results they had obtained with copper and gold therapy of RA and other degenerative diseases. For RA of less than 1 year's duration, copper therapy (particularly Cupralene) was more effective and less toxic than gold. If gold therapy was tolerated, it was preferred for patients who had had RA for more than 1 year. However, copper complexes were far superior to gold in treating chronic polyarticular synovitis with effusion. This definite clinical entity, associated with a high ESR and sometimes marked anemia, was found to be resistant to gold therapy. In 80% of the patients treated, effusions entirely or nearly cleared after two or three series of injections.

Thirty-seven patients with chronic polyarticular gout responded favorably to small doses of copper complexes, as they did also to gold therapy. Of these, 21 given i.m. injections in two to three series enjoyed remarkably rapid elimination of signs of gouty polyarthritis; 8 others showed improvement in four to five series. In 8 others, which were more recent cases, i.v. or i.m. therapy was also successful.

The first attempts by Forestier and others with ankylosing spondylitis were less successful. Increasing the dose and keeping their patients on a very strict regime of injections, however, gave improved results.

Psoriasis arthropathica generally responsed rapidly to copper therapy, which was far superior to gold therapy and brought about astonishing improvements in both skin and joint lesions.

Benefit from copper therapy in infectious arthritis of known or unknown origin or in rheumatic fever was less certain. A decrease in ESR was found to correspond with clinical improvement over a period of 2 or 3 months.

In summary, copper therapy administered according to recommended doses was effective and did not, as a rule, cause untoward reactions. The onset of improvement was immediate, definite, and progressive in most patients. Twelve patients were followed for over 2 years and showed no relapse. This period was considered long enough to make it unlikely that these were spontaneous remissions. Erythrocyte sedimentation rates returned to normal faster in patients with RA of less than 1 year's duration than in those with chronic disease. Of 59 patients, 51 experienced no ill effects attributable to the medication. Only minor side effects, such as malaise, nausea, and slight albuminuria, were noted in 8 patients; there were no signs of marrow depression in this group. Copper compounds used to continue therapy in patients who were resistant to or intolerant of gold were found to be advantageous since they could be used without ill effects in the event of gold-therapy-induced dermatites (rashes or stomatitis) or nephritis (albuminuria).

Results of a study using copper-morrhuate therapy for RA were also reported in 1950 by Graber-Duvernay and Van Moorleghem (1950). This mixture of copper complexes, formed with the fatty acids obtained from cod liver oil, was given i.v. as a colloidal suspension along with novacaine. Of 117 RA patients so treated, only 30% were benefited.

In 1950 Tyson et al. (1950) reported their evaluation of Cupralene in what were described as 27 typical RA patients. The first group of 20 treated in this study were reported to be unquestionably severe RA patients who had had their disease for 2 months to 15 years. These patients were either resistant to or intolerant of gold therapy. In the first course of treatment, doses were used as recommended by Forestier for patients with disease of less than 1 year's duration. Only 2 patients, who had had the disease for 2 months to 1 year, obtained subjective improvement according to criteria of the American Rheumatism Association (ARA) for marked improvement or remission. All patients tolerated the total course of therapy without toxicity.

In correlating duration of disease with subjective improvement, it was obvious that patients who had had the disease for shorter durations benefited most. Consequently 7 patients with unquestionable RA of less than 1 year's duration were treated in addition to the original 20. Individual and total doses of Cupralene given to this group were increased to nonrecommended levels. The therapeutic results obtained in this study were reported to have been essentially similar to those with the smaller doses. However, it was found that none of these patients, no matter how long or how short the duration of his or her disease, showed any significant improvement throughout the average follow-up period of 1 year. Although Forestier had recommended multiple shorter series of injections before evaluation, Tyson et al. were forced to discontinue therapy after one prolonged series when severe toxicity was encountered with the larger doses. From these effects it seems clear why Forestier recommended small single doses and smaller total doses per series of injections of this copper complex.

In some instances, when a small quantity of Cupralene escaped from the vein into subcutaneous tissues, a very severe and painful cellulitis resulted. In no instance, however, did suppuration or necrosis occur, and in 3 to 10 days the cellulitis subsided.

Whereas during the early course of Cupralene treatment, with smaller doses administered i.v. twice a week, toxicity was rare, in the later stage using larger doses toxicity was the rule. On discontinuing copper therapy all symptoms of toxicity disappeared, and within a month blood counts returned to normal.

In the following year Kuzell et al. (1951) published their study of copper therapy for a variety of rheumatoid and degenerative diseases, using the criteria of Steinbrocker et al. (1949) and the ARA. Dicuprene and Cupralene were used to treat patients who had RA, RA with psoriasis, Reiter's syndrome, ankylosing spondylitis, and chronic gouty arthritis.

Of the 31 RA patients treated, only 6 progressed to Grades I and II. Therefore only 20% appreciably improved, and that was interpreted to be less than satisfactory, or apparently without benefit. However, patients who had improved had less severe stages of RA. Five patients or 56% of those treated who had less severe or early stage psoriasis with RA also responded with Grade I or II improvement without skin lesion progression or exacerbation. Slightly more improvement in these patients than in arthritics without psoriasis was suggested to merit further study. Two of the 3 patients with ankylosing spondylitis showed responses of Grade I and III. However, the fact that these patients received other drugs in addition to copper complexes prevented any conclusion concerning the efficacy of copper therapy. Three patients with Reiter's syndrome were reported to have been "apparently cured" with Dicuprene. In each of these, positive cultures of pleuropneumonia-like organisms were obtained from conjunctiva, and in the male from penile sores and urethral discharge. Only a limited Grade III response was observed with Dicuprene for 1 patient with disseminated lupus erythematosis. Of 18 patients with chronic gouty arthritis, 13 had a Grade I,

II, or III response to copper therapy, and this result was also viewed as worthy of study.

According to Kuzell et al. (1951), the toxicity of these substances was slight, infrequent, and mild and transitory in nature, in contrast to the toxicities observed by Tyson et al. (1950). If Cupralene escaped from the vein, intense pain and necrosis followed. Some of the lack of beneficial effect and/or of toxicity reported by Tyson et al. (1950) and Kuzell et al. (1951) has been attributed to drug decomposition before administration (Sorenson and Hangarter, 1977; Sorenson, 1978). Dicuprene produced pain on i.m. injection, so that in the later phases of the study this material was given i.v. Three patients developed nongravid amenorrhea while taking Dicuprene, but menstruation resumed after cessation of medication, and one became pregnant.

Consistent with an earlier report by Forestier and Certonciny (1949) that the LD_{50} values determined in rats for Cupralene and Dicuprene were 160 and 120 mg/kg, respectively (route of administration not given), Kuzell et al. (1951) reported rat LD_{50} values of 160 mg/kg intraperitoneally and 376 mg/kg i.m. for Cupralene, and 126 mg/kg following i.m. administration of Dicuprene.

Studies of the mechanism of action of Cupralene were subsequently reported by Wintrobe et al. (1953). They found that 250 mg of Cupralene (50 mg Cu) given i.v. to normal adults caused a rapid increase in plasma and RBC copper, which did not occur with oral administration, followed by a decrease to normal levels in 4 hr. It was suggested that the disappearance from plasma was associated with uptake by the reticuloendothelial system before erythropoiesis. This was consistent with the earlier observations of Heilmeyer and Stuwe (1938) and Heilmeyer et al. (1941) that serum copper increased in patients with RA but returned to normal with remission of their arthritic disease and associated anemia, as well as with the antianemic effects of Cupralene reported by Fenz (1941). These observations and suggestions were consistent also with the known requirements of copper for iron utilization in hemoglobin synthesis (Hart et al., 1928).

Although Cupralene and Dicuprene were shown to be effective and in some instances superior to gold therapy for rheumatoid and other degenerative connective tissue diseases, nothing more was published concerning these copper complex drugs after 1955.

That year O'Reilly (1955) reported results of a single study with i.m. sodium *m*-(*N*-allylcuprothiocarbamide)-1-benzoate (Alcuprin) (**8**), an analogue of

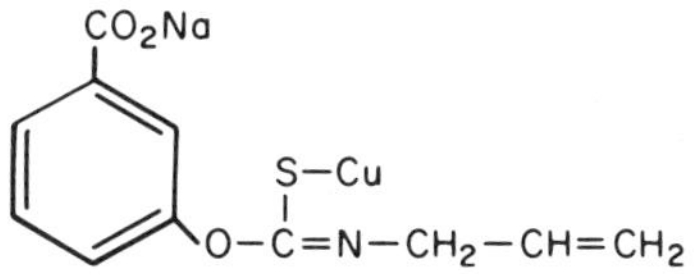

CO_2Na

S−Cu

O−C=N−CH_2−CH=CH_2 **8**

Cupralene, and Cuprimyl (**7**) in 10 patients with well-established RA of 1 to 13 years' duration. Four patients had had crysotherapy. All patients had two courses of treatment in a single series. Each course consisted of 10 doses of Alcuprin or Cuprimyl. Although no toxic effects were observed, only 2 patients experienced slight improvement. Two patients in the group had psoriasis, but they received no benefit from the copper therapy. Even though it was recognized that greater beneficial effects might have been expected in a clinical trial where cases of short duration were not excluded, the results of a suggested further trial using patients with early disease have not been reported.

In 1950, however, Hangarter had begun his research with a new copper preparation, Permalon. The following is an account of Hangarter's recognition of the therapeutic potential of copper and the results of his work at Bad Oldesloe Hospital, which continued until 1971, when he retired and the manufacture of Permalon was discontinued.

Historically, the therapeutic potential of copper for the treatment of rheumatic diseases was independently recognized by Hangarter (Hangarter and Lubke, 1952) in 1939, when he learned that Finnish copper miners were unaffected by rheumatism as long as they stayed with the mining industry. This was particularly striking, since rheumatism was widespread in Finland and workers in other industries and other towns had more rheumatic disease than these copper miners. Hangarter attributed this observation to copper, a view that was consistent with observations that serum copper increased in patients with rheumatic disease and infections and returned to normal with remission (Heilmeyer and Stuwe, 1938; Heilmeyer et al., 1941). Therapeutic results with inorganic copper alone were found by Hangarter (1974) to be comparable with those for gold therapy, although copper treatment was associated with fewer side effects. On the basis of these results and his experience with i.v. salicylic acid therapy for RA, he and Dr. Reiser of the Albert Chemical Co. developed Permalon, an aqueous solution of $CuCl_2$ and sodium salicylate. One 20-ml ampoule contained 2.0 g (12.5 mol) of sodium salicylate and 2.5 mg (39 μmol) of copper. It is likely that Cu(II)-$(\text{salicylate})_2$ (**9**) was a component of this solution.

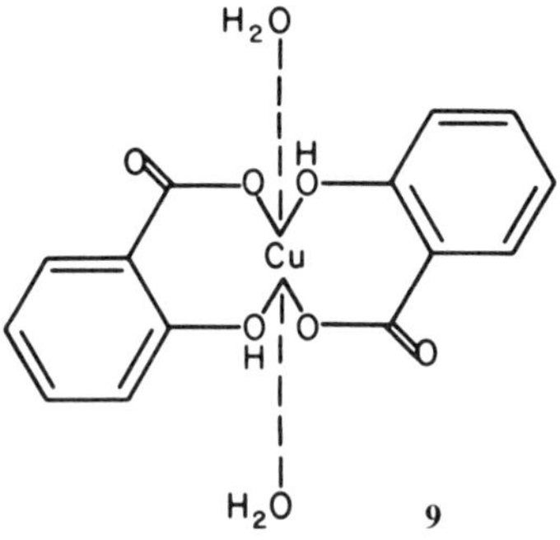

9

From 1950 to 1954, Permalon was administered by i.v. injection. Daily injections of one ampoule over an average period of only 8 to 14 days produced beneficial results. In cases of obvious therapeutic success according to the criteria of the ARA and of Steinbrocker et al. (1949), that is, remission of fever, alleviation of pain, increased mobility, inhibition of exudative joint effusions, and a decrease in the ESR in various stages of rheumatic diseases, the initial period of medication was followed by one or several days without treatment (Hangarter and Lubke, 1952). Sudden discontinuation was not advised, since prolonged administration guaranteed a better and longer lasting effect.

Unlike results obtained with Cupralene and Dicuprene in the treatment of patients with acute rheumatic fever (ARF), Permalon produced dramatic improvement. A single injection of Permalon (20 ml) markedly decreased pain, reddening, and swelling of affected joints. These ARF patients were completely free of pain and fever and had increased mobility after one or two additional administrations. Because erythrocyte sedimentation rates decreased more slowly, treatment was continued until normal ESR values were achieved. No relapses were recorded during the observation period. An average of 20 injections (50 mg Cu) was necessary to achieve a Symptom-Free* classification for 100% of patients.

Effects achieved by i.v. therapy with copper-salicylate in cases of rheumatic carditis were also remarkable. After transient, highly inflammatory acute articular episodes accompanied by high fever, a typical endocarditis or myocarditis had developed in the original group of 11 patients. In these severe cases all signs of articular inflammation completely subsided after three or four injections (7.5 to 10.0 ml Cu) at intervals of 12 hr. Cardiovascular performance was restored from both the clinical and the electrocardiographic point of view. Symptoms of endocarditis also diminished from day to day. Contrary to the favorable objective clinical course of the disease, ESR values improved but treatment was continued until normal rates were achieved.

The broadest clinical experience with Permalon therapy was gained in the treatment of RA. Of the original group of 27 patients, 13 (48%) progressed to a Symptom-Free classification, 13 (48%) advanced to the Improved† classification, and only 1 patient was unaffected‡ by Permalon therapy. Many of these patients had previously been ineffectually treated with a variety of antirheumatic drugs, including salicylates, gold compounds, and corticoids, but they responded to Permalon in a step-by-step fashion.

The initial effect after the first injection was rapid alleviation of pain as well as improved articular mobility, even in patients with severe deformities, due to

* Symptom-free: The absence of articular inflammation, disappearance of nonarticular inflammation, return of articular mobility—deformation only as a result of irreversible changes, normal ESR, no radiological evidence of progression. Free from pain and fever.

† Improved: ESR still elevated, articular swelling—though only slight—still present, disturbances in articular mobility with only little sign of activity still evident, no increase in deformities, no radiological evidence of progression. Arthralgia only occasionally. No fever.

‡ Unaffected: General condition unchanged, painful, no change in inflammatory signs, radiological evidence of progression, elevated ESR, restricted mobility, deformation and fever of varying degrees, but not significantly decreased for evaluation as improved.

decreased articular swelling resulting from extensive joint effusions. Erythrocyte sedimentation rates were characterized as dropping slowly but steadily, even though the objective clinical picture indicated a rapid, complete remission. In many cases complete normalization of the ESR was observed only after discharge from hospital at out-patient follow-up evaluations. These results were confirmed by Fahndrich (1952) and Broglie (1954).

Four patients with sarcoidosis responded in a similar manner to Permalon therapy, but they only achieved the status of Improved.

Permalon therapy of five patients with arthrosis deformans was least successful. Pain and mobility were influenced to such a slight degree that this therapy could not be considered superior to conventional forms of treatment, although three of the patients progressed to the Improved category.

Two patients suffering from a classical erythema nodosum accompanied by very painful arthralgia were treated with Permalon; both achieved complete remissions. This was particularly significant since experience had shown that this type of rheumatic disease was very resistant to treatment. After two or three injections these patients were completely free of pain. At the same time, body and local temperatures dropped to normal, along with an astonishingly rapid absorption of skin infiltrates. After 12 days of Permalon therapy ESR values were normal. No relapses were recorded during the observation period, and complications were not observed. Compared to oral sodium salicylate treatment (not given at the same time), Permalon was much more effective and acted more rapidly.

Rheumatic neuritis (sciatica and facial) also responded to Permalon therapy with rapid alleviation of pain, and all five of the treated patients reached the Symptom-Free classification. However, as far as motor impairment and its duration were concerned, the effects of Permalon did not differ from those of other therapeutic methods.

It is of interest to point out here that Hangarter had recognized the importance of copper in his preparation of copper-salicylate. It was known that in order to achieve the same therapeutic success with only i.v. sodium salicylate therapy, a serum level of at least 25 mg % was necessary. To reach this serum concentration with Permalon a patient had to receive a total dose of 12 g of salicylic acid, six Permalon injections, divided into equal administrations throughout a 24-hr period. Since this much Permalon had not been given and the salicylate level in the blood reached a peak value of only 20 to 24 mg % and dropped to 5 to 8 mg % within 24 hrs, Hangarter concluded that the effect of Permalon was not due to salicylic acid alone. It is reasonable to suggest that the marked therapeutic efficacy of Permalon was due to the presence of a copper-salicylate complex or complexes.

Hangarter found no systemic toxic or noxious side effects in these initial studies with Permalon, even on long-term i.v. administration. Analyses of blood components and kidney and liver functions (thymol test, Takata-Ara, Weltmann, and elimination of bile pigments) gave no evidence of pathological changes or reactions. Blood serum levels of all components remained normal. Tolerance was good even if two injections per day were given. It was particularly significant that gastrointestinal disturbances and cerebral toxic reactions, which were investigated, were not found. A few patients experienced transient sciatic pain during i.v. injection of the first Permalon preparations, but this effect was no longer observed after subsequent pharmaceutical development of the preparation. However, occasionally i.v. administration was accompanied by pain or transient injection site reddening if the therapy was long-term and the veins were in poor condition or the contents of the 20-ml ampoule injected too rapidly.

For this reason Permalon was administered to all patients by i.v. infusion from 1954 to 1971. This mode of administration achieved equally good results with fewer treatments and without causing venous irritation. All patients received 500 ml of normal saline containing three to four ampoules of Permalon per infusion (7.5 to 10.0 mg Cu). To avoid intimal irritation with slow infusions, a 2-ml ampoule containing 0.4 mg of Novocain was routinely added to the infusion solution. This dose of Novocain was much less than the usual therapeutic dose of 4.0 mg/kg body weight and was not considered to be a systemic analgesic dose.

Hangarter pointed out that the number of infusions necessary to achieve therapeutic success depended upon onset and degree of disease. This was similar to results obtained with Cupralene and Dicuprene, which were found to be more effective in diseases of shorter duration. However, Permalon was effective in diseases of all durations if given in the prescribed manner. In general, six to eight infusions (45.0 to 80 mgCu) at intervals of 2 to 4 days sufficed. These infusions were well tolerated despite the high dosage level. A transient nausea accompanied by tinnitus occurred occasionally but was considered to be a minor effect in view of the rapid regression of usually severe symptoms and signs of disease.

The absence of gastrointestinal functional disturbances usually observed at much lower oral salicylic acid doses was again striking. With infusion therapy the blood chemistry, kidney and liver functions, blood sugar, serum electrophoresis, electrolyte metabolism, ECG, and rheumatic serology were evaluated. No pathological changes or abnormal reactions were found, nor were any cerebral, respiratory, or circulatory toxicities observed.

The clinical results of i.v. infusion therapy corresponded to the very successful clinical results previously described for patients treated with daily i.v. injections of single 20-ml ampoules of Permalon. With infusion therapy, 78 ARF patients or 100% become Symptom-Free.* Acute symptoms subsided, in many cases almost immediately, after one or occasionally two to three infusions (20 to 30 mgCu). The average duration of treatment was only 14 to 18 days. This therapy produced remission of fever, increased articular mobility, and decreased swellings, as well as normalization of ESR and absence or disappearance of cardiac manifestations.

From 1954 to 1971, 620 patients in all stages (Steinbrocker et al., 1949) of

RA received Permalon by i.v. infusion (Hangarter, 1974). Of these, 403 (65%) achieved Symptom-Free status, 143 (23%) gained Improved* status, and 74 (12%) remained in the Unaffected* category when the supply of Permalon was discontinued. On an average, the number of Permalon infusions ranged between six (45 to 60 mg Cu) and a maximum of ten (75 to 100 mg Cu). With RA as well as agressive forms of polyarthritis, patients progressed to the Symptom-Free classification in a step-by-step manner. The initial effect, evident after one or two infusions, was a marked alleviation of pain and improved mobility, which was readily seen in cases of severe deformities. Just as in ARF, infusions brought about a remission of fever and a constant regression of articular swellings and associated articular exudations. The erythrocyte sedimentation rate, which was the most important indicator of all objective pathological processes, was characterized by a slow and steady drop, parallel to the decrease in rheumatic serology, even though the objective clinical picture indicated very rapid complete remission. In many cases a normal ESR was reached only after hospital discharge in out-patient follow-up studies. Latex tests remained positive longer than elevated ESR values.

No serious toxic disturbances were recorded in association with this high dose of Permalon. There was a toxic dose limit where nausea and tinnitus appeared, but in view of the rapid alleviation of pain these toxicities were usually considered as minor side effects by the patients.

Patients with cervical spine-shoulder syndrome, including shoulder bursitis (frozen shoulder, calcareous tendonitis, and Duplay's disease), as well as lumbar spine syndrome, were also successfully treated with Permalon (Hangarter, 1974). These patients generally received infusions as described for patients with RA. However, if these patients suffered from marked impairment of mobility accompanied by severe nocturnal pain, they were given an average of four to eight infusions (40 to 80 mg Cu) per day at intervals of 2 to 3 days. Of the 162 patients with these syndromes who underwent therapy, 95 (57%) became Symptom-Free, 52 (32%) were classified as Improved, with complete relief from pain, and only 18 (11%) experienced no beneficial effect.

Two hundred and eighty patients who had the diagnosis of sciatic neuritis or neuralgia with or without intervertebral disc changes underwent in-patient treatment with Permalon (Hangarter, 1974). Most of these patients were seriously affected and suffered from long-lasting pain, particularly nocturnal pain, and considerable impairment of mobility with regard to walking and changing position. Objective symptoms were accompanied by true neurological and trophic disturbances, as well as highly acute sciatica associated with severe shooting lumbagoid pain. Some patients presented with a highly positive Lasegue's sign, without reflexes or sensitivity to stimuli. With a majority of sciatic cases there were no signs of neuritis, and etiology appeared to be due to a slipped disc. Most patients included in this Permalon study had already undergone several years

* See definitions on p. 111.

of unsuccessful orthopedic and conventional antirheumatic therapy. Therapy designed to treat neuritis and neuralgia was also unsuccessful.

Sciatic patients received between six and eight Permalon infusions (60 to 80 mgCu) every 3 to 4 days, depending on duration and severity of the disease. A total of 120 sciatica patients without lumbar involvement experienced the following results: 76 (63%) became Symptom-Free,* 38 (32%) reached the Improved† classification, and the remainder, 6 (5%), obtained Slightly Improved‡ results. Of the 160 sciatic patients with lumbar involvement 95 (59%) progressed to the Symptom-Free category, 39 (24%) obtained Improved status, 10 (6%) reached the Slightly Improved status, and 16 (11%) remained Unchanged.§

In spite of the above-mentioned diverse etiology of these diseases, Permalon therapy was very effective. Even patients who had undergone previous unsuccessful treatment with conventional preparations were rapidly and persistently relieved from pain and experienced an overall alleviation of their condition. Far advanced abnormalities in posture as a result of spinal or neurological lesions were ameliorated. Impairment of mobility in regard to walking and inability to change position subsided within a short period. Spinal scoliosis was corrected. Neurological symptoms provoked by the disease, absence of or differences in reflexes—a positive Lasegue's sign, and disturbances in sensitivity and trophism were normalized very rapidly. It was particularly interesting that 12 of the 16 patients who were therapy resistant had serious slipped disc problems that required neurosurgical repair.

Apart from tinnitus, sweating, and transient nausea, no serious toxic disturbances were observed. Blood chemistry, kidney and liver functions, serum electrophoresis, and electrolyte metabolism revealed no pathological changes or abnormal clinicochemical reactions.

More recently, through Dr. Thomas L. Schulte's efforts, a new anti-inflammatory copper-dependent metalloenzyme has been developed for drug use. This drug, Orgotein, has been shown to be safe and effective for the treatment of established osteoarthritis when given intra-articularly into knee and hip joints in single or multiple doses (Lund-Olesen and Menander, 1974). Administration of 2 to 20 mg, given in one to ten injections with intervals of 3 to 64 days between injections, brought about long-lasting decreases in disease activity. Of 19 patients treated, 16 experienced a decrease in pain, reduced joint effusion, and increased mobility lasting for more than 90 days after termination of Orgotein treatment. Stimulation of polymorphonuclear leukocyte chemotaxis, stabilization of ly-

* Symptom-Free: Disappearance of (subjective) symptoms, Lasegue's sign negative, normal reflexes with equal quality on both sides, no disturbances in sensitivity, no tenderness on pressure, and mobility restored.

† Improved: Not completely relieved from (subjective) symptoms or persistance of one or several symptoms listed in the Symptom-Free classification, yet no longer any impairment of mobility.

‡ Slightly Improved: Impairment of walking ability still demonstrable, though only moderate, with persistence of one or several symptoms.

§ Unchanged: No response at all to treatment.

sosomal membranes, reduction of lysosomal membrane permeability, and superoxide dismutase activity were implicated as modes of action for Orgotein.

After a recent disclosure that copper complexes had anti-inflammatory activity Walker and his colleagues evaluated the clinical efficacy of the copper bracelet in a single blind cross-over study. Their preliminary results demonstrated that, for a significant number of arthritic patients, wearing a copper bracelet appeared to have some therapeutic value (Walker and Griffin, 1976; Walker and Keats, 1976; Walker et al., 1977). Significantly ($p < .01$) more of the patients who differentiated between the copper bracelet and a placebo (anodized aluminum bracelet) perceived the copper bracelet as being more effective. Also, previous wearers of the copper bracelet were significantly ($p < .02$) worse when not wearing it.

Publications by Walker and Griffin (1976) and Walker and Keats (1976) also provided evidence that, while the copper bracelets were worn, the average weight loss was 12 mg/month. Equilibrations of metallic copper turnings with human sweat also demonstrated that the components in sweat could solubilize this metal and possibly facilitate its absorption. Subsequently Walker et al. (1977) demonstrated that a copper complex, which was likely to be formed while wearing a copper bracelet, could be absorbed by perfusing cat skin with $[^{64}Cu(II)](glycine)_2$. The ^{64}Cu was found in all layers of the skin, and after 6 to 7 hr a steady-state subcutaneous concentration was reached. During a 24-hr period 1 mg of the complex had perfused the intact skin.

In light of this and other physiologic evidence which supports the observed clinical effectiveness of the copper bracelet Whitehouse and Walker (1977) have suggested that it would be appropriate for the medical profession to reconsider the validity of the copper bracelet in the treatment of arthritics.

7.1. Physiology of Copper Complexes in Rheumatoid Arthritis

Copper is one of a number of metals known to be "essential" because they, like essential amino acids, essential fatty acids, and essential cofactors, are required for normal metabolism but are not synthesized *de novo.* The essentiality of copper was established by Hart, Steenbock, Waddell, and Elvehjem (1928) when they demonstrated that it was required for the synthesis of hemoglobin. Since that time copper has been shown to be required for growth, keratinization, pigmentation, bone formation, reproduction, fertility, development of the central and peripheral nervous systems, cardiac function, cellular respiration, nerve function, extracellular connective tissue formation, and regulation of monoamine concentrations (Underwood, 1977). Tissue requirements for copper are believed to be met and controlled in a homeostatic fashion based upon availability, absorption, storage, utilization, and excretion (Evans, 1973). All normal tissues contain this element in the form of copper-dependent components, including metalloproteins, metalloenzymes, and lower molecular weight complexes of biologic importance (Osterberg, 1974).

Copper has received considerable attention with regard to its presence in normal blood, plasma, and serum components. Perusal of this literature (Sorenson, 1978) revealed that total serum (TS) copper concentrations were most commonly determined and reported through the years to be between 94 and 133 $\mu g/100$ ml. Variation in these values is in part due to geographic location, which may affect dietary intake, but sample contamination and analytical error have been cited to account for marked deviations from the currently accepted norm of 110 ± 12 $\mu g/100$ ml for males and 123 ± 16 $\mu g/100$ ml for females (Underwood, 1977; Cartwright, 1950). Larger than usual mean TS copper concentrations may also be the result of including in a "normal" population individuals who have an infection or inflammation of one sort or another, since these abnormalities are associated with an elevation in TS copper concentrations (Heilmeyer and Stuwe, 1938; Heilmeyer et al., 1941; Cartwright and Wintrobe, 1949; Vikbladh, 1950, and 1951; Gubler et al., 1952a; Brendstrup, 1953a; Pekarek and Beisel, 1972; Beisel et al., 1976). Total serum copper concentrations have also been reported to increase slightly with age in adults (Harman, 1965; Zackheim and Wolf (1972)).

Total plasma or serum copper has three principal forms. Ceruloplasmin (Cp) (0.34% Cu), an α_2-globulin with oxidase activity, is synthesized in the liver and increases in serum as an acute-phase reactant. It accounts for 80 to 90% of serum copper (Gubler et al., 1952a). The remaining non-Cp copper is also an acute-phase reactant and is released from liver stores as albumin and amino acid complexes which serve in its transportation and utilization (Caruthers et al., 1966; Osterberg, 1974).

Sex-related variation in serum copper has been attributed to observations that estrogens increase TS copper values (Markowitz et al., 1955; Clemetson, 1968; Halstead et al., 1968). This view is supported by two more recent reports that normal women had significantly higher serum copper concentrations than did normal males (Zackheim and Wolf, 1972; Barnett and Brozovic, 1975).

Over the last 40 years studies of tissues from individuals with rheumatoid arthritis have revealed that copper concentrations change in this disease state. Results reported from 1938 to 1953 are presented in Table 1. Heilmeyer and Stuwe (1938) were the first to report a total plasma copper increase in men and women with active RA; the level returned to normal with remission. These changes in TS copper were not related to sex but were most pronounced in active febrile disease and correlated with a decrease in hemoglobin and an increase in erythrocyte sedimentation rate (Heilmeyer et al., 1941). This arthritic-disease-related alteration of serum copper was confirmed by Evers (1952) and Van Ravesteijn (1945), who also found that RBC copper decreased in spite of increased whole blood copper.

Subsequently Brendstrup (1953a, 1953b) correlated the TS copper increase

Table 1. Alterations of Copper Concentrations[a] in Rheumatoid Arthritis Reported from 1938 to 1952

Sex	Copper		RBC[b] Copper		Whole Blood Copper		Reference
	Normal	Arthritic	Normal	Arthritic	Normal	Arthritic	
?	105 (80–140)	210					Heilmeyer and Stuwe (1938)
M	106 ± 16 (15)						Cartwright (1950)
F	107 ± 17 (15)						
M, F		173 (23) (118–216)					
M	123 (12) (85–152)	179 (2) (140–217)	129 (7) (96–169)	100 (2) (60–140)	129 (8) (90–160)	149 (2) (117–181)	Van Ravesteijn (1945)
F	131 (12) (103–152)	192 (6) (110–250)	120 (2) (118–122)	89 (4) (88–182)	131 (2) (122–140)	175 (4) (125–195)	
M, F		147 (122) (81–254)					Evers (1952)

[a] Mean or mean ± standard deviation in micrograms per 100 ml of serum, plasma, RBC or whole blood (number of individuals in study)/(Range of values found).
[b] RBC = red blood cells.

in febrile patients and the return to normal in remission with other parameters associated with disease activity. As shown in Table 2, active disease was associated with increased TS copper, which was directly related to decreased mobility and hemoglobin (Hb) values as well as to increased ESR and duration. The mean value of TS copper in patients with normal mobility differed relatively little from the mean value found for patients with inactive joint lesions. Patients with active RA who showed clinical improvement had a pronounced fall toward normal TS copper and ESR values, but these parameters remained relatively unchanged in patients showing no improvement.

Brendstrup's data also confirmed that the normal sex-related serologic difference in TS copper was no longer apparent in active disease states because of marked increases in the concentration of this metal in both men and women, as shown in Table 3. In addition, these serologic changes were most pronounced in patients with more active rheumatic disease. This led Brendstrup to suggest that the pathologic activity of RA could be evaluated by determining the magnitude of the TS copper concentration. Alternatively, the rate of remission was related to the rate at which the TS copper concentration returned to normal.

In summary, the results published from 1938 to 1953 clearly demonstrated that, compared to normal healthy individuals, patients with RA had higher mean TS copper concentrations which were directly related to disease severity or ac-

Table 2. Mean Mobility Indices (MI), Erythrocyte Sedimentation Rate (ESR), Duration, Hemoglobin (Hb), Total Copper (Cu), and Disease Activity for Patients with Chronic Rheumatoid Arthritis[a]

MI[b]	ESR[c]	Duration[d]	Hb (%)	Cu[e]	Activity[f]	
1.0 (25)	10 (25)	5 (23)	92 (25)	135 (24)	19i	6a
	(1–36)	(1–16)	(77–110)	(86–243)		
0.97 (31)	24 (31)	6 (29)	92 (31)	152 (25)	16i	15a
(1–0.95)	(1–90)	(0.5–26)	(73–111)	(82–319)		
0.93 (24)	23 (24)	6 (21)	91 (24)	173 (21)	8i	16a
(0.95–0.90)	(4–51)	(1–31)	(81–114)	(117–231)		
0.87 (30)	39 (30)	8 (29)	87 (30)	175 (26)	4i	26a
(0.90–0.83)	(1–116)	(1–26)	(62–115)	(91–282)		
0.80 (23)	43 (23)	8 (20)	82 (23)	203 (19)	3i	20a
(0.83–0.75)	(3–91)	(1–17)	(57–100)	(117–261)		
0.61 (18)	50 (18)	18 (18)	81 (18)	193 (18)	0i	18a
(0.75–0.07)	(10–127)	(3–43)	(65–104)	(121–292)		

[a] Brendstrup (1953a).
[b] Mobility index (number of individuals)/(Range of values included in group).
[c] Millimeters per hour (number of individuals)/(Range of values in group).
[d] Years (number of individuals)/(Range of values in group).
[e] See footnote *a*, Table 1.
[f] Number of individuals with inactive (i) and active (a) disease.

Table 3. Disease Activity, Sex, Mean Erythrocyte Sedimentation Rate (ESR), Duration of Illness, Hemoglobin (Hb), Total Copper (Cu), and Mobility Indices (MI) in Chronic Rheumatoid Arthritis[a]

Activity[b]	Sex	ESR[c]	Duration[d]	Hb (%)	Cu[e]	MI[f]
i	F	6 (29)	6 (28)	90 (29)	133 (28)	0.96 (29)
i	M	3 (21)	8 (18)	102 (21)	132 (17)	0.95 (21)
a	F	45 (69)	8 (63)	83 (69)	187 (62)	0.86 (69)
a	M	43 (101)	9 (94)	85 (101)	188 (88)	0.84 (101)

[a] Brendstrup (1953a).
[b] See footnote *f* to Table 2.
[c] See footnote *c* to Table 2.
[d] See footnote *d* to Table 2.
[e] See footnote *e* to Table 2.
[f] See footnote *b* to Table 2.

tivity as measured by increased corporal and local temperatures, immobility, duration of disease, pain, edema, and ESR, as well as diminished strength and decreased Hb values. Small sex-related differences in normal individuals were obscured by marked increases found for both male and female patients. The TS copper, which increased in association with the onset and persistence of active disease, returned to normal with remission.

It is now known from animal studies that the rise in TS copper is accompanied by a fall in liver copper concentrations. Since serum copper-containing components are synthesized in the liver and appear in serum after the onset of disease, it seems reasonable to suggest that this is a physiologic response to RA.

The low or normal serum copper concentrations shown in the range of values obtained for some patients with RA deserve some comment. These values may be the result of a failure of this aspect of the physiologic response due to depletion of liver copper stores. Depletion may be due to increased turnover, resulting in copper excretion (Wiesel, 1959; Koskelo et al., 1967), and failure to replete these stores because of either loss of appetite or inadequate diet. Failure of serum copper to maximally increase as a physiologic response to disease could lead to chronic disease.

After the reports of Brendstrup (1953a, 1953b) little was published concerning the changes in blood essential metal concentrations associated with RA until 1965, when interest in this topic was renewed. This interest has continued to the present, and results obtained during this period are presented in Table 4.

In 1965 Plantin and Strandberg (1965) reported results which confirmed the report by Van Ravesteijn (1945) that total whole blood copper was elevated in RA. Analyses of whole blood for copper were shown to give values that were essentially the same regardless of disease activity. Thus analysis of whole blood for copper did not provide useful diagnostic or prognostic information concerning the activity of this disease.

Earlier observations concerning copper in RA were extended by Neidermeier (1965) with the report that the increase in TS copper was due to an increase in Cp and non-Cp copper concentrations. Since Cp contains 0.34% Cu, the observed mean increase in Cp represented an increase of 34 μg/100 ml, whereas the mean non-Cp copper increased only 11 μg/100 ml serum. The increased synthesis of Cp in response to RA was subsequently confirmed by Koskelo et al. (1966) and shown to be related to disease activity (Koskelo et al., 1967). Patients with severe to moderately active RA had significantly ($p < .05$) accelerated daily Cp turnover rates. The highest turnover rate was found for the patient with the highest Cp concentration and severely active RA. Moderately active RA was associated with lower concentrations of Cp and moderate turnover rates. It was suggested that these results indicated that increased daily turnover was accompanied by an increased rate of synthesis and, in the steady state, a corresponding increase in the elimination rate. Turnover studies with γ-globulin also revealed an accelerated synthesis of this protein in the majority of patients with RA (Koskelo et al., 1967).

Lorber et al. (1968) were the first to report a statistically significant elevation of TS copper concentrations in RA when these patients were compared with age-matched normal individuals. However, they found that almost the entire TS copper content of normal individuals was bound to Cp. This was corroborated in a second group of normal individuals with the demonstration that TS copper contained only a small amount of the non-Cp element (7 μg/100 ml), much less than reported by Neidermeier (1965) (24 $\pm$ 16 μg/100 ml). In addition Lorber et al. (1968) found that their population of RA patients had a very large mean non-Cp copper concentration (103 μg/100 ml), much higher than reported by Niedermeier (1965) (35 $\pm$ 25 μg/100 ml). Based upon these results it seems reasonable to interpret the large increase in non-Cp copper as an integral part of the physiologic response to active disease in RA patients.

The results of Lorber et al. (1968) were questioned by Sternlieb et al. (1969) on the basis of their analyses of sera from RA patients, which gave a mean non-Cp copper concentration of only 9 μg/100 ml. Lorber (1969) suggested that this observation was due to the low TS copper concentration found by Sternlieb et al. (1969) and attributed the latter to the methodology used to obtain their result.

However, the different analytic procedures used by Neidermeier (1965), Lorber et al. (1968), and Sternlieb et al. (1969), although perhaps accounting for small differences, do not seem to fully explain the large differences in the three sets of data. Alternatively it is suggested here that it may be possible to accommodate the observed differences by taking into account disease activity in the three populations studied. According to reports published up to 1953, the second group of RA patients studied by Lorber et al. (1968) seem to have had the most active disease, since they had the highest TS copper concentration (248 μg/100 ml). Since the patients studied by Neidermeier (1965) and Sternlieb

Table 4. Altered Copper Concentrations in Rheumatoid Arthritis Reported since 1965.

Sex	Copper[a] Normal	Copper[a] Arthritic	Ceruloplasmin[b] Normal	Ceruloplasmin[b] Arthritic	Non-Cp Copper[a] Normal	Non-Cp Copper[a] Arthritic	Reference
M, F	99[c] (70)	174[c] (46)					Plantin and Strandberg (1965)
M, F	148 ± 25 (19)	192 ± 34 (21)	36 ± 9 (19)	46 ± 11 (21)	24 ± 16 (19)	35 ± 25 (21)	Niedermeier (1965)
M, F			37 ± 5 (58) (24–50)	65 ± 18 (31)[d] (41–131)			Koskelo et al. (1966)
M, F			40 ± 6 (13) (27–48)	59 ± 14 (7)[d] (49–88)			Koskelo et al. (1967)
?	96 (89) (60–132)	168 (10) (109–276)					Bonebrake et al. (1968)
M, F	139 (13) (117–161)	181[e] (20) (161–201)					Lorber et al. (1968)
M, F	119 (4) (107–137)	248 (5) (210–338)	33 (4) (28–42)	41 (5) (38–52)	7 (4) (0–12)	103 (5) (75–156)	Lorber (1969)
?		150 (7) (132–205)		46 (7) (42–61)		9 (7) (0–20)	Sternlieb et al. (1969)
M, F	97 ± 32 (105)	110 ± 29[f] (105)					Neidermeier and Griggs (1971)
M	99 (9) (73–171)	96 (14) (77–122)	27 (9) (18–52)	26 (14) (20–43)			Bajpayee (1975)
F	107 (13) (73–146)	107 (14) (73–146)	31 (13) (20–47)	31 (14) (20–44)			
M, F	104 ± 11 (1)	266 ± 85[e] (27) (66–392)					Sorenson and DiTommaso (1976)
M		254 ± 10 (7)					
F		271 ± 79 (20)					

Sex	Synovial Fluid Copper[g]		Synovial Fluid Ceruloplasmin[h]		Synovial Fluid Non-Cp Copper[g]		References
	Normal	Arthritic	Normal	Arthritic	Normal	Arthritic	
M, F	50 ± 14 (6)	107 ± 22^{i} (23)	4 ± 2 (6)	24 ± 7^{i} (23)	35 ± 14 (6)	26 ± 15 (23)	Neidermeier et al. (1962), Neidermeier (1965), Neidermeier and Griggs (1971)
M, F	28 ± 16^{j} (50)	85 ± 21^{d} (50)					
?	34^{j} (7–73)						Bonebrake et al. (1968)
		98 (12) (66–149)					
?	43 (6) (22–59)						

[a] Mean or mean ± standard deviation in micrograms per 100 ml of serum (number of individuals in study)/(Range of values found).
[b] Mean or mean ± standard deviation in milligrams per 100 ml of serum ceruloplasmin (number of individuals in study)/(Range of values found).
[c] Mean in micrograms per 100 g or 100 ml of whole blood (number of individuals in study).
[d] $p < .001$.
[e] $p < .05$.
[f] $p = .002$.
[g] Mean or mean ± standard deviation in micrograms per 100 g of synovial fluid (number of individuals in study)/(Range of values found).
[h] Mean or mean ± standard deviation in micrograms per 100 g of synovial fluid (number of individuals in study).
[i] $p < .01$.
[j] Sample taken postmortem.

et al. (1969) had TS copper concentrations of 192 and 150 $\mu g/100$ ml, respectively, these patients had less active disease. All three groups of patients had about the same Cp concentration, which appears to be 15 to 20% larger than normal values. From the increases in non-Cp copper it appears that the patients in the second group studied by Lorber et al. (1968) had more active disease than those studied by Sternlieb et al. (1969). This last group either had inactive disease, were in remission, or were no longer physiologically responding to the disease processes.

Lorber et al. (1968) concluded that the elevation in TS copper associated with RA was primarily due to an elevation of non-Cp copper and suggested that this copper might serve as a deliterious sulfhydryl group oxidizing agent, denaturing Hb, disrupting RBC and lysosomal membranes, causing the release of tissue lytic enzymes. This last speculation is consistant with the observation of Chuapil et al. (1972) that copper destabilized lysosomal membranes. They further speculated that the beneficial effects of certain chelating agents such as penicillamine could be attributed to their copper chelating action, which presumably was believed to promote the excretion of the metal. This hypothesis was used to account for the beneficial effects of penicillamine, resulting in lowered TS copper, increased serum sulfhydryl content, and decreased rheumatoid factor titers in patients receiving this drug. However, the observation by Chayen et al. (1969) that a lower concentration of copper stabilizes lysosomal membranes, perhaps via sulfhydryl group oxidation, and as a result decreases the ratio of free versus bound lysosomal enzymes is contradictory. Also, it has been suggested that the mechanism of action of penicillamine and other drug chelating agents that appear to lower serum copper by bringing about remission is the result of copper-drug complex formation *in vivo,* which facilitates remission by promoting tissue utilization of copper rather than by promoting excretion of this element (Schubert, 1960, 1966; Sorenson, 1976a, 1976b). Based upon these considerations it seems reasonable to interpret the large increase in non-Cp copper observed by Lorber et al. (1968) as an integral part of the physiologic response to active disease, which facilitates remission rather than causing adverse effects via sulfhydryl oxidation.

In a subsequent study of TS copper Niedermeier and Griggs (1971) found concentrations that were unusually small and very close to what is generally accepted as normal. This raises the question of whether many of the patients in this group had less active disease, were in remission, or were no longer able to respond to the disease process and had chronic rheumatic disease. These suggestions are based upon the results of Brendstrup (1953a) and the recent findings that RA patients at or near the peak activity of their disease had 2 to 3 times the normal TS copper levels (Sorenson and DiTommaso, 1976). These recent data also provide confirmation of the absence of a significant difference between the levels found for men and for women with active RA.

Recently Bajpayee (1975) questioned previous reported increases of copper and Cp concentrations in sera of patients with RA. His data with plasma from both male and female patients who had a diagnosis of RA demonstrated that total plasma copper and Cp concentrations were the same as those found for his normal male and female populations. Based upon these results it was suggested (Sorenson and DiTommaso, 1976) that the patients in his study also had inactive disease, were in remission, or were no longer able to respond to the disease process and had chronic rheumatic disease.

Bajpayee (1975) suggested that the well-known sex difference (female: male ratio 2:1) in populations of arthritic patients may have accounted for elevated levels of copper and Cp because of the estrogen-mediated increase in plasma copper concentration in the female subgroup of his RA population. However, his data did not show that women who were not known to be on estrogen therapy had significantly different levels of copper or Cp than did men. Also, the TS copper concentrations for patients with active disease (Brendstrup, 1953a; Sorenson and DiTommaso, 1976) were found to be much greater than the reported (Bajpayee, 1975) estrogen-induced increase in TS copper.

Also shown in Table 4 are data collected to show the concentration changes of copper in synovial fluid (SF) of RA patients. After the qualitative observation by Schmid and McNair (1956) that Cp was present in SF, Niedermeier et al. (1962), Niedermeier (1965), and Niedermeier and Griggs (1971) quantitated its presence and alteration in RA. A significant increase in total SF copper was shown to be due to a large increase in SF Cp, while the non-Cp copper concentration decreased. Increases in Cp concentration appeared to approximately parallel the duration of disease.

This increase in SF copper was confirmed by Bonebrake et al. (1968), who found that, in addition to an increase in TS copper and total SF copper, SF from RA patients contained two copper-containing components, as compared to only one in SF from noninflamed joints. Ceruloplasmin was identified in all SF samples; the other, found only in inflamed joints, was suggested to be a copper-albumin complex.

In summary, the results reported since 1965 have increased our understanding of essential metal metabolism associated with RA. Marked increases in TS copper were confirmed in patients with active disease. The increased rate of synthesis and the accelerated turnover rate of Cp were found to be directly related to disease activity. Copper was also found to increase in SF of RA patients, and it was suggested that Cp, which accounted for most of the rise in SF copper, increased with increasing duration of disease. Accelerated rates of Cp synthesis and turnover, marked increases in TS copper, and an accumulation of Cp copper in SF can all be interpreted as part of a multifaceted physiologic response to inflammatory disease.

Unfortunately, some of the work done since 1965 did not take into account the results reported from 1938 to 1953 or the possibility that these results had physiologic significance. Consequently, the changes in copper metabolism were interpreted as pathologic. It is suggested here that, if these changes in copper-requiring tissue components are at least considered to be physiologic and correlated with disease activity, duration, severity, pain, ESR, Hb levels, edema, and immobility, it may be possible to achieve a better understanding of rheumatic disease.

Progress is being made with regard to our understanding of the copper complex equilibria required to produce the forms of copper necessary for its role(s) as an acute-phase reactant. Pioneering efforts in this regard were begun by Perrin (1965, 1969) and his colleagues (Hallman et al., 1971; Perrin and Agarwal, 1973) with the development of a computer simulation model for the study of complex formation between low molecular weight complexing agents such as amino acids and copper in plasma. This approach has been extended by Williams and his colleagues (Williams, 1972; May et al., 1976, 1977; May and Williams, 1977), who included more competing metals and a larger number of complexing agents, including plasma peptides, in their computerized model. With this technique they have been able to suggest the various complexes of copper present in blood as well as the relative amounts of each. In a recent series of papers Williams and his colleagues (Jackson et al., 1978, 1978a; Fiabane and Williams, 1978; Fiabane et al., 1978; Micheloni et al., 1978; Arena et al., 1978) considered the role of increased amounts of copper complexes in the blood of patients with rheumatoid arthritis. With this type of information available it may be possible to develop a better pharmacologic approach to the treatment of arthritic and other degenerative diseases.

7.2. Pharmacologic Activities of the Antiarthritic or Anti-inflammatory Copper Complexes

Before 1950 the anti-inflammatory activities of copper complexes had been exclusively studied in human beings for the treatment of tuberculoses and rheumatic diseases. In 1950 Kuzell and Gardner (1950) reported that i.m. administration of Cupralene (**1**) favorably influenced polyarthritis produced in rats with a pleuropneumonia-like organism. However, a year later, using the same model of experimental arthritis, Kuzell et al. (1951) reported that both Cupralene and Dicuprene (**7**) were ineffective in preventing or "curing" the rather variable inflammation in this model. They felt that these results were consistent with their clinical results, which could have been viewed more favorably but were not. This interpretation is now somewhat clouded since it seems that the Cupralene they used either had decomposed or was impure material (Sorenson and Hangarter, 1977).

A more favorable interpretation was offered by Vykydal et al. (1956). In summarizing their results with Formalin-induced rat paw edema, they suggested that the mechanism of action of parenterally administered Cupralene was similar to that of adrenalcorticotrophic hormone. It was suggested that this favorable interpretation was consistent with the clinical results of Fenz (1941, 1951) as well as of Forestier and his colleagues (1944, 1946, 1948, 1949, 1950).

Subsequently Schubert (1960, 1966) published his accounts of the hypothermic effect of i.v.-administered sodium salicylate, $CuCl_2$, and copper-salicylate (**9**). Copper salicylate was more effective in lowering yeast-induced rat hyperthermia than either $CuCl_2$ or sodium salicylate. The *in vivo* formation of a copper complex of an analogue of salicylic acid was invoked by Schubert to account for the lowering of rat body temperature in either the normothermic or the hyperthermic state. These observations provide mechanistic insight into Hangarter's (1952, 1974) observed rapid reduction of fever in febrile patients with acute rheumatic and degenerative connective tissue diseases when they were treated with his copper salicylate preparation (Permalon). These results also account for the mild to severe chills observed by Tyson et al. (1950) when larger doses of Cupralene were used in their clinical studies.

In 1958 Sutter et al. (1958) reported that a large oral dose of cuprous iodide, 250 mg/kg body weight (mpk), caused a 60% reduction of the granuloma and almost completely eliminated exudation in the rat granuloma-pouch (GP) model of inflammation. However, Adams and Cobb (1967) found that even larger oral doses of up to 320 mpk did not delay the development of ultraviolet (UV)-light-induced erythema in guinea pigs, and the activity of this compound was no longer pursued.

In 1969 the patent of Laroche (1969) disclosed that copper complexes of acetic, lauric, oleic, caprylic, butyric, sebasic, lipoic, and cinnamic acids were orally effective in rats as anti-inflammatory agents at doses corresponding to 18 and 60 mpk Cu in the carrageenan paw edema (CPE) and croton oil granuloma models of inflammation, respectively. It was claimed that mixtures of these complexes with clinically used antiarthritic drugs would be useful in treating a variety of rheumatic diseases. However, this approach was never applied in clinical therapy.

Also in 1969, Bonta (1969) reported that the mineral basic cupric carbonate $[Cu(OH)_2CuCO_3]$ (malachite) failed to cause a 50% inhibition of kaolin-induced rat paw edema (KPE) at 20 mpk when given orally but did so at 20 mpk when administered subcutaneously (s.c.). In addition to this substance Bonta reported that a series of copper chelates, which were not specified, were only weakly active when given orally but had pronounced anti-inflammatory activity when given s.c. in the KPE, cotton wad granuloma (CWG), and GP models of inflammation.

Because the anti-inflammatory effects were thought to be regularly associated with marked tissue irritation at the injection site, Bonta suggested that these copper compounds might not have specifically induced the anti-inflammatory effects, and that the anti-inflammatory activity was due to nonspecific tissue irritation at the sites of injection, which were remote from the site of induced inflammation. In a follow-up study of malachite, he (1969) found a contrasting profile of activity in the guinea pig UV-erythema model of inflammation. Unlike the results obtained by Adams and Cobb (1967), malachite caused a 50% inhibition of UV-induced erythema at 25 mpk when given orally but was inactive when administered s.c. at the same dose. In the guinea pig, as in rats, this compound produced marked irritation at the s.c. injection site. If in the rat edema test the effect of this compound is indirect and is due to the s.c. irritation, the latter factor did not seem to play a role in inhibiting guinea pig UV-erythema. Though the gastric mucosa of guinea pigs treated orally with malachite showed harsh irritation, there was no evidence at that time as to whether or not that provided a basis for an irritant-induced antierythema effect. Bonta (1969) suggested that one could not rule out the possibility that in the guinea pig gastric but not s.c. irritation evoked a remote anti-inflammatory effect. Alternatively he pointed out that he had no argument to exclude the possibility that in guinea pigs the copper-containing substance, after being absorbed from the gastrointestinal tract, acted by virtue of its anti-inflammatory effect. This possibility seems to be just as likely in rats.

However, Bonta's impression was that the compounds that were not irritating were devoid of anti-inflammatory effect. In addition, it appeared that adrenalectomy abolished their anti-inflammatory activity. Tissue irritation might have stressed the rats to such as extent as to activate the hypophysyl-adrenal axis to discharge endogenous corticosteroids. This in turn may have exerted the anti-inflammatory effect. However, Bonta believed that this was not very likely, since corticosterone—the main glucocorticoid produced by the rat adrenal—had a particularly weak anti-inflammatory effect.

Bonta (1969) also pointed out that the possibility of an indirect corticoid-induced anti-inflammatory effect had not been ruled out. Nevertheless he chose to follow the philosophical view that local tissue irritation and remote anti-inflammatory effect may have been casually connected with each other.

Support for this idea was obtained when rats were treated intraperitoneally with phenylquinone (an irritant that causes writhing) 30 min before subplantar injection of kaolin. There was marked inhibition of the swelling usually caused by kaolin. This indicated that the remote inhibitory effect of tissue irritation was similar to the action of corticosteroids and nonsteroidal anti-inflammatory drugs. It was considered conceivable that tissue irritation with phenylquinone may have released some factor(s) which, being discharged into the blood stream, caused a suppression of the inflammatory response to kaolin. Bonta (1969) succeeded in transferring the postulated tissue factor in lyophilized peritoneal exudate to other rats. In subsequent experiments, serum from phenylquinone-treated rats was also shown to exert an anti-inflammatory effect in the KPE test; even serum from normal animals did so, though to a lesser extent.

In an effort to correlate all of these experimental observations it is pointed out here that one of the tissue factors responsible for irritant-induced anti-inflammatory activity may be the increase in serum Cp, as well as albumin- and amino acid-copper complexes. These are known to increase in the serum of human beings and other animals in response to irritants (Heilmeyer and Stuwe, 1938; Heilmeyer et al., 1941; Wintrobe et al., 1953), inflammation (Underwood, 1977; Evans, 1973; Cartwright and Wintrobe, 1949; Karabelas, 1972), and infection (Heilmeyer et al., 1938, 1941; Wintrobe et al., 1953; Markowitz, 1955) as a primary response to the etiologic agents. The increase of these acute-phase reactants in serum is consistent with the observation that inorganic copper and copper complexes have anti-inflammatory activity if tissue distribution is assured by parenteral administration, and with the idea that this activity is physiologic in nature. This aspect of the anti-inflammatory activity of copper complexes is considered in greater detail in Section 7.1.

Copper acetate [$Cu(II)_2(acetate)_4$] was also found to have potent anti-inflammatory activity in the CPE model of inflammation (Sorenson, 1974). This observation, along with the recognition that all of the antiarthritic drugs could form copper complexes, suggested that these complexes might be the active forms of the antiarthritic drugs (Sorenson, 1976b). The rationale for this suggestion was based upon two very likely possibilities. First, giving $Cu(II)_2(acetate)_4$ could result in the formation of copper complexes *in vivo* with complexing agents such as the amino acids. Second, it was known that serum copper concentrations increase in arthritic disease, so that giving a complexing agent such as an antiarthritic drug could result in the formation of a copper complex of that drug *in vivo*.

At the outset the hypothesis was tested with copper complexes prepared with complexing agents that had no known anti-inflammatory activity of their own. This was done to avoid observing activity resulting from an active anti-inflammatory compound, in the event that dissociation of the complex occurred to give what would be a known anti-inflammatory drug. If activity observed for $Cu(II)_2(acetate)_4$ was due to the amount of copper being given, then giving copper complexes with less of the element in them would not be as effective as giving $Cu(II)_2(acetate)_4$. Initially, then, copper complexes prepared with inactive complexing agents and containing decreased quantities of the metal were used to test the hypothesis that copper complexes had anti-inflammatory activity.

To test these complexes and reduce or minimize the risk of their loss by dissociation in the stomach they were administered s.c. To further reduce or minimize the risk of losing the copper in these complexes they were routinely given as Tween 80 suspensions. Suspending agents such as carboxymethylcellulose, tragacanth, and acacia were avoided since these are polycarboxylic acids which might have removed the copper from these complexes or formed mixed complexes. Administration of these compounds under these conditions of reduced risk of loss by dissociation throughout the course of this study gave satisfactory biologic results for many copper complexes (Sorenson, 1976a). Results obtained for only two of the complexes synthesized from biologically inactive complexing agents are presented here.

The anti-inflammatory activities of $Cu(II)_2(acetate)_4$, anthranilic acid, 3,5-diisopropylsalicylic acid, $Cu(II)(anthranilate)_2$ (**10**, X = NH_2, R = H),

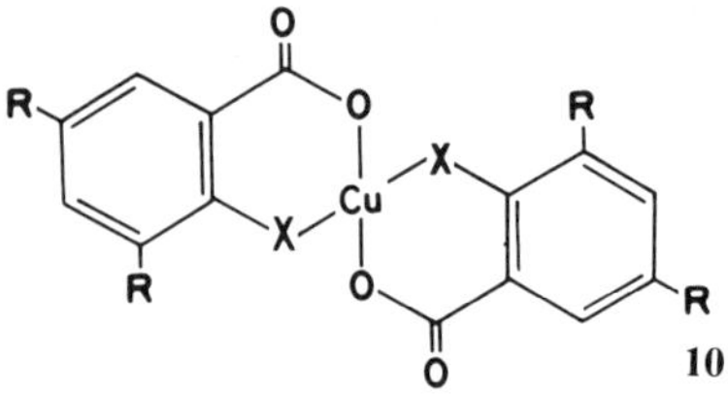

10

and $Cu(II)(3,5\text{-diisopropylsalicylate})_2$ [$Cu(II)(3,5\text{-dips})_2$] [**10**, X = OH, R = $(CH_3)_2CH$—] are presented in Table 5. As mentioned above, cupric acetate had been found to be active (A) in the CPE model of inflammation at 8 mpk, but with subsequent follow-up testing it proved to be inactive (I) at the initial screening doses of 100 and 30 mpk, respectively, in the CWG and adjuvant arthritis (AA) models of inflammation. Anthranilic acid and 3,5-dips were found to be inactive, as expected, at the large initial screening doses of 200 and 30 mpk in the CWG and AA tests, respectively. However, $Cu(II)(anthranilate)_2$ and $Cu(II)(3,5\text{-dips})_2$ were very potent in all three models of inflammation.

The difference in effectiveness of these compounds does not appear to be related to the amount of copper in them. Cupric acetate, which was effective only in the CPE model, contained nearly 32% Cu, whereas $Cu(II)(anthranilate)_2$ and $Cu(II)(3,5\text{-dips})_2$, which were active in all three models of inflammation, contained 18.9 and 12.5% Cu, respectively. Therefore, to account for the observed anti-inflammatory activity, it is necessary to invoke the existence of the intact complex since the observed activity cannot be attributed to the activity of anthranilic acid or 3,5-dips or to the amounts of copper in the complexes.

In addition the rat LD_{50} data clearly indicated that $Cu(II)(anthranilate)_2$ and $Cu(II)(3,5\text{-dips})_2$ were safe to use. This safety was quantitated by calculating the therapeutic index (TI), which is given in Table 5 as the $LD_{50}:ED_{95}$ ratio, for each of these two compounds. Since an ideal TI is suggested to be 10, where the lethal dose is 10 times the effective dose, the TI data obtained for these copper complexes demonstrate that they are unusually safe to use.

These two copper complexes were also evaluated with regard to their production of signs of central nervous system toxicity. Subcutaneous injection of 20, 40, 80, and 320 mpk failed to produce signs of CNS stimulation or depression in mice. This was consistent with the lack of observed CNS effects in all of the anti-inflammatory testing of these two complexes in rats, although $Cu(II)_2(acetate)_4$ had produced CNS toxicity of unknown pathogenesis at all active doses in rats. It was also found that mouse LD_{50} values for both of these compounds were greater than 320 mpk, the largest dose given.

To investigate possible liver damage in association with these two copper

Table 5. Antiinflammatory Activities, LD_{50} Data, and Therapeutic Indices (TI) Obtained for Copper Complexes of Acetic, Anthranilic, and 3,5-Diisopropylsalicylic Acids[a,b]

Compound	Carrageenan Paw Edema	Cotton Wad Granuloma	Adjuvant Arthritis	Copper (%)	LD_{50}	TI	
						CPE	PA
Cupric acetate	A at 8	I at 100	I at 30	31.8	350	70	
Anthranilic acid	I at 200	NT[c]	I at 30	13.5			
3,5-Dips acid	1 at 200	NT	I at 30				
$Cu(II)(anthranilate)_2$	A at 8	A at 25	A at 1.2	18.9	750 ± 106	94	625
Cu(II)(3,5-dips)	A at 8	A at 5	A at 1.2	12.5	240 ± 33	30	200

[a] Sorenson (1976a).

[b] All compounds were given by s.c. injection and expressed as milligrams per kilogram of body weight.

[c] Not Tested.

complexes, 20 mpk (1 to 20 times the effective doses) were given for 10 days. The usual liver function regimen was 10 mpk for 4 days, so that giving twice as much for 2.5 times as long was not an experiment weighted in favor of the two complexes. In addition, two doses of $Cu(II)_2(acetate)_4$, 20 mpk and 10 mpk, were tested in the same protocol to distinguish its toxicity from the toxicities of the two complexes. Neither $Cu(II)(anthranilate)_2$ nor $Cu(II)(3,5\text{-dips})_2$ caused any liver function or liver cell damage, although $Cu(II)_2(acetate)_4$ produced some liver cell damage at both doses (Sorenson, 1976a).

Having demonstrated an increase in anti-inflammatory activity for the copper complexes of anthranilic acid and 3,5-dips, as well as their safe use, it was decided to test the hypothesis that copper complexes of the clinically used antiarthritic drugs might be more effective as anti-inflammatory agents than the parent drugs. For this purpose copper complexes were prepared using acetylsalicylic acid (aspirin) (**11**), 2[3-(trifluoromethyl)phenyl]aminonicotinic acid (niflumic acid) (**12**), D-penicillamine (D-pen) (**13**), 1-phenyl-5-aminotetrazole (fenamole) (**14**), and salicylic acid.

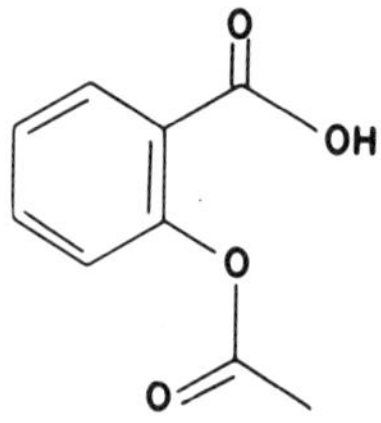

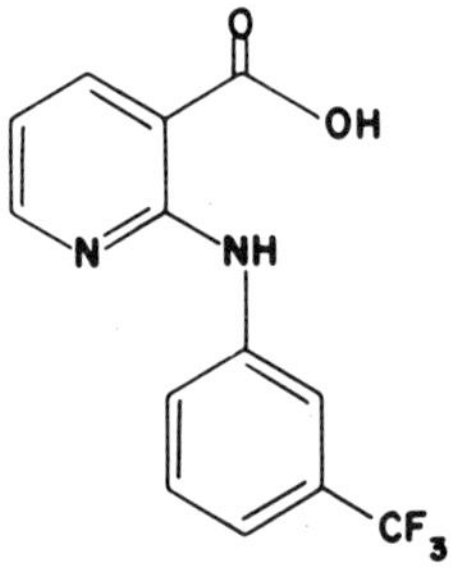

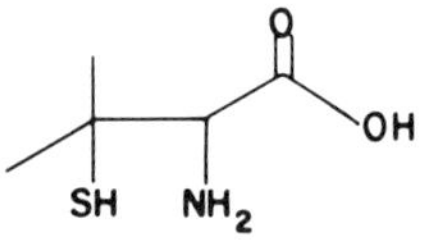

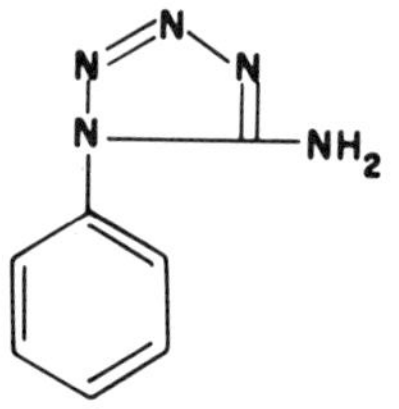

Only the copper complexes of aspirin [$Cu(II)_2(aspirinate)_4$] (**15**) and the copper salicylates have been structurally characterized (Manojlovic-Muir, 1967; Kato et al., 1964). The binuclear structure of $Cu(II)_2(aspirinate)_4$, like the structures of $Cu(II)_2(acetate)_4$ and other carboxylic acid copper complexes, is well known. A more detailed discussion of this octahedral bipyrimidal conformation had been presented elsewhere (Sorenson, 1976a), along with plausible copper complex structures of the other antiarthritic drugs cited above.

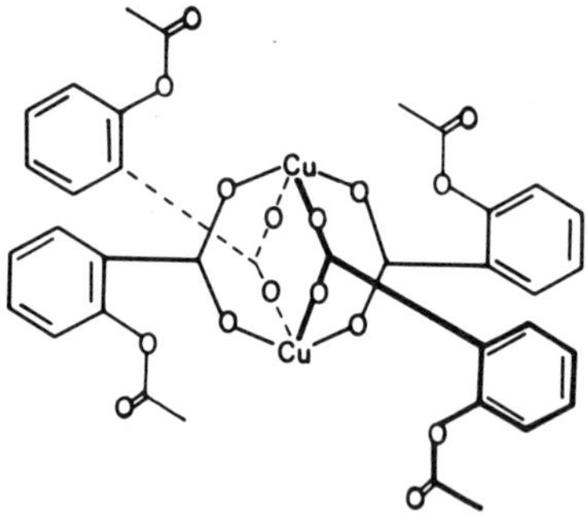

The results of tests on the CPE, CWG, and AA models of inflammation are presented in Table 6. Comparing the activity observed for aspirin and its copper complex demonstrated that Cu(II)$_2$(aspirinate)$_4$ was more active than aspirin. In the CPE model the complex was active at one-eighth the lowest active dose of aspirin. No s.c. data are available for aspirin in the CWG model, so that comparison here is based upon oral data for aspirin. In the AA model the anti-inflammatory activity of Cu(II)$_2$(aspirinate)$_4$ was found to be 5 times greater than the activity observed for aspirin. In addition, Cu(II)$_2$(aspirinate)$_4$ was active at 30 mpk when given i.v., whereas aspirin was inactive. Intravenous testing was done because it was suggested that a compound which was active because it was an irritant would not be active if given i.v.

The copper chelate of niflumic acid [Cu(II)$_n$(niflumate)$_{2n}$] was also found to be active in all three models. Comparison here is somewhat lacking, since one cannot directly compare the activities of two compounds administered by different routes. However, the following comparison is viewed as helpful with regard to disclosure of potentially useful information. The parent compound was active at 40 and 25 mpk orally in the CPE and CWG models, while the chelate was active at 8 and 10 mpk s.c. in these two models. The lowest active dose in the AA model was 6 mpk orally for niflumic acid and 1.2 mpk s.c. for Cu(II)$_n$(niflumate)$_{2n}$. When both Cu(II)$_n$(niflumate)$_{2n}$ and niflumic acid were tested i.v. in the adjuvant arthritic rat, the lowest active dose of niflumic acid was only 30 mpk, whereas its copper complex was active at 6 mpk.

Clinical use of D-penicillamine (D-pen) as an antiarthritic compound is interesting since it is well known to be inactive in these three animal models of inflammation at the large initial screening doses used. However, two copper coordination compounds obtained with D-pen, one a polymeric cuprous complex and the other suggested to be a cupric disulfide complex, were found to be active in these models. The cupric bis-D-pen compound was inactive at the initial screening dose in the CFE model and was not tested in the subsequent follow-up models of inflammation.

The activities of both coordination compounds prepared from fenamole were just as remarkable. The parent compound was inactive at the initial screening dose in the CPE model, active only at the initial screening dose in the CWG model, and active only at 4 times the usual initial screening dose in the AA model of inflammation. However, the two fenamole copper complexes were found to be active at low doses in the CPE and CWG models. In the AA model only one of the two complexes was tested, but it proved to be active. The two copper-salicylate complexes were also found to be potent anti-inflammatory agents in the CPE and CWG models of inflammation. These results are particularly significant in light of Hangarter's successful therapeutic results with his preparation of copper salicylate (see Section 7).

The anti-inflammatory activities of copper complexes synthesized with inactive and active anti-inflammatory complexing agents, along with similar results

Table 6. Anti-inflammatory Activities, LD_{50} Data, and Therapeutical Indices (TI) of Some Antiarthritic Drugs and Their Copper Complexes[a,b]

Compound	Carrageenan Paw Edema	Cotton Wad Granuloma	Adjuvant Arthritis	Copper (%)	LD_{50}	TI	
						CPE	PA
Aspirin	A at 64	A at 200 i.g.	A at 6		1500 i.g. 790 r.t.[c]		
$Cu(II)_2(aspirinate)_4$	A at 8	A at 10	A at 1.2	15.0	760 ± 100	95	633
Niflumic acid	A at 40 i.g.	A at 25 i.g.	A at 6 i.g.		370 ± 25 i.p.	9	62
$Cu(II)_n(niflumate)_{2n}(H_2O)_n$	A at 8	A at 10	A at 1.2	10.1	650 ± 80	81	542
D-Penicillamine	I at 200	I at 100	I at 30				
$Cu(I)$D-pen$(H_2O)_{1.5}$	A at 8	A at 10	NT	26.7			
$Cu(II)_n$(D-pen)$_{2n}(H_2O)_{2n}$	I at 200	NT[d]	NT	16.1			
$Cu(II)$(D-pen disulfide)$(H_2O)_2$	A at 8	A at 25	A at 30	15.4			
Fenamole	I at 200	A at 100 i.g.	A at 100 i.g.				
$Cu(II)_n(fenamole)_n(acetate)_{2n}$	A at 8	A at 25	A at 30	18.5			
$Cu(II)_n(fenamole)_{2n}(HCl)_{2n}$	A at 16	A at 10	NT	13.9			
$Cu(II)_2(salicylate)_4(Na_4)$	A at 16	A at 10	NT	16.4			
$Cu(II)(salicylate)_2 4H_2O$	A at 8	A at 25	NT	15.4			

[a] Sorenson (1976a).

[b] All compounds were given by s.c. injection unless indicated as intragastric (i.g.) and expressed as milligrams per kilogram of body weight.

[c] Rectal.

[d] Not Tested.

obtained with 12 other amino acid, carboxylic acid, amine, and corticoid copper complexes (Sorenson, 1976a), supported the hypothesis that copper coordination compounds which could be formed *in vivo* may account for the clinical usefulness of the antiarthritic drugs. The suggestion that this beneficial activity is due to the complex is also supported by the lack of direct correlation between anti-inflammatory activity and the amount of copper in these compounds.

Comparison of the acute toxicity data obtained for some of these complexes and their parent compounds suggests that the copper chelates are less toxic. Oral and rectal LD_{50} values for aspirin have been reported to be 1500 and 790 mpk, respectively (Christensen, 1973). The LD_{50} for $Cu(II)_2(aspirinate)_4$ was found to be 760 mpk when given s.c. Since it is quite likely that the LD_{50} for aspirin is less than 760 mpk when given s.c., it seems safe to suggest that the complex is less acutely toxic. The case concerning niflumic acid is more certain. The LD_{50} for this chelate, 650 mpk s.c., was much higher than the value obtained for the parent compound given by the oral route, 370 mpk. The LD_{50} for the niflumic acid given intraperitoneally was found to be 155 ± 9 mpk, which would indicate that the LD_{50} value if obtained by the s.c. route would be between 155 and 370 mpk. Based upon these data it seems reasonable to suggest that, in addition to being more potent as anti-inflammatory agents, these copper complexes are less toxic than the parent compounds. A consequence of this is that the TI values, which are a measure of safety, shown in Table 6 are larger for the complexes than for the parent compounds.

Of even greater interest than this marked reduction in toxicity was the observation that these complexes were potent *antiulcer* agents. Since it is well known that clinically used antiarthritic drugs cause ulcers and gastrointestinal distress, the observed antiulcer activity further distinguishes these coordination compounds from their parent compounds as being safer and potentially much more therapeutically useful than the currently used drugs.

The antiulcer activities of these complexes, presented in Table 7, were determined in the Shay ulcer model after *intragastric* or *oral* dosing. Cupric acetate was found to be only very weakly active as an antiulcer compound; activity at 225 mpk is viewed as a very low order of potency. However, $Cu(II)(anthranilate)_2$, $Cu(II)(3,5\text{-dips})_2$, $Cu(II)_2(aspirinate)_4$, $Cu(II)_n(niflumate)_{2n}$, $Cu(II)_n(\text{D-pen})_{2n}$, Cu(II)(D-pen disulfide), $Cu(II)_n(fenamole)_n(acetate)_{2n}$, $Cu(II)_n(fenamole)_{2n}(HCl)_{2n}$, and $Cu(II)(salicylate)_2 \cdot 4H_2O$, along with 27 amino acid, carboxylic acid, amine, and corticoid-21-phosphate copper complexes proved to be potent antiulcer compounds (Sorenson, 1976a, 1976b). These results are particularly significant since a common reason for withdrawal from aspirin and D-penicillamine therapy of arthritic diseases is the gastrointestinal irritation these drugs cause. The data also support Hangarter's observation that patients treated with his preparation of copper salicylate did not suffer from gastrointestinal distress.

In addition to the Shay antiulcer activity, a number of these chelates were

Table 7. Antiulcer Activity of Some Copper Complexes[a]

Compound	Shay Antiulcer Activity[b]
$Cu(II)_2(acetate)_4(H_2O)_2$	A at 225
$Cu(II)(anthranilate)_2$	A at 4.5
$Cu(II)(3,5\text{-dips})_2$	A at 2.3
$Cu(II)_2(aspirinate)_4$	A at 11.3
$Cu(II)_n(niflumate)_{2n}$	A at 4.5
$Cu(II)_n(\text{D-pen})_{2n}$	A at 4.5
$Cu(II)(\text{D-pen disulfide})\cdot 3H_2O$	A at 4.5
$Cu(II)_n(fenamole)_n(acetate)_{2n}$	A at 4.5
$Cu(II)_n(fenamole)_{2n}(HCl)_{2n}$	A at 4.5
$Cu(II)(salicylate)_2 \cdot 4H_2O$	A at 4.5

[a] Sorenson (1976a).

[b] All doses were given intragastrically and expressed as milligrams per kilogram of body weight.

also shown to have antiulcer activity in the corticoid-induced ulcer model (Sorenson, 1976a). Phenomenologically it is of interest that preliminary studies of $Cu(II)_{3n}(\text{hydrocortisone-21-phosphate})_{2n}(H_2O)_{9n}$ demonstrated that this complex inhibited ulcers in the corticoid-induced ulcer model at doses as low as 18 mpk.

All of the copper complexes that were active antiulcer compounds were also shown to decrease gastrointestinal secretions. Both acid and pepsin in these secretions were decreased, 5 hr after ligation, in the Shay rat. The possibility that these copper complexes decreased pepsin activity by inhibiting pepsin was studied *in vitro,* but no pepsin inhibition was found. The observed antisecretory activity could not be explained as anticholinergic activity since the copper coordination compounds studied failed to possess any significant ganglionic or postganglionic cholinergic blocking activity, as shown in Table 8. This lack of anticholinergic activity or blockade of the autonomic nervous system was consistent with the observations that most of these compounds failed to affect psychomotor behavior in treatments with six doses of complex ranging from 5 to 320 mpk. Acute toxicity was not observed in this dose range when the compounds were given s.c.

It was simultaneously suggested by Whitehouse et al. (1975) that D-penicillamine might be a precursor drug and that the active species formed *in vivo* might be a copper–D-penicillamine complex since they had found a copper complex of D-penicillamine to be effective in the AA, CPE, urate paw edema (UPE), and oleyl alcohol paw edema models of inflammation. These observations were extended by Whitehouse (1976) in his discussion of the role of copper in inflammatory disorders and with the report by Denko and Whitehouse (1976)

Table 8. Anticholinergic Activity, Qualitative Mouse Behavioral Effects, and Estimations of LD_{50} Values in Mice of Some Copper Complexes[a]

Compound	Anticholinergic Activity[b] TEA[c] (%)	Anticholinergic Activity[b] Atropine (%)	Qualitative Behavioral Change[d]	LD_{50}[d]
Cu(II)(anthranilate)$_2$	<11	<1	I at 320	>320
Cu(II)(3,5-dips)$_2$	<20	<1	I at 320	>320
Cu(II)$_2$(aspirinate)$_4$	<20	<1	I at 320	>320
Cu(II)$_n$(niflumate)$_{2n}$(H$_2$O)$_n$	NT[e]	NT	I at 320	>80 <320
Cu(I)$_n$D-pen(H$_2$O)$_{1.5n}$	NT	NT	I at 320	>80 <320
Cu(II)$_n$(D-pen)$_{2n}$(H$_2$O)$_{2n}$	NT	NT	Depressant at 320	>80 <320
Cu(II)$_n$(D-pen disulfide)$_n$ (H$_2$O)$_{3n}$	NT	NT	I at 320	>80 <320
Cu(II)$_n$(fenamole)$_n$(acetate)$_{2n}$	<7	NT	I at 320	>320
Cu(II)$_n$(fenamole)$_{2n}$(HCl)$_{2n}$	<7	NT	I at 320	>320
Cu(II)$_2$(salicylate)$_4$(Na)$_4$	NT	NT	I at 320	>320
Cu(II)(salicylate)$_2$(H$_2$O)$_4$	<20	<1	I at 320	>320

[a] Sorenson (1976a).
[b] < indicates inactivity; values were not different from control values.
[c] Tetraethylammonium bromide.
[d] All doses given subcutaneously and expressed as milligrams per kilogram of body weight.
[e] Not Tested.

that copper complexes of glycine and citric acid had anti-inflammatory activity in a model of calcium pyrophosphate-induced inflammation. Whitehouse and Walker (1976) subsequently reported that Cu(II)(D-penicillamine)$_2$, Cu(I)–D-penicillamine, $Na_5Cu(I)_8Cu(II)_6$–D-penicillamine$_{12}$Cl, $Na_2Cu(I)$-thiomalate, Cu(I)-dithiodiglycol, $Na_2Cu_2(S_2O_3)_2 \cdot H_2O$, Cu(I)(thioacetamide)$_4$Cl, Cu(I)$(Cu_3CN)_4CO_4$, Cu(I)Cl-dimethylsulfoxide, bisglycinatocopper(II), and Cu(II)-ascorbate were effective in CPE, UPE, KPE, and/or AA models of inflammation after s.c. dosing. A comparison of copper, gold, and silver thiomalate and thiosulfate complexes in these models of inflammation revealed that the copper complexes were effective in preventing inflammation and compared favorably in this respect with proven anti-inflammatory drugs, whereas the gold and silver complexes were virtually inactive, even at higher doses. These data support the results obtained by Forestier in his clinical comparisons of copper and gold complexes.

After the report (Sorenson, 1974) that Cu(II)$_2$(aspirinate)$_4$ had more anti-inflammatory activity than aspirin in the CPE and AA models of inflammation when they were given s.c., Williams et al. (1976) reported that Cu(II)$_2$(aspirinate)$_4$ was only as active as aspirin in CPE inflammation but about twice as active as aspirin in AA when each was given orally. Rainsford and Whitehouse

(1976b) also reported that $Cu(II)_2(aspirinate)_4$ and a preparation of Cu(II)-salicylate had the same activity as aspirin and salicylic acid in CPE and UPE when they were given orally. Boyle et al. (1976) found that copper complexes of aspirin (**11**), clopirac (**16**), ketoprofen (**17**), (+)-naproxen (**18**), niflumic acid (**12**), and indomethacin (**19**) were only as active as the parent drugs when given orally in CPE.

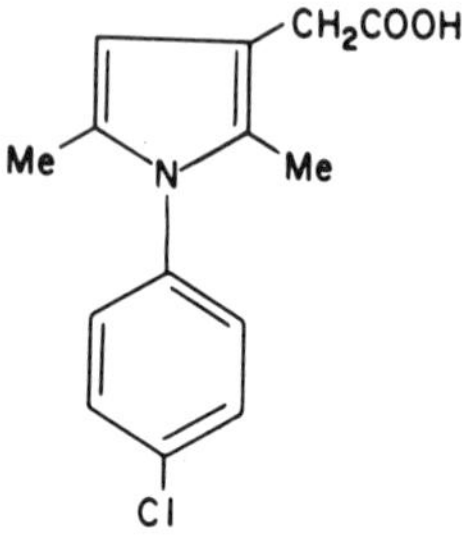

16

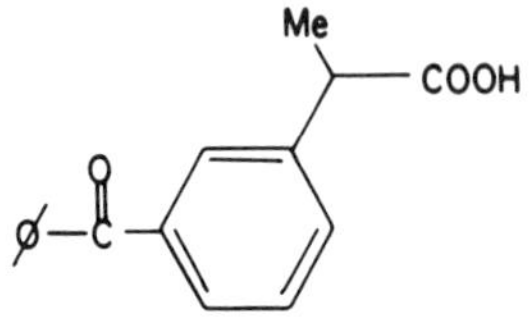

17

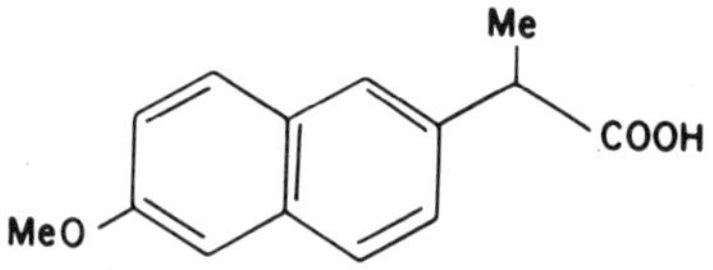

18

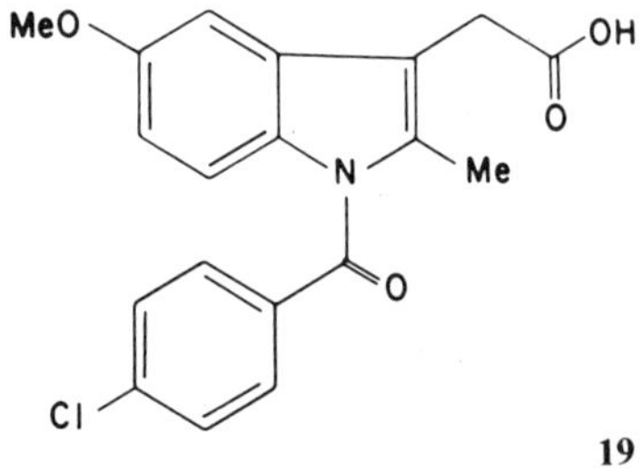

19

It has been suggested (Sorenson, 1977) that the reasons for the reduced oral anti-inflammatory activity found by Williams et al. (1976) and Rainsford and Whitehouse (1976b) were in part gastric acid destruction and the use of ionic suspending agents, tragacanth and acacia, which are capable of either removing copper from the complexes in question or forming quaternary complexes. If copper were removed from the complex, only the parent drug would be available for absorption. Or, if a quaternary complex were formed, it is likely that only a small amount of complex would be absorbed. However, methylcellulose was used as the suspending agent in the Boyle et al. (1976) study, so it appears that gastric acidity may be the primary cause for the reduced activity found after oral dosing of these animals. Nevertheless, ionic suspending agents such as tragacanth, acacia, and carboxymethylcellulose should always be avoided since they may affect the results when gastric acidity is of no concern. A third reason for reduced anti-inflammatory activity of these copper complexes after oral dosing may be a decreased rate of absorption in the stomach. If there is a slower rate of absorption and the complexes pass into the small intestine, they are no longer likely to be absorbed since copper complexes undergo olation in basic media and yield CuO. If this occurs, neither the complex nor the copper is available for absorption since CuO is very insoluble in basic media.

When $Cu(II)_2(aspirinate)_4$ and Cu(II)-salicylate were given by s.c. injection,

Rainsford and Whitehouse (1976b) found them to be 5 to 6 times more active against CPE and UPE than when orally administered, providing additional support for the results obtained by Hangarter (1974). Boyle et al. (1976) also confirmed the observation that $Cu(II)_2(acetate)_4$ was more active in the CPE model when given s.c. than following oral administration. It was suggested that this increased activity after s.c. administration was an irritant-induced anti-inflammatory effect (Rainsford and Whitehouse, 1976b). However, reported data (Sorenson, 1976a) which were in part obtained in consideration of this possibility do not support that suggestion.

The compound $Cu(II)_2(acetate)_4$, which caused marked irritation on s.c. injection, was active only in the CPE model of inflammation and had no anti-inflammatory effect in the CWG and AA models, even though it was just as irritating in the animals used in these studies (Sorenson, 1976a). In contrast to this experience many copper complexes were found to be active in all three models of inflammation after s.c. administration and were not noted to cause irritation. A very large dose of $Cu(II)(tryptophan)_2$, 200 mpk, caused some irritation at the site of injection but was inactive s.c. as an anti-inflammatory agent in the CPE model, although it was active orally in this model and not only lacked oral irritating properties but also had antiulcer activity at a dose as high as 225 mpk. Since rats used in the CWG model of inflammation were adrenalectomized, compounds that were active because they were irritants and stimulated adrenal production of the corticoids could not be active in this model. Nevertheless, all of the copper complexes studied in the CWG model were shown to be active.

It was also suggested that compounds which caused irritation at the site of injection and as a result were anti-inflammatory would be inactive if given i.v. The copper complexes which were given by this route were active in the AA model of inflammation, although sometimes the parent, clinically used drugs were inactive. Irritation resulting in stimulation or depression of the CNS was also suggested as a mechanism of irritant-induced anti-inflammatory activity. However, studies with large doses of copper complexes failed to reveal any CNS stimulation or depression. As an alternative to the toxic-irritant mechanism of action it is suggested here that the increase in anti-inflammatory activity of the copper complexes given s.c. is the result of achieving greater concentrations of the active species in the inflamed tissues. The active species may facilitate the physiologic processes required for the return to normal tissue status.

Some of the more recent studies of copper complexes have been done with very large doses to evaluate their gastric irritation or ulcerogenic properties since it is well known that all of the currently used antiarthritic drugs cause gastrointestinal distress and/or ulcers. Williams et al. (1976) were the first to report that $Cu(II)_2(aspirinate)_4$ caused no gastrointrointestinal erosions in rats with doses up to 1380 mpk of the complex, whereas aspirin produced 50% or greater incidence of erosions at much lower doses of 100 to 300 mpk. Rainsford and Whitehouse (1976b) found that their preparation of Cu(II)-salicylate and

$Cu(II)_2(aspirinate)_4$ either reduced or prevented gastric damage in Rainsford's cold-stress rat ulcer model. Even a large dose of Cu(II)-salicylate caused no gastric damage in rats with AA. It was suggested that the mechanism of this protective effect was a stimulation of gastric mucous effusion by Cu(II) ions (Rainsford and Whitehouse, 1976a, 1976b). However, the unlikelihood that the ionic species of copper exists in any significant concentration in this biologic system, and the lack of significant antiulcer activity for $Cu(II)_2(acetate)_4$ (Sorenson, 1974, 1976a), which is readily dissociable, suggest that the existence of copper complexes must be invoked to account for the variation in antiulcer activity after treatment with various complexes. It has also been suggested that the antiulcer activity observed by Rainsford and Whitehouse (1976a, 1976b) might have been greater had they not acidified the copper solutions before administration, since acid is likely to at least partially dissociate these metal complexes (Sorenson, 1977).

Boyle et al. (1976) found that the copper complexes of ketoprofen (**17**), (+)-naproxen (**18**), and indomethacin (**19**) were as ulcerogenic as their parent compounds but the copper complexes of aspirin (**11**), niflumic acid (**12**), and clopirac (**16**) failed to cause the gastric damage produced by the parent drugs. The ulcerogenic copper complexes were shown to be as effective as their parent compounds in inhibiting "total" prostaglandin synthesis, but the nonulcerogenic complexes failed to cause the inhibition of "total" prostaglandin synthesis observed for the parent drugs. These effects on prostaglandin synthesis were suggested to account for the observed ulcerogenicity. When $Cu(II)_2(aspirinate)_4$ was studied with regard to its effect on prostaglandin E_2 (PGE_2) and prostaglandin $F_{2\alpha}$ ($PGF_{2\alpha}$) synthesis, it was found that this complex caused a concomitant increase in $PGF_{2\alpha}$ and a decrease in PGE_2 synthesis. Although the relative changes in synthesis of these prostaglandins are consistant with data in the literature, the magnitudes of the changes are somewhat in doubt since the *in vitro* test system used in these studies contained EDTA, which may compete for the copper in these complexes, and the tests were done at pH 8.0, a value at which insoluble polynuclear coordination complexes are formed, causing the copper to be less available for biochemical reaction processes.

A question has been raised concerning the toxicity associated with the use of a copper-containing substance in therapy, even though the complexes were found to be less toxic than the parent compounds, including the clinically used drugs (Sorenson, 1976a), to lack gastrointestinal irritation properties (Williams et al., 1976; Boyle et al., 1976), and to possess antiulcer activity (Sorenson, 1976; Rainsford and Whitehouse, 1976a, 1976b). With this evidence it seems worthwhile to at least consider the possibility that therapy of rheumatoid diseases may require copper in a complexed form to counteract the pathogenic processes associated with these diseases so that remission can be achieved. In this context, if a copper complex is less toxic and more effective than existing drugs, the fact that the complex contains copper should be viewed positively.

7.3. Biochemistry of Anti-inflammatory Copper Complexes

During the last few years important gains have been made in understanding the possible biochemical mechanisms of action for these copper complexes. At least five plausible biochemical mechanisms can be invoked to account for the anti-inflammatory and antiulcer activities of copper complexes: induction of lysyl oxidase, modulation of prostaglandin syntheses, induction of superoxide dismutase or superoxide dismutase mimetic activity, stabilization of lysosomal membranes, and modulation of the activity of histamine.

It is well known that repair of sites of inflammation, including ulcers, requires the cross-linking and extracellular maturation of the connective tissue components collagen and elastin. The enzyme responsible for this, lysyl oxidase, is copper dependent (Chou et al., 1969; Carnes, 1971; O'Dell, 1976). As a result, this aspect of wound or tissue repair assumes particular significance with regard to a role for copper complexes having both anti-inflammatory and antiulcer activity. Recently Harris (1976) has demonstrated that lysyl oxidase activity is induced in copper-deficient chickens with $CuSO_4$. He has also been able to induce lysyl oxidase activity in these animals with Cu(II)(anthranilate)$_2$, Cu(II)(3,5-dips)$_2$, and Cu(II)$_2$(aspirinate)$_4$ (personal communication).

In light of these findings the absence of ulcerogenicity (Williams et al., 1976) and the observation of antiulcer activity with these and other copper complexes merit special consideration. If it is supposed that at least one possible chemical mechanism for ulcerogenesis is the loss of copper from the copper-dependent enzyme lysyl oxidase, which is required to maintain the integrity of the extracellular connective tissue in the mucosa and submucosa, then the ulcerogenicity of aspirin and other anti-inflammatory compounds may be due to their ability to form copper complexes, resulting in enzyme loss as a direct consequence of their coordination-chemical reactivity. It follows, then, that existing ulcers can be successfully treated with copper complexes which can induce lysyl oxidase activity and promote tissue repair.

The suggestion that these compounds promote tissue repair is consistent with the observation that Cu(II)(L-tryptophan)$_2$-treated rats with surgically induced ulcers (or wounds) (Townsend, 1961) appeared to heal at a markedly increased rate compared to nontreated controls. Townsend found that, at day 5 postulcer induction, treated animals that were given 25 mpk Cu(II)(L-tryptophan)$_2$ beginning on day 1 postsurgery were a full 5 days ahead of the nontreated control animals with regard to healing of these ulcers, and they remained ahead of the nontreated controls throughout the course of the 20-day experiment (personal communication). In this same ulcer model 25 mpk of Cu(II)$_2$(aspirinate)$_4$ also brought about the same increase in rate of ulcer repair (Townsend and Sorenson, 1978). In addition, visual comparison of nontreated and Cu(II)(L-tryptophan)$_2$- or Cu(II)$_2$(aspirinate)$_4$-treated rats revealed that gastric adhesions had formed only in the nontreated rats.

Histochemical studies revealed that the regression adjacent to the lesion seen at 5 days postsurgery in nontreated rats was absent in the $Cu(II)_2(aspirinate)_4$-treated rats and the size of the ulcer or wound was much smaller as a result of increased re-epethelialization and gland formation (Townsend and Sorenson, 1978). On day 10 postsurgery, re-epithelialization was completed in only one of the nontreated rats, new glands were highly irregular, and the mucocollagenous band joining the cut ends of the muscularis mucosae was discontinuous and fenestrated. In contrast, re-epithelialization was complete in all treated rats, new glands were regularly shaped, and the mucocollagenous band was continuous and thicker. In the nontreated rats, collagen in newly formed mucosa and submucosa at the base of the lesion was very dense and composed of thick, wavy, disorientated bundles, whereas treated rats had normally orientated, fine collagen fibers in the same location. The quantity and the quality of the replaced collagen in the treated animals were superior to those found in the nontreated rats. While the desired collagenous changes were interpreted as being the result of lysyl oxidase induction, the earliest observed anti-inflammatory effects were attributed in part to modulation of prostaglandin synthesis.

Modulation of prostaglandin synthesis by copper complexes may result in a decrease in the synthesis of the proinflammatory (vasodilator) prostaglandin, PGE_2, and a concomitant increase in the synthesis of the anti-inflammatory prostaglandin (vasoconstructor), $PGF_{2\alpha}$ (Sorenson, 1974). This suggestion was consistent with the report of Lee and Lands (1972), which has been confirmed by Maddox (1973), who found a depression in PGE_2 synthesis and an increase in $PGF_{2\alpha}$ after the addition of $CuSO_4$ or $CuCl_2$ to seminal vesicle homogenates. Additional support for these observations has been provided by Vargraftig et al. (1975), who used $CuCl_2$ and found copper to be required for platelet synthesis of prostaglandins from arachindonic acid and suggested that this accounted for the observed anti-inflammatory activity of copper complexes. This suggestion has been supported by Boyle et al. (1976), who demonstrated that $Cu(II)_2$-$(aspirinate)_4$ brought about a concomitant decrease in PGE_2 and increase in $PGF_{2\alpha}$ synthesis. These workers also confirmed the results of Lee and Lands (1972) and Maddox (1973) by showing that $CuCl_2$ caused a decrease in the synthesis of PGE_2 while increasing the synthesis of $PGF_{2\alpha}$.

McCord (1974) has suggested that rheumatoid arthritis may be the result of a deficiency or lack of superoxide dismutase enzyme activity. This suggestion is of interest, because it is known that human superoxide dismutase contains copper, which is required for its dismutase activity (Fridovich, 1975). In addition, Weser and his colleagues (Brigelius et al., 1974, 1975; Paschen and Weser, 1975) reported that the copper chelates $Cu(II)(lysine)_2$, $Cu(II)(glycylhistadine)_2$, $Cu(II)(glycylhistadylleucine)_2$, $Cu(II)(tryptophan)_2$, and $Cu(II)(histadine)_2$ have superoxide dismutase-like activity. Since it had been shown that Cu(II)-$(tryptophan)_2$ and $Cu(II)(lysine)_2$ have anti-inflammatory and antiulcer activity, it seemed reasonable to suggest that other anti-inflammatory copper chelates have superoxide dismutase mimetic activity (Sorenson, 1976b). This suggestion has recently been supported by the reports of De Alvare et al. (1976), Weser et al. (1978), and Younes and Weser (1977) that $Cu(II)(salicylate)_2 \cdot 4H_2O$, $Cu(II)(p\text{-aminosalicylate})_2$, $Cu(II)(dips)_2$, $Cu(II)_2(aspirinate)_4$, $Cu(II)_2$-(niflumate)$_4$, and $Na_5Cu(I)_8Cu(II)_6$-D-penicillamine$_{12}$Cl have superoxide dismutase mimetic activity.

The superoxide dismutase mimetic activity observed for $Cu(II)(tryptophan)_2$ and $Cu(II)_2(aspirinate)_4$ may account for the observation that mucosal and submucosal tissues of rats with surgically induced ulcers did not undergo autolylic destruction along the borders of the original incision when they were treated with these two complexes. This dismutase activity may also account for the absence of hard liver, spleen, pancreas, and stomach adhesions in these rats, which were found in the nontreated controls (Townsend and Sorenson, 1977).

A fourth possible biochemical mechanism of action for copper complexes is based upon the report of Chayen et al. (1969) that copper was important for redox control of human synovial lysosomes. They found that copper decreased the permeability of lysosomes and, in addition, lowered the ratio of free versus bound lysosomal enzymes. This is an alternative biochemical mechanism which may be used to account for the absence of autolytic destruction and adhesions associated with the use of copper complexes in the treatment of surgically induced ulcers (Townsend and Sorenson, 1977). The possibility that the anti-inflammatory activity of the copper complexes was due to specific lysosomal enzyme inhibition has been partially ruled out by McAdoo et al. (1973), who found that many complexes failed to inhibit cathepsin-D, a lysosomal proteinase.

Biochemical modulation of the physiologic effects of histamine (**20**) may also

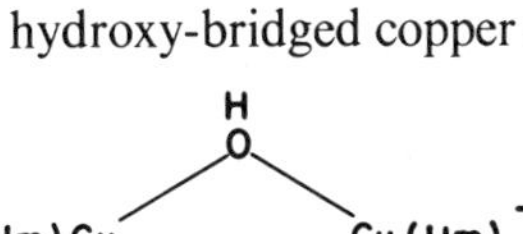

$CH_2CH_2NH_2$

N NH

20

be an important biochemical role for copper complexes. Walker et al. (1973, 1975) and Walker and Reeves (1977) considered the coordination chemistry of histamine *in vivo* and provided an abundance of evidence to support the suggestion that a binuclear, hydroxy-bridged copper complex (**21**) was the active

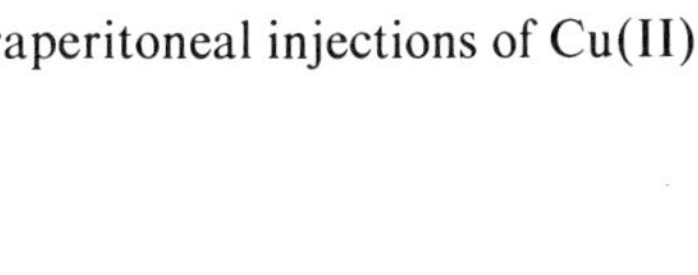

$$\left[(Hm)Cu(\mu\text{-}OH)_2Cu(Hm)\right]^{2+}$$

21

form of histamine. Intraperitoneal injections of $Cu(II)(histamine)Cl_2$ at one-fiftieth the dose of histamine produced the same subliminal anaphylactic symptoms as histamine. Injection of $CuSO_4$ 2 min before the injection of a nonlethal dose of histamine caused death by anaphylactic shock in all mice within 10 min. The anaphylactic LD_{50} dose for Cu(II)(histamine)Cl_2 was found to be one-sixteenth that for histamine. These data support the suggestion that histamine-induced vascular permeability as an acute-phase inflammatory reaction may be mediated by a copper-histamine complex after degranulation of mast cells in the connective tissue spaces.

With the recognition that a copper complex of histamine was responsible for its histaminic activity Walker and Reeves (1977) suggested that competition for this copper accounted for the familiar observation that salicylic acid has antihistamine activity. They then demonstrated that salicylic acid prevented lethality in mice given an LD_{50} dose of histamine. Since an increase in vascular permeability is an important physiologic response in inflamed tissues, these observations merit in-depth study.

7.4. Summary

Copper complexes constitute a unique class of antiarthritic drugs for two important reasons. First, animal studies have revealed that they are potent antiulcer agents, in contrast to currently used ulcerogenic drugs. This is consistent with the lack of gastrointestinal irritation reported by Hangarter. Second, treatment of human rheumatoid diseases brought about long-lasting remissions with short-term copper complex therapy, in contrast to the try-and-hope approach with existing drugs. These two unique features of the biologic activity of copper complexes, along with their demonstrated anti-inflammatory activity in many animal models of inflammation, suggest that copper complex therapy may be more beneficial than existing drug therapy. Copper complex therapy is also consistent with what has been interpreted as the normal physiologic response to rheumatoid diseases.

Serum from patients with rheumatoid disease have been shown to contain increased concentrations of Cp and low molecular weight copper complexes. Since these increases occur before remission, and low molecular weight copper complexes are known to have anti-inflammatory activity, it is suggested that the increase in serum copper-containing components is a physiologic response to these diseases. This physiologic response then facilitates the biologic processes required to prevent further tissue destruction and to promote tissue repair, which leads to remission.

Recent pharmacologic and biochemical studies have provided evidence for five plausible copper-dependent mechanisms of action which may account for the observed biologic activity of copper complexes: induction of lysyl oxidase or lysyl oxidase mimetic activity, induction of superoxide dismutase or superoxide dismutase mimetic activity, modulation of prostaglandin synthesis, modulation of histamine activity, and lysosomal membrane stabilization. Ongoing research may provide evidence for still other copper-dependent processes which may more fully explain the biologic activity of copper complexes.

The possibility that rheumatoid diseases are associated with deficits in copper-dependent processes also suggests that the etiology and/or chronicity of these diseases may result from a frank or quasi copper deficiency. As Elms (1974) has suggested, it seems possible that collaboration between workers in the field of trace element metabolism and those in experimental therapeutics would prove fruitful.

8. ANTINEOPLASTIC ACTIVITY OF COPPER COMPLEXES

Brada and Altman (1978) have pointed out that inorganic copper salts have been known for some time to protect against tumors in rats. Sharples (1946) was the first to observe that increasing the amount of copper in the diet of rats fed 4-dimethylaminoazobenzene lengthened the hepatic tumor induction time. This observation has been confirmed and extended in subsequent studies, as reviewed by Brada and Altman (1978). Inorganic copper was also found to be effective in preventing rodent liver tumors due to ethionine (Brada and Altman, 1978) and a variety of other animal carcinomas (Donath and Putnoky, 1969; Savitskii, 1970).

However, none of these treatments with inorganic copper was as effective as therapy with copper complexes. A single 5-mpk dose of Cu(II)(dimethylglyoxime)$_2$ (**22**) increased the life span of mice with Ehrlich ascites and sarcoma

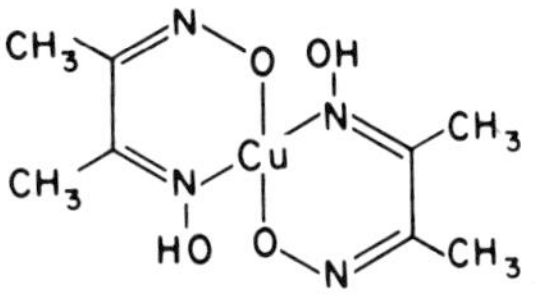

CH$_3$ N O OH N CH$_3$ Cu CH$_3$ N HO O N CH$_3$ 22

180 tumors from 2 to 3 times that of nontreated controls (Takamiya, 1960). Other copper complexes that have been reported to have similar rodent antitumor activities are Cuphen (**6**) (Dwyer et al., 1965), Cu(II)[2-keto-3-ethoxybutyraldehyde bis(thiosemicarbazone)] (**23**; R = —CH(CH$_3$)(OC$_2$H$_5$) (Petering

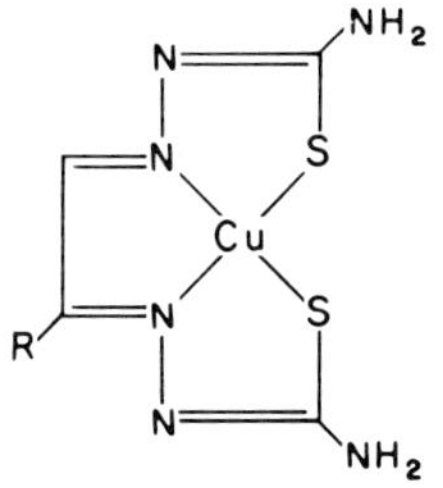

NH$_2$ N N S Cu N S R N NH$_2$ 23

and Van Giessen, 1966; Petering and Petering, 1975; Coats et al., 1976; Petering, 1978), Cu(II)[pyruvaldehyde bis(thiosemicarbazone)] (**23;** R = CH_3) (Cappuccino et al., 1967), and copper complexes of 2-formyl pyridine and 1-formylisoquinoline thiosemicarbazones (**24a** and **24b,** respectively) (Sartorelli and Creasy, 1969; Agrawal et al., 1974), as well as copper-bleomycin (Pietsch, 1975).

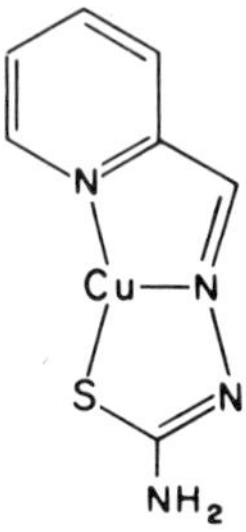

24a 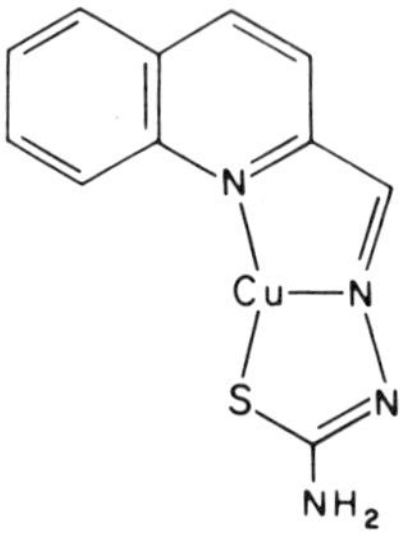 **24b**

In addition to demonstrating the antitumor activity of these copper complexes, the authors cited above have provided a great deal of information concerning their possible mechanisms of action. One reasonable conclusion to be drawn from this information is that there are copper complexes which serve in a homeostatic fashion to prevent the development of neoplasms. At some point in time this physiologic response may fail, thus allowing tumors to grow. Since the classical approach to cancer chemotherapy has not been successful, there seems to be little risk in pursuing the possibility that a therapeutically useful copper complex can be found.

ACKNOWLEDGMENTS

I am grateful to the International Copper Research Association for financial assistance in the writing of this chapter and to Mrs. Betty Stearns, Miss Cathy Sullivan, and Mr. Lonnie Jackson for help in the preparation of the manuscript. Drs. Ulf Bergquist, William R. Collie, Jim Camakaris, and David M. Danks offered many helpful comments, which are also greatly appreciated.

REFERENCES

Adadevoh, B. K. and Dada, O. A. (1973). "Effect of Intrauterine Copper on the Uptake of Estradiol-^{14}C by Rat Tissues," *Fertil. Steril.,* **24,** 54–59.

Adams, P. C., Strond, D. R., Bresnan, M. J., and Lucky, A. W. (1974). "Kinky Hair Syndrome: Serial Study of Radiological Findings with Emphasis on the Similarity to the Battered Child Syndrome," *Radiology,* **112,** 401–407.

Adams, S. S. and Cobb, R. (1967). "Non-Steroidal Antiinflammatory Drugs," *Prog. Med. Chem.,* **5,** 59–138.

Adelstein, S. J. and Vallee, B. L. (1962). "Copper." In C. L. Lomar and F. Bronner, Eds., *Mineral Metabolism*, Vol. 2. Academic Press, New York, pp. 371–401.

Aedo, A. R. and Zipper, J. (1973). "Effect of Copper Intrauterine Devices (IUD's) on Estrogen and Progesterone Uptake by the Rat Uterus," *Fertil. Steril.*, **24**, 345–348.

Agrawal, K., Booth, B., Michaud, R., Moore, E., and Sartorelli, A. (1974). "Comparative Studies of Antineoplastic Activity of 5-Hydroxy-2-formylpyridine Thiosemicarbazone and Its Selenosemicarbazone Quanylhydrazone and Semicarbazone Analogs," *Biochem. Pharmacol.*, **23**, 2421–2429.

Aguilar, M. J., Chadwick, D. L., Okuyama, K., and Kamoshita, S. (1966). "Kinky Hair Disease. I: Clinical and Pathologic Features," *J. Neuropathol. Exp. Neurol.*, **25**, 507–522.

Albert, A. (1961). "Design of Chelating Agents for Selected Biological Activity," *Fed. Proc.*, **20**(II), Suppl. 10, 137–147.

Albert, A., Gibson, M. I., and Rubbo, S. D. (1953). "The Influence of Chemical Constitution on Antibacterial Activity. VI: The Bactericidal Action of 8-Hydroxyquinoline (Oxine)," *Br. J. Exp. Pathol.*, **34**, 119–130.

Alexander, F. W. (1974). "Copper Metabolism in Children," *Arch. Dis. Child.*, **49**, 589–590.

Alexander, F. W., Clayton, B. E., and Delves, H. T. (1974). "Mineral and Trace Metal Balances on Children Receiving Normal and Synthetic Diets," *Quant. J. Med.*, **43**, 89–111.

Al-Rashid, R. A. and Spangler, J. (1971). "Neonatal Copper Deficiency," *N. Engl. J. Med.*, **285**, 841–843.

Al-Rashid, R. A. and Spangler, J. (1972). Letter to the Editor in reply to J. R. Seely, G. B. Humphrey, and B. J. Matter (1972), "Copper Deficiency in a Premature Infant Fed an Iron-Fortified Formula," *N. Engl. J. Med.*, **286**, 110.

An Onlooker's Notebook. (1974). "Copper on the Up and Up?" *Pharm. J.*, Oct. 5, pp. 318–319.

Arena, G., Kavu, G., and Williams, D. R. (1978). "Metal-Ligand Complexes Involved in Rheumatoid Arthritis. V: Formation Constants for Calcium(II)-, Magnesium(II)- and Copper(II)-Salicylate and Acetylsalicylate Interactions," *J. Inorg. Nucl. Chem.*, **40**, 1221–1226.

Arnold, C. (1940). "Uber die Behandlung von Lugen- und Ehlkopftuberkulose mit Kupfersalzen (Ebesol)," *Beitr. Klin. Tuberk.*, **95**, 112–119.

Ashkenazi, A., Levin, S., Djaldetti, M., Fishel, E., and Benvenisti, D. (1973). "The Syndrome of Neonatal Copper Deficiency," *Pediatrics*, **52**, 525–533.

Auld, D., Kawaguchi, H., Livingston, D., and Vallee, B. (1974). "RNA-Dependent DNA Polymere (Reverse Transcriptase) from Avian Mycloblastosis Virus: a Zinc Metalloenzyme," *Proc. Natl Acad. Sci.*, **71**, 2091–2095.

Bajpayee, D. P. (1975). "Significance of Plasma Copper and Caeruloplasmin Concentrations in Rheumatoid Arthritis," *Ann. Rheum. Dis.*, **34**, 162–165.

Barnett, M. D. and Brozovic, B. (1975). "A Simple Automated Micromethod for Measuring Serum Copper," *Clin. Chem. Acta*, **58**, 295–298.

Beisel, W. R., Pekarek, R. S., and Wannemacher, R. W., Jr. (1976). "Homeostatic

Mechanism Affecting Plasma Zinc Levels in Acute Stress." In A. S. Prasad and D. Oberleas, Eds., *Trace Elements in Human Health and Disease,* Vol. 1: *Zinc and Copper.* Academic Press, New York, pp. 87–106.

Bickel, L. (1972). Cited Louis Pasteur's "Advice to his students." In *Rise Up to Life: Lord Florey's biography.* Angus and Robertson, Sidney.

Billings, D. M. and Degnan, M. (1971). "Kinky Hair Syndrome: A New Case and a Review," *Am. J. Dis. Child.,* **121,** 447–449.

Biological Concepts, Inc. (1973). "Method of Inhibiting the Fertility of the Reproduction Organs of Mammals." Derwent Publications, London. Patent 1,338,493.

Bonebrake, R. A., McCall, J. T., Hunder, G. C., and Polley, H. L. (1968). "Copper Complexes in Synovial Fluid," *Arthritis Rheum.,* **11,** 95.

Bonta, I. L. (1969). "Microvascular Lesions as a Target of Anti-inflammatory and Certain Other Drugs," *Acta Physiol. Pharmacol. Neerl.,* **15,** 188–222.

Boyle, E., Freeman, P. C., Goudie, A. C., Magan, F. R., and Thomson, M. (1976). "Role of Copper in Preventing Gastrointestinal Damage by Acidic Anti-inflammatory Drugs," *J. Pharm. Pharmacol.,* **28,** 865–868.

Brada, Z. and Altman, N. H. (1978). "The Inhibitory Effect of Copper on Ethionine Carcinogenesis." In G. N. Schrauzer, Ed., *Inorganic and Nutritional Aspects of Cancer.* Plenum Press, New York, pp. 193–206.

Bray, P. F. (1965). "Sex-Linked Neurodegenerative Disease Associated with Monilethrix," *Pediatrics,* **36,** 417–420.

Brendstrup, P. (1953a). "Serum Copper, Serum Iron and Total Iron-Binding Capacity of Serum in Patients with Chronic Rheumatoid Arthritis," *Acta Med. Scand.,* **146,** 384–392.

Brendstrup, P. (1953b). "Serum Copper, Serum Iron and Total Iron-Binding Capacity of Serum in Acute and Chronic Infections," *Acta Med. Scand.,* **145,** 315–325.

Brigelius, R., Spottl, R., Bors, W., Lengfelder, E., Saran, M., and Weser, U. (1974). "Superoxide Dismutase Activity of Low Molecular Weight Cu_2 Plus-Chelates Studied by Pulse Radiolysis," *FEBS Lett.,* **47,** 72–75.

Brigelius, R., Hartman, H. J., Bors, W., Saran, M., Lengfelder, E., and Weser, U. (1975). "Superoxide Dismutase Activity of Cu(Tyr)2, and Cu,Co-Erythrocuprein," *Hoppe-Seylers Z. Physiol. Chem.,* **356,** 739–745.

Broglie, M. (1954). "Die Therapie des entzundichen Elenkrheumatismus," *Dtsch. Med. Wochenschr.,* **79,** 769–773, 816 and 825–829.

Bucknall, W. E., Haslam, R. H. A., and Holtzman, N. A. (1973). "Kinky Hair Syndrome: Response to Copper Therapy," *Pediatrics,* **52,** 653–657.

Butler, H. M., Laver, J. C., Shulman, A., and Wright, R. D. (1970). "The Use of Phenanthroline Metal Chelates for the Control of Topical Infections Due to Bacteria, Fungi, and Protozoa," *Med. J. Aust.,* **2,** 309–314.

Buu-Hoi, N. P. and Xuong, N. D. (1953). "The Thiourea Type of Tuberculostatic Compounds and Their Mechanism of Action," *C. R. Acad. Sci.,* **237,** 498–500.

Cappuccino, J. G., Banks, S., Brown, G., Searge, M., and Tarnowski, G. S. (1967). "The Effect of Copper and Other Metal Ions on the Antitumor Activity of Pyruvaldehyde Bisthiosemicarbazone," *Cancer Res.,* **27,** 968–973.

Carl, E. and Marquardt, P. (1949). "Kupferkomplexbildung und tuberkulostatische Chemotherapeutica," *Z. Naturforsch.*, **4,** 280–283.

Carlton, W. W. and Kelly, W. A. (1969). "Neural Lesions in the Offspring of Female Rats Fed a Copper Deficient Diet," *J. Nutr.*, **97,** 42–52.

Carnes, W. H. (1971). "Role of Copper in Connective Tissue Metabolism," *Fed. Proc.*, **30,** 995–1000.

Cartwright, G. E. (1950). "Copper Metabolism in Human Subjects," In W. D. McElroy and B. Glass, Eds., *Symposium on Copper Metabolism.* Johns Hopkins Press, Baltimore, Md., pp. 283, 284.

Cartwright, G. E. and Wintrobe, M. M. (1949). "Chemical, Clinical, and Immunological Studies on the Products of Human Plasma Fractionation; Anemia of Infection; Studies on Iron-Binding Capacity of Serum," *J. Clin. Invest.*, **28,** 86–98.

Caruthers, M. E., Hobbs, C. B., and Warren, R. L. (1966). "Raised Serum Copper and Caeruloplasmin Levels in Subjects Taking Oral Contraceptives," *J. Clin. Pathol.*, **19,** 498–500.

Caton, M. P. L. (1971). "The Prostaglandins," *Prog. Med. Chem.*, **8,** 317–376.

Chayen, J., Bitensky, L., Butcher, R. G., and Poulter, L. W. (1969). "Redox Control of Lysosomes in Human Synovia," *Nature,* **222,** 281–282.

Chou, W. S., Savage, J. E., and O'Dell, B. L. (1969). "Role of Copper in Biosynthesis of Intramolecular Cross-Links in Chick Tendon Collagen," *J. Biol. Chem.*, **244,** 5785–5789.

Christensen, H. E. (1973). *The Toxic Substances List.* U.S. Department of Health, Education, and Welfare, U.S. Public Health Service, NIOSH, Rockville, Md., pp. 33.

Chvapil, M., Ryan, J. N., and Zukoski, C. F. (1972). "The Effect of Zinc and Other Metals on the Stability of Lysosomes," *Proc. Soc. Exp. Biol. Med.*, **140,** 642–646.

Citron, K. (1972). "Tuberculosis—Chemotherapy," *Br. Med. J.*, **1,** 426.

Clemetson, C. A. B. (1968). "Caeruloplasmin and Green Plasma," *Lancet,* **2,** 1037.

Coats, E. A., Milstein, S. R., Holbein, G., McDonald, J., Reed, R., and Petering, H. G. (1976). "Comparative Analysis of the Cytotoxicity of Substituted [Phenylglyoxal Bis(4-methyl-3-thiosemicarbazone)] Copper(II) Chelates," *J. Med. Chem.*, **19,** 131–135.

Cordano, A., Baertl, J. M., and Graham, G. G. (1964). "Copper Deficiency in Infancy," *Pediatrics,* **34,** 324–336.

Cymerman-Craig, J., Willis, D., Rubbo, S. D., and Edgar, J. (1955). "Mode of Action of Isonicotinic Hydrazide," *Nature,* **176,** 34–35.

Cymmerman-Craig, J., Rubbo, S. D., Edgar, J., Vaughn, G. N., and Willis, D. (1956). "Mode of Action of Isonicotinic Hydrazide," *Nature,* **177,** 480.

Daniel, E. E. (1964). "Effect of Cocaine and Adrenaline on Contractures and Downhill Ion Movements Induced by Inhibitors of Membrane ATPase in Rat Uteri," *Can. J. Physiol. Pharmacol.*, **42,** 497–526.

Danks, D. M. (1975). "Steely Hair, Mottled Mice and Copper Metabolism," *N. Engl. J. Med.*, **293,** 1147–1149.

Danks, D. M. (1977). "Copper Transport and Utilization in Menkes' Syndrome and in Mottled Mice," *Inorg. Perspect. Biol. Med.,* **1,** 73–100.

Danks, D. M., Campbell, P. E., Stevens, B. J., Mayne, V., and Cartwright, E. (1972a). "Menkes' Kinky Hair Syndrome: An Inherited Defect in Copper Absorption with Widespread Effects," *Pediatrics,* **50,** 188–201.

Danks, D. M., Stevens, B. J., Campbell, P. E., Gillespie, J. M., Walker-Smith, J., Blomfield, J., and Turner, B. (1972b). "Menkes' Kinky-Hair Syndrome," *Lancet,* **1,** 1100–1103.

Danks, D. M., Cartwright, E., Stevens, B. J., and Townley, R. R. W. (1973a). "Menkes' Kinky Hair Disease: Further Definition of the Defect in Copper Transport," *Science,* **179,** 1140–1142.

Danks, D. M., Cartwright, E., and Stevens, B. J. (1973b). "Menkes' Steely-Hair (Kinky-Hair) Disease," *Lancet,* **1,** 891.

Danks, D. M., Camakaris, J., and Stevens, B. J. (1978). "The Cellular Defect in Menkes' Syndrome and in Mottled Mice." In M. Kirchgessner, D. A. Roth-Maier, H.-P. Roth, F. J. Schwarz, and E. Weigand, Eds., *Proceedings of the Third International Symposium on Trace Element Metabolism in Man and Animals.* Technische Universitaet Munchen, Freising-Weihenstephan, F. R. Germany, pp. 401-404.

De Alvare, L. R., Goda, K., and Kimura, T. (1976). "Mechanism of Superoxide Anion Scavenging Reaction by Bis(salicylato)-Copper(II) Complex," *Biochem. Biophys. Res. Commun.,* **69,** 687–694.

Deane, R. S., Mills, E. L., and Hamel, A. J. (1970). "Antibacterial Action of Copper in Respiratory Therapy Apparatus," *Chest,* **58,** 373–377.

Dekaban, A. S., O'Reilly, S., Aamodk, R., and Rumble, W. F. (1974). "Study of Copper Metabolism in Kinky Hair Disease (Menkes' Disease) and in Hepatolenticular Degeneration (Wilson's Disease) over Utilizing ^{67}Cu and Radioactivity Counting in the Total Body and Various Tissues," *Trans. Am. Neurol. Assoc.,* **99,** 106–109.

Denko, C. W. and Whitehouse, M. W. (1976). "Experimental Inflammation Induced by Naturally Occurring Microcrystalline Calcium Salts," *J. Rheum.,* **3,** 54–62.

Di Paolo, R. V. and Newberne, P. M. (1972). "Copper Deficiency and Myelination in Newborn Rat," *Proc. Fed. Am. Soc. Exp. Biol.,* **31,** 699.

Di Paolo, R. V. and Newberne, P. M. (1973). "Copper Deficiency and Myelination in the Central Nervous System of the Newborn Rat: Histological and Biochemical Studies." In D. D. Hemphill, Ed., *Trace Substances in Environmental Health*—VII. University of Missouri, Columbia, pp. 225–232.

Doluizio, J. T. and Martin, H. N. (1963a). "Metal Complexation of Tetracycline Hydrochlorides," *J. Med. Chem.,* **6,** 16–20.

Doluizio, J. T. and Martin, H. N. (1963b). "Binding of Tetracycline Analogs to Conalbumin in Absence and Presence of Cupric Ions," *J. Med. Chem.,* **6,** 20–23.

Donath, I. and Putnoky, G. (1969). "Experimental Data on the Cyteostatic Effect of Copper," *Magy. Onkol.,* **13,** 247–250 (*Cancer Chemother. Abstr.,* 10 (1969), No. 71).

Dury, A. and Bradbury, J. T. (1941). "Copper-Induced Pseudopregnancy in the Adult Estrous Rat," *Am. J. Physiol.*, **135,** 587–590.

Dwyer, F. P., Mayhew, E., Roe, E. M. F., and Shulman, A. (1965). "Inhibition of Landschutz Ascites Tumor Growth by Metal Chelates Derived from 3,4,7,8-Tetramethyl-1,10-phenanthroline," *Br. J. Cancer,* **19,** 195–199.

Elms, M. E. (1974). "Anti-inflammatory Drugs and Tissue Copper," *Lancet,* Nov. 30, pp. 1329–1330.

Evans, G. W. (1973). "Copper Homeostasis in the Mammalian System," *Physiol. Rev.*, **53,** 535–570.

Evers, A. (1952). "Das Verhalten des Serumfupfers bei primar chronischer Polyarthritis," *Z. Rheumaforsch.*, **11,** 164–176.

Everson, C. J., Schrader, R. E., and Wang, T. I. (1968). "Chemical and Morphological Changes in Brains of Copper Deficient Guinea Pigs," *J. Nutr.*, **96,** 115–125.

Fahndrich, W. H. (1952). "Die primer chronische Polyarthritis und ihre Behandlung," *Medizinische,* **46,** 1450–1454.

Fenz, E. (1941). "Le Cuivre dans le Rhumatisme Articulaire," *Munch. Med. Wochenschir.*, **18,** 398–403.

Fenz, E. (1951). "Kupfer, ein neues Mittel gegen chron und subakuten Gelenkrheumatismus," *Munch. Med. Wochenschr.*, **41,** 1101–1106.

Fevold, H. L., Hisaw, F. L., and Greep, R. O. (1936). "Augmentation of Gonad Stimulating Action of Pituitary Extracts by Inorganic Substances, Particularily Copper Salts," *Am. J. Physiol.*, **117,** 68–74.

Fiabane, A. M. and Williams, D. R. (1978). "Metal-Ligand Complexes Involved in Rheumatoid Arthritis. II: Acidic Drug-Serum Albumin-Copper(II) Interactions Investigated Using Visible Spectrophotometry and Molecular Filtration," *J. Inorg. Nucl. Chem.*, **40,** 1195–1200.

Fiabane, A. M., Touche, M. L. D., and Williams, D. R. (1978). "Metal-Ligand Complexes Involved in Rheumatoid Arthritis. III: Bovine Serum Albumin-Copper(II), -Zinc(II), and -Lead(II) Interactions Investigated Using Potentiometric Analysis and Molecular Filtration," *J. Inorg. Nucl. Chem.*, **40,** 1201–1207.

Fiscina, B., Oster, G. K., Oster, G., and Swanson, J. (1973). "Gonococcicidal Action of Copper *in vitro*," *Am. J. Obstet. Gynecol.*, **116,** 86–90.

Forestier, J. (1944). "Les Sels Organiques de Cuivre dans le Traitement des Rhumatismes Chroniques," *Bull. Acad. Med.*, **2,** 22–24.

Forestier, J. (1949a). "Comparative Results of Copper Salts and Gold Salts in Rheumatoid Arthritis," *Am. Rheum. Dis.*, **8,** 132–134.

Forestier, J. (1949b). "La Chryso- et la Cuprotherapies dans le Traitment de la Polyarthritic Chroniques Evolution," *Am. Rheum. Dis.*, **8,** 27–29.

Forestier, J. and Certonciny, A. (1946). "Le Traitement des Rhumatismes Chronique par les Sels Organiques des Cuivre," *Presse Med.*, **64,** 884–885.

Forestier, J. and Certonciny, A. (1949). "La Goutte Polyarticulaire Chronique et Son Traitment par les Metaux Lourds," *Bull. Acad. Med.*, **133,** 243–245.

Forestier, J., Jacqueline, F., and Lenoir, S. (1948). "La Cuprotherapie Intramusculaire dans les Rhumatismes Chroniques Inflammatores," *Presse Med.*, **56,** 351–352.

Forestier, J., Certonciny, A., and Jacqueline, F. (1950). "Therapeutic Value of Copper Salts in Rheumatoid Arthritis," *Stanford Med. Bull.*, **8,** 12–13.

Foye, W. O. (1955). "Metal Chelates and Antitubercular Activity. II: *p*-Aminosalicylic Acid," *J. Pharm. Sci.*, **44,** 415–418.

Foye, W. O. and Duvall, R. N. (1958). "Metal Chelates and Antitubercular Activity. III: *p*-Aminosalicylic Acid: Chelate vs. Complex," *J. Pharm. Sci.*, **47,** 282–285.

Foye, W. O., Winthrop, E. L., Swintosky, J. V., Chamberlain, R. E., and Guarini, J. R. (1955). "Metal Chelates of Streptomycin," *J. Pharm. Sci.*, **44,** 261–263.

French, J. H., Sherard, E. S., Lubell, H., Brotz, M., and Moore, C. L. (1972). "Trichopolidystrophy," *Arch. Neurol.*, **26,** 229–244.

Fridovich, I. (1975). "Superoxide Dismutases," *Ann. Rev. Biochem.*, **44,** 147–159.

Gallagher, C. H. and Reeve, V. E. (1971). "Copper Deficiency in the Rat: Effect on Adenine Nucleotide Metabolism," *Aust. J. Exp. Biol. Med. Sci.*, **49,** 445–451.

Gallagher, C. H., Judah, J. D., and Rees, K. R. (1956). "The Biochemistry of Copper Deficiency. II: Synthetic Processes," *Proc. R. Soc.*, **B145,** 195–205.

Garnica, A., Frias, J., and Rennert, O. (1973). "Studies of Ceruloplasmin in Menkes' Kinky Hair Disease," *Pediatr. Res.*, **7,** 387.

Gershon, H. (1974). "Antifungal Activity of Bis-chelates of 5-, 7-, and 5,7-Halogenated 8-Quinolinols with Copper(II)—Determination of Approximate Dimensions of Long and Short Axes of Pores in the Fungal Spore Wall," *J. Med. Chem.*, **17,** 824–827.

Glasser, A. C. and Doughty, R. M. (1962). "Substituted Heterocyclic Thioureas. I: Antitubercular Activity," *J. Pharm. Sci.*, **51,** 1031–1033.

Goka, T. J., Stevenson, R. E., Hefferan, P. M., and Howell, R. R. (1976). "Menkes' Disease: A Biochemical Abnormality in Cultured Human Fibroblasts," *Proc. Natl. Acad. Sci.*, **73,** 604–606.

Goralewski, G. (1940). "Das Kupfer in der Behandburg der Lungentuberkulose," *Z. Tuberk.*, **84,** 313–319.

Graber-Duvernay, J. and Van Moorleghem, G. (1950). "Le Morrhuate de Cuvire dans la Therapeutique des Polyarthritis Chroniques," *J. Lyons Med.*, **183,** 113–116.

Graham, G. C. and Cordano, A. (1969). "Copper Depletion and Deficiency in the Malnourished Infant," *John Hopkins Med. J.*, **124,** 139–150.

Griscom, N. T., Craig, J. N., and Neuhauser, E. B. D. (1971). "Systemic Bone Disease Developing in Small Premature Infants," *Pediatrics*, **48,** 883–895.

Grover, W. D. and Henkin, R. I. (1978). "Trichopoliodystrophy (TPD)—Menkes' Steely Hair Disease: New Aspects of Pathology and Treatment." In M. Kirchgessner, Ed., *Proceedings of the Third International Symposium on Trace Element Metabolism in Man and Animals* (in press).

Grover, W. D. and Scrutton, M. C. (1975). "Copper Infusion Therapy in Trichopolidystrophy," *J. Pediatr.*, **86,** 216–220.

Grover, W. D., Johnson, W. C., and Henkin, R. I. (in press). "Clinical and Biochemical Aspects of Trichopoliodystrophy."

Gubler, C. J., Lahey, M. E., Cartwright, G. E., and Wintrobe, M. M. (1952a). "Studies on Copper Metabolism. X: Factors Influencing the Plasma Copper Level of the Albino Rat," *Am. J. Physiol.*, **171**, 652–658.

Gubler, C. J., Lahey, M. E., Chase, M. S., Cartwright, B. E., and Wintrobe, M. M. (1952b). "Studies on Copper Metabolism. III: The Metabolism of Iron in Copper Deficient Swine," *Blood,* **7**, 1075–1092.

Gubler, C. J., Lahey, M. E., Cartwright, G. E., and Wintrobe, M. M. (1953). "Studies on Copper Metabolism. IX: The Transportation of Copper in Blood," *J. Clin. Invest.*, **32**, 405–414.

Hallman, P. S., Perrin, D. D., and Watt, A. E. (1971). "The Computed Distribution of Copper(II) and Zinc(II) Ions among Seventeen Amino Acids Present in Human Blood Plasma," *Biochem. J.*, **121**, 549–555.

Halsted, J. A., Hackley, B., and Smith, J. C. (1968). "Plasma-Zinc and Copper in Pregnancy and after Oral Contraceptives," *Lancet,* **2**, 278–279.

Hangarter, W. (1974). *Die Salicysaure und Ihre Abkommlinge—Ursprung, Wirkung und Anwendung in der Medizine.* F. K. Schattauer, Verlag, New York, pp. 312–316, 326–330.

Hangarter, W. and Lubke, A. (1952). "Über die Behandlung rheumatischer Erkuankungen mit einer Kupfer-natrium-salicylat-komplerverbindung (Permalon)," *Dtsch. Med. Wochenschr.*, **77**, 870–872.

Harman, D. (1965). "The Free Radical Theory of Aging: Effect of Age on Serum Copper Levels," *J. Gerontol.*, **20**, 151–153.

Harris, E. D. (1976). "Copper-Induced Activation of Aortic Lysyl Oxidase *in vivo*," *Proc. Natl. Acad. Sci.*, **73**, 371–374.

Hart, E. B., Steenbock, H., Waddell, J., and Elvehjem, C. A. (1928). "Iron in Nutrition; Copper as Supplement to Iron for Hemoglobin Building in Rat," *J. Biol. Chem.*, **77**, 797–812.

Heilmeyer, L. and Stuwe, G. (1938). "Der Eisen-kupferantagonismus im Blutplasma beim Infektionsgeschehen," *Klin. Wochenschr.*, **17**, 925–926.

Heilmeyer, L., Keiderling, W., and Stuwe, G. (1941). *Kupfer und Eisen als Korpereigne Wirkstoffe und Ihre Bedeutung beim Krankheitsgeschehen.* Verlag, Jena, pp. 59.

Henkin, R. I., Schulman, J. D., Schulman, C. B., and Bronzert, D. A. (1973). "Changes in Total, Nondiffusible, and Diffusible Plasma Zinc and Copper during Infancy," *J. Pediatr.*, **82**, 831–837.

Heydorn, K., Damsgaard, E., Horn, N., Mikkelsen, M., Tygstrup, I., Vestermark, S., and Weber, J. (1975). "Extra-Hepatic Storage of Copper—Male Fetus Suspected of Menkes' Disease," *Human-genetik.*, **29**, 171–175.

Hiroi, M., Stevens, K., and Gorski, R. (1969). "Effects of Intravenous Cupric Sulfate on Hypophysial Activity in the Female Rat," *J. Reprod. Fertil.*, **18**, 439–444.

Horn, N. (1976). "Copper Incorporation Studies on Cultured Cells for Prenatal Diagnosis of Menkes' Disease," *Lancet,* May 29, pp. 1156–1158.

Howell, J. McC. and Davidson, A. N. (1959). "The Copper Content and Cytochrome Oxidase Activity of Tissues from Normal and Swayback Lambs," *Biochem. J.*, **72**, 365–368.

Hunt, D. M. (1974). "Primary Defect in Copper Transport Underlies Mottled Mutants in the Mouse," *Nature,* **249,** 852–854.

Hunt, D. M. (1976). "A Study of Copper Treatment and Tissue Copper Levels in the Murine Congenital Copper Deficiency, Mottled," *Life Sci.,* **19,** 1913–1920.

Itoh, S., Inuzuka, K., and Suzuki, T. (1970). "New Antibiotics Produced by Bacteria Grown on *n*-Paraffin (Mixture of C_{12}, C_{13}, and C_{14} Fractions)," *J. Antibiot.,* **23,** 542–545.

Jackson, G. E., May, P. M., and Williams, D. R. (1978). "Metal-Ligand Complexes Involved in Rheumatoid Arthritis. I: Justifications for Copper Administration," *J. Inorg. Nucl. Chem.,* **40,** 1189–1194.

Jackson, G. E., May, P. M., and Williams, D. R. (1978a). "Metal-Ligand Complexes Involved in Rheumatoid Arthritis. VI: Computer Models Simulating the Low Molecular Weight Complexes Present in Blood Plasma for Normal and Arthritic Individuals," *J. Inorg. Nucl. Chem.,* **40,** 1227–1234.

Johnson, W. S. (1966). "Chelating Properties of Some *N-N′*-Disubstituted Thioureas Having Antitubercular Activity." Thesis, University of Kentucky Press, Lexington.

Jordon, F. T. W. and Nassar, T. J. (1971). "The Influence of Copper on the Survival of Infectious Bronchitis Vaccine Virus in Water," *Vet. Res.,* **89,** 609–610.

Josephs, H. W. (1931). "Treatment of Anemia in Infants with Iron and Copper," *Bull. Hopkins Hosp.,* **49,** 246–258.

Kamamoto, Y., Makiura, S., Sugihara, S., Hiosa, Y., Arai, M., and Ito, N. (1973). "The Inhibitory Effect of Copper on DL-Ethionine Carcinogenesis," *Cancer Res.,* **33,** 1129–1135.

Karabelas, D. S. (1972). *Copper Metabolism in the Adjuvant Induced Arthritic Rat.* University Microfilms Limited, Ann Arbor, Mich.

Karpel, J. T. and Peden, V. H. (1972). "Copper Deficiency in Long-Term Parenteral Nutrition," *J. Pediatr.,* **80,** 32–36.

Kato, M., Jonassen, H. B., and Fanning, J. C. (1964). "Copper(II) Complexes with Subnormal Magnetic Moments," *Chem. Rev.,* **64,** 99–128.

Kelly, W. A., Kesterson, J. W., and Carlton, W. W. (1974). "Myocardial Lesions in the Offspring of Female Rats Fed a Copper Deficient Diet," *Exp. Mol. Pathol.,* **20,** 40–56.

Koskelo, P., Kekki, M., Virkkunen, M., Lassus, A., and Somer, R. (1966). "Serum Ceruloplasmin Concentrations in Rheumatoid Arthritis, Ankylosing Spondylitis, Psoriasis, and Sarcoidosis," *Acta Rheum. Scand.,* **12,** 261–266.

Koskelo, P., Kekki, M., Nikkila, E. A., and Virkkunen, M. (1967). "Turnover of ^{131}I-Labeled Ceruloplasmin in Rheumatoid Arthritis," *Scand. J. Clin. Lab. Invest.,* **19,** 259–262.

Krivis, A. F. and Rabb, J. M. (1969). "Cuprous Complexes Formed with Isonicotinic Acid Hydrazide," *Science,* 164, 1064–1065.

Kuzell, W. C. and Gardner, G. M. (1950). "ACTH, Pregnenolone, Glutathione, and Gonadotropins in Experimental Polyarthritis," *Stanford Med. Bull.,* **8,** 83–88.

Kuzell, W. C., Shaffarzick, R. W., Mankle, E. A., and Gardner, G. M. (1951). "Copper

Treatment of Experimental and Clinical Arthritis," *Ann. Rheum. Dis.*, **10,** 328–336.

Laroche, M. J. (1969). *Nouveaux Medicaments Anti-inflammatoires.* Societe d'Exploration des Laboratories BOTTU, S. A. Bull. Off. Propr. Ind. 3. French Patent 6518M.

Lee, G. R., Nach, S., Lukens, J. N., and Cartwright, G. E. (1968). "Iron Metabolism in Copper-Deficient Swine," *J. Clin. Invest.*, **47,** 2058–2069.

Lee, G. R., Williams, D. M., and Cartwright, G. E. (1976). "Role of Copper in Iron Metabolism and Heme Biosynthesis." In A. S. Prasad and D. Oberleas, Eds., *Trace Elements in Human Health and Disease,* Vol. 1: *Zinc and Copper.* Academic Press, New York, pp. 373–390.

Lee, R. E. and Lands, W. E. M. (1972). "Cofactors in Biosynthesis of Prostaglandins F1-alpha and F2-alpha," *Biochem. Biophys. Acta,* **260,** 203–211.

Levinson, W., Faras, A., Woodson, B., Jackson, J., and Bishop, J. M. (1973). "Inhibition of RNA-Dependent DNA Polymerase of Rous Sarcoma Virus by Thiosemicarbazones and Several Cations," *Proc. Natl. Acad. Sci.,* **70,** 164–168.

Levinson, W., Mikelens, P., Jackson, J., and Kaska, W. (1978). "Anti-tumor Virus Activity of Copper-Binding Drugs." In G. N. Schrauzer, Ed., *Inorganic and Nutritional Aspects of Cancer.* Plenum Press, New York, pp. 161–178.

Levinson, W., Rohde, W., Mikelens, P., Jackson, J., Antony, A., and Ramakrishnan, T. (in press). "Inactivation and Inhibition of Rous Sarcoma Virus by Copper Binding Ligands: Thiosemicarbazones, 8-Hydroxyquinolines, and Isonicotinic Acid Hydrazide."

Lorber, A. (1969). "Communication in Response to Sternlieb et al., regarding Nonceruloplasmin Copper in Rheumatoid Arthritis," *Arthritis Rheum.,* **12,** 458–460.

Lorber, A., Cutler, L. S., and Chang, C. C. (1968). "Serum Copper Levels in Rheumatoid Arthritis: Relationship of Elevated Copper to Protein Alterations," *Arthritis Rheum.,* **11,** 65–71.

Lott, I. T., DiPaolo, R., Schwartz, D., Janowska, S., and Kanfer, J. N. (1975). "Copper Metabolism in the Steely-Hair Syndrome," *N. Engl. J. Med.,* **292,** 197–199.

Lou, H. C., Holmer, G. K., Reske-Nielsen, E., and Vagn-Hansen, P. (1974). "Lipid Composition in Gray and White Matter of the Brain in Menkes' Disease," *J. Neurochem.,* **22,** 377–381.

Lund-Olesen, K. and Menander, K. B. (1974). "Orgotein-New Anti-inflammatory Metalloprotein Drug—Preliminary Evaluation of Clinical Efficacy and Safety in Degenerative Joint Disease," *Curr. Ther. Res.,* **16,** 706–717.

McAdoo, M. H., Dannenberg, A. M., Jr., Hayes, C. J., James, S. P., and Sanner, J. H. (1973). "Inhibition of Cathepsin D-Type Proteinase of Macrophages by Pepstatin, a Specific Pepsin Inhibitor, and Other Substances," *Infect. Immunol.,* **7,** 655–665.

McCord, J. M. (1974). "Free Radicals and Inflammation: Protection of Synovial Fluid by Superoxide Dismutase," *Science,* **185,** 529–531.

Maddox, I. S. (1973). "Role of Copper in Prostaglandin Synthesis," *Biochem. Biophys. Acta,* **306,** 74–81.

Mallekaer, A. M. (1974). "Kinky Hair Syndrome," *Acta Paediatr. Scand.,* **63,** 289–296.

Manojlovic-Muir, L. (1967). "Copper(II) Aspirinate," *Chem. Commun.,* 1057–1058.

Markowitz, H., Gubler, C. J., Mahoney, J. P., Cartwright, G. E., and Wintrobe, M. M. (1955). "Studies on Copper Metabolism. XIV: Copper, Ceruloplasmin and Oxidase Activity in Sera of Normal Human Subjects, Pregnant Women, and Patients with Infection, Hepatolenticular Degeneration and the Nephrotic Syndrome," *J. Clin. Invest.,* **34,** 1498–1508.

Marston, H. R., Allen, S. H., and Swaby, S. L. (1971). "Iron Metabolism in Copper Deficient Rats," *Br. J. Nutr.,* **25,** 15–30.

May, P. M. and Williams, D. R. (1977). "Computer Simulation of Chelation Therapy: Plasma Mobilizing Index as a Replacement for Effective Stability Constant," *FEBS Lett.,* **78,** 134–138.

May, P. M., Linder, P. W., and Williams, D. R. (1976). "Ambivalent Effect of Protein Binding on Computed Distributions of Metal Ions Complexed by Ligands in Blood Plasma," *Experientia,* **32,** 1492–1493.

May, P. M., Linder, P. W., andWilliams, D. R. (1977). "Computer Simulation of Metal-Ion Equilibria in Biofluids: Models for the Low-Molecular-Weight Complex Distribution of Calcium(II), Magnesium(II), Manganese(II), Iron(III), Copper(II), Zinc(II), and Lead(II) Ions in Blood Plasma," *J. Chem. Soc., Dalton Trans.,* pp. 588–595.

Menkes, J. H., Alter, M., Steigleder, G. K., Weakley, D. R., and Sung, J. H. (1962). "A Sex-Linked Recessive Disorder with Retardation of Growth, Peculiar Hair, and Focal Cerebral and Cerebellar Degeneration," *Pediatrics,* **29,** 764–779.

Menkes, J. H. (1972). "Kinky Hair Disease," *Pediatrics,* **50,** 181–182.

Micheloni, M., May, P. M., and Williams, D. R. (1978). "Metal-Ligand Complexes Involved in Rheumatoid Arthritis. IV: Formation Constants and Species Distribution Considerations for Copper(II)-Cystinate, -Oxidised Penicillaminate and -Oxidised Glutathionate Interactions and Considerations of the Action of Penicillamine In Vivo," *J. Inorg. Nucl. Chem.,* **40,** 1209–1219.

Michez, B. and Ortegal, A. (1945). "Propos du Traitment dans les Polyarthritis Chroniques Evolutives par un Sel de Cuvire," *Brux. Med.,* cited by Forestier et al. (1948).

Mikelens, P. E., Woodson, B. A., and Levinson, W. E. (1976). "Association of Nucleic Acids with Complexes of *N*-Methylisatin-β-thiosemicarbazone and Copper," *Biochem. Pharmacol.,* **25,** 821–827.

Mills, C. F. and Williams, R. B. (1962). "Copper Concentration and Cytochrome Oxidase and Ribonuclease Activities in Brains of Copper-Deficient Lambs," *Biochem. J.,* **85,** 629–632.

Neumann, P. S. and Sass-Kortsak, A. (1967). "The State of Copper in Human Serum: Evidence for an Amino Acid Bound Fraction," *J. Clin. Invest.,* **46,** 646–658.

Niedermeier, W. (1965). "Concentration and Chemical State of Copper in Synovial Fluid and Blood Serum of Patients with Rheumatoid Arthritis," *Ann. Rheum. Dis.*, **24,** 544–548.

Niedermeier, W. and Griggs, J. H. (1971). "Trace Metal Composition of Synovial Fluid and Blood Serum of Patients with Rheumatoid Arthritis," *J. Chronic Dis.*, **23,** 537–536.

Niedermeier, W., Creitz, E. E., and Holley, H. L. (1962). "Trace Metal Composition of Synovial Fluid from Patients with Rheumatoid Arthritis," *Arthritis Rheum.*, **5,** 439–444.

O'Brien, J. S. and Sampson, E. L. (1966). "Kinky Hair Disease. II: Biochemical Studies," *J. Neuropathol. Exp. Neurol.*, **25,** 523–530.

O'Dell, B. L. (1976). "Biochemistry and Physiology of Copper in Vertebrates." In A. S. Prasad and D. Oberleas, Eds., *Trace Elements in Human Health and Disease,* Vol. 1: *Zinc and Copper,* Academic Press, New York, pp. 391–413.

O'Reilly, T. J. (1955). "Treatment of Rheumatoid Arthritis with Organic Copper Compounds," *Br. Med. J.*, **1,** 150.

Osaka, K., Sata, N., Matsumoto, S., Ogino, H., Kodama, S., Yokoyama, S., and Sugiyama, T. (1977). "Congenital Hypocupremia Syndrome with and without Steely Hair: Report of Two Japanese Infants," *Dev. Med. Child. Neurol.*, **19,** 62–68.

Oster, G. and Salgo, M. P. (1975). "The Copper Intrauterine Device and Its Mode of Action," *N. Engl. J. Med.*, **293,** 432–438.

Oster, G. and Salgo, M. P. (1977). "Copper in Mammalian Reproduction." In S. Garattini, A. Goldin, F. Hawking, and I. J. Kopin, Eds., *Advances in Pharmacology and Chemotherapeutics,* Vol. 14. Academic Press, New York, pp. 327–409.

Osterberg, R. (1974). "Models for Copper-Protein Interaction Based upon Solution and Crystal Structure Studies," *Coord. Chem. Rev.*, **12,** 309–347.

Osterberg, R., Sjoberg, B., and Branegard, B. (1975). *A Critical Review of Copper in Medicine.* Final Report, INCRA Project 234. International Copper Research Association, New York, pp. 1–70.

O'Sullivan, D. and Sadler, P. (1961). "Agents with High Activity against Type Z Poliovirus," *Nature,* **192,** 341–343.

Paschen, E. and Weser, U. (1975). "Problems concerning the Biological Action of Superoxide Dismutase (Erythrocuprein)," *Hoppe-Seylers Z. Physiol. Chem.*, **356,** 727–737.

Pederson, T. C. and Aust, S. D. (1973). "The Role of Superoxide and Singlet Oxygen in Lipid Peroxidation Promoted by Xanthine Oxidase," *Biochem. Biophy. Res. Commun.*, **52,** 1071–1078.

Pekarek, R. S. and Beisel, W. R. (1972). "Redistribution and Sequestering of Essential Trace Elements during Acute Infection." In *Proceedings of the 9th International Congress on Nutrition,* Vol. 2. Karger, Basel, pp. 193–198.

Perrin, D. D. (1965). "Multiple Equilibria in Assemblages of Metal Ions and Complexing Species: a Model for Biological Systems," *Nature,* **206,** 170–171.

Perrin, D. D. (1969). "Distribution of Metal Ions of Mixture of Complexing Agents—a Model for Biological Systems," *Suomen Kem.*, **42,** 205.

Perrin, D. D. and Agarwal, R. P. (1973). "Multimetal-Multiligand Equilibriums. Model for Biological Systems." In H. Sigel, Ed., *Metal Ions in Biological Systems,* Vol. 2. Marcel Dekker, New York, pp. 167–206.

Petering, D. H. (1978). "Reaction of Copper Complexes with Erlich Cells." In G. N. Schrauzer, Ed., *Inorganic and Nutritional Aspects of Cancer.* Plenum Press, New York, pp. 179–191.

Petering, D. H. and Petering, H. G. (1975). "Metal Chelates of 3-Ethoxy-2-oxobutyraldehyde Bis(thiosemicarbazone) H_2KTS." In A. C. Sartorelli and D. G. Johns, Eds., *Antineoplastic and Immunosuppressive Agents,* Vol. II. Springer-Verlag, New York, pp. 841–849.

Petering, H. G. and Van Giessen, G. J. (1966). "The Essential Role of Cupric Ion in the Biological Activity of 3-Ethoxy-2-oxobutyraldehyde Bisthiosemicarbazone, a New Antitumor Agent." In J. Peisach, P. Aisen, and W. E. Plumberg, Eds., *The Biochemistry of Copper.* Academic Press, New York, pp. 197–209.

Peterson, E. K., Gillespie, D. C., and Cook, F. E. (1966). "A Wide-Spectrum Antibiotic Produced by a Species of *Sporgangium,*" *Can. J. Microbiol.,* **12,** 221–230.

Pietsch, P. (1975). "Phleomycin and Bleomycin." In A. C. Sartorelli and D. G. Johns, Eds., *Antineoplastic and Immunosuppressive Agents,* Vol. II. Springer-Verlag, New York, pp. 850–876.

Plantin, L. O. and Strandberg, P. O. (1965). "Whole Blood Concentrations of Copper and Zinc in Rheumatoid Arthritis Studied by Activation Analysis," *Acta Rheum. Scand.,* **11,** 30–34.

Prins, H. W. and Van den Hamer, C. J. A. (1978). "Copper Content of the Liver of Young Mice in Relation to a Defect in the Copper Metabolism." In M. Kirchagessner, D. A. Roth-Maier, H.-P. Roth, F. J. Schwarz, and E. Weigand, Eds., *Proceedings of the 3rd International Symposium on Trace Element Metabolism in Man and Animals,* Technische Universitaet Munchen, Freising-Weihenstephan, F. R. Germany, pp. 397–400.

Prohaska, J. R. and Wells, W. W. (1974). "Copper Deficiency in the Developing Rat Brain: A Possible Model for Menkes' Steely-Hair Disease," *J. Neurochem.,* **23,** 91–98.

Rainsford, K. D. and Whitehouse, M. W. (1976a). "Gastric Mucous Effusion Elicited by Oral Copper Compounds—Potential Antiulcer Activity," *Experientia,* **32,** 1172–1173.

Rainsford, K. D. and Whitehouse, M. W. (1976b). "Concerning the Merits of Copper Aspirin as a Potential Antiinflammatory Drug," *J. Pharm. Pharmacol.,* **28,** 83–86.

Rasul, A. R. and Howell, J. McC. (1973). "Further Observations on the Response of the Peripheral and Central Nervous System of the Rabbit to Sodium Diethyldithiocarbamate," *Acta Neuropathol.,* **24,** 161–173.

Reiber, M. and Bemski, G. (1969). "On the Mycrobactericidal Effect of the Interaction Isoniazid-Copper," *Arch. Biochem. Biophys.,* **131,** 655–658.

Reske-Nielsen, E., Lou, H. C., Anderson, P., and Vagn-Hansen, P. (1973). "Brain-Copper Concentrations in Menkes' Disease," *Lancet,* **1,** 613.

Reske-Nielsen, E., Lou, H. C., and Vagan-Hansen, P. L. (1974). "Menkes' Disease—A Hypothesis with Recommendations for Future Investigations," *Acta Neuropathol.,* **28,** 361–363.

Roth, W., Zuber, F., Sorkin, E., and Erlenmeyer, H. (1951). "Uber die Eigenschaften einiger Metallkomplexe der *p*-Aminosalicylsaure," *Helv. Chim. Acta,* **43,** 430–433.

Salgo, M. P. and Oster, G. (1973). "Copper Stimulation and Inhibition of the Rat Uterus," *Fertil. Steril.,* **25,** 113–120.

Sartorelli, A. and Creasy, W. (1969). "Cancer Chemotherapy," *Ann. Rev. Pharmacol.,* **9,** 51–72.

Savitskii, I. V. (1970). "Complex Action of Some Heavy Metal Compounds and High Temperature on the Development of Experimental Tumors in Rats," *Gig. Tr.,* 42–49 (*Chem. Abstr.,* **74,** 51582b).

Schmid, K. and MacNair, M. B. (1956). "Characterization of the Proteins of Human Synovial Fluid in Certain Disease States," *J. Clin. Invest.,* **35,** 814–824.

Schubert, J. (1960). "The Relationship of Metal-Binding to Salicylate Action." In M. J. Seven and L. A. Johnson, Eds., *Metal-Binding in Medicine.* J. B. Lippincott, Philadelphia, pp. 325–328, 348.

Schubert, J. (1966). "Chelation in Medicine," *Sci. Am.,* **214,** 40–50.

Seelenfreund, M. H., Gartner, S., and Vingar, P. F. (1968). "The Ocular Pathology of Menkes' Disease (Kinky Hair Disease)," *Arch. Ophthalmol.,* **80,** 718–720.

Seely, J. R., Humphery, G. B., and Matter, B. J. (1972). "Copper Deficiency in a Premature Infant Fed an Iron-Fortified Formula," *N. Engl. J. Med.,* **286,** 109–110.

Sharples, G. R. (1946). "The Effects of Copper on Liver Tumor Induction by *p*-Dimethylaminoazobenzene," *Fed. Proc.,* **5,** 239–240.

Shirahata, K., Deguchi, T., Hayaski, T., Matsubara, I., and Suzuki, T. (1970). "The Structures of Fluoropsins C and F," *J. Antibiot.,* **23,** 546–550.

Singh, S. and Bresnan, M. J. (1973). "Menkes' Kinky-Hair Syndrome (Trichopoliodystrophy): Low Copper Levels in Blood, Hair and Urine," *Am. J. Dis. Child.,* **125,** 572–578.

Sorenson, J. R. J. (1974). "The Antiinflammation Activity of Some Copper Chelates." In D. D. Hemphill, Ed., *Trace Substances in Environmental Health*—VIII: *A Symposium.* University of Missouri, Columbia, pp. 305–311.

Sorenson, J. R. J. and DiTommaso, D. J. (1976). "The Significance of Increased Plasma and Serum Copper Concentrations in Rheumatoid Arthritis," *Ann. Rheum. Dos.,* **35,** 186–188.

Sorenson, J. R. J. (1976a). "Some Copper Chelates and Their Antiinflammatory and Antiulcer Activities," *Inflammation,* **1,** 317–331.

Sorenson, J. R. J. (1976b). "Copper Chelates as Possible Active Forms of the Anti-arthritic Agents," *J. Med. Chem.,* **19,** 135–148.

Sorenson, J. R. J. (1977). "The Evaluation of Copper Complexes as Potential Anti-arthritic Drugs," *J. Pharm. Pharmacol.,* **29,** 450–452.

Sorenson, J. R. J. (1978). "Copper Complexes, a Unique Class of Antiarthritic Drugs," *Prog. Med. Chem.,* **15,** 211–260.

Sorenson, J. R. J. and Hangarter, W. (1977). "Treatment of Rheumatoid and Degenerative Diseases with Copper Complexes: A Review with Emphasis on Copper-Salicylate," *Inflammation,* **2,** 217–238.

Sorkin, E., Roth, W., and Erlenmeyer, H. (1952). "Metal Ions and Biological Activity. VII: The Effect of Cupric Ion on Tuberculostasis," *Helv. Chim. Acta,* **35,** 1736–1741.

Spellacy, W. N., Hiser, B. J., and Birk, S. A. (1974). "The Effect of Copper Intrauterine Devices on Endocervical Gonococcal Cultures," *Fertil. Steril.,* **25,** 772–773.

Steinbrocker, O., Trager, C. H., and Batterman, R. C. (1949). "Therapeutic Criteria in Rheumatoid Arthritis," *J. Am. Med. Assoc.,* **140,** 659–662.

Sternlieb, I., Sandson, J. I., Morell, A. G., Karotkin, E., and Scheinberg, I. H. (1969). "Nonceruloplasmin Copper in Rheumatoid Arthritis," *Arthritis Rheum.,* **12,** 458–460.

Sturgeon, P. and Brubaker, C. (1956). "Copper Deficiency in Infants: A Syndrome Characterized by Hypocupremia, Iron Deficiency Anemia and Hypoproteinemia," *Am. J. Dis. Child.,* **92,** 254–265.

Sutter, M. D., Adjarian, R., and Haskell, A. R. (1958). "Observations on the Therapeutic Activity of Cuprous Iodide," *J. Am. Vet. Med. Assoc.,* **132,** 279–280.

Takamiya, K. (1960). "Anti-Tumor Activities of Copper Chelates," *Nature,* **185,** 190–191.

Tipton, I. H. and Cook, M. J. (1963). "Trace Elements in Human Tissue. II: Adult Subjects from the United States," *Health Phys.,* **1,** 103–145.

Townsend, S. F. (1961). "Regeneration of Gastric Mucosa in Rats," *Am. J. Anat.,* **109,** 133–147.

Townsend, S. F. and Sorenson, J. R. J. (1978). "Effect of Copper Aspirinate on Regeneration of Gastric Mucosa Following Surgical Lesion." In M. Kirchgessner, D. A. Roth-Maier, H.-P. Roth, F. J. Schwarz, and E. Weigand, Eds., *Proceedings of the 3rd International Symposium on Trace Element Metabolism in Man and Animals,* Technische Universitaet Munchen, Freising-Weihenstephan, F. R. Germany, pp. 370–372.

Tuchler, K. and Razenhofer, H. (1940). "Cuprion ein neuartiges Schwermetallsalz zur Behandlung der Tuberkulose," *Wien. Med. Wochenschr.,* **90,** 115–119.

Tyson, T. L., Holmes, H. H., and Ragan, C. (1950). "Copper Therapy of Rheumatoid Arthritis," *Am. J. Med. Sci.,* **220,** 418–420.

Ueno, T. (1956b). "Influence of Metal Ions on the *in vitro* Activity of Antituberculosis Agents and Its Relation to Metal Complex Formation. I: Influence of Metal Ions on the Activity of Antituberculosis Agents," *J. Pharm. Soc. Jap.,* **76,** 825–830 (*Chem. Abstr.,* **50,** 15712i).

Underwood, E. J. (1977). *Trace Elements in Human and Animal Nutrition,* 3rd ed. Academic Press, New York, pp. 57–115.

Vagn-Hansen, P., Reske-Nielsen, E., and Lou, H. C. (1973). "Menkes' Disease—a New Leucodystrophy: A Clinical and Neuropathological Review Together with a New Case," **25,** 103–119.

Van den Hamer, C. J. A. and Prins, H. W. (1978). "Results of ^{64}Cu-Loading Test Applied to Patients with Inherited Defect in Their Copper-Metabolism (Menkes' Disease)." In M. Kirchgessner, D. A. Roth-Maier, H.-P. Roth, F. J. Schwarz, and E. Weigand, Eds., *Proceedings International Symposium on Trace Element Metabolism in Man and Animals,* Technische Universitaet Munchen, Freising-Weihenstephan, F. R. Germany, pp. 394–396.

Van Ravesteijn, A. H. (1945). "Over de Koperstofwisseling by den Mensch en over het Kopergehalte van het Bloed by normale n zieke Personen." M.D. thesis, P. Den Boer, Utrecht.

Vargaftig, B. B., Tranier, Y., and Chignard, M. (1975). "Blockade by Metal Complexing Agents and by Catalase of Effects of Archidonic Acid on Platelets—Relevance to Study of Antiinflammatory Mechanisms," *Eur. J. Pharmacol.,* **33,** 19–29.

Vikbladh, I. (1950). "Studies on Zinc in Blood—I," *Scand. J. Clin. Lab. Invest.,* **2,** 143–148.

Vikbladh, I. (1951). "Studies on Zinc in Blood—II," *Scand. J. Clin. Lab. Invest.,* Suppl. 2, **3,** 1–74.

Volpintesta, E. J. (1974). "Menkes' Kinky Hair Syndrome in a Black Infant," *Am. J. Dis. Child.,* **138,** 244–246.

Voyatzakis, V. A. E., Vasilikiotis, G. S., Karageorgiou, G., and Kassapoglou, I. (1968). "Influence of Metallic Ions on the Antituberculosis Activity of Isonicotinoyl Hydrazones," *J. Pharm. Sci.,* **57,** 1255–1257.

Vykydal, M., Klabusay, L., and Truavsky, K. (1956). "Zum Mechanismus der Wirkung von Schwermetallen auf die experimentelle Arthritis," *Arzneim. Forsch.,* **6,** 568–569.

Walker, W. R. and Griffin, B. J. (1976). "Solubility of Copper in Human Sweat," *Search,* **7,** 100–101.

Walker, W. R. and Keats, D. M. (1976). "Investigation of Therapeutic Value of Copper Bracelet-Dermal Assimilation of Copper in Arthritic-Rheumatoid Conditions," *Agents Actions,* **6,** 454–459.

Walker, W. R. and Reeves, R. (1977). "Salicylates and Copper: Ternary Copper(II) Complexes Containing Salicylate and Nitrogenous Chelates Such as Histamine." In K. D. Rainsford, K. Brune, and M. W. Whitehouse, Eds., *Aspirin Related Drugs: Their Actions and Uses.* Birkhauser Verlag, Basel, pp. 109–117.

Walker, W. R., Reeves, R., and Kay, D. J. (1975). "Role of Cu^{2+} and Zn^{2+} in Physiological-Activity of Histamine in Mice," *Search,* **6,** 134–135.

Walker, W. R., Shaw, Y. H. L., and Li, N. C. (1973). "Nature of Copper(II) Interaction with Thyroxine Analogs," *J. Chem. Soc.,* **95,** 3015–3017.

Walker, W. R., Reeves, R. R., Brosnan, M., and Coleman, G. D. (1977). "Perfusion of Intact Skin by a Saline Solution of Bis(glycenate)copper(II)," *Bioinorg. Chem.,* **7,** 271–276.

Walker-Smith, J. A., Turner, B., Bloomfield, J., and Wise, G. (1973). "Therapeutic Implications of Copper Deficiency in Menkes' Steely-Hair Syndrome," *Arch. Dis. Child.,* **48,** 958–962.

Wesenberg, R. L., Gwinn, J. S., and Barnes, G. R., Jr. (1969). "Radiological Findings in the Kinky-Hair Syndrome," *Radiology,* **92,** 500–506.

Weser, U., Richter, C., Wendel, A., and Younes, M. (1978). "Reactivity of Antiinflammatory and Superoxide Dismutase Active Cu(II)-Salicylates," *Bioinorg. Chem.,* **8,** 201–213.

Whitehouse, M. W. (1976). "Ambivalent Role of Copper in Inflammatory Disorders," *Agents Actions,* **6,** 201–206.

Whitehouse, M. W. and Walker, W. R. (1977). Letter to the Editor: "The 'Copper Bracelet' for Arthritis," *Med. J. Aust.,* **1,** 938.

Whitehouse, M. W. and Walker, W. R. (1978). "Copper and Inflammation," *Agents Actions,* **8,** 85–95.

Whitehouse, M. W., Field, L., Denko, C. W., and Ryall, R. (1975). "Is Penicillamine a Precursor Drug?" *Scand. J. Rheumat.,* **4,** Suppl. 8 (Abstr. 183).

Wiesel, L. L. (1959). "Metal Chelation in the Mechanism of Action of Glucogenic Corticosteroids," *Metabolism,* **8,** 256–264.

Williams, D. A., Walz, D. T., and Foye, W. O. (1976). "Synthesis and Biological Evaluation of Tetrakis-μ-(acetysalicylato)-dicopper(II)," *J. Pharm. Sci.,* **65,** 126–128.

Williams, D. R. (1972). "Metals, Ligands and Cancer," *Chem. Rev.,* **72,** 203–213.

Winick, M. (1970). "Fetal Malnutrition," *Clin. Obstet. Gynecol.,* **13,** 526–541.

Wintrobe, M. M., Cartwright, G. E., and Gubler, C. J. (1953). "Studies of the Function and Metabolism of Copper," *J. Nutr.,* **50,** 395–419.

Younes, M. and Weser, U. (1977). "Superoxide Dismutase Activity of Copper-Penicillamine: Possible Involvement of Cu(I) Stabilized Sulfur Radical," *Biochem. Biophys. Res. Commun.,* **78,** 1247–1253.

Zackheim, H. S. and Wolf, P. (1972). "Serum Copper in Psoriasis and Other Dermatoses," *J. Invest. Dermatol.,* **58,** 28–32.

Zelkowitz, M., Cote, L., Miranda, A., Eastwood, A., Hays, A., and DiMauro, S. (1976). *Trichopoliodystrophy: Ultrastructure and Biochemical Studies in Muscle.* Child Neurology Society, Monterey, Calif.

Zipper, J., Medel, M., and Prager, R. (1969). "Supression of Fertility by Intrauterine Copper and Zinc in Rabbits: A New Approach to Intrauterine Contraception," *Am. J. Obstet. Gynecol.,* **105,** 529–534.

6

COPPER HOMEOSTASIS AND METABOLISM IN MAMMALIAN SYSTEMS

Gary W. Evans

U.S. Department of Agriculture, Science and Education Administration, Human Nutrition Laboratory, Grand Forks, North Dakota

1. FLUCTUATIONS IN COPPER BALANCE

The accompanying chapters of this book amply describe the deleterious effects that occur when human beings and other animals consume diets that contain suboptimal levels, as well as toxic levels, of copper. If the dietary copper contents of people and animals were constant and neither deficient nor excessive, studies on copper homeostasis would be purely academic. However, both human beings and animals are subjected to fluctuations in the copper content of the diet. Animals are often shifted back and forth from feedlot to forage land, and the contents of the diets differ between the two. Human beings, from the day of birth, constantly encounter fluctuations in dietary copper content. Infant diets that contain suboptimal levels of copper have been described (Cordano et al., 1964; Josephs, 1931). Moreover, evidence has been obtained which suggests that commonly consumed diets in the United States contain less than an adequate quantity of copper (Klevay, 1975; Wolf et al., 1977). At the other extreme, excess copper can be ingested when acidic foods are eaten after having been stored in metallic containers (Hopper and Adams, 1958; McMullen, 1971). Excess levels of copper have been ingested by people drinking whiskey sours from a copper shaker at a cocktail party (McMullen, 1971).

Stresses on the copper homeostatic mechanisms of the body are certainly not limited to dietary deficiencies or excesses. Deleterious effects have been associated with the copper in various types of prothesis (Barranco, 1972; Frykholm et al., 1969; Trachtenberg, 1972). Copper-containing fungicides have been linked to pulmonary problems in vineyard workers in France, Portugal, and Italy (Pimentel and Marques, 1969). Copper deficiency has been observed in patients receiving total parenteral alimentation (Dunlap et al., 1974; Karpel and Peden, 1972), and copper toxicity in patients on hemodialysis (Blomfield, 1969). Copper metabolism can also be affected by physiological changes that occur within the body. Marked alterations in copper balance accompany pregnancy, inflammation, disease, and ingestion of oral contraceptives (Evans, 1973).

With a few exceptions, humans beings and animals have evolved with the capability to cope with external and internal stresses on copper balance. Most mammalian creatures possess specific cellular mechanisms that can conserve copper during periods of want or excrete the metal if an excess of copper enters the body. If the periods of deficiency or excess are not long lasting, the mechanisms that regulate copper balance can successfully prevent the occurrence of severe abnormalities. This chapter describes our current knowledge of the mechanisms involved in the regulation of copper homeostasis.

2. COPPER-BINDING LIGANDS INVOLVED IN COPPER HOMEOSTASIS

Whether or not copper is bound by a specific endogenous ligand in the intestinal lumen and subsequently transported through the outer membrane of the intestinal epithelial cells remains unknown. Kirchgessner and Grassmann (1970) suggested that amino acids facilitate copper absorption from the diet. However, Marceau et al. (1970) observed no significant effect on copper absorption when amino acids were administered orally with copper.

The absorption of dietary copper may be regulated in part by a sulfhydryl-rich protein inside the intestinal epithelial cells. Starcher (1969) first identified a copper-binding protein in the intestinal mucosa from chicks. The intestinal protein described by Starcher had an apparent molecular weight of 10,000 daltons. Later, Evans et al. (1970) isolated a copper-binding protein from bovine intestine and suggested that the protein was similar to the metallothioneins, the class of proteins that contain approximately 30% cysteine but are devoid of aromatic amino acids (Pulido et al., 1966). Subsequently, Evans and LeBlanc (1976) isolated a copper-binding protein from rat intestine, and the results suggested that the intestinal protein was not a metallothionein. However, more recent experiments in our laboratory (Evans and Johnson, 1978) indicate that the protein described by Evans and LeBlanc was probably an artifact resulting from uncontrolled oxidation of copper-thionein. When the copper-binding protein from rat intestine was purified by the use of completely anaerobic conditions, the copper was associated with a cysteine-rich thionein.

In an attempt to delineate the role of copper-thionein in copper absorption, we have examined the effects of protein synthesis inhibitors on copper absorption and thionein synthesis (Evans and Johnson, 1978). When rats were given an oral dose of cycloheximide followed by ^{64}Cu and [^{14}C]lysine, the absorption of ^{64}Cu was significantly increased whereas the incorporation of [^{14}C]lysine into thionein was significantly decreased. In addition, the incorporation of [^{14}C]lysine into thionein was significantly greater in rats given an oral dose of 20 μg Cu than in rats given an oral dose of 1.0 μg Cu. These results suggest that copper-thionein may function in regulating the quantity of copper that passes from the mucosal cell into the blood.

After passing through the intestinal epithelial cells, copper is transported through the blood as a histidine-copper-albumin complex (Lau and Sarkar, 1971). Albumin from most species contains a specific first binding site for copper. This specific binding site consists of the α-amino nitrogen, the imidazole nitrogen from histidine in position 3, and two other peptide bond nitrogen atoms (Bradschaw et al., 1968; Dixon and Sarkar, 1974). Dog serum albumin does not contain the important histidine-3 residue (Dixon and Sarkar, 1974), and this species is extremely susceptible to copper toxicity (Goresky et al., 1968).

Hepatic cells have a key role in the metabolism of copper, and several copper-binding ligands are involved in the processes. When copper enters the hepatocytes, the metal is initially bound to a protein that has a molecular weight of approximately 10,000 daltons (Terao and Owen, 1973). The exact nature of this low molecular weight copper-binding protein in the liver has been the subject of controversy, but the majority of evidence indicates that the protein is a thionein, as suggested by this author in a previous article (Evans, 1973). Copper-thionein has been identified in the liver cytosol from neonatal rats (Evans et al., 1975), fetal calves (Hartmann and Weser, 1977), adult rats (Bremner and Young, 1976a), adult pigs (Bremner and Young, 1976b), and mature beef (Evans et al., 1970). In addition, copper-thionein has been isolated from the particulate fraction of neonatal calf liver (Porter, 1974).

Some controversy arose regarding the exact nature of the hepatic copper-binding protein after the reports of Winge et al. (1975), Riordan and Gower (1975), and Evans et al. (1975), each of whom described copper-binding proteins that differed from the metallothioneins. As pointed out by Bremner and Young (1976a) and Hartmann and Weser (1977), isolation of pure copper-thionein requires extreme care under anaerobic conditions. Thus incomplete purification and/or uncontrolled oxidation has undoubtedly given rise to erroneous interpretations of the results obtained from the isolation of copper-binding proteins.

The experiments of Premakumar et al. (1975) suggest that synthesis of the hepatic copper-binding protein can be induced by copper. When rats were injected with copper and [^{3}H]lysine, incorporation of the labeled amino acid into a copper-binding protein was markedly increased over that in control animals. In addition, actinomycin D prevented *de novo* synthesis of the copper-binding protein. These observations suggest that the hepatic copper level regulates the transcription of the RNA template for the copper-binding protein, thionein. Thus hepatic copper-thionein may function in detoxification as well as intracellular transport.

Within hepatic cells, copper that was initially bound to a copper-binding protein eventually appears in ceruloplasmin, copper-dependent enzymes, and bile components (Terao and Owen, 1973). The fraction of copper that is associated with ceruloplasmin is thought to be destined for distribution to the extrahepatic organs. Several years ago, Broman (1964) first suggested that ceruloplasmin is a copper transport protein. Later, Owen (1965) demonstrated that intravenously injected radioactive copper did not appear in extrahepatic tissues until after the emergence of [^{64}Cu]ceruloplasmin. After intravenous injection of [^{67}Cu]ceruloplasmin, Marceau and Aspin (1972) and Owen (1971) observed a decrease in plasma radioactivity commensurate with an increase in tissue radioactivity. In addition, Marceau and Aspin (1973) demonstrated that radioactive copper is incorporated into hepatic and brain cytochrome *c* oxidase following injection of ^{67}Cu-labeled plasma. More recently, Hsieh and Frieden (1975) completed a series of experiments which demonstrated that ceruloplasmin copper is utilized in the biosynthesis of cytochrome oxidase. Rats were fed a copper-deficient diet to deplete the cytochrome *c* oxidase and then injected intravenously with ceruloplasmin, copper-albumin, copper-histidine, or $CuCl_2$. Cytochrome *c* oxidase activity then was measured in several tissues. Of the copper complexes tested, ceruloplasmin was the most efficacious in restoring cytochrome oxidase activity, an observation which indicates that ceruloplasmin is a copper transport protein from which copper atoms are transferred to cytochrome *c* oxidase and possibly to other copper proteins.

Copper is secreted from the hepatic cells into the bile bound to very low molecular weight ligands (Evans and Cornatzer, 1971; Gollan, 1975; Lewis, 1973; Terao and Owen, 1973); but as the metal passes along the bile ducts, very high molecular weight complexes are formed (Gollan, 1975; Lewis, 1973; McCullars et al., 1977). Evans and Cornatzer (1971) have suggested that the low molecular weight fraction of biliary copper is made up of copper-amino acid complexes. Lewis (1973) has presented experimental evidence which suggests that copper complexes with bile acids, particularly tauro-chenodeoxycholic acid. Lewis has also suggested that the high molecular weight fraction of copper in bile is composed of micelles of bile acid-copper complexes. However, Gollan (1975) has demonstrated that the high molecular weight fraction of biliary copper is a macromolecule rather than a micelle. Moreover, McCullars et al. (1977) have recently presented evidence which suggests that the macromolecular moiety that binds copper in the bile is a conjugated bilirubin. These observations suggest that copper is initially secreted into bile bound to amino acids and/or bile acids. As the copper passes through the ducts, the metal becomes complexed with bilirubins, and in this form the copper is unavailable for reabsorption.

The foregoing discussion is summarized in the schematic drawing in Figure 1.

3. METABOLIC DISORDERS ARISING FROM INBORN ERRORS OF METABOLISM WITHIN THE COPPER HOMEOSTATIC SYSTEM

The deleterious effects that result from malfunctioning copper homeostatic systems can best be appreciated by a brief review of the symptoms that result from two inborn errors of metabolism that affect copper homeostasis.

3.1. Menkes' Steely Hair Syndrome (Trichopoliodystrophy)

In 1962 Menkes and his colleagues (1962) wrote a detailed description of a degenerative disease of the central nervous system. The condition was inherited as a sex-linked recessive trait, and the affected males were both physically and mentally retarded and had peculiar white, stubbly hair. Later, O'Brien and Sampson (1966), who first coined the name "kinky hair disease," examined the fatty acid levels in brain tissue from Menkes' patients and found a decreased quantity of docosahexanoic acid, a highly unsaturated fatty acid. At that time the significance of the peroxidation of lipids was not understood.

In the early 1970s David Danks and his colleagues in Australia began examining infants suffering from Menkes' kinky hair disease. Acting on the suggestion of Dr. J. M. Gillespie of the Division of Protein Chemistry, C.S.I.R.O., in Melbourne, Danks and associates (1972a) examined the copper status of these infants, and a new era in copper metabolism began. In a series of publications Danks et al. (1972a, 1972b, 1973b) demonstrated that infants with Menkes' disease absorb copper poorly and have a decreased concentration of the element

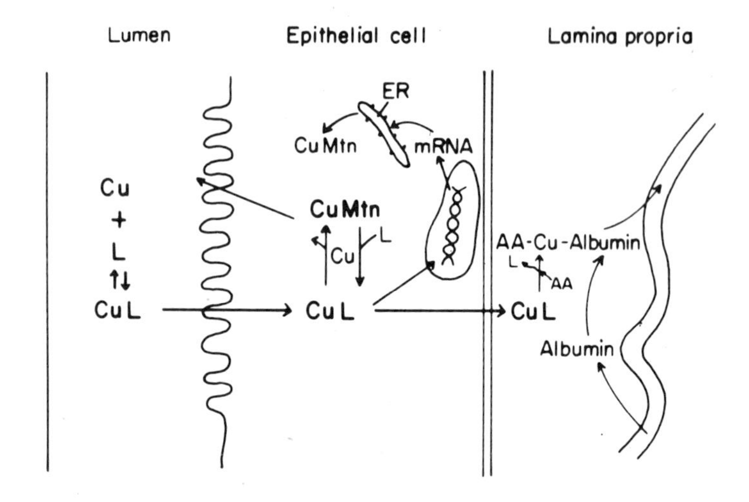

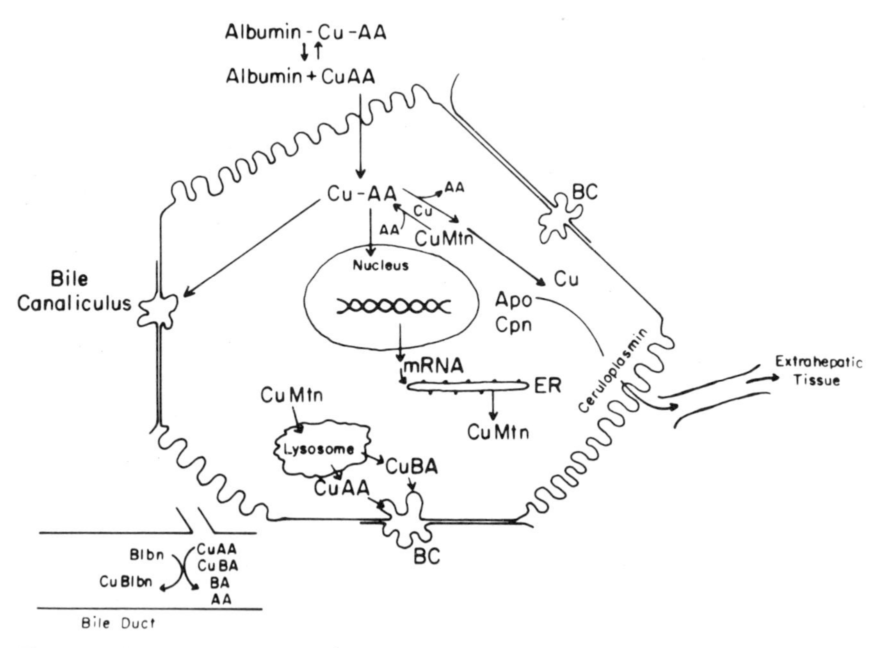

Figure 1. Hypothetical illustration of the functions of various copper-binding ligands involved in the maintenance of copper homeostasis. Abbreviations: CuL = copper-binding ligand; CuMtn = copper-thionein; mRNA = messenger ribonucleic acid; ER = endoplasmic reticulum; AA = amino acid; Apo Cpn = apoceruloplasmin; BA = bile acid; BC = bile canaliculus; Blbn = bilirubin.

in the plasma and liver but an elevated concentration of copper in the intestinal epithelium. These observations have been verified by Walker-Smith et al. (1973) and Dekaban et al. (1975). In addition, Goka et al. (1976) have demonstrated that cultured skin fibroblasts from patients with Menkes' disease contain an elevated concentration of copper.

With the exception of anemia, all of the pathological abnormalities associated with copper deficiency in experimental animals have been observed in patients with Menkes' syndrome (Danks, 1975). In fact, the similarity between wool from copper-deficient sheep and hair from Menkes' patients prompted Danks et al. (1973a) to suggest that the term "steely hair" be used to describe the hair of these affected infants.

Therapy (see also Sorenson, this volume)

Since oral copper is poorly absorbed by infants afflicted with Menkes' syndrome, parenteral copper administration has been used as a therapeutic measure, but the results have been disappointing (Danks, 1975). Although copper therapy restores ceruloplasmin and hepatic copper levels to normal, the symptoms of the disease usually are not alleviated. This may be so because most of the cases reported to date were at least 3 months old before they were identified and treatment was begun. At this late stage of development the tissue damage caused by the malabsorption syndrome may be so extensive that it is irreversible. Alternatively, the mutant gene involved in Menkes' syndrome may alter the accessibility of serum copper to tissue cells (Danks, 1975).

Until research establishes exactly how copper is taken up by tissue cells, the methods now being used for copper therapy in Menkes' syndrome infants will probably be uneffective. When methods are developed to circumvent the metabolic block of copper uptake by the cells, therapeutic programs will have to be carefully analyzed and designed because Menkes' syndrome is a complex anomaly. For example, some patients exhibit symptoms at birth (Danks et al., 1972a), whereas in others, clinical manifestations first appear months after birth (Walker-Smith et al., 1973). These observations suggest the possibility that placental copper transport may be affected, but in varying degrees, by the mutant gene that causes Menkes' disease. If the placenta is affected by the mutant gene, heterozygous carrier females will have to be identified and therapy begun to ensure adequate transport of copper to the developing fetus.

Genetic Heterogeneity

The apparent complexity of the steely hair syndrome may be due in part to genetic heterogeneity. Although in most affected infants examined to date, hepatic copper uptake is normal during infusion therapy, in some patients hepatic copper uptake is abnormal. Garnica and Fletcher (1975) observed an abnormally high rate of urinary copper excretion in one patient both during and after copper infusion therapy. Since excess copper is normally taken up by the liver and excreted through the bile, this observation suggests that hepatic copper uptake may be affected in some infants with steely hair syndrome. In addition, Horn et al. (1975) examined copper distribution in a male fetus suspected of this inborn error and discovered that the liver was the only tissue that contained less copper than specimens from control subjects. The copper concentrations of the kidneys, spleen, pancreas, and placenta of the diseased fetus were significantly higher than those in controls. Thus the gene mutation that occurs in the steely hair syndrome may affect the copper transport system of organs other than the intestine.

Animal Model for Menkes' Syndrome

Research on the steely hair syndrome has been greatly facilitated by the observations of Hunt (1974), who discovered that the sex-linked inherited disorder in mottled mutant (Mo^{br}) mice is nearly identical to the steely hair syndrome in human beings. Hunt found decreased concentrations of copper in livers and brains from Mo^{br} male mice, but the copper concentration of the intestines from the mutant mice was much higher that normal. Ceruloplasmin activity was significantly decreased in the Mo^{br} mice.

Experiments in our laboratory have confirmed the observations of Hunt and also demonstrate that hepatic copper uptake is impaired in Mo^{br} mice (Evans and Reis, 1978). In Mo^{br} mice suckling ^{64}Cu-labeled dams, 72% of the total radioactivity in the pups was in the intestinal cells; the liver contained only 3% of the radioactivity. In homozygous littermates, however, less than 25% of the radioactivity was found in the intestine, and the liver contained 50% of the radioactivity. These results demonstrate that copper accumulates in the intestine of Mo^{br} mice and suggest that the copper absorbed from the intestine of these mutants is diverted to extrahepatic tissues.

During our experimentation with Mo^{br} mice, we also examined copper metabolism in the Mo^{br} heterozygous females. The kidney copper concentration of heterozygous females was significantly higher than that of homozygous females, but there was no difference in liver and brain copper concentrations. In addition, we have observed a marked decrease in copper absorption in heterozygous females (Evans and Reis, 1978). The observations discussed above demonstrate that Mo^{br} mice will have an important role in elucidating the biochemical defect that arises from the lethal mutant gene present in infants affected with Menkes' steely hair disease.

3.2. Wilson's Disease (Hepatolenticular Degeneration)

The inborn error of copper metabolism known as Wilson's disease was first described in 1912 (Wilson, 1912). Since that time, volumes of literature have appeared describing various aspects of the disorder. To adequately cover the subject of Wilson's disease would require a chapter in itself. Therefore the comments in this section will be brief and directed toward informing the uninformed regarding the pathogenesis, pathology, and treatment. For more comprehensive discussions of Wilson's disease, the reader is referred to several excellent reviews (Bearn, 1972; Sass-Kortsak, 1965; Scheinberg and Sternlieb, 1975; Walshe, 1966).

The mutant gene in individuals with Wilson's disease produces a yet undiscovered biochemical defect that results in excess retention of hepatic copper. Most affected patients have a decreased concentration of plasma ceruloplasmin, and excretion of copper into the bile is impaired. Thus the defect apparently involves the mechanisms regulating the passage of copper into the ceruloplasmin-synthesizing and biliary excretion pathways.

After several years of accumulating copper, the capacity of the liver is eventually exceeded and copper begins diffusing into the plasma and fluids of extrahepatic tissues. At this point in the progression of the disease, necrosis can be detected in the hepatic cells and the first clinical symptoms begin to appear in the form of liver dysfunction.

If the untreated patient survives the liver disease, pathological changes resulting from excess copper eventually appear in the central nervous system, kidneys, and cornea. At this stage, neurologic and psychiatric symptoms may appear and renal function is impaired. If untreated, the manifestations of copper toxicity result in death at an early age.

Fortunately, the manifestations of Wilson's disease can be arrested and prevented by drug therapy. The drug now being used most successfully is D-penicillamine (β,β-dimethylcysteine), which produces a marked increase in the urinary excretion of copper. Treatment with penicillamine results in dramatic recovery of the affected patients, and continuous therapy expands the life spans of individuals who possess this mutant gene, which alters copper homeostasis in the liver.

4. CONCLUSION

The foregoing discussion of Menkes' steely hair syndrome and Wilson's disease illustrates the deleterious effects that result from malfunctioning copper homeostatic mechanisms. The discussion also illustrates our lack of knowledge regarding the basic biochemical mechanisms that regulate copper absorption and excretion. As described in previous paragraphs, several investigations have demonstrated that mammals absorb and excrete copper according to body requirements. In addition, some of the ligands involved in copper metabolism have been isolated. However, several aspects of copper metabolism remain to be explained; little is known about copper transport across cell membranes or about the exchange of copper between ligands. Until these processes are explained, our knowledge of copper homeostasis will not be complete.

REFERENCES

Barranco, V. P. (1972), "Eczematous Dermatitis Caused by Internal Exposure to Copper," *Arch. Dermatol.,* **106,** 386.

Bearn, A. G. (1972). "Wilson's Disease." in J. B. Stanbury, J. C. Wyngaarden, and D. S. Fredrickson, Eds., *The Metabolic Basis of Inherited Disease,* 3rd ed. McGraw-Hill, New York, p. 1033.

Blomfield, J. (1969). "Copper Contamination of Exchange Transfusion," *Lancet,* **1,** 731.

Bradshaw, R. A., Shearer, W. T., and Gurd, F. R. N. (1968). "Sites of Binding of Copper(II) Ion by Peptide (1–24) of Bovine Serum Albumin," *J. Biol. Chem.,* **243,** 3817.

Bremner, I. and Young, B. W. (1976a). "Isolation of (Copper,Zinc)-Thioneins from Pig Liver," *Biochem. J.,* **155,** 631.

Bremner, I. and Young, B. W. (1976b). "Isolation of (Copper,Zinc)-Thioneins from the Livers of Copper-Injected Rats," *Biochem. J.,* **157,** 517.

Broman, L. (1964). "Chromatographic and Magnetic Studies on Human Ceruloplasmin," *Acta Soc. Med. Upsalien,* **69,** Suppl. 7, 1.

Cordano, A., Baertl, J. M., and Graham, G. G. (1964). "Copper Deficiency in Infancy," *Pediatrics,* **34,** 324.

Danks, D. M. (1975). "Steely Hair, Mottled Mice, and Copper Metabolism," *N. Engl. J. Med.,* **293,** 1147.

Danks, D. M., Campbell, P. E., Stevens, B. J., Mayne, V., and Cartwright, E. (1972a). "Menkes's Kinky Hair Syndrome: An Inherited Defect in Copper Absorption with Widespread Effects," *Pediatrics,* **50,** 188.

Danks, D. M., Campbell, P. E., Walker-Smith, J., Stevens, B. J., Gillespie, J. M., Blomfield, J., and Turner, B. (1972b). "Menkes' Kinky-Hair Syndrome," *Lancet,* **1,** 1100.

Danks, D. M., Cartwright, E., and Stevens, B. (1973a). "Menkes' Steely-Hair (Kinky Hair) Disease," *Lancet,* **1,** 891.

Danks, D. M., Cartwright, E., Stevens, B. J., and Townley, R. R. W. (1973b). Menkes' Kinkey Hair Disease: Further Definition of the Defect in Cooper Transport," *Science,* **179,** 1140.

Dekaban, A. S., Aamodt, R., Rumble, W. F., Johnston, G. S., and O'Reilly, S. (1975). "Kinky Hair Disease: Study of Copper Metabolism with Use of ^{67}Cu," *Arch. Neurol.,* **32,** 672.

Dixon, J. W. and Sarkar, B. (1974). "Isolation, Amino Acid Sequence and Copper(II)-Binding Properties of Peptide (1-24) of Dog Serum Albumin," *J. Biol. Chem.,* **249,** 5872.

Dunlap, W. M., Jones, G. W., III, and Hume, D. M. (1974). "Anemia and Neutropenia Caused by Copper Deficiency," *Ann. Intern. Med.,* **80,** 470.

Evans, G. W. (1973). "Copper Homeostasis in the Mammalian System," *Physiol. Rev.,* **53,** 535.

Evans, G. W. and Cornatzer, W. E. (1971). "Biliary Copper Excretion in the Rat," *Proc. Soc. Exp. Biol. Med.,* **136,** 719.

Evans, G. W. and Johnson, P. E. (1978). "Copper and Zinc Binding Ligands in the Intestinal Mucosa." In M. Kirchgessner, Ed. *Trace Element Metabolism in Animals—3.* Institut fur Ernahrungsphysiologie, Freising-Weihenstephan, W. Germany, p. 98.

Evans, G. W. and LeBlanc, F. N. (1976). "Copper-Binding Protein in Rat Intestine: Amino Acid Composition and Function," *Nutr. Rep. Int.,* **14,** 281.

Evans, G. W. and Reis, B. L. (1978). "Impaired Copper Homeostasis in Neonatal Male and Adult Female Brindled (MoBr) Mice," *J. Nutr.,* **108,** 554.

Evans, G. W., Majors, P. F., and Cornatzer, W. E. (1970). "Mechanism for Cadmium and Zinc Antagonism of Copper Metabolism." *Biochem. Biophys. Res. Commun.,* **40,** 1142.

Evans, G. W., Wolentz, M. L., and Grace, C. I. (1975). "Copper-Binding Proteins in the Neonatal and Adult Rat Liver Soluble Fraction," *Nutr. Rep. Int.,* **12,** 261.

Frykholm, K. O., Frithiof, L., Fernstrom, A. I. B., Moberger, G., Blohm, S. G., and Bjorn, E. (1969). "Allergy to Copper Derived from Dental Alloys as a Possible Cause of Oral Lesions of Lichen Planus," *Acta Dermato-Venereol.,* **49,** 268.

Garnica, A. D. and Fletcher, S. R. (1975). "Parenteral Copper in Menkes' Kinkey-Hair Syndrome," *Lancet,* **2,** 659.

Goka, T. J., Stevenson, R. E., Hefferan, P. M., and Howell, R. R. (1976). "Menkes' Disease: A Biochemical Abnormality in Cultured Human Fibroblasts," *Proc. Natl. Acad. Sci.,* **73,** 604.

Gollan, J. L. (1975). "Studies on the Nature of Complexes Formed by Copper with Human Alimentary Secretions and Their Influence on Copper Absorption in the Rat," *Clin. Sci. Mol. Med.,* **49,** 237.

Goresky, C. A., Holmes, T. H., and Sass-Kortsak, A. (1968). "The Initial Uptake of Copper by the Liver in the Dog," *Can. J. Physiol. Pharmacol.,* **46,** 771.

Hartmann, H. and Weser, U. (1977). "Copper-Thionein from Fetal Bovine Liver," *Biochim. Biophys. Acta,* **491,** 211.

Hopper, S. H. and Adams, H. S. (1958). "Copper Poisoning from Vending Machines," *Public Health Rep.,* **73,** 910.

Horn, N., Mikkelsen, M., Heydorn, K., Damsgaard, E., and Tygstrup, I. (1975). "Copper and Steely Hair," *Lancet,* **1,** 1236.

Hsieh, H. S. and Frieden, E. (1975). "Evidence for Ceruloplasmin as a Copper Transport Protein," *Biochem. Biophsy. Res. Commun.,* **67,** 1326.

Hunt, D. M. (1974). "Primary Defect in Copper Transport Underlies Mottled Mutants in the Mouse," *Nature,* **249,** 852.

Josephs, H. W. (1931). "Treatment of Anemia in Infants with Iron and Copper," *Bull. Johns Hopkins Hosp.,* **49,** 246.

Karpel, J. T. and Peden, V. H. (1972). "Copper Deficiency in Long-Term Parenteral Nutrition," *J. Pediatr.,* **80,** 32.

Kirchgessner, M. and Grassmann, E. (1970). "The Dynamics of Copper Absorption." In C. F. Mills, Ed., *Trace Element Metabolism in Animals,* Livingstone, Edinburgh, pp. 277–287.

Klevay, L. M. (1975). "The Ratio of Zinc to Copper of Diets in the United States," *Nutr. Rep. Int.,* **11,** 237.

Lau, S. and Sarkar, B. (1971). "Ternary Coordination Complex between Human Serum Albumin, Copper(II), and L-Histidine," *J. Biol. Chem.*, **246**, 5938.

Lewis, K. O. (1973). "The Nature of the Copper Complexes in Bile and Their Relationship to the Absorption and Excretion of Copper in Normal Subjects and in Wilson's Disease," *Gut*, **14**, 221.

Marceau, N. and Aspin, N. (1972). "Distribution of Ceruloplasmin-Bound ^{67}Cu in the Rat," *Am. J. Physiol.*, **222**, 106.

Marceau, N. and Aspin, N. (1973). "The Intracellular Distribution of the Radiocopper Derived from Ceruloplasmin and from Albumin," *Biochim. Biophys. Acta*, **328**, 338.

Marceau, N., Aspin, N., and Sass-Kortsak, A. (1970). "Absorption of Copper-64 from Gastrointestinal Tract of the Rat," *Am. J. Physiol.*, **218**, 377.

McCullars, G. M., O'Reilly, S., and Brennan, M. (1977). "Pigment Binding of Copper in Human Bile," *Clin. Chim. Acta*, **74**, 33.

McMullen, W. (1971). "Copper Contamination of Soft Drinks from Bottle Pourers," *Health Bull. (Edinburgh)*, **29**, 94.

Menkes, J. H., Alter, J., Steigleder, G. K., Weakley, D. R., and Sung, J. H. (1962). "A Sex-Linked Recessive Disorder with Retardation of Growth, Peculiar Hair, and Focal Cerebral and Cerebellar Degeneration," *Pediatrics*, **29**, 764.

O'Brien, J. S. and Sampson, E. L. (1966). "Kinky Hair Disease. II: Biochemical Studies," *J. Neuropathol. Exp. Neurol.*, **25**, 523.

Owen, C. A., Jr. (1965). "Metabolism of Radiocopper (Cu^{64}) in the Rat," *Am. J. Physiol.*, **209**, 900.

Owen, C. A., Jr. (1971). "Metabolism of Copper 67 by the Copper-Deficient Rat," *Am. J. Physiol.*, **221**, 1722.

Pimentel, J. C. and Marques, F. (1969). "Vineyard Sprayer's Lung: A New Occupational Disease," *Thorax*, **24**, 678.

Porter, H. (1974). "The Particulate Half-Cystine-Rich Copper Protein of Newborn Liver: Relationship to Metallothionein and Subcellular Localization in Non-mitochondrial Particles Possibly Representing Heavy Lysosomes," *Biochem. Biophys. Res. Commun.*, **56**, 661.

Premakumar, R., Winge, D. R., Wiley, R. D., and Rajagopalan, K. V. (1975). "Copper-Induced Synthesis of Copper-Chelatin in Rat Liver," *Arch. Biochem. Biophys.*, **170**, 267.

Pulido, P., Kagi, J. H. R., and Vallee, B. L. (1966). "Isolation and Some Properties of Human Metallothionein," *Biochemistry*, **5**, 1768.

Riordan, J. R. and Gower, I. (1975). "Purification of Low Molecular Weight Copper Proteins from Copper Loaded Liver," *Biochem. Biophys. Res. Commun.*, **66**, 678.

Sass-Kortsak, A. (1965). "Copper Metabolism," *Adv. Clin. Chem.*, **8**, 1.

Scheinberg, I. H. and Sternlieb, I. (1975). "Wilson's Disease." In G. E. Gaull, Ed., *Biology of Brain Dysfunction*, Vol. 3. Plenum, New York, p. 247.

Starcher, B. C. (1969). "Studies on the Mechanism of Copper Absorption in the Chick," *J. Nutr.*, **97**, 321.

Terao, T. and Owen, C. A., Jr. (1973). "Nature of Copper Compounds in Liver Supernate and Bile of Rats: Studies with ^{67}Cu," *Am. J. Physiol.,* **224,** 682.

Trachtenberg, D. I. (1972). "Allergic Response to Copper—Its Possible Gingival Implications," *J. Periodontol.,* **43,** 705.

Walker-Smith, J. A., Turner, B., Blomfield, J., and Wise, G. (1973). "Therapeutic Implications of Copper Deficiency in Menkes's Steely-Hair Syndrome," *Arch. Dis. Child.,* **48,** 958.

Walshe, J. M. (1966). "Wilson's Disease, a Review." In J. Peisach, P. Aisen, and W. E. Blumberg, Eds., *Biochemistry of Copper.* Academic Press, New York, p. 475.

Wilson, S. A. K. (1912). "Progressive Lenticular Degeneration: A Familial Nervous Disease Associated with Cirrhosis of the Liver," *Brain,* **34,** 295.

Winge, D. R., Premakumar, R., Wiley, R. D., and Rajagopalan, K. V. (1975). "Copper-Chelatin: Purification and Properties of a Copper-Binding Protein from Rat Liver," *Arch. Biochem. Biophys.,* **170,** 253.

Wolf, W. R., Holden, J., and Greene, F. E. (1977). "Daily-Intake of Zinc and Copper from Self-Selected Diets," *Fed. Proc.,* **36,** 1175.

7

THE USE OF RADIOCOPPER TO TRACE COPPER METABOLIC TRANSFER AND UTILIZATION

Normand Marceau

Laboratory of Medical Biophysics, Laval University Hospital, Ste-Foy, Quebec, Canada

1. INTRODUCTION

Much of the work reported on the metabolic transport of copper using radiocopper has been done on rats. A search through the literature revealed that very few data have appeared for animals during the last few years. In human beings the recent interest in the use of radiocopper has been focused on measurements of total body copper retention by means of whole-body counting after an intravenous injection of ^{67}Cu. Such metabolic studies in human beings have been motivated by the reliablity of the method to discriminate between normal individuals and heterozygous and homozygous patients of Wilson's disease, the most important aberration of human copper metabolism.

Although this chapter aims at reviewing the work on the use of radioactive copper (^{64}Cu and/or ^{67}Cu) in order to follow, in qualitative and quantitative manner copper transport in the body, it is of interest to summarize first some of the related information on the utilization of copper in various body compartments, the end point in biological copper transport. From data obtained on various animal species, it has been established that only trace amounts accumulate in mammals. In fact, 70 mg Cu with a range of 50 to 120 mg is present in the normal human adult, the highest concentrations being found in the liver, kidneys, heart, and bone marrow (Sass-Kortsak, 1965; Underwood, 1962). Human blood contains 1 to 1.23 μg Cu/ml with similar concentrations in plasma and red blood cells (Sass-Kortsak, 1965). In rats around 1.25 μgCu/ml is found in whole blood with 1.44 μg/ml in plasma (Sass-Kortsak, 1965). Other animal species yield blood copper concentrations that differ greatly from those observed in human beings (Evans and Weideranders, 1967). The relevance of these data will be appreciated when the effect of a copper load on metabolic rate measurements is considered in a later section.

The distribution of copper within a given tissue varies considerably, depending on the type of cells present and the type of copper. This element is normally incorporated in proteins in the brain, liver, and erythrocytes, as well as in enzymes such as tyrosinase and cytochrome *c* oxidase (Marceau, 1971). In plasma, 95% of the total copper is present in ceruloplasmin, the rest being bound to albumin and amino acids (Sass-Kortsak et al., 1967). All these proteins have been isolated, and many of their physical chemical properties characterized. Such information emphasizes the essential role of copper as a functional cellular constitutent. Moreover, the close correlation between copper and cellular activity and its apparent importance in maintaining health (Evans, 1973) imply that elaborate metabolic pathways are present to ensure its hemostatic level in the body (our knowlege on this subject is reviewed in details in the preceding chapter by Evans).

The present chapter concerns work not only on the whole body transport of copper after ingestion of a radiocopper isotope but also on its transfer from plasma to organ compartments and its temporal distribution at the cellular and molecular levels after an intravenous injection of ^{64}Cu and/or ^{67}Cu bound to albumin or ceruloplasmin. Although much of the information has been obtained in rats, the results can probably be advantageously extrapolated to human beings.

2. PRODUCTION OF RADIOCOPPER ELEMENTS

There are two useful radioactive isotopes of copper, namely, ^{64}Cu and ^{67}Cu. Certain properties of these are shown in Table 1. Copper-64 is readily available commercially, but its half-life of 12.8 hr has limited its use in lengthly experiments. Copper-64 of a medium degree of specific activity is produced by bombarding ^{63}Cu with thermal neutrons to obtain ^{64}Cu by the following reaction:

$$^{63}Cu(\eta,\gamma)^{64}Cu$$

However, ^{64}Cu of very high specific activity can be produced by bombarding ^{64}Zn with fast neutrons (Frize, 1964). The reaction is given by

$$^{64}Zn(\eta,p)^{64}Cu$$

This requires chemical separation of the copper from zinc (Frize, 1964). Lots of 25 to 30 mCi can be produced by this method.

Three methods for the production of ^{67}Cu in millicurie quantities have been described. Sternlieb et al. (1961b) used the reaction:

$$^{64}N(\alpha,p)^{67}Cu$$

to produce 2 mCi of ^{67}Cu after irradiation of enriched ^{64}N by 40-MeV alpha-particles. However, in their method, carrier copper was added to the irradiated target in order to achieve efficient chemical separation of the product, which greatly reduced the specific activities of the preparation. O'Brien (1967) prepared ^{67}Cu by the reaction

$$^{67}Zn(\eta,p)^{67}Cu$$

Irradiation of 100 mg of enriched ^{67}Zn for 5 days in a flux of 7×10^{14} n/cm^2·sec

Table I. Properties of Useful Radioactive Copper Isotopes[a]

Property	Cu^{64}	Cu^{67}
Half-life (hr)	12.8	58.5
Energy of principal gamma emission (MeV)	0.511 (from β^+ decay)	0.184 0.093
Maximum energy of beta emission (MeV)	β^- 0.57	β^- 0.40
	β^+ 0.66	0.48 0.58

[a] Data from Sass-Kortsak (1965).

produced approximately 12 mCi of the required nuclide. Marceau et al. (1970b) have produced millicurie quantities of high specific activity ^{67}Cu by the reaction

$$^{68}\mathrm{Zn}(\gamma,\mathrm{p})^{67}\mathrm{Cu}$$

after irradiation of natural zinc in the bremsstrahlung beam from a linear accelerator. Up to 15 mCi of ^{67}Cu with a specific activity of at least 1 mCi/μg Cu was produced after separation of the ^{67}Cu from the zinc by distillation and ion exchange resin purification.

The work to be described made use of either or both of these radioelements to trace the copper transfer from its site of absorption in the gastrointestinal tract to its site of utilization in the various organ compartments.

3. ABSORPTION OF COPPER FROM THE GASTROINTESTINAL TRACT

Balance studies have been used to measure the total absorption of copper from the gastrointestional (GI) tract (Tu et al., 1965). These measurements, however, did not yield the net absorption or the rate of copper absorption. Whole-body counting techniques have also been used to estimate the absorption of an oral dose of ^{67}Cu (Farren and Mistilis, 1967); but, just as in balance studies, these methods could hardly discriminate between variations in the actual transfer from GI tract to blood and the degree of body retention.

The "input discovery" technique has been used to measure the rate and extent of net absorption of copper from GI tract in rats (Marceau et al., 1970a). The results obtained for the rates of absorption by varying the amount of ingested copper are reproduced in Figure 1. When the amount of cold copper in the ^{64}Cu ingested dose varied from 0 to 12 μg, the rate of absorption was seen to increase rapidly to reach a maximum 0.5 hr after administration. Thereafter the rate of absorption dropped rapidly and leveled off at later times. As the oral copper load was increased to 18 and 36 μg, the peak value of the absorption rate decreased progressively and was shifted to later times. However, the changes in the absorption rates with time after ingestion of the dose were quantitatively similar over the range of doses of copper tested. The integral net absorption of copper was calculated from these curves, and the results are shown in Figure 2. The main observation is that the absorption of copper from the GI tract is linear from trace amount to 12 μg. At higher doses it would seem that the transfer sites in the GI tract are oversaturated. Although such values may seem at first to be relatively small, their meaning becomes more apparent when the actual amount of copper found in the body is considered. To our knowledge, no comparable studies have been reported for human beings. The effect of the dietary form of copper was also determined in rats, and the results are shown in

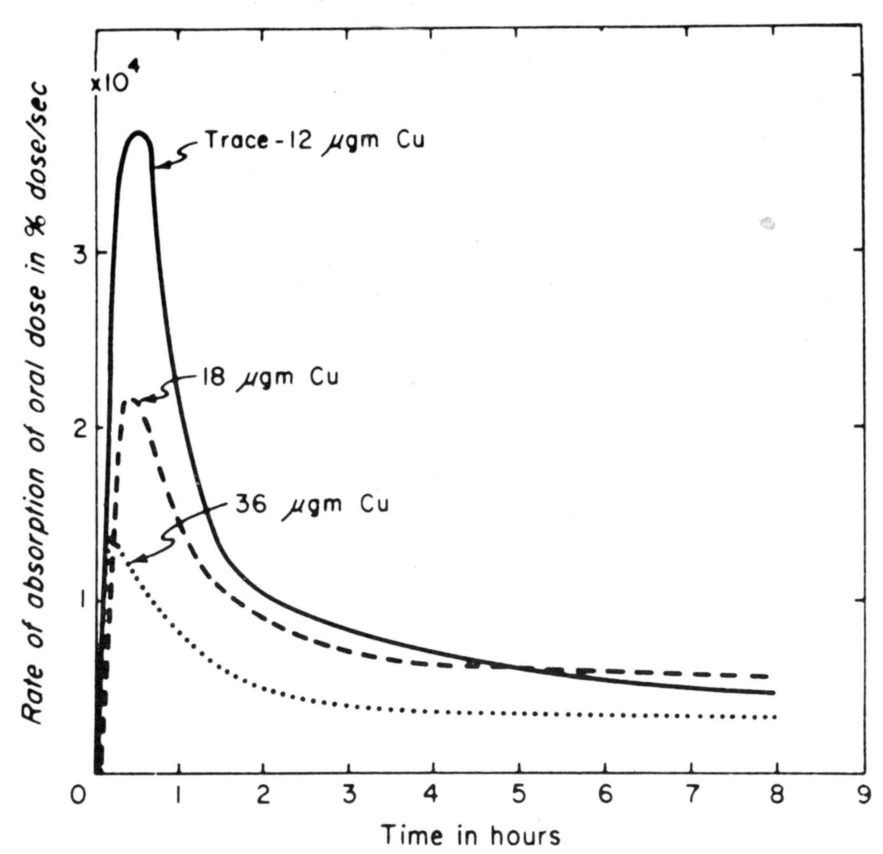

Figure 1. The rate of absorption of ^{64}Cu activity as a function of time after oral administration. Calculated curves are shown for doses of trace to 12, 18, and 36 μg Cu. From Marceau et al. (1970a).

Figure 3. The percentages of the copper dose appearing in plasma after ingestion as ionic copper and as copper bound to amino acid (histidine), mixed with milk, or tightly bound to ceruloplasmin were compared. The data presented in Figure 3 demonstrate that the absorption of copper from the GI tract was independent of the binding form of the ingested copper (Marceau et al., 1970a).

The influence of molybdenum and zinc on copper absorption has been determined (Marceau, 1968). When the oral dose of cold copper was 6 μg, no effects were observed for the amount absorbed in plasma unless the molybdenum or the zinc dose was increased to 50 times that of the copper, indicating that these elements do not interfere with the intestinal absorption of copper except at very high and unphysiological doses (see, however, Kirchgessner et al., this volume).

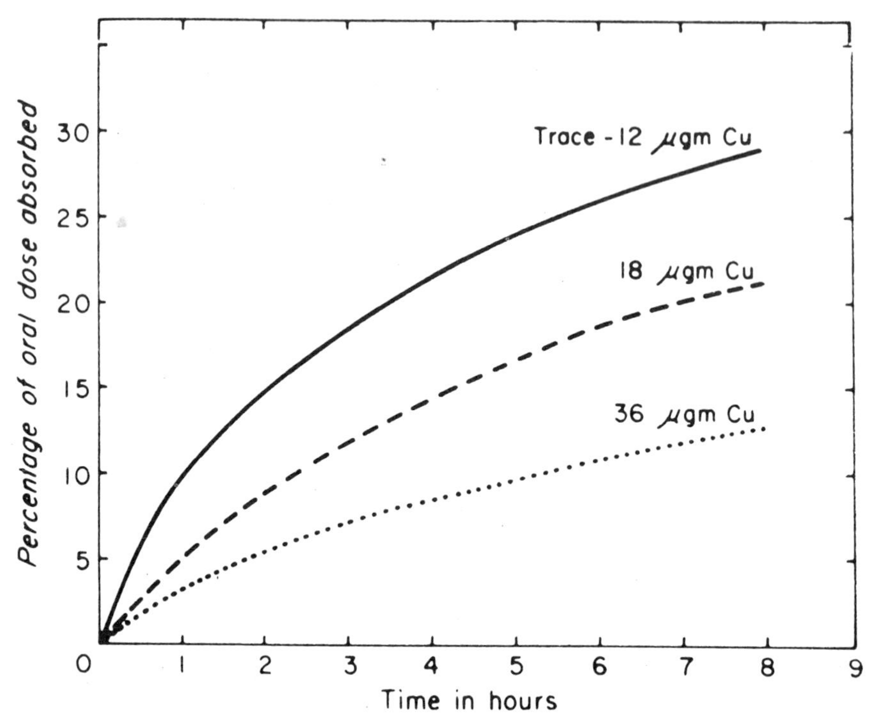

Figure 2. Percentage of ^{64}Cu activity absorbed as a function of time after oral administration. Calculated curves are shown for doses of trace to 12, 18, and 36 μg Cu. From Marceau et al. (1970a).

4. DISAPPEARANCE OF RADIOCOPPER IN BLOOD

It is well established that, when radioactive copper first enters the plasma compartment after an oral or an intravenous administration, the radioactivity is found mainly in the albumin fraction (Owen, 1964; Owen and Hazelrig, 1965). Copper administered intravenously as a single dose disappears extremely fast from the blood, that is, the plasma compartment. In fact, in normal human beings half of a 0.1 to 1 μg dose of copper disappears from the plasma department in less than 10 min (Sass-Kortsak, 1965; Owen and Hazelrig, 1965; Marceau et al., 1970a). However, the disappearance does not follow first-order kinetics, and the percentage of dose as a function of time can be approximated by a sum of at least three exponentials, as obtained in rats (Marceua et al., 1970a). Figure 4 shows the effect of a copper load on the kinetics of ^{64}Cu disappearance and its reappearance in the ceruloplasmin newly secreted by the liver. These results indicate that the response of the copper transfer system is linear up to a dose of at least 10 μg. Moreover, the appearance of ceruloplasmin-bound ^{64}Cu reaches a maximum by 3 to 6 hr; at 6 hr more than 98% of the radioactive copper found

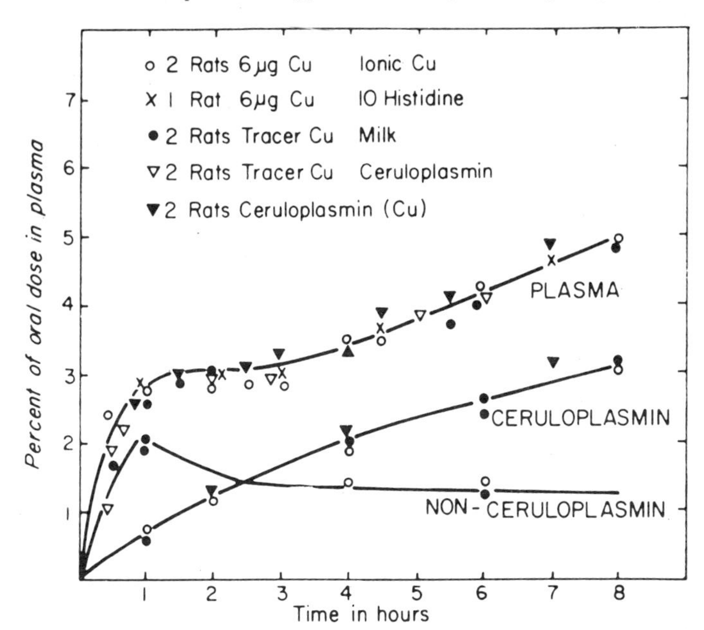

Figure 3. Percentage of ^{64}Cu activity in total plasma, ceruloplasmin, and nonceruloplasmin fractions as a function of time after oral administration. Values are plotted for doses of copper given as ionic copper, with histidine, with milk, with ceruloplasmin, and incorporated into ceruloplasmin. From Marceau et al. (1970a).

in plasma is bound to this copper protein, whereas the remaining 2% represents the fraction still bound to albumin (Marceau et al., 1970a). It is important to realize that on the basis of these data, which have been confirmed by others (Owen and Hazelrig, 1965), essentially all of the copper utilized by tissues and transfered from albumin is restricted to the few hours that followed its injection in plasma. Thereafter the transport of copper in plasma is assumed by ceruloplasmin.

5. TISSUE UPTAKE OF COPPER AFTER ITS INJECTION AS [^{64}CU]-ALBUMIN

A large fraction of the radiocopper that disappears from the plasma is taken up by the liver (Sass-Kortsak, 1965; Owen and Hazelrig, 1965; Aspin and Sass-Kortsak, 1966). Between 65 and 90% of the dose is detected in this organ 1 or 2 hr after injection, the amount taken up being a function of the injected copper load. After 2 hr there is a slow decrease in the activity in the liver, which reflects the excretion of copper via the bile and the release into the blood of labeled ceruloplasmin. The existence of an enterohepatic circulation of copper has been

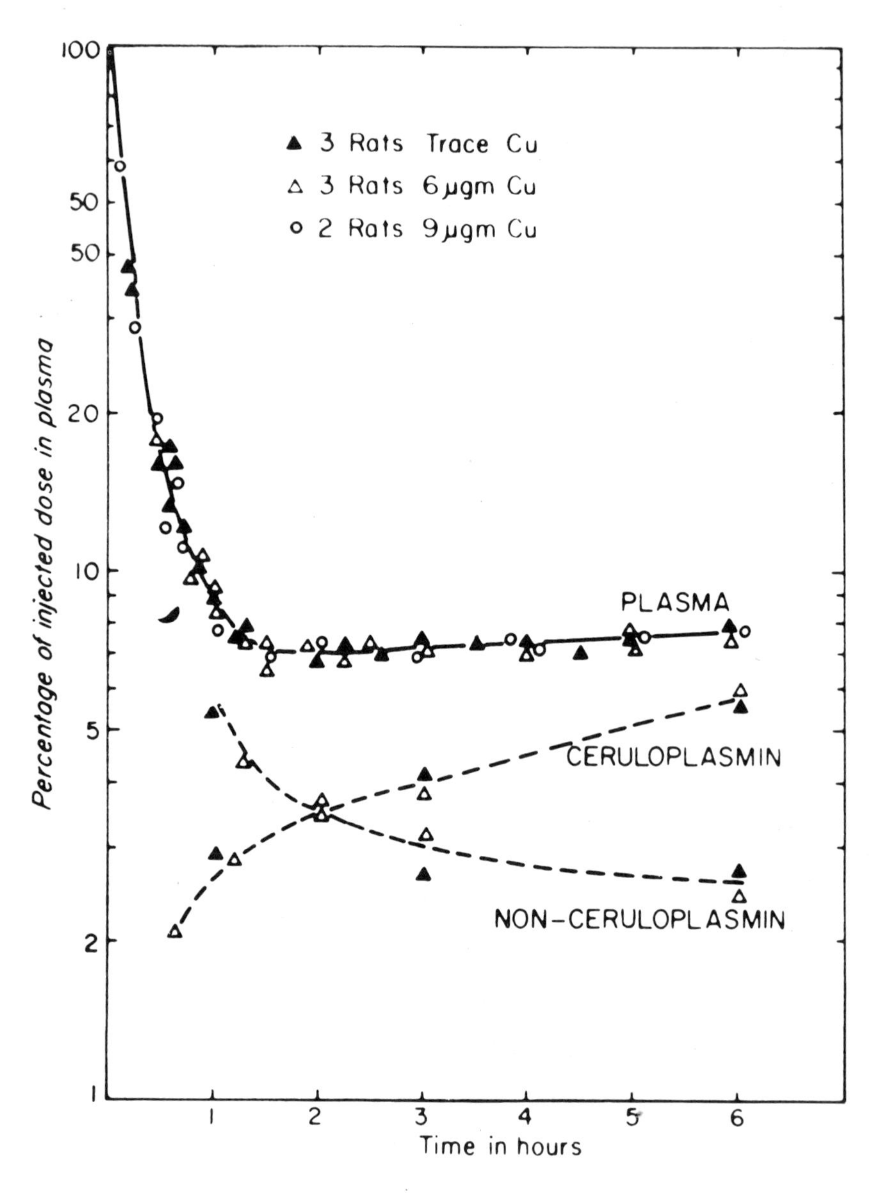

Figure 4. Percentage of ^{64}Cu activity in total plasma, ceruloplasmin, and nonceruloplasmin fractions as a function of time after intravenous injection. Values are plotted for the injection of trace and 6 and 9 μg Cu. From Marceau et al. (1970a).

demonstrated (Jennet et al., 1962), but biliary copper was found to be poorly reabsorbed (Farren and Mistilis, 1967). These pathways have been demonstrated *in vitro* using an isolated rat liver preparation (Owen and Hazelrig, 1966).

The uptake of albumin-bound ^{64}Cu by tissues other than the liver has been determined (Owen and Hazelrig, 1965). The main observation has been a rapid and nonselective diffusion of the radioelement throughout the tissues (Goresky et al., 1968). It is also of interest to note that this tissue copper is exchangeable with that of the plasma compartment (Goresky et al., 1968; Marceau and Aspin, 1972).

6. TISSUE TRANSFER OF CERULOPLASMIN-BOUND ^{64}CU OR ^{67}CU

The fact that after an intravenous injection of copper as [^{64}Cu]albumin most of the radioactivity reappeared in plasma as radioactive copper tightly bound to ceruloplasmin suggested that this glycoprotein has a role in the transport copper to tissues (Marceau and Aspin, 1973b; Frieden, this volume). Ceruloplasmin-bound ^{67}Cu has been injected intravenously into rats, and the radioactivity has been measured in the whole body, spleen, kidneys, gastrointestinal tract, liver, lungs, heart, brain, bones, bone marrow, skin muscles, and plasma during a period of 140 hr. As shown in Figure 5, a whole-body half-life of 145

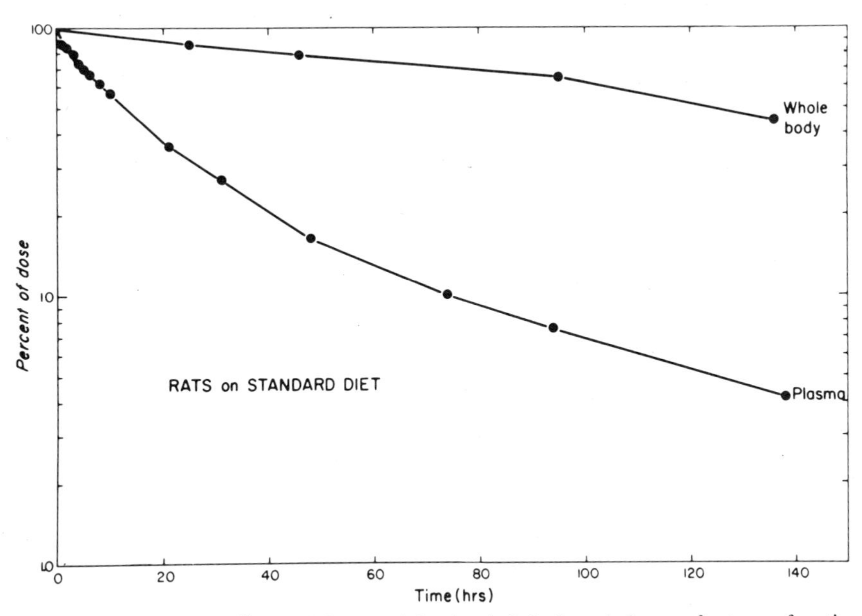

Figure 5. Percentage of ^{67}Cu activity remaining in whole body and plasma of rats as a function of time after an intravenous injection of ceruloplasmin-bound ^{67}Cu. Data points represent means of four or five measurements. From Marceau and Aspin (1972).

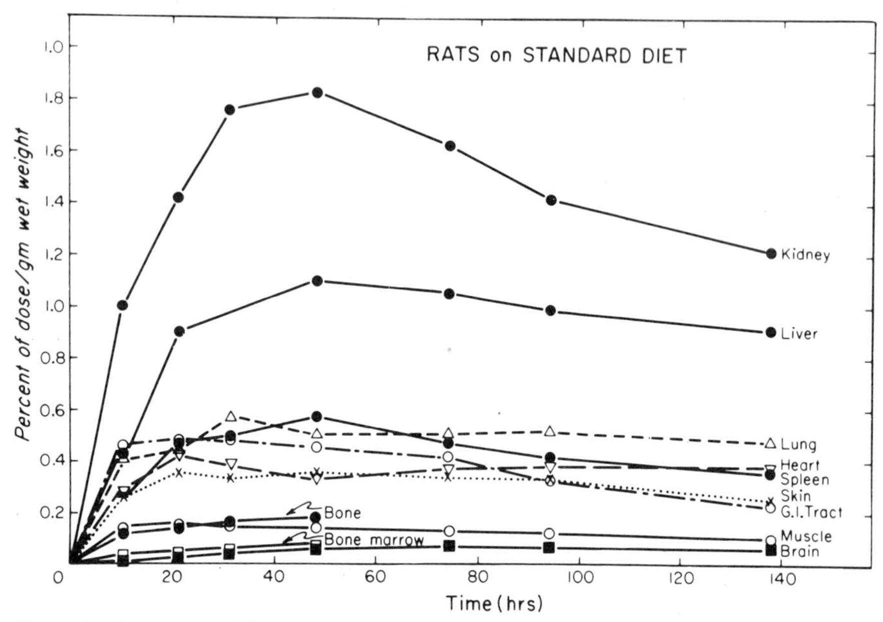

Figure 6. Percentage of ^{67}Cu activity remaining in whole body and plasma as a function of time after intravenous injection of ceruloplasmin-bound ^{67}Cu in rats fed a copper-deficient diet (broken line) and a control (solid line). Each data point represents the mean of three to five measurements. From Marceau and Aspin (1972).

hr was found for ceruloplasmin copper, whereas in plasma half of the radioactivity disappeared in 17 to 18 hr. This activity further decreases to 10% of its original value in 75 hr. The levels of total ^{67}Cu activity in the dissected organs were also measured, and after correction of the data for the ^{67}Cu activity trapped in plasma an uptake of ^{67}Cu was observed for all the tissues assayed. As shown in Figure 6, where the results are expressed as percentage of dose per gram of wet weight, the radioactivity rose to a maximum at about 40 hr, after which it declined slightly. Kidneys and liver showed the greatest degree of uptake; brain and bone marrow, the least. However, the relative distribution of the dose looked very different when the uptake was calculated on the basis of total organ weight rather than on a unit weight basis. Since the muscle comprised 47% of the rat body weight (Marceau and Aspin, 1972), the uptake in this tissue became a significant fraction of the total dose, even though the fraction taken up per unit weight was very small. Other studies on the influence of a copper load on the disappearance of ceruloplasmin-bound copper in plasma have shown that this form of copper in tissues is not exchangeable with the copper in plasma (Marceau and Aspin, 1972). The use of ceruloplasmin double-labeled with ^{67}Cu and [^{3}H]leucine has demonstrated that the protein breakdown occurs concomitantly with the transfer of the copper to the tissues (Marceau and Aspin, 1973a).

7. INTRACELLULAR TRANSFER OF COPPER

Our studies on copper transport have been extended to the temporal transfer of albumin- and ceruloplasmin-bound copper to subcellular organelles and macromolecular elements. Because copper is so predominant in liver, we looked at the relative distribution of radioactive copper in liver nuclei, lysosomes, mitochondria, microsomes, and the soluble fraction in the organ (Marceau and Aspin, 1973b). At day 1 after an intravenous injection of ceruloplasmin-bound ^{67}Cu, most of the radioactivity was detected in the soluble fraction. However, at day 6 there was a relative decrease in the soluble fraction radioactivity which could be matched with an equivalent relative increase of ^{67}Cu in the mitochondrial fraction. Similar studies in the first hours that followed the injection of albumin-bound ^{64}Cu, that is, during the period of massive transfer to the liver, revealed that most of the radioactivity was present in the soluble fraction with little in the mitochondria (Marceau and Aspin, 1973b). These studies at the subcellular level have shown that, although copper from albumin and copper from ceruloplasmin both appeared primarily in the soluble fraction of the liver cells, the rates at which they were transferred seem very different. The copper from albumin appeared in the first 2 hr that followed its injection in plasma, whereas the process by which the copper from ceruloplasmin is gradually transferred from plasma to mitochondria via the solbule fraction took place within a matter of days.

8. IDENTIFICATION OF THE PROTEINS INVOLVED IN COPPER TRANSFER AND UTILIZATION IN TISSUES

The identity of the cuproproteins, that is, albumin and ceruloplasmin, in plasma has been well established, and our interest turned to the nature of those present in the soluble fraction and mitochondria and their involvement in the transfer of copper from albumin or ceruloplasmin. Accordingly, the distribution within the soluble fraction of rat liver of the copper from albumin and ceruloplasmin was studied as a function of time after an intravenous injection (Marceau and Aspin, 1973b). Both ceruloplasmin-bound ^{67}Cu and albumin-bound ^{64}Cu were injected into the same rat at 56 hr and 30 min, respectively, before sacrifice. Fractionation by gel filtration on Sephadex G-100 of the soluble fraction obtained from the homogenate revealed that the ^{67}Cu activity from ceruloplasmin eluted as a single peak, distant from that of ^{64}Cu from albumin. The estimated molecular weight of the ^{67}Cu protein in the cytoplasm was 30,000 to 40,000 daltons, whereas that of the ^{64}Cu protein was 10,000 to 15,000 daltons (Marceau and Aspin, 1973b). Others have shown that at 24 hr after an intravenous injection of albumin-bound ^{67}Cu most of the radioactivity is found in the 30,000 to 40,000 dalton protein (Terao and Owen, 1973), confirming the above observation that the transfer of albumin copper to this protein occurs via ceruloplasmin. Fractionation of polyacrylamide gel electrophoresis also demonstrated that the two intracellular proteins detected in this manner differed not only in molecular weight but also in charge (Marceau and Aspin, 1973b).

The same results were obtained for the intracellular distribution of ceruloplasmin- and albumin-bound copper in rat brain and spleen, except that in these tissues the radioactive copper from albumin was detected in a number of protein fractions, including the 30,000-molecular-weight fraction obtained for copper from ceruloplasmin (Marceau and Aspin, 1973b).

The observation that ^{67}Cu from ceruloplasmin appeared in the mitochondrial fraction of the liver suggested its association with the copper protein cytochrome *c* oxidase (Marceau and Aspin, 1973b). Extraction of the membrane enzyme with nonionic detergents at 6 days after an intravenous injection of ceruloplasmin-bound ^{67}Cu and further separation by polyacrylamide gel electrophoresis showed that ^{67}Cu correlated with a specific stain for the oxidase activity. Such results demonstrated that the copper from ceruloplasmin indeed appeared in cytochrome c oxidase, and other studies using chelating agents demonstrated that this copper is tightly bound to the enzyme (Marceau and Aspin, 1973b). Comparative studies with radiocopper obtained from albumin demonstrated that, although this copper may appear in cytochrome *c* oxidase, it can be removed with chelating agents (Marceau and Aspin, 1973b). All these results could be explained in the light of data on the physicochemical properties of cytochrome *c* oxidase, establishing that two types of copper are present in the enzyme, one intrinsic and the other extraneous (Lemberg, 1969). This indeed strongly supports the above results on the radiocopper transfer measurements, which show that the major fraction of the extraneous copper associated with the enzyme is derived from albumin, whereas ceruloplasmin copper appears as intrinsic fractional copper.

Comparison of the results obtained on the fractionation of the radiolabeled proteins in the soluble fraction with those reported by others (Morell et al., 1961; Evans, 1973) demonstrated that the protein of the liver soluble fraction that receives copper from albumin has properties very similar to those of a copper storage protein, called L-6-D or metallothionein. Moreover, the protein that receives copper from ceruloplasmin showed properties very similar to those of cytocuprein, also known as superoxide dismutase (Carrico and Deutsch, 1969; McCord and Fridovich, 1969).

Furthermore, metallothionein is believed to be the soluble receptor in the liver for copper coming from albumin in plasma. Since this copper is either excreted in bile or incorporated into ceruloplasmin, it is evident that metallothionein has to play an important role in this branching metabolic pathway. Studies done by Linquist (1967) on the biliary excretion of a copper load revealed that the excess copper is contained in numerous discrete granules arranged around bile caniculi. These granules are rich in acid phosphatase, and such granules are considered to be lysosomes (Linquist, 1968). Moreover, Porter et al. (1962) isolated from neonatal human and bovine livers a cuproprotein that they named "mitochondrocuprein," since it was extracted from a mitochondrial subfraction (Porter et al., 1964). More recently, Porter (1974) has re-evaluated this work and concluded that this substance was instead a cuproprotein from lysosomes. Its copper content is only 0.02% in human liver, compared to 3.5% for the neonatal protein (Porter, 1974). Thus it seems likely that this cuproprotein is involved in the transfer of metallothionein copper in bile.

As for the site of transfer of copper from metallothionein to ceruloplasmin, it seems reasonable at first to suggest that the incorporation of copper into the protein occurs on microsomes. However, considering the subunit structure of ceruloplasmin and the fact that it is a glycoprotein, it may very well be that the copper is transferred at the level of the Golgi apparatus. This proposal is quite appealing. In view of the fact that the hepatic copper is eliminated only by the bile or the newly synthesized ceruloplasmin and also that lysosomes are apparently formed from the Golgi vesicles (De Duve, 1969), it is proposed here that metallothionein transfers a fraction of its copper to ceruloplasmin at the level of the Golgi apparatus and the rest becomes trapped in lysosomes, from which it is excreted in bile. On the basis of this hypothesis, "metallothionein" and "mitochondrocuprein" seem to be the same protein.

9. A SCHEMATIC MODEL FOR THE TRANSPORT AND UTILIZATION OF COPPER IN MAMMALIAN ORGANISMS

All the metabolic transfer pathways discussed above are schematically represented in Figure 7. In this scheme we find three types of copper proteins: those considered to be true enzymes, whose function depends on the presence of copper (i.e., tyrosinase, amine oxidases, and cytochrome *c* oxidase); those that serve primarily transfer purposes (i.e., metallothionein, mitochondrocuprein, and cytocuprein); and those whose role is mainly transport (i.e., albumin and ceruloplasmin). Albumin is a vehicle that transports copper from the gastrointestinal tract mainly to the liver. This copper can be transferred through metallothionein (L-6-D), to be either excreted in bile via mitochrondrocuprein or conveyed back into plasma within ceruloplasmin. Ceruloplasmin is the specfic vehicle in plasma for carrying copper via cytocuprein (superoxide dismutase) to cytochrome *c* oxidase.

10. COPPER TRANSFER IN HUMAN BEINGS

Much of the interest in studying copper utilization in human beings is related to a relatively rare disease which is characterized by copper intoxication: Wilson's

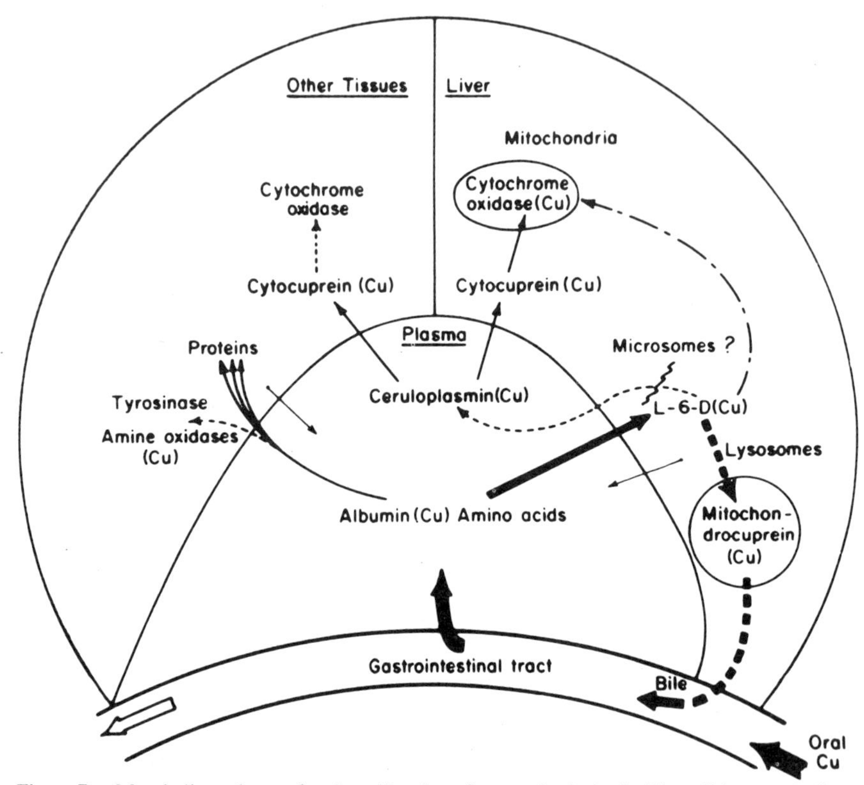

Figure 7. Metabolic pathways for the utilization of copper in the body. The solid arrows indicate the pathways, the existence of which is supported by solid experimental evidence. The broken arrows correspond to those postulated in the present chapter. The dotted-broken line represents a nonspecific transfer of copper.

disease or hepatolenticular degeneration (Sass-Kortsak, 1965, 1975). This disease, which expresses itself predominantly in young people, results from an "inborn error of metabolism," the exact nature of which is not clearly defined. Genetic studies have revealed that the condition is inherited in an autosomal recessive fashion (Bearn, 1960). Evaluations of the copper content in tissues have shown abnormal accumulations of copper in the liver, brain, kidneys, and cornea (Cumings, 1959; Boudin and Pépin, 1959). Finally, it has also been repeatedly shown that the concentration of copper in plasma is reduced as a result of the low level of ceruloplasmin (Walshe, 1967).

Radioactive copper has been used to follow the copper transport in these subjects (Jensen and Kamin, 1957; Walshe, 1967). The rate of absorption of copper from the GI tract is no different in patients with Wilson's disease from that in normal individuals (Aspin and Sass-Kortsak, personal communication).

Nevertheless, after an intravenous injection of radioactive copper the activity remaining in plasma is higher in these patients than in controls (Earl et al., 1954). Radioactive copper which enters the liver of patients from the plasma does not appear to be incorporated in ceruloplasmin, since this labeled protein cannot be found in the blood (Scheinberg and Gitlin, 1952; Sass-Kortsak et al., 1959; Maytum et al., 1961). The low fecal excretion of the radiocopper in the bile by the patient has been used as evidence of low biliary excretion of copper in this disease condition (Bearn and Kundel, 1954; Osborn et al., 1963). In other words, copper that enters the liver of a patient becomes trapped there, eventually resulting in copper levels that are toxic to the liver cells. Patients with marked liver damage show a lower uptake of ^{64}Cu than do asymptomatic patients or control individuals (Matthews, 1954; Osborn and Walshe, 1961; Walshe, 1963). This could be the result of a high liver copper concentration preventing the entry of more copper, or of the destruction of many liver cells by copper poisoning.

Heterozygous individuals with Wilson's disease present very complex metabolic features. They may or may not accumulate copper in tissues (Sternlieb et al., 1961c). The non-ceruloplasmin-bound copper in plasma is normal, and ceruloplasmin levels are intermediate between those of the normal and the homozygous abnormal subjects (Scheinberg and Gitlin, 1952; Bush et al., 1955). When intravenous injection of radioactive copper is given, its rate of disappearance from plasma is normal as a result of an uptake in the liver comparable to that seen in normal subjects (Aspin and Sass-Kortsak, 1966). The liver does not discharge radioactive-copper-labeled ceruloplasmin at a normal rate, resulting in a lower radioactive copper activity level in plasma than in control subjects (Osborn et al., 1963; Aspin and Sass-Kortsak, 1966). Finally, whole-body counting techniques with ^{67}Cu have been used to differentiate between homozygous and heterozygous patients of Wilson's disease and normal subjects (Willvonseder et al., 1974; Tauxe et al., 1974). In spite of the slight disturbances in copper metabolism mentioned above, it is a fact that heterozygotes of Wilson's disease are free of the ailment throughout their lives (Sass-Kortsak, 1975).

The primary defect in Wilson's disease which causes the copper intoxication is not yet known (Thompson et al., 1977), but there are some indications of anomalies in a "transfer" protein. First, microscopic studies by Goldfisher and Sternlieb (1968) on the liver of these patients have demonstrated that during the progression of the disease there is a massive displacement of copper from the cytoplasm to lysosomes in liver cells. Second, at the molecular level, Morrell et al. (1961) and Evans (1973) have found that the copper content of metallothionein is increased in these patients. Third, Porter (1968) observed that the excess of copper is found in the "mitochondocuprein" fraction. All these observations suggest that in these patients the "metallothionein" and, of course, the "mitochondrocuprein" are modified, and that this anomaly leads to the development of the disease.

ACKNOWLEDGMENTS

The work described in this chapter was supported by a grant from the Hartford Foundation, New York, to Professor Andrew Sass-Kortsak. I am most thankful to Mrs. Monique Bournatzis for her secretarial assistance in preparing the text.

REFERENCES

Aspin, N. and Sass-Kortsak, A. (1966). "Radiocopper Studies on a Family with Wilson's Disease." In J. Peisach, P. Aisen and W. E. Blumberg, Eds., *Biochemistry of Copper.* Academic Press, New York, pp. 503–513.

Bearn, A. G. (1960). "Wilson's Disease." In J. B. Stanbury, J. B. Wyngaarden, and D. S. Frederickson, Eds., *The Metabolic Basis of Inherited Disease.* McGraw-Hill, New York, pp. 809–838.

Bearn, A. G. and Kundel, H. G. (1954). "Abnormalities of Copper Metabolism in Wilson's Disease and Their Relationship to the Aminoaciduria," *J. Clin. Invest.,* **33,** 400–409.

Boudin, G. and Pépin, B. (1959). Dégénérescence Hépatolenticulaire. Masson, Paris.

Bush, J. A., Mahoney, J. P., Markowitz, H., Gubler, C. J., Cartwright, G. E., and Wintrobe, M. M. (1955). "Studies on Copper Metabolism. XVI: Radioactive Copper Studies in Normal Subjects and in Patients with Hepatolenticular Degeneration," *J. Clin. Invest.,* **36,** 1766–1778.

Carrico, R. J. and Deutsch, H. F. (1969). "Isolation of Human Hepatocuprein and Cerebrocuprein: Their Identity with Erythrocuprein," *J. Biol. Chem.,* **244,** 6087–6094.

Cumings, J. N. (1959). *Heavy Metals and the Brain.* Thomas, Springfield, Ill.

De Duve, C. (1969). "The Lysosome in Retrospect." In J. T. Dingle and H. B. Fell, Eds., *Lysosome in Biology and Pathology.* North-Holland, Amsterdam, pp. 3–40.

Earl, C. N., Moulton, M. J., and Silverstone, B. (1954). "Metabolism of Copper in Wilson's Disease and in Normal Subjects: Studies with Cu^{64}," *Am. J. Med.,* **16,** 205–213.

Evans, G. N. and Weideranders, R. E. (1967). "Blood Copper Variations among Species," *Am. J. Physiol.,* **213,** 183–184.

Evans, G. W. (1973). "Copper Homeostasis in the Mammalian System," *Physiol. Rev.,* **53,** 535–571.

Farren, P. and Mistilis, S. P. (1967). "Absorption of Exogenous and Endogenous Biliary Copper in the Rat," *Nature,* **213,** 291–292.

Frize, K. (1964). "Preparation of High Specific Activity ^{64}Cu," *Radiochim. Acta,* **3,** 166–167.

Goldfisher, S. and Sternlieb, I. (1968). "Changes in the Distribution of Hepatic Copper in Relation to the Progression of Wilson's Disease," *Am. J. Pathol.,* **52,** 833–891.

Goresky, C. A., Holmes, T. H., and Sass-Kortsak, A. (1968). "The Initial Uptake of Copper by the Liver in the Dog," *Can. J. Physiol. Pharmacol.*, **46,** 771–784.

Jennet, F., Richterich, R., and Aebig, H. (1962). "Bile et Céruloplasmine. Etude *in vitro* à l'Aide du Foieisolé," *J. Physiol. (Paris)*, **54,** 729–737.

Jensen, W. N. and Kamin, H. (1957). "Copper Transport and Excretion in Normal Subjects and in Patients with Laenic's Cirrhosis and Wilson's Disease: A Study with Cu^{64}," *J. Lab. Clin. Med.*, **49,** 200–210.

Lemberg, M. R. (1969). "Cytochrome Oxydase," *Physiol. Rev.*, **49,** 48–123.

Lindquist, R. R. (1967). "Studies on the Pathogenesis of Hepatolenticular Degeneration. 1: Acid Phosphatase Activity in Copper Loaded Rat Livers," *Am. J. Pathol.*, 471–482.

Lindquist, R. R. (1968). "Studies on the Pathogenesis of Hepatolènticular Degeneration. 3: The Effect of Copper on Rat Liver Lysosomes," *Am. J. Pathol.*, **53,** 903–912.

Marceau, N. (1968). "The Absorption of Copper from the Gastrointestinal Tract of the Rat." M.Sc. thesis, University of Toronto, Toronto, Ontario.

Marceau, N. (1971). "Biophysical Studies of Ceruloplasmin in Relation to Other Copper Proteins." Ph.D. thesis, University of Toronto, Toronto, Ontario.

Marceau, N. and Aspin, N. (1972). "Distribution of Ceruloplasmin-Bound ^{67}Cu in the Rat," *Am. J. Physiol.*, **222**(1), 106–110.

Marceau, N. and Aspin, N. (1973a). "The Association of the Copper Derived from Ceruloplasmin with Cytocuprein," *Biochim. Biophys. Acta,* **328,** 351–358.

Marceau, N. and Aspin, N. (1973b). "The Intracellular Distribution of the Radiocopper Derived from Ceruloplasmin and from Albumin," *Biochim. Biophys. Acta,* **293,** 338–350.

Marceau, N., Aspin, N., and Sass-Kortsak, A. (1970a). "Absorption of Copper-64 from the Gastrointestinal Tract of the Rat," *Am. J. Physiol.*, **218,** 337-383.

Marceau, N., Kruck, T. P. A., McConnell, D. B., and Aspin, N. (1970b). "The Production of Copper-67 from Natural Zinc Using a Linear Accelerator," *Int. J. Appl. Radiat. Isot.*, **21,** 667–669.

Matthews, W. B. (1954). "The Absorption and Excretion of Radiocopper in Hepatolenticular Degeneration (Wilson's Disease)," *J. Neurol. Neurosurg. Psychiatr.*, **17,** 242–246.

Maytum, W. J., Goldstein, N. P., McGuchin, W. F., and Owen, C. A., Jr. (1961). "Copper Metabolism in Wilson's Disease, Laennec's Cirrhosis and Hemachromatic: Studies with Radiocopper (^{64}Cu)," *Proc. Staff Meet Mayo Clin.*, **36,** 641–660.

McCord, J. M. and Fridovich, L. (1969). "Superoxyde Dismutase, an Enzymatic Function for Erythrocuprein," *J. Biol. Chem.*, **244,** 6049–6055.

Morell, A. G., Shapiro, J. R., and Scheinberg, I. A. (1961). "Copper Binding Protein in Human Liver." In J. M. Walshe and J. N. Cumings, Eds., *Wilson's Disease, Some Current Concepts.* Thomas, Springfield, Ill., pp. 36–41.

O'Brien, H. A., Jr. (1967). "The Preparation of ^{67}Cu from ^{67}Zn in a Nuclear Reactor," *Int. J. Appl. Radiat. Isot.*, **20,** 121–124.

Osborn, S. B. and Walshe, J. M. (1961). "Copper Uptake by the Liver: Study of a Wilson's Disease Family." In J. M. Walshe and J. N. Cumings, Eds., *Wilson's Disease, Some Current Concepts.* Thomas, Springfield, Ill., pp. 141–150.

Osborn, S. B., Roberts, C. N., and Walshe, J. M. (1963). "Uptake of Radiocopper by the Liver: A Study of Patients with Wilson's Disease and Various Control Groups," *Clin. Sci.,* **24,** 13–22.

Owen, C. A., Jr. (1964). "Absorption and Excretion of ^{64}Cu-Labeled Copper by the Rat," *Am. J. Physiol.,* **207,** 1203–1206.

Owen, C. A., Jr. and Hazelrig, J. B. (1965). "Metabolism of Radiocopper (^{64}Cu) in the Rat," *Am. J. Physiol.,* **209,** 900–904.

Owen, C. A., Jr. and Hazelrig, J. B. (1966). "Metabolism of ^{64}Cu-Labeled Copper by the Isolated Rat Liver," *Am. J. Physiol.,* **210,** 1059–1064.

Porter, H. (1974). "The Partiulate Half-Cystine-Rich Copper Protein of Newborn Liver: Relationship to Metallothioneine and Subcellular Localization in Heavy Lysosomes," *Biochem. Biophys. Res. Commun.,* **56,** 661–668.

Porter, H. (1968). "Copper Proteins in Brain and Liver." In *Wilson's Disease.* National Foundation, March of Dimes, New York, pp. 23–28.

Porter, H., Johnston, J., and Porter, E. (1962). "Neonatal Hepatic Mitochoncuprein. I: Isolation of a Protein Fraction Containing More than 4% Copper from Mitochondria of Immature Bovine Liver," *Biochim. Biophys. Acta,* **65,** 66-74.

Porter, H., Sweeney, M., and Porter, E. (1964). "Neonatal Hepatic Mitochondrocuprein. II: Isolation of the Copper-Containing Subfraction from Mitochondria of Newborn Human Liver," *Arch. Biochem.,* **104,** 97–106.

Sass-Kortsak, A. (1965). "Copper Metabolism," *Adv. Clin. Chem.,* **8,** 1–67.

Sass-Kortsak, A. (1975). "Wilson's Disease: A Treatable Liver Disease in Children," *Pediatr. Clin. North Am.,* **22,** 963–984.

Sass-Kortsak, A., Chernian, M., Geiger, D. W., and Slater, R. J. (1959). "Observations on Ceruloplasmin in Wilson's Disease," *J. Clin. Invest.,* **38,** 1672–1682.

Sass-Kortsak, A., Clarke, R., Harris, D. I. M., Neumann, P. Z., and Sarkar, B. (1967). "The Biological Tranport of Copper." In *Proceedings of the 2nd International Congress of Neuro-Genetics and Neuro-Ophthalmology,* Montreal, pp. 625–631.

Scheinberg, I. A. and Gitlin, D. (1952). "Deficiency of Ceruloplasmin in Patients with Hepatolenticular Degeneration (Wilson's Disease)," *Science,* **116,** 484–485.

Sternlieb, I., Morell, A. G., Bauer, C. D., Combes, B., DeBobes-Sternberger, S., and Scheinberg, I. H. (1961a). "Detection of the Heterozygous Carrier of the Wilson's Disease Gene," *J. Clin. Invest.,* **40,** 707–715.

Sternlieb, I., Morell, A. G., Tucker, W. D., Greene, M. W., and Scheinberg, I. H. (1961b). "The Incorporation of Copper into Ceruloplasmin *in vivo:* Studies with ^{64}Cu and ^{67}Cu," *J. Clin. Invest.,* **40,** 1834–1841.

Sternlieb, I., Morell, A. G., and Scheinberg, I. H. (1961c). "Homozygotes and Heterozygotes in Wilson's Disease." In J. M. Walshe and J. N. Cumings, Eds., *Wilson's Disease, Some Current Concepts.* Thomas, Springfield, Ill., pp. 133–140.

Tauxe, W. N., Willvonseder, R., and Goldstein, N. P. (1974). "Body Retention of Injected ^{67}Cu in Patients with Homozygous or Heterozygous Wilson's Disease and in Normal

Subjects: Comparison of Whole-Body Counting Systems," *Mayo Clin. Proc.*, **49,** 382–386.

Terao, T. and Owen, C. A. Jr. (1973). "Nature of Copper Compounds in Liver Supernate and Bile of Rats: Studies with ^{67}Cu," *Am. J. Physiol.*, **224**(3), 682–686.

Thompson, W. G., Hyslop, P. S., Barr, R., and Sass-Kortsak, A. (1977). "Wilson's Disease: A Common Liver Disorder," *Can. Med. Assoc. J.*, **117**(1), 45–48.

Tu, J. B., Blackwell, P. Q., and Watten, R. H. (1965). "Copper Balance Studies in Wilson's Disease," *Metabolism,* **14,** 653-666.

Underwood, E. J. (1962). *Trace Elements in Human and Animal Nutrition,* 2nd ed. Academic Press, New York, pp. 48–99.

Walshe, J. M. (1963). "Copper Metabolism and the Liver," *Postgrad. Med. J.*, **39,** 183–192.

Walshe, J. M. (1967). "The Physiology of Copper in Man and Its Relation to Wilson's Disease," *Brain,* 149–176.

Willvonseder, R., Goldstein, N. P., and Tauxe, W. N. (1974). "Long-Term Body Retention of Radiocopper (^{67}Cu) and the Diagnosis of Wilson's Disease," *Mayo Clin. Proc.*, **49,** 387–393.

8

CHEMISTRY OF COPPER PROTEINS AND ENZYMES

Ulrich Weser

Lutz M. Schubotz

Maged Younes

Anorganische Biochemie, Physiologisch Chemisches Institut der Universität Tübingen, Tübingen, Federal Republic of Germany

1. INTRODUCTION

The history of copper proteins is almost as old as our knowledge of the chemical properties of the protein ligands involved. In 1847 Harless described the presence of copper in the oxygen-binding protein hemocyanin, prepared from molluscs. At that time interest in this metal was essentially of an analytical and/or academic nature. In 1938 the importance of copper in biological oxidation was successfully shown with the discovery that cytochrome oxidase is a copper protein (Keilin and Hartree, 1938). From the same laboratory many other copper proteins were prepared, including tyrosinase (Keilin and Mann, 1938), the cupreins (Mann and Keilin, 1938), and laccase (Keilin and Mann, 1939).

An overwhelming number of studies on other copper proteins were reported from 1940 onward and are comprehensively documented in many symposia and reviews (McElroy and Glass, 1950; Peisach et al., 1966; Malkin and Malmström, 1970; Vänngard, 1972; Malkin, 1973; Weser, 1973; Fridovich, 1974a, 1974b, 1976; Fee, 1975; Malmström et al., 1975; Michelson et al., 1977). Therefore this chapter will not undertake to resummarize these conveniently available data. Instead, an attempt will be made to focus on the biochemical reactivity of the liganded copper not only in polypeptides but also in low molecular weight chelates.

Many structures of iron proteins were deduced from X-ray diffraction measurements. Ironically it took some 130 years to elucidate from X-ray diffraction studies the first structure of a copper protein, the 2Cu,2Zn superoxide dismutase (Richardson et al., 1975, 1976). The structure-function correlation of this protein-bound copper (Figure 1) will be considered in more detail.

Specific attention will be paid also to a new class of copper proteins designated in the author's laboratory as copper-sulfur proteins. This type of copper protein would nicely fill an evolutionary gap in conjunction with the iron-sulfur proteins. Unfortunately the functional side of these unique copper-thiolate-rich proteins remains obscure.

2. CLASSIFICATION OF COPPER FORMS

The actual charge density as a consequence of chelation and the different

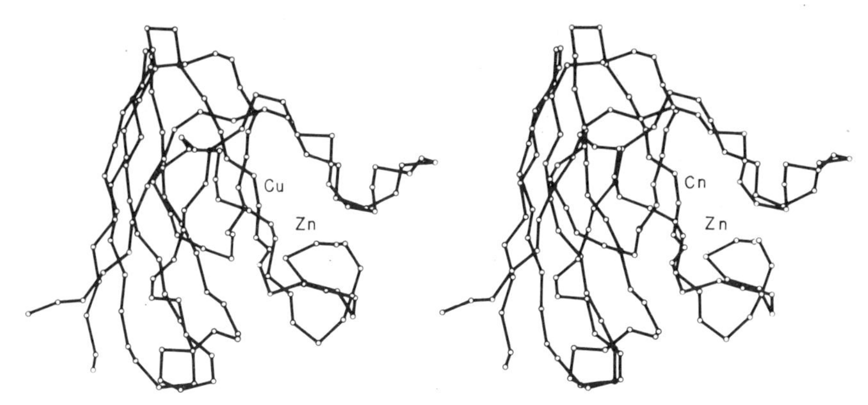

Figure 1. Stereo drawing of the α-carbon backbone for a subunit of Cu,Zn superoxide dismutase, obtained by X-ray diffraction at 3 Å resolution. Richardson et al. (1976), reproduced with the kind permission of the authors.

chemical environments found in many copper proteins greatly affects the binding situation of the copper. Owing to distinct differences in some physicochemical properties, including electron absorption, chiroptical behavior, electron paramagnetism, and magnetic susceptibility, it has been possible to define three different forms of bound Cu(II) (Malmström et al., 1975, and references therein).

2.1. Type 1 Copper

Type 1 Cu(II) causes the extraordinary deep-blue color seen in many copper-containing oxidases. A remarkable multibanded electron absorption near 600 nm is always recorded. The molar absorption coefficient (Malkin and Malmström, 1970) is approximately two orders of magnitude higher than the constants obtained using most of the common copper complexes. The electron paramagnetic resonance parameters are unique in that unusually low *g* values and hyperfine splitting constants are measured (Malmström and Vänngard, 1960; Malmström et al., 1970; Lee and Dawson, 1973; Deinum et al., 1974).

The strong electron absorption cannot be assigned exclusively to *d-d* transitions. Possible ligand transitions (Falk and Reinhammar, 1972) and, especially, charge-transfer bands involving copper-liganded sulfur have been proposed (Miskowski et al., 1975). The latter proposal is strongly supported by X-ray photoelectron spectrometric studies on copper-sulfur bondings (Rupp and Weser, 1976a, 1976b, 1976c, 1978; Younes and Weser, 1977; Weser et al. 1979, Younes et al., 1979). It was demonstrated that copper having the formal oxidation state +2 and bound in a square planar geometrical environment had an actual charge density of +1 when a thiolate-sulfur ligand was involved. The thiolate-sulfur may possibly develop into a sulfur radical which is stabilized by Cu(I). Thus it is concluded that the sulfur radical contributes, to a substantial degree, to the deep blue color of copper oxidases.

2.2. Type 2 Copper

A second type of copper is present in many multicopper oxidases. The electron absorption properties and the electron paramagnetic resonance parameters are quite similar to those obtained from most simple Cu(II) complexes. The reactivity of this copper is unique as it binds, rather strongly, many anionic inhibitors, including cyanide, azide, and halides. However, it should be emphasized that this type of copper is still different from that found in copper proteins of low electron absorption (e.g., the cupreins). Erroneous conclusions result when this type of copper is studied in an isolated manner; it must be examined in conjunction with the other types of copper in the multicopper oxidases (Fee, 1975; Malmström et al., 1975, and references therein).

2.3. Type 3 Copper

Initially all copper not detectable by electron paramagnetic resonance was attributed to Cu(I). However, the uptake of electrons by this type of copper suggested the presence of Cu(II), being antiferromagnetically coupled. On the other hand, examples of the presence of Cu(I) have recently been shown in the case of copper-thionein (Weser et al., 1977a). In this copper-sulfur protein, Cu(I) was monitored by X-ray photoelectron spectrometry. In addition to these types of copper, possible Cu(I)-Cu(III) pairs and $(Cu^{2+})_2$RSSR are discussed.

3. GENERAL CHARACTERISTICS OF BIOCHEMICALLY ACTIVE COPPER

In cellular biochemistry many transition metals play a dominant role in oxygenation and oxidation reactions. These metals, having themselves unpaired electrons, form complexes with triplet oxygen, and the respective orbitals of the two components overlap. We have to consider all unpaired electrons on such a complex as an entity, the assignment of the exact number of unpaired electrons being very difficult. According to Hamilton (1974), such a system can react with a singlet molecule. Spin-allowed processes become possible, yielding oxidized singlet products.

In the biochemistry of oxygen we cannot look at the reactivities of transition metals in isolation. The reactions of many different excited oxygen species have to be considered at the same time. In biological systems a controlled reactivity of these oxygen species is observed. In other words, the overwhelming number of interconversion reactions is minimized. This task is readily accomplished by the concerted action of a macromolecular protein portion bound, for example, to $3d^9$ copper. The specificity of the desired reaction is dictated by the protein portion. The following are examples of specific copper-dependent pathways of oxygen metabolism:

O_2^0: reversible oxygenation (hemocyanin)

$\cdot O_2^-$: superoxide dismutation (the cupreins)

O_2^{2-}: peroxide metabolism (copper-containing oxidases)

O^{2-}: formation of water (cytochrome oxidase, blue oxidases (Malmström et al., 1975) including ascorbate oxidase, ceruloplasmin, and laccase)

The major data on the copper proteins presently known are summarized in Table 1. As stated at the beginning, the two copper proteins designated as cuprein and copper-sulfur protein will be discussed in more detail.

4. CUPREIN (2CU,2ZN SUPEROXIDE DISMUTASE)

In the biochemistry of oxygen, the reactivity of metal proteins dealing with reversible oxygenation, the metabolism of peroxides, and the formation of water are fairly well understood. By way of contrast, we are only now beginning to understand the metabolism of superoxide, including all possible side reactions which lead to many interesting oxygen species. In this context a critical summary on what we know at present about the biochemical action of Cu,Zn superoxide dismutase will be helpful.

Almost 40 years have elapsed since Mann and Keilin (1938) isolated a class of mammalian copper proteins from both erythrocytes and liver. They called these proteins cupreins. At this stage the functional aspect of this type of metal protein was obscure. It was McCord and Fridovich (1969) who reported the successful isolation of an enzyme called superoxide dismutase, which turned out to be identical with the cupreins. In addition to the copper content and the superoxide dismutase activity, the identity of the different cellular cupreins was shown by Carrico and Deutsch (1969), using human tissues; and, surprisingly, 2 additional g-atoms of zinc was found to be bound to the protein portion (Carrico and Deutsch, 1970; Weser et al., 1971). Within the past few years, dramatic progress has been made in elucidating the structure of this Cu,Zn protein, which had its climax in X-ray diffraction studies (Richardson et al., 1975, 1976).

4.1. Specificity of Protein-Bound Copper

The apparently simple picture of the enzymatic reaction of the cupreins became

Table 1. Major Data on Presently Known Copper Proteins

Copper Protein	Molecular Weight	Total Copper (g-atoms per Mole Protein) (blue/non-blue/EPR non-detectable)	Function	Source	References
Azurin	14,000	1 (1/0/0)	Electron transport	*Bordetella pertussis*	Sutherland and Wilkinson (1963)
Plastocyanin	21,000	2 (2/0/0)		Spinach chloroplasts	Peisach et al. (1976)
Stellacyanin	17,000	1 (1/0/0)		*Rhus vernicifera*	Blumberg and Peisach (1966)
Mung bean Blue protein	22,000	1 (1/0/0)		Mung bean	Shichi and Hackett (1963)
Umecyanin	15,000	1 (1/0/0)		Horseradish root	Stigbrand (1971)
Hemocyanin	7–8 × 10^6 (2500–7500 per subunit)	2 (0/10%/90% Cu I)	O_2 transport	Blood of molluscs, cephalopods, gastropods, decapods	Lontie and Witters (1973)
Ceruloplasmin	151,000	8 (2/2/4)	Oxidation of Fe^{2+}	Animal serum	Frieden (1971)
2Cu,2Zn superoxide-dismutase	32,000	2 (0/2/0) + 2Zn	$\cdot O_2^-$ dismutation	Eucaryotes	Weser (1973), Fridovich (1974a) Michelson et al. (1977)

Tyrosinase (poly-phenol-oxidase)	119,000	4 (0/0/4)	Oxygenase	Mushroom, mammalian tissue	Bouchilloux et al. (1963)
Dopamine-β-hydroxylase	290,000	2 (0/2?/0)		Bovine adrenal medulla	Goldstein (1966)
Laccase	120,000	4 (1/1/2)	Blue oxidases catalyze the reduction of O_2 to H_2O	*Rhus vernicifera*	Levine (1966)
Laccase	141,000/6000	4 (1/1/2)		*Polyporus versicolor*	Mosbach (1963)
Ascorbate oxidase	130,000	8 (2?/2?/4?)		Cucumber Squash	Dawson et al. (1975)
D-Galactose oxidase	75,000	1 (0/1?/0)	Nonblue oxidases catalyze the reduction of O_2 to H_2O	*Dactylium deuderides*	Amaral et al. (1963)
Monoamine-oxidase	120,000	1 (0/1/0)		Bovine plasma	Achee et al. (1968)
Diamine oxidase	96,000	1 (0/1/0)		Pea seedlings	Hill and Mann (1964)
Diamine oxidase	190,000	2 (0/2/0)		Pig kidney	Yamada et al. (1967)
Uricase	120,000	1 (0/1/0)		Mammalian liver	Mahler (1963)
Cytochrome-*c*-oxidase	~200,000	2 (0/1/1)-(+2heme *a* +2heme a_3)	Terminal oxidase	Mitochondriae	Wharton (1973)
Yeast copper thioneins	10,000[a] ± 1000	1–8 (0/0/1-8 Cu (I))	?	*Saccharomyces cerevisiae*	Weser et al. (1977a)
Fetal copper thionein	10,000[a] ± 1000	8 (0/8%/92% Cu (I)) + 2Zn	?	Fetal bovine liver	Hartmann and Weser (1977)

[a] Determined by gel filtration.

Table 2. Second-Order Rate Constants for the Reaction of $\cdot O_2^-$ and Different Copper Chelates in the Presence and Absence of Serum Albumin

Copper Complex	$10^{-9} \times k_{245}$ M^{-1} sec^{-1} Serum Albumin Omitted	$10^{-9} \times k_{245}$ M^{-1} sec^{-1} Serum Albumin Added
Cu(II)aquohydroxo complex	2.70 ± 0.20	0.25 ± 0.05
2Cu,2Zn superoxide dismutase	1.30 ± 0.10	1.30 ± 0.10
2Cu,2Co superoxide-dismutase	1.30 ± 0.10	1.30 ± 0.10
$Cu(tyr)_2$	1.00 ± 0.10	0.90 ± 0.10
$Cu(gly\text{-}his)_2$	0.29 ± 0.02	0.15 ± 0.05
$Cu(gly\text{-}his\text{-}leu)_2$	0.21 ± 0.02	0.10 ± 0.05

rather unclear (Brigelius et al., 1974, 1975). In pulse radiolytic studies, $CuSO_4$ alone proved twice as active ($k_{245} = 2.7 \pm 0.20 \times 10^9$ M^{-1} sec^{-1}) as the protein-bound copper of native superoxide dismutase ($k_{245} = 1.30 \pm 0.10 \times 10^9$ M^{-1} sec^{-1}) at neutral pH (Klug-Roth and Rabani, 1976). The catalysis of superoxide radical dismutation at acidic pH values is even four times faster in the presence of simple copper salts (Rabani et al., 1973). It was argued that the protein portion of native cuprein is required to protect the Cu(II) coordination site for superoxide radical binding from accidental and undesired coordination by all sorts of biogenic chelators. Whereas the catalytic action of $CuSO_4$ was dramatically diminished in the presence of serum albumin, no such inhibition was seen when the somewhat hydrophobic $Cu(tyr)_2$ complex was used. The rate constants for superoxide dismutation in the presence of serum albumin remained constant at 1.0×10^9 M^{-1} sec^{-1} and accorded fully with the corresponding constant calculated per equivalent of enzyme-bound Cu(II) (Brigelius et al., 1974; 1975; Younes et al., 1978).

Apart from the Cu,Zn protein, there are other transition metals, including iron and manganese, which are strongly bound in certain proteins. These metal proteins display superoxide dismutase activity as well (Yost and Fridovich, 1973; Lumbsden and Hall, 1974). As in the case of low molecular weight copper chelates, Fe-EDTA* was shown to be capable of marked superoxide dismutation (Halliwell, 1975; McClune et al., 1977).

In fact, it is expected that the catalysis of superoxide dismutation can be considered a common reaction of most transition metals. Thus we have to ask: Why should nature create high molecular weight superoxide dismutases? Small transition metal chelates are much better suited for transport and incorporation

* EDTA = ethylenediaminetetraacetic acid.

into membranes and subcellular portions of the living cell where superoxide radicals with longer lifetimes are known to be generated.

4.2. Indirect Assays of Superoxide Dismutase Activity

The low molecular weight copper complexes showed nearly the same activity toward superoxide as the protein-bound Cu(II) of native Cu,Zn superoxide dismutase in pulse radiolysis experiments, even in the presence of the natural chelator, serum albumin (Brigelius et al., 1974, 1975). However, their ability to inhibit the reduction of nitroblue tetrazolium (Beauchamp and Fridovich, 1971) was three orders of magnitude lower than the inhibitory effect of cuprein-Cu(II) (Younes and Weser, 1976a) (Table 3). This leads to considerable doubt that the generation of blue formazan in the widely used assay for superoxide dismutase activity is exclusively due to the effect of the superoxide anion. Similar doubts were raised by Amano et al. (1975), who observed that proteins other than the known superoxide dismutase were able to suppress formazan generation during the phagocytosis of latex particles by polymorphonuclear leukocytes with the same effectiveness as the cupreins. Using the xanthine-xanthine oxidase system, we observed a higher specificity for the inhibition of formazan generation by native erythrocuprein, compared to the effect of small copper chelates, than when potassium superoxide was used as the reducing agent. The reactivity of $Cu(tyr)_2$ was raised threefold, and that of $Cu(lys)_2$ fivefold, when the tetrazolium salt was reduced by potassium superoxide. In the case of the xanthine-xanthine oxidase system, many more reactive species could be generated than with potassium superoxide. Massey et al. (1971) proposed that the adducts of oxygen and flavins in flavoenzymes could yield different species,

Table 3. Inhibition of Formazan Generation by Cuprein-Cu(II), $Cu(II)(tyr)_2$, and $Cu(II)(lys)_2$

Inhibitor	Cu(II) Concentration for 50% Inhibition (n*M*)	[Cu(II) Chelate] × $[Cu(II)\text{-}Cuprein]^{-1}$
Xanthine oxidase system		
Cuprein-Cu(II)	40	–
$Cu(II)(tyr)_2$	45 000	1125
$Cu(II)(lys)_2$	86 000	2150
KO_2-system		
Cuprein-Cu(II)	60	–
$Cu(II)(tyr)_2$	22 000	376
$Cu(II)(lys)_2$	25 000	417

depending on the surrounding protein moiety. Even electrons might be released, thus accounting for part of the nitroblue tetrazolium reduction.

Another example of the problems with an apparent superoxide-mediated reaction involves the oxidation of adrenaline. In many laboratories the superoxide anion radical was regarded as the agent responsible for the generation of adrenochrome (Valerino and McCormack, 1969; Misra and Fridovich, 1972). Forman and Fridovich (1972) took the adrenochrome formation as general evidence that electrolytically produced superoxide radicals could migrate in aqueous solutions to react with small molecules. Pulse radiolytic studies revealed that adrenochrome formation occurs via a complicated mechanism and is not a simple superoxide-mediated oxidation (Bors et al., 1975). As we have demonstrated that low molecular weight copper chelates are as good superoxide dismutases as the native Cu,Zn protein itself (Brigelius et al., 1974, 1975), no detectable difference between the two types of "superoxide dismutase" should be expected. However, a contrary observation was made: all employed copper chelates accelerated the rate of adrenochrome generation in glycine buffer. The observed acceleration rate proved to be a logarithmic function of the copper concentration (Table 4) (Younes and Weser, 1976b).

Trapping agents of both hydroxyl radicals and singlet oxygen had no effect. Surprisingly, hydrogen peroxide inhibited the oxidation of adrenaline completely. No inhibition was seen, however, in the presence of catalase. This effect is not easily understood. Some very reactive species might be produced during the destruction of peroxide in the presence of catalase, even if they were bound to the enzyme, for example, the superoxide radical. The Cu(II)-induced adrenochrome formation can be explained by the ability of Cu(II) to withdraw an electron from adrenaline-bound oxygen. Thus a semiquinone is formed which is immediately oxidized to yield adrenochrome (Rapp et al., 1973).

Provided that there is a specific reaction of cuprein on the autooxidation of adrenaline, the formation of Cu(I) should be demonstrated. Furthermore a genuine enzymatic reaction requires the involvement of the macromolecular protein portion bound to the copper. Both the reduction of Cu(II) to Cu(I) and

Table 4. Acceleration of Adrenaline Oxidation by Copper Chelates

Copper Complex	Concentration Required to Double the Velocity of Adrenochrome Formation (μM): Autooxidation	Enzymically produced $\cdot O_2$
$Cu(gly)_2$	11.2	17.4
$Cu(salicylate)_2$	11.2	16.0
$Cu(lys)_2$	9.6	10.5
$Cu(tyr)_2$	5.3	8.6

the participation of the protein moiety should be detectable by circular dichroism.

Concentrations of 50 μM cuprein allowed the convenient determination of chiroptical properties during adrenaline autooxidation (Weser and Schubotz, 1978). The reduction of cuprein Cu(II) to Cu(I) was observed by a decrease in the 610-nm Cotton band. From marked and characteristic changes in the protein spectrum between 260 and 400 nm the formation of a ternary complex of cuprein, adrenaline, and oxygen is proposed. The Cotton effects observed do not originate from the addition of the adrenaline spectrum to the cuprein spectrum (Figure 2). It is interesting to note that the overall protein conformation, as seen from the constancy of the 210-nm Cotton trough, remained unchanged throughout the reaction. The constancy of the overall protein structure during the binding with intermediate radical forms of adrenaline supports the observation that during the reaction of a radical substrate with the corresponding enzyme virtually no activation energy is measured. A distinct entatic state, as observed for many metalloenzymes by Vallee and Williams (1968), is not re-

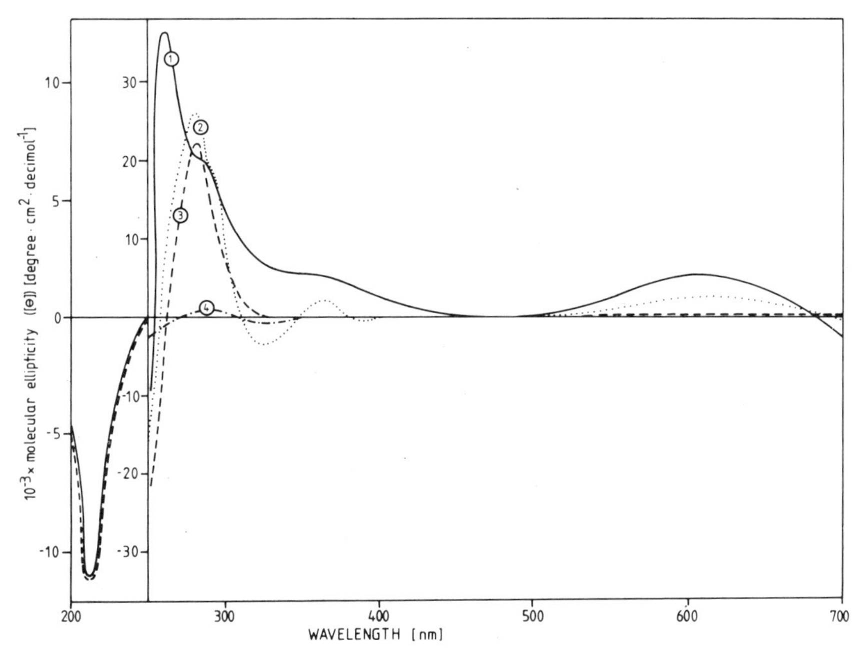

Figure 2. Circular dichroism of cuprein and adrenaline. 1, 50 μM cuprein in 0.1 M carbonate buffer at pH 10; 2, cuprein as in 1 after the addition of 0.5 mM adrenaline; 3, cuprein as in 1 but reduced by excess dithionite; 4, 0.5 mM adrenaline in 0.1 M carbonate buffer, pH 10, within 10 min of autooxidation. Molecular ellipticities for protein spectra below 25 nm were calculated on the basis of the mean residue weights of the amino acids.

quired in the case of cupreins, thereby minimizing conformational changes of the protein moiety.

By way of contrast, such a conformational change cannot be seen when the apoprotein or the heat-denatured enzyme is added (Figure 3). In these cases the total spectra of adrenaline oxidation products and the employed proteins must be taken into consideration.

The highly superoxide-dismutase-active copper chelate $Cu(tyr)_2$ was also unable to exert this phenomenon, although the reduction of Cu(II) was complete. It should be mentioned that the broad Cotton band in the 600-nm region appeared to be negative when $Cu(tyr)_2$ was examined (Figure 4).

5. COPPER-SULFUR PROTEINS

Proteins containing metal-coordinated sulfur are much more abundant than originally thought. The iron-sulfur proteins were supposed to be the sole representatives of this class of proteins. However, Zn, Cd, Hg and Cu have to be included by virtue of their coordination with thiolate-sulfur. Equine horse liver alcohol dehydrogenase contains two tetrahedrally coordinated zinc ions. One

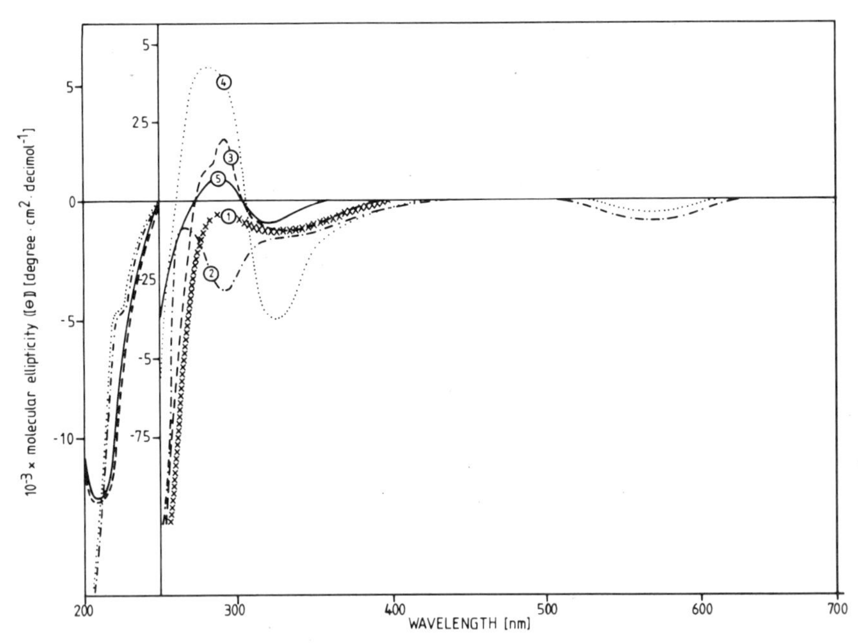

Figure 3. Circular dichroism of apocuprein, heat-denatured cuprein, and adrenaline. 1, 50 μM apocuprein in 0.1 M carbonate buffer, pH 10; 2, 50 μM heat-denatured cuprein in 0.1 M carbonate buffer, pH 10; 3, apocuprein as in 1 and 0.5 mM adrenaline; 4, heat-denatured cuprein as in 2 and 0.5 mM adrenaline; 5, adrenaline as 4 in Figure 2.

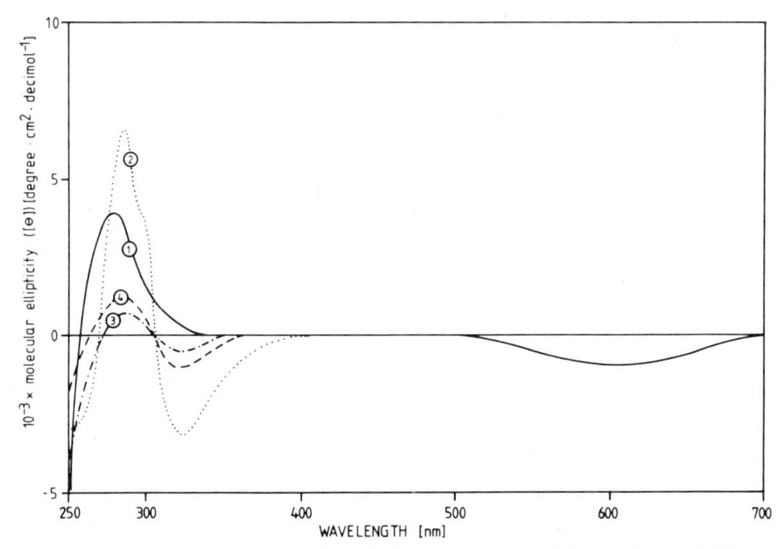

Figure 4. Circular dichroism of $Cu(tyr)_2$ and adrenaline. 1, 100 μM $Cu(tyr)_2$ in 0.1 *M* carbonate buffer, pH 10; 2, $Cu(tyr)_2$ as in 1 and 0.5 m*M* adrenaline; 3, 0.5 m*M* adrenaline in carbonate buffer, pH 10; 4, $Cu(tyr)_2$ as in 1 reduced by excess dithionite.

is coordinated to four cysteine sulfurs formally representing the "rubredoxin type" and is known to be "inactive" with regard to the catalysis of alcohol dehydrogenation (Brändén et al., 1975). Even the enzymically active zinc is bound to two cysteine sulfurs and one histidine nitrogen, with a vacant site for substrate binding. The distorted tetrahedral structure of either zinc ion is maintained.

It is very tempting to regard the tetrahedrally coordinated metal-sulfur cluster (Figure 5) as a general unit in biochemistry. For evolutionary reasons the sulfur ligand belongs to the ligands reaching farthest backward. Tetrahedral sulfur coordination is present in both the plant-type 2Fe-ferredoxins and adrenodoxin, and four distorted iron-sulfur tetrahedra form the cubane-like cluster of microbial iron-sulfur proteins (Brändén et al., 1975; Lovenberg, 1973).

X-ray diffraction measurements of iron-sulfur proteins and liver alcohol dehydrogenase are available. Our knowledge of the metal-sulfur chromophore of copper-thioneins is restricted to spectrometric data. The metal:sulfur ratio is close to 1:3 in the case of zinc- and cadmium-thionein (Kägi et al., 1974; Weser et al., 1977a), and 1:2 in the copper-thioneins (Hartmann and Weser, 1977). The tetrahedral thiolate-zinc chromophore in alcohol dehydrogenase and examples of low molecular weight metal-sulfur complexes, including Zn_2(dimethyldithiocarbamate)$_4$ (Klug, 1966) and Cu(I)-thiourea (Okaya and Knobler, 1964), led to the proposal of tetrahedral metal-thiolate clusters in those ubiquitous metallothioneins. It is suggested that the zinc-thioneins contain one three-coordinated sulfur atom, while two of such three-coordinated sulfur atoms

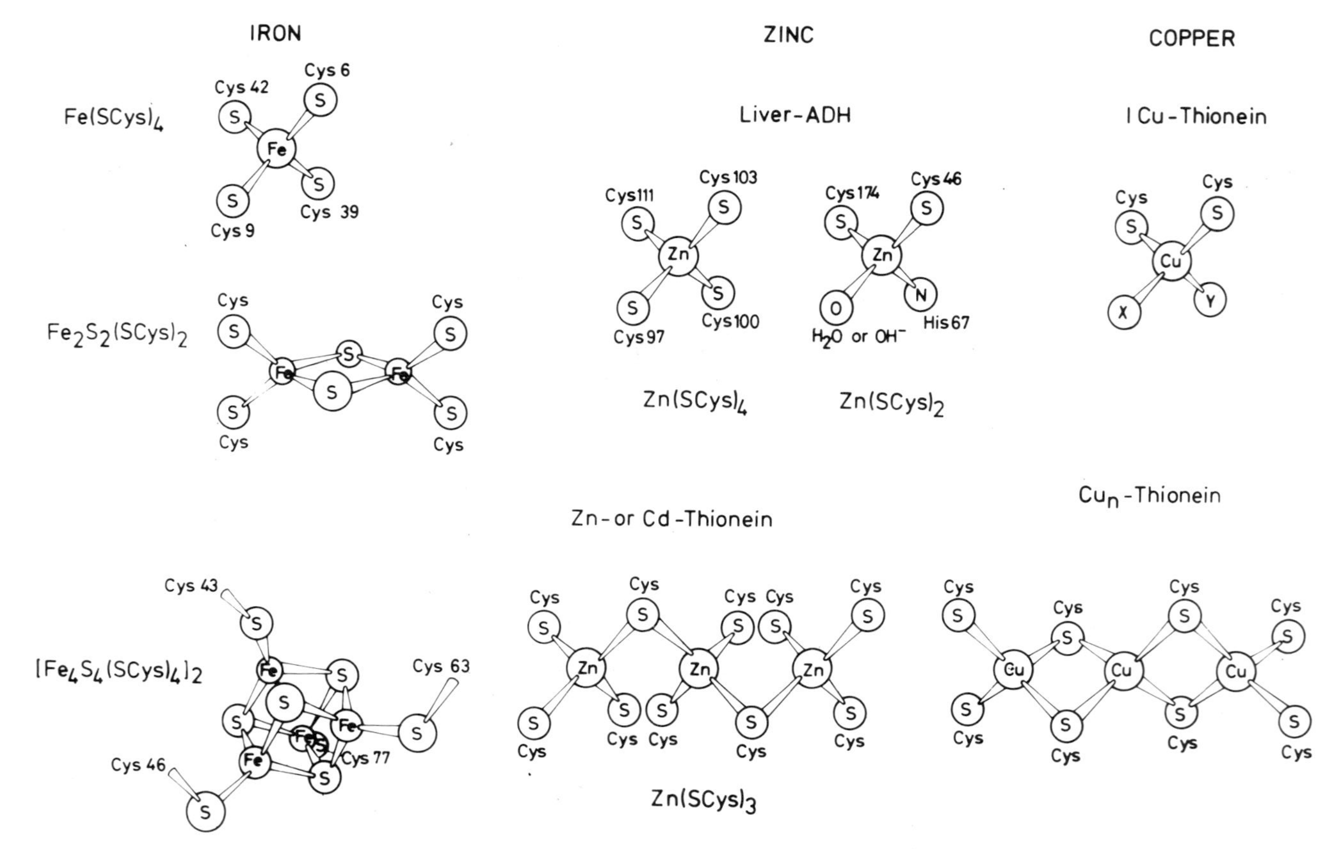

Figure 5. Metal clusters of metal-sulfur proteins.

are assumed to be present in the copper-thioneins. This proposal is made in light of the physicochemical measurements, including electron absorption spectrometry, chiroptical properties, magnetic resonance, and X-ray photoelectron spectrometry.

5.1. The Copper(thiolate)$_2$Chromophore

Electron absorption spectrometry has been successfully used to characterize the copper-sulfur chromophores. Because of differences in the copper content and the absence of disturbing aromatic amino acid residues, it was very convenient to demonstrate the ultraviolet spectrometric properties per mole of copper (Figure 6). Regardless of the biological origin, a similar absorption profile was detected. The ϵ_{Cu} values of copper-thioneins isolated from animals remained constant. A significant rise in ϵ_{Cu} values, however, was measured for the yeast copper-thionein (Table 5).

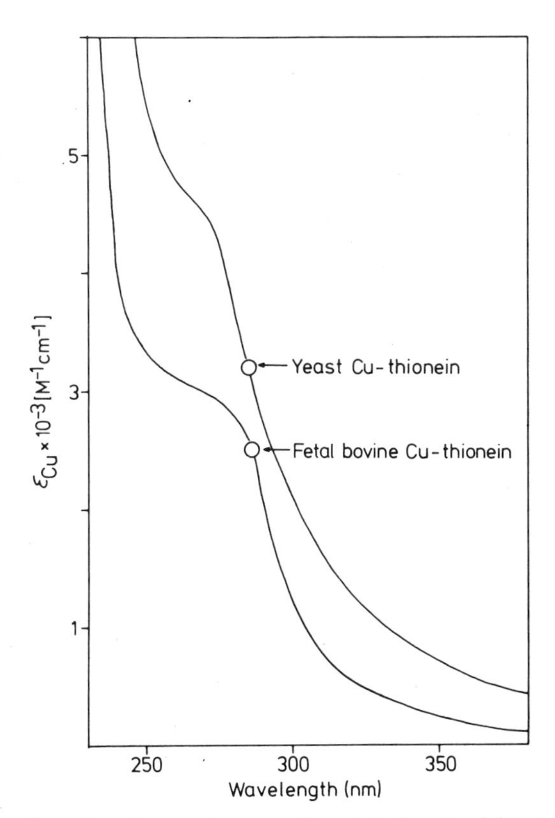

Figure 6. Molar absorption of copper in copper-thionein in the ultraviolet region.

Table 5. Molar Absorption Coefficients of Copper in Different Copper-Thioneins

Wavelength (nm)	Bovine Fetus	ϵ_{Cu} (M^{-1} cm^{-1})		
		Chicken	Rat[a]	Yeast
300	1210	1390	1030	2040
280	2870	2240	2220	3650
250	3320	3580	3170	5210

[a] Calculated from D. R. Winge, R. Premakumar, R. D. Wiley, and K. V. Rajagopalan, *Arch. Biochem. Biophys.*, **170**, 253–266, 1975.

The similarity in the chemical environment of the copper chromophore is more distinct in circular dichroism measurements. Although the magnitudes of the Cotton extrema vary in the different copper-thioneins, their positions remained essentially at the same wavelength, regardless of the copper-thionein studied (Table 6). The Cotton extrema of the microbial copper-thioneins were highest in resolution (Figure 7). Thus all subsequent measurements were carried out on these copper-thioneins. The magnitude of the observed Cotton bands paralleled both the cysteine and the copper content. It is tempting to conclude that these Cotton bands are attributable to coordination of the copper with the two cysteine sulfurs in conjunction with the overall protein conformation. As in the ultraviolet absorption, the θ values correlated linearly with the copper content (Figure 8).

Additional proof of the copper-sulfur coordination was obtained when the thionein-copper was successfully displaced either by protons (Figure 9*a*) or by *p*-chloromercuribenzoate (Figure 10). The remarkable pH stability should be noted. No changes occurred throughout the pH region examined except for the sharp drop of θ values within the pH range of 0.5 to 0.2 (Figure 9*b*). The *p*-chloromercuribenzoate titration supported nicely the conclusion of a 1:2 stoichiometry between copper and sulfur. Both copper displacement studies showed clearly that the copper-thiolate chromophore had been cleaved (Weser et al., 1977a; Rupp and Weser, 1978).

Table 6. Circular Dichroism Extrema of Copper-Thioneins

Origin	Wavelength (nm)					
Fetal bovine liver	370 (+)			285 (s−)	275 (−) 255	220 (s−)
Chicken liver	350 (+)		310 (−)	290 (−)	260 (+)	220 (s−)
Cu(I) titrated Yeast	360 (+)	330 (+)	300 (s−)	280 (−)	250 (+)	215 (s−)

s = shoulder.

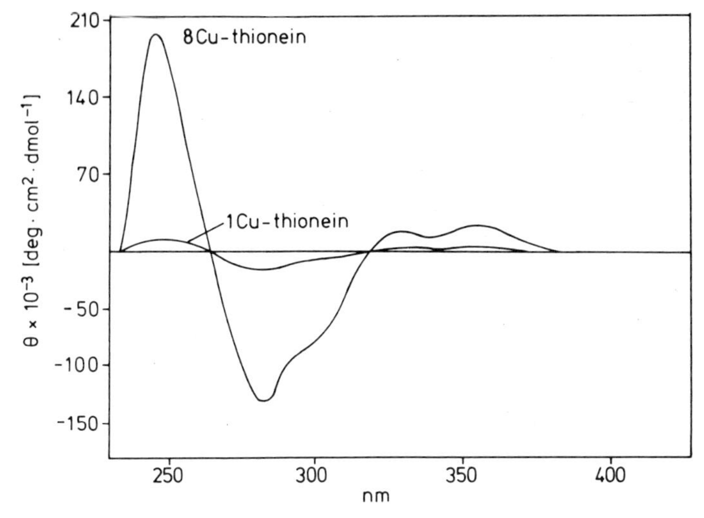

Figure 7. Circular dichroism of yeats copper-thionein.

5.2. The Redox State of Thionein-Bound Copper

In many X-ray photoelectron spectrometric studies on copper chelates of biochemical interest (Rupp and Weser, 1976a, 1976b, 1976b; Weser et al., 1979; Younes et al., 1979), we were able to distinguish between $3d^{10}$ Cu(I) and $3d^9$ Cu(II). When Cu(I)-containing compounds are measured, a single homogeneous copper $2p_{1/2}$ or copper $2p_{3/2}$ signal is detectable. The binding energy levels of the copper $2p_{3/2}$ core electrons are substantially different when Cu(II) is present: the main signal has been shifted by 2 eV, and a more or less split satellite from 940 ± 2 to 942 ± 2 eV appears. This satellite structure proved characteristic for the presence of Cu(II). Antiferromagnetic coupling of unpaired electrons did not affect the satellite structure observed. Upon measuring compounds containing spin-coupled Cu(II), including Cu(1,3-diphenyltriazene)$_4$, Cu(HCOO)$_2$, and Cu(succinate)$_2$·4H_2O, the satellite structure was fully detectable.

When erythrocuprein was measured, the copper $2p_{3/2}$ satellites were obscured by a high signal:noise ratio (Weser, 1973). Eight-copper-thionein contains about 10 times more copper, allowing the recording of a good-quality copper $2p_{3/2}$ spectrum. A large and homogeneous copper $2p_{3/2}$ signal appeared at 932.3 eV (Weser et al., 1977a), attributable to the exclusive presence of Cu(I). As no satellites were observed, antiferromagnetic coupling Cu(II) could be excluded. Treatment of the copper-thionein samples with hydrogen peroxide caused a blue shift of the main signal to 933.9 eV. In addition, two satellites at 940.0 and 942.0 eV were clearly detectable. It should be emphasized that this is the first time that both the existence of Cu(I) and the conversion of Cu(I) into Cu(II) were

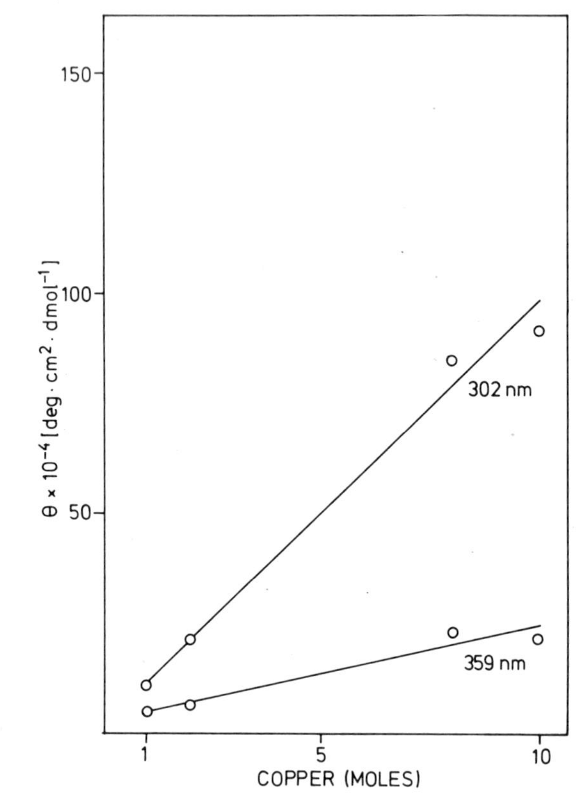

Figure 8. Correlation of copper content and dichroic amplitude, using different copper-thioneins.

unequivocally shown in a native copper protein by X-ray photoelectron spectrometry.

The binding energy values of the core electrons of Cu(I) and Cu(II) are paralleled by the corresponding data obtained for the sulfur 2*p* levels. A single sulfur core electron signal appears at 162.7 ± 0.1 eV when freshly prepared eight-copper-thionein is used. This numerical value can be assigned to the presence of thiolate-sulfur coordinated to Cu(I). Upon the sole treatment with hydrogen peroxide a dramatic shift of the sulfur 2*p* signal to 167.9 eV was seen, suggesting that all sulfur was oxidized to RSO_3^-. During the course of the hydrogen peroxide titration intermediate sulfur redox states were recorded at 163.7 and 166.6 eV, which were attributed to RSSR and RSO_2^-.

As in the circular dichroism studies, it can be taken as fact that the hydrogen peroxide treatment causes an irreversible breakdown of the copper-thiolate chromophore. In the newly formed Cu(II) protein, the Cu(II) may be assumed to be preferentially coordinated to double-bonded oxygen or nitrogen in a way similar to that of the copper-biuret complex (Rupp and Weser, 1976a).

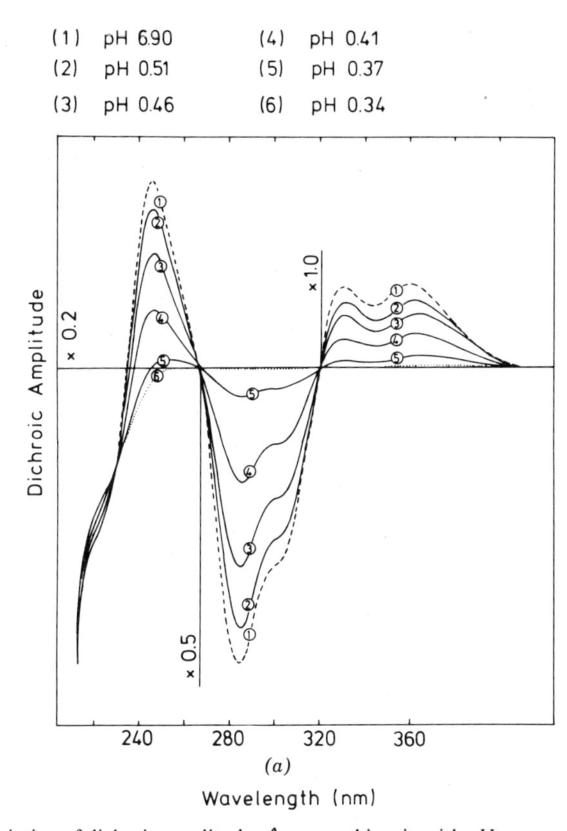

Figure 9a. Variation of dichroic amplitude of copper-thionein with pH.

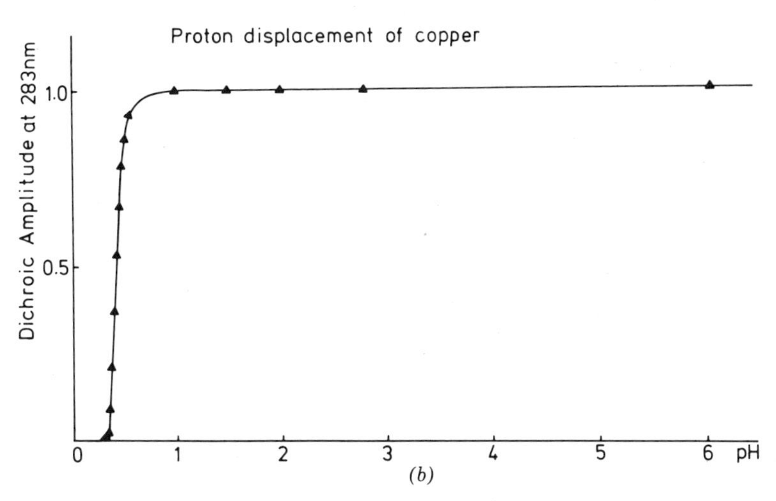

Figure 9b. Proton displacement of copper in copper-thionein.

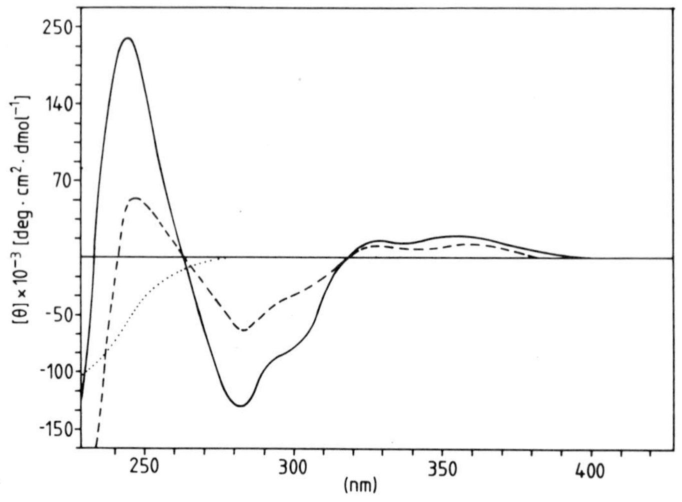

Figure 10. *p*-Chloromercuribenzoate titration of eight-copper-thionein (—), 1.6 μM (- - -) and 2.6 μM (· · ·) *p*-CMB.

6. LOW MOLECULAR WEIGHT COMPLEXES AS MODELS OF COPPER PROTEINS AND ENZYMES

One approach in studying the structures and functions of proteins and enzymes, which are complex macromolecules, is to examine simpler molecules which exhibit some of the features that appear to be essential characteristics of the macromolecules. As the model compounds are normally easier to study, the information gained can be extrapolated to the parent molecule. However, structural model compounds of metalloenzymes are not easy to obtain. Knowledge that the active sites exhibit unusual stereochemical features led to the conclusion that these sites are in an "entatic state" (Vallee and Williams, 1968). The macromolecular ligand appears to induce such an unusual geometry. Nevertheless, many model systems have been synthesized that mimic various characteristics of many metalloproteins and enzymes. We have to be aware, however, that in all cases we are dealing with a model and not with a replica (Malmström, 1970).

Many iron-sulfur clusters have been synthesized that exhibit structural features of various iron-sulfur proteins (for a review see Holm, 1975). Synthetic oxygen carriers were used to study the main aspects of reversible oxygenations (Martell and Taqui Khan, 1973). In a recent review, Freeman (1973) demonstrated the usefulness of the information gained from a study of zinc complexes in understanding the structure of the active site of carboxypeptidase. *N,N′*-Ethylenebis(trifluoracetylacetoniminato)Cu(II) was proposed as a model for the equatorial coordination of Cu(II) in galactose oxidase (Giordano and

Bereman, 1974). The electron paramagnetic resonance parameters differed only slightly in g_{zz} and A_{zz}, suggesting the presence of a strong π-bonding axial ligand in the enzyme. The *d-d* transitions were calculated in both cases and were in agreement with the experimental data.

Other model chelates have been used that possess catalytic activities comparable to these of metal enzymes, but normally the enzymic reactions may be up to 10^{11} times faster than the model ones (Hughes, 1972). Levitzki et al. (1965) examined the catalysis of the oxidation of ascorbic acid and *p*-hydroquinone by different Cu(II) complexes and observed Michaelis-Menten characteristics with Cu(II)-poly-L-histidine ($K_m = 1.5 \times 10^{-5}$ *M*). Sharma and Schubert (1969) were able to demonstrate that Cu(II)-imidazole, as well as Cu(II)-histidine (Sharma et al., 1970), exhibited catalytic activity in the neutral pH region. They suggested a mechanism in which the copper complexes react with the HOO^- anion. Upon reduction of Cu(II), the O-Cu-O bonding is increased and rupture of the O-O bond is facilitated. Sigel (1969) studied the catalase-like activity of various Cu(II) chelates with polyamines, peptides, native RNA, DNA, 2,2′-bipyridyl, and glycinamide and deduced that the catalatic activity depends upon the number of coordinated ligand groups. In all these cases, however, the concentrations of the chelates needed to exert a catalytic action comparable to that of the native enzymes were very high.

We shall focus our interest on structural and functional model chelates of 2Cu,2Zn superoxide dismutase (cuprein), as well as the group of proteins containing type 1 copper centers.

6.1. Model Complexes of 2Cu,2Zn Superoxide Dismutase

Structure Models

Many X-ray photoelectron spectrometric studies on native and modified bovine erythrocuprein (2Cu,2Zn superoxide dismutase), as well as different Cu-, Zn-, Co-, and Cd-amino acid chelates, have been carried out in our laboratory to gain information regarding the charge densities and the coordinations of the metals in the native enzyme (Jung et al., 1972, 1973a, 1973b, 1973c). It could be deduced that copper, which had a charge density characteristic of Cu(II) throughout, is not bound to sulfur. Marked differences between the binding energies of ionic- and complex-bound metals and their ligands were found, the binding energy of the copper in native cuprein indicating an ionic nature (Jung et al., 1972, 1973c).

On the basis of electron paramagnetic resonance studies of bovine superoxide dismutase, which revealed that the two evuivalent copper sites are of rhombic geometry (Rotilio et al., 1972), Morpurgo et al. (1973) suggested that crystalline Cu-ethylenetriamine diformate was a good structural model compound for the active site of the protein. They found that an aqueous solution of Cu-ethylenetriamine^{2+} cannot account for the properties of the protein-bound copper. The electron paramagnetic resonance spectrum of the complex solution itself has axial symmetry, and ligand binding does not greatly affect this feature, as is the case with cuprein. The redox behavior also proved different. A close fit, however, was observed with crystalline Cu-diethylenetriaminediformate: a rhombic crystal field was found, resulting in a three-g-value spectrum (g_1 = 2.0341, g_2 = 2.0939, g_3 = 2.2437; bovine superoxide dismutase: g_1 = 2.023, g_2 = 2.108, g_3 = 2.265) and a main absorption at 650 nm. The rhombicity was related to substantial displacement of the oxygen from a formate group coplanar to the three nitrogen atoms, while in the case of cuprein the rhombic distortion was assigned to displacement of a water molecule from the square planar position due to hydrogen bonding to some vicinal groups. When these conditions were altered (Rotilio et al., 1972) by binding of cyanide, fluoride, or azide or by achieving pH values below 4 or above 9, the electron paramagnetic resonance spectrum became axial, as was the case with aqueous Cu-diethylenetriamine^{2+} ions. The similarity in the optical properties of bovine superoxide dismutase and the complexes was assumed to require a similarity in the chemical nature of the ligands (Morpurgo et al., 1973).

Superoxide Dismutase-Active Copper Chelates

The ability of a great variety of Cu(II)-amino acid and peptide chelates to catalyze the spontaneous dismutation of the superoxide radical anion was examined in our laboratory, using two assay systems: the cytochrome *c* reductase and the chemiluminescence test. The xanthine-xanthine oxidase system was employed as superoxide radical source. The Cu(II) chelates of tyrosine, lysine, and histidine were the most active catalysts of superoxide dismutation (Joester et al., 1972), as were some di- and tripeptides thereof. The activities were very high compared to the value for the native enzyme (up to 7.5%).

This finding prompted us to measure the second-order rate constants of the reactions of some crystalline low molecular weight copper complexes with pulse radiolytically produced superoxide ions (Brigelius et al., 1974, 1975; Younes et al., 1978). The results of these studies were noted earlier in this chapter (see Section 4.1). Klug-Roth and Rabani (1976) reported the second-order rate constants of the reactions between some copper complexes of other amino acids and superoxide radicals to lie around $10^6\ M^{-1}\ sec^{-1}$. These chelates, however, were not available in defined crystalline form. Recently it could be shown that hydrated electrons reacted with Cu(II)-peptide complexes at a relatively slow rate, an organic portion being the first site of attack (Faraggi and Leopold, 1976). This implies that the major portion of these primary radicals produced during the pulse radiolysis experiments reacts with oxygen to produce superoxide radicals, as this reaction is a very rapid one ($k = 2 \times 10^{10}\ M^{-1}\ sec^{-1}$) (Gordon et al., 1963). Therefore no side reactions of the primary radicals with the chelates we investigated could have occurred. In addition to the fact that low molecular weight copper chelates showed nearly the same superoxide dismutase activity as native cuprein, it was found that many other complexes, such as Cu(phenanthroline)$_2$(ClO_4)$_2$ (Valentine and Curtis, 1975), some iron-porphyrin systems (Kovács and Matkovics, 1975), and Fe(II)- and Fe(III)-EDTA chelates (Halliwell, 1975; McClune et al., 1977), exhibited such activity. This rendered it doubtful that the basic physiological function of the cupreins is merely to catalyze the spontaneous dismutation of superoxide.

On the other hand, while native cuprein was able to prevent the production of singlet oxygen in different systems (Weser and Paschen, 1972; Paschen and Weser, 1973; Weser et al., 1975), the low molecular weight superoxide dismutase-active copper chelates failed to do so. This was found throughout, whether singlet oxygen was produced enzymatically (Weser and Paschen, 1972), in a totally inorganic system (K_3CrO_8) (Paschen and Weser, 1973), or photochemically (Weser et al., 1975). In the case of the CrO_8^{3-} decay, it was confirmed that, aside from superoxide and hydroxyl radicals, singlet oxygen was produced (Paschen and Weser, 1975). Native 2Cu,2Zn superoxide dismutase was able to suppress the chemiluminescence arising from excited oxygen species specifically, even in the absence of a chemiluminescence-mediating agent and in the presence of Tris buffer, a potent hydroxyl-radical scavenger. No such specificity was seen, however, when Cu(tyr)$_2$, Cu(lys)$_2$, or Cu(his)$_2$ was used instead (Paschen and Weser, 1975). The same results were obtained when singlet oxygen was generated photochemically in heavy water solutions in the presence of methylene blue (Weser et al., 1975).

Considering that the Cu(II)-amino acid chelates, such as Cu(tyr)$_2$ and Cu(lys)$_2$, had nearly the same superoxide dismutase activity as native cuprein, one would expect them to inhibit the reactions that are normally suppressed by the copper protein, if these reactions were mediated by superoxide radicals. In many investigations the participation of superoxide radicals in a reaction was postulated whenever a cooxidation of adrenaline or a coreduction of nitroblue tetrazolium was observed, which could be inhibited by cuprein. The different reactivities of the superoxide dismutase-active low molecular weight copper chelates in both the reduction of nitroblue tetrazolium (Younes and Weser, 1976a) and the autooxidation of adrenaline (Younes and Weser, 1976b) have already been reported and show that these reactions do not occur only because of the presence of superoxide anion radicals. Therefore, although it is tempting to regard the inhibition of various reactions by 2Cu,2Zn superoxide dismutase as evidence for the participation of superoxide radicals in these reactions, such conclusions should not be drawn until the substrate specificity of this enzyme is definitely established.

One of the major advantages of the low molecular weight copper chelates which exhibit superoxide dismutase activity is that they can exert their activity at or near the active sites of enzymes that might produce superoxide radicals during their catalytic action. Because of its molecular dimensions the native superoxide dismutase fails to show activity. This was demonstrated to be true in the case of the hepatic microsomal demethylation of *p*-nitroanisol and aminopyrine, which was inhibited by $Cu(tyr)_2$ but not by native cuprein, Cu^{2+} ions, or $Cu(lys)_2$. This effect was related to the superoxide dismutase activity of the complex, exerted in a scarcely accessible, low-polarity environment (Richter et al., 1976). Furthermore it was found that $Cu(tyr)_2$ was able to suppress, in addition, the dealkylation of 7-ethoxycoumarin by induced rat microsomes. Oxygen uptake by the microsomes was also inhibited by the copper chelate. On the other hand, the complex failed to inhibit 7-ethoxycoumarin dealkylation induced by 3-chlorobenzoic acid (Richter et al., 1977). A double reciprocal plot of the rate of *p*-nitrophenol formation as a function of *p*-nitroanisol concentration in the presence and in the absence of $Cu(tyr)_2$ (Figure 11) demonstrates the inhibitory action of the chelate on the hepatic microsomal dealkylation.

These observations indicate that superoxide radicals are formed by cytochrome P-450 during its multifunctional oxidase action. The $Cu(tyr)_2$ complex seems to exert its superoxide-dismutating activity at the level of a cytochrome P-450 intermediate species to which superoxide radical is coordinated, as shown in Figure 12.

6.2. Anti-inflammatory Activity

Recently Sorenson (1976a and this volume) reported that a great number of

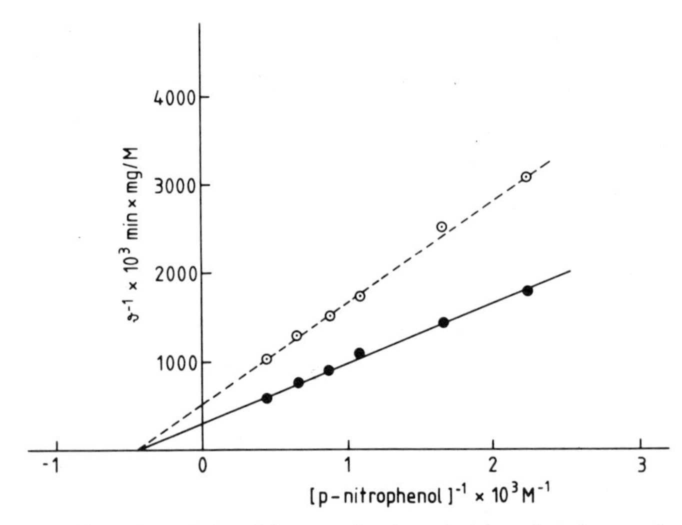

Figure 11. Double reciprocal plot of the rate of *p*-nitroanisol demethylation as a function of its concentration in the absence (-•-) and in the presence (-o-) of 25 μM $Cu(tyr)_2$. The incubation mixture was composed of 150 m*M* KCl, 50 m*M* Tris-HCl of pH 7.75, 10 m*M* $MgCl_2$, 7 m*M* isocitrate, 10 m*M* nicotinamide, 0.06 mU of isocitrate dehydrogenase, 1.5 mg/ml microsomal protein, and 1 m*M* *p*-nitroanisol.

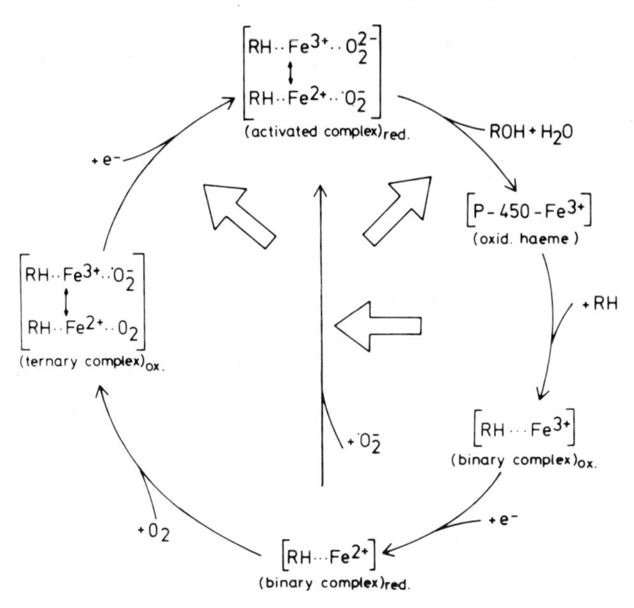

Figure 12. Possible sites of action of superoxide dismutase-active Cu(II) chelates during the catalytic cycle of the cytochrome P-450 hydroxylase system as proposed by H. W. Strobel and M. J. Coon, *J. Biol. Chem.*, **246,** 7826–7829 (1971). These sites are indicated by ⇒.

copper complexes of antiarthritic agents were more active as anti-inflammatory agents than were their parent compounds. Furthermore, he stated that Cu(II) chelates of some inactive chelating agents were very potent anti-inflammatory compounds (Sorenson, 1976b). These complexes were less toxic and had, in addition, antiulcer activity, as was reported elsewhere (Rainsford and Whitehouse, 1976; Boyle et al., 1976). Sorenson postulated that copper chelates might be the active metabolites of the clinically used antiarthritic agents:

$$\text{Agent} \xrightarrow{Cu^{2+}} \text{copper chelate} \rightarrow \text{antiarthritic activity}$$

As McCord (1974) had postulated that hydroxyl radicals produced via the reaction of superoxide with hydrogen peroxide were responsible for the inflammatory type of arthritis, and Oyanagui (1976) had found that superoxide dismutase administrated intravenously to rats completely suppressed the prostaglandin phase swelling of carrageenan foot odema, the possibility that the anti-inflammatory-active copper chelates exerted their action via the suppression of superoxide production had to be considered. It was shown that the (bissalicylato)Cu(II) complex had a superoxide dismutase activity (De Alvare et al., 1976). The authors observed the electron paramagnetic resonance spectral changes during the reaction of the chelate with potassium superoxide in dimethyl sulfoxide and proposed a mechanism for this reaction (Figure 13).

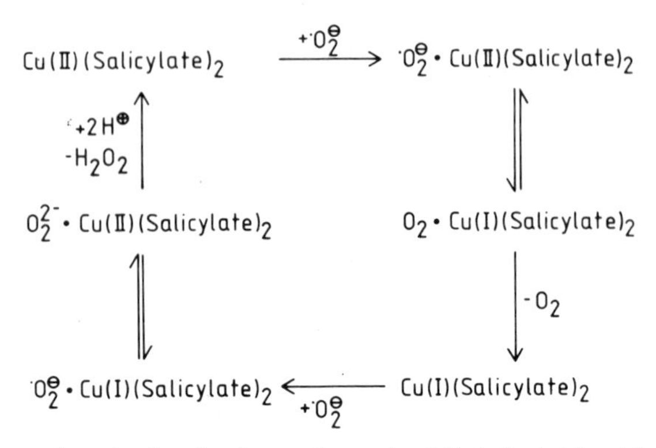

Figure 13. Proposed mechanism for the reaction cycle of Cu(salicylate)$_2$ with superoxide.

In our laboratory the activity of chelated Cu(II) with four different aspirin-like drugs in various superoxide dismutase assays was examined (Weser et al., 1977b). The oxidation state of the copper involved had been measured by X-ray photoelectron spectrometry and was found to be +2 ($3d^9$) throughout. All copper complexes were able to suppress the xanthine-xanthine oxidase-mediated reduction of both cytochrome *c* and nitroblue tetrazolium, as well as formazan formation by potassium superoxide in a specific manner. Furthermore, the hydroxylation of benzo-[α]-pyrene and the dealkylation of 7-ethoxycoumarin by induced hepatic rat microsomes were successfully inhibited by the copper chelates employed.

In this context the copper-penicillamine complex appeared to be of special interest in two respects. First, penicillamine itself proved inactive as an anti-inflammatory agent, whereas the copper chelate was a potent inhibitor of the models of inflammation that Sorenson (1976a) had investigated. Second, it was found that the copper was present as Cu(I) in the red-violet copper-penicillamin complex, the sulfur probably being present as a radical (Rupp and Weser, 1976a, 1976c):

$$RS^- + Cu(II) \rightleftharpoons RS^-Cu(II) \rightleftharpoons \text{-}[RSCu(I)]_n$$

This complex proved also able to inhibit various reactions mediated by superoxide and perhaps other excited oxygen species. X-ray photoelectron spectrometry revealed that the actual charge density of copper before and after the reaction with superoxide radicals was +1, while the binding energy of the sulfur remained constant. This indicated that in this case the sulfur radical, present as Cu(I)·SR˙, is the species actually undergoing the redox cycle (Younes and Weser, 1977; Lengfelder et al., 1979):

$$Cu(I)\cdot SR + \cdot O_2^- \rightleftharpoons Cu(I)^-SR + O_2$$

$$Cu(I)^-SR + \cdot O_2^- \rightleftharpoons Cu(I)\cdot SR + O_2$$

6.3. Models for the Blue Copper Centers

The characteristics of the blue copper centers were briefly discussed in Section 2. The exact chemical environment of the bound "blue" copper, however, is not fully known. In a resonance-Raman study employing some blue copper proteins, a band near 270 cm^{-1} was assigned to a copper-sulfur stretching mode, and a $S(\pi) \rightarrow Cu$ charge transfer was deduced (Miskowski et al., 1975; Siiman et al., 1976). In the case of stellacyanin, Rist et al. (1970) were able to show by means of electron nuclear double resonance that more than one set of nitrogens contribute to the double resonance signal. The blue copper site geometry was found to be between planar, the usual structure of four-coordinated Cu(II), and tetrahedral, which is characteristic for four-coordinated Cu(I) complexes (Holm and O'Connor, 1971). From data obtained by low-temperature absorption, circular dichroism, and magnetic circular dichroism of some blue copper proteins and ligand field calculations based on a tetrahedral structure distorted about 6° toward a square plane, Solomon et al. (1976) suggested a structural model for the blue copper centers. The characteristic absorption and circular dichroism bands were assigned to ligand-to-metal transitions $S(\pi) \rightarrow d_{x^2-y^2}$, $S(\delta) \rightarrow d_{x^2-y^2}$, and $N^*(\pi) \rightarrow d_{x^2-y^2}$, and a CuN_2N^*S unit was thought to represent the active centers of type 1 copper proteins (Figure 14). A distorted $CuSN_3$ core had previously been suggested by Williams (1973).

Much work has been done on protein complexes and low molecular weight copper chelates which exhibit some of the unusual spectral features shown by the blue copper centers. The characteristic absorption bands and their intensities found for some of these model chelates are listed in Table 7.

Morpurgo et al. (1976) studied the spectral properties of several anion complexes of Cu(II)- and Co(II)-bovine carbonic anhydrase, the crystal structures of which are well known. In the case of the sulfide and the 2-mercaptoethanol complexes of both metal-substituted enzymes, two intense transitions in the

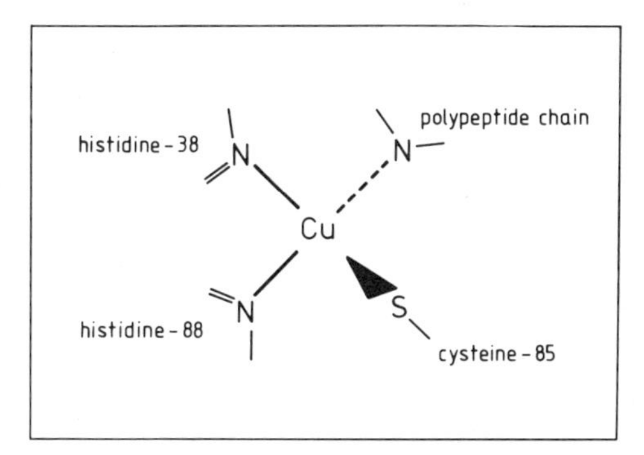

Figure 14. Proposed structural model for the blue copper center of bean plastocyanin. Solomon et al. (1976).

Table 7. Absorption Bands and Molar Absorption Coefficients of Copper Chelates

Group of Ligands	Complex	λ_{max}, ~400-nm Band (nm)	ϵ_{400} (M^{-1} cm^{-1})	λ_{max}, ~600-nm Band (nm)	ϵ_{600} (M^{-1} cm^{-1})	Reference
Cyclic polythioethers	Cu-12-ane-S_4[a]	387	6.0×10^3	675	2.0×10^3[g]	
	Cu-13-ane-S_4[b]	390	6.0×10^3	627	1.8×10^3[g]	
	Cu-14-ane-S_4[c]	390	8.2×10^3	570	1.9×10^3[g]	Jones et al. (1975a)
	Cu-15-ane-S_4[d]	414	8.0×10^3	565	1.1×10^3[g]	
Open-chain polythioethers	Cu-et_2-TTU[e]	410	6.8×10^3	612	1.1×10^3[g]	Jones et al. (1975a)
Sulfur-containing peptides	2-Mercaptopropionylglycine-Cu (green)	400	Shoulder	605	2.6×10^2	Sugiura et al. (1975)
	3-Mercaptopropionylglycine-Cu	400	Shoulder	630	1.0×10^2	
	2,5-Dimercaptopropionylglycine-Cu	400	Shoulder	605	2.8×10^2	
	2-Mercaptopropionyl-L-cysteine-Cu	405	7.0×10^2	580	2.8×10^2	Sugiura and Hirayama (1977)
	N-Mercaptoacetyl-L-histidine-Cu	438	9.7×10^2	598	8.3×10^2	

Phosphinate-bridged dithiocompounds	Bis(imidotetraphenyldithiodiphosphino-*S*,*S'*)-Cu	407	4.1×10^3	575	3.6×10^{3h}	Bereman et al. (1976)
Imidazole derivatives	1-Methyl-2-mercaptoimidazole-Cu(A)	—	—	620	6.3×10^{2i}	Dobry-Duclaux and Périchon (1976)
	1-Methyl-2-mercaptoimidazole-Cu(B)	420	9.6×10^2	620	3.7×10^{3j}	
N_3SR ligands	$Cu[HB(3,5\text{-}me_2pz)_3](p\text{-}NO_2C_6H_4S)^f$	—	—	588	3.9×10^{3k}	Thompson et al. (1977)

[a] $\hat{=}$ Cu-1,4,7,10-tetrathiocyclododecane.
[b] $\hat{\Delta}$ Cu-1,4,7,10-tetrathiocyclotridecane.
[c] $\hat{=}$ Cu-1,4,8,11-tetrathiocyclotetradecane.
[d] $\hat{=}$ Cu-1,4,8,12-tetrathiocyclopentadecane.
[e] $\hat{=}$ Cu-3,6,10,13-tetrathiopentadecane.
[f] $\hat{=}$ Cu-[hydrotris(3,5-dimethyl-1-pyrazolyl)borate]·(*p*-nitrobenzenethiolate).
[g] In 80% methanol-20% water.
[h] In methylene chloride.
[i] In ethanol.
[j] In acetonitrile.
[k] At −78°C in tetrahydrofuran.

optical absorption spectra were observed, suggesting charge-transfer interactions between sulfur and an acceptor group of the protein. A great similarity between the Co(II)-bovine carbonic anhydrase-sulfides and Co(II)-stellacyanin was detected, whereas the corresponding Cu(II) enzymes showed marked differences. The authors suggest that the Cu(II) is pentacoordinated in native stellacyanin, unlike Cu(II)-bovine carbonic anhydrase-sulfides and the Co(II) enzymes, while tetrahedral Co(II)-stellacyanin was proposed as a model for the reduced copper site.

Markedly positive potentials are among the unusual features exhibited by blue copper proteins. Patterson and Holm (1975) made polarographic measurements of 37 copper chelates containing primarily salicylaldiimine, β-ketoamine, β-aminoamine, and pyrrole-2-aldiimine ligand systems in order to account for these unusual features. Nonplanar bischelate complexes were found to be easier to reduce than their planar analogues. Complexes differing only in donor atoms could be more readily reduced in the order $N_4 < N_2O_2 < N_2S_2$. If one takes into account the potentials of the complexes these workers studied, and assumes four-coordinated Cu(I) in the models and reduced proteins, a potentially suitable ligand structure of blue copper proteins would include one thiolate-sulfur, with some or all of the remaining ligands being histidyl groups, and an entire ligand structure constrained toward nonplanar coordination in both redox forms.

6.4. Copper-Thioether Chelates

Copper-thioether complexes were suggested as simple low molecular weight models for blue copper centers in proteins as they exhibit spectral characteristics and redox potentials similar to those of the native macromolecules (Jones et al., 1975a; Dockal et al., 1976). Kinetic measurements had revealed that both the dissociation rate constants and the stability constants depended on the size of the ring which was thought to be associated with the binding of the third and fourth donor atoms and associated conformational changes (Jones et al., 1975 b). Copper(II)-14-ane-S_4 [$\hat{=}$ Cu(II)-1,4,8,11-tetrathiocyclotetradecane] was found to be the most stable ($K_m = 3.1 \times 10^3\ M$, $k_f = 2.8 \times 10^4\ M^{-1}\ \text{sec}^{-1}$). X-ray crystallographic data for this chelate showed the Cu(II) to be symmetrically centered within the plane of the four sulfur atoms. High molar absorptions were observed with the chelates near 600 nm, which were thought to result from sulfur-to-Cu(II) charge transfers, as is supposed to be the case with blue proteins (Jones et al., 1975a). The authors also found that the copper complexes of 12-ane-S_4 ($\hat{=}$ 1,4,7,10-tetrathiocyclododecane) and 13-ane-S_4 ($\hat{=}$ 1,4,7,10-tetrathiocyclotridecane), which are too small to permit planar coordination to Cu(II), as is the case with 14-ane-S_4, exhibited molar absorption coefficients near 600 nm, which are as high as the value for the Cu(II) chelate of the latter ligand.

It was concluded that no specific coordination geometry to the Cu(II) is required for the generation of the charge-transfer band. Dockal et al. (1976) found that the coordination to either mercaptide-sulfur or thioether-sulfur might influence the Cu(II)-Cu(I) potential in such a way as to render it more positive. The potentials did not appear to be primarily dependent on coordinative distortion. With the thioether-copper chelates, positive redox potentials comparable to those of native blue proteins were observed. On the basis of these observations it was suggested that the thioether-sulfurs of methionine groups represented the ligating sulfur donor atom(s) in the copper proteins (Jones et al., 1975a).

Miskowski et al. (1976) studied the ligand-to-metal charge-transfer absorptions of some Cu(II)-thioether complexes and found that, relative to the $S(\delta) \rightarrow Cu(II)$ ligand-to-metal charge-transfer absorptions, those that could be assigned in part to $S(\pi) \rightarrow Cu(II)$ ligand-to-metal charge transfers were considerably less intense, red-shifted, and not well separated in energy from interfering ligand field absorptions.

6.5. Copper Complexes with Sulfur-Containing Peptides

Sugiura et al. (1975) described the characteristics of a green complex of 1-mercaptopropionylglycine as a model for the active center of blue copper proteins. Very stable and broad absorption peaks were observed at 400 and 605 nm, and the magnetic circular dichroism curve consisted of two negative bands at 450 and 600 nm. The spin-Hamiltonian parameters of the electron paramagnetic resonance spectrum of the complexes ($g_{\parallel} = 2.259$, $g_{\perp} = 2.040$, and $A_{\parallel} = 82$ G), as well as the bonding parameters, were found to be comparable to those of the chromophores in the copper proteins. The authors observed a large degree of covalency of the copper-sulfur bonding in the complex, which would suggest a decrease in the unpaired electron density in the Cu(II) atom. They considered this fact to be one of the main reasons for the unusual characteristics of the chromophore. In fact, using X-ray photoelectron spectrometry, which permits an exact determination of the actual charge density of copper (Rupp and Weser, 1976a, 1976b, 1976c), it could be shown that the binding energy of the copper in this chelate was characteristic of Cu(I), suggesting that a sulfur-copper charge transfer actually occurs (Rupp and Weser, 1976b).

Several Cu(II) and Ni(II) complexes of sulfhydryl- and imidazole-containing peptides were synthesized and characterized by visible absorption, circular dichroism, proton nuclear magnetic resonance, and electron paramagnetic resonance spectra. The 1:1 *N*-mercaptoacetyl-L-histidine–Cu(II) complex showed unique spectral characteristics similar to those of blue copper proteins: an intense absorption at 598 nm ($\epsilon = 830\ M^{-1}\ cm^{-1}$) and a small copper hyperfine coupling constant ($A_{\parallel} = 93$ G) (Sugiura and Hirayama, 1977). The 600-nm band was assigned to the $S(\delta) \rightarrow Cu(d)$ charge-transfer transitions. The data obtained with this model complex suggested that blue copper sites involve cysteine-sulfhydryl and histidine-imidazole coordinations. When the complexes were reduced by dithionite (1:1), the respective Cu(I) chelates were obtained. From proton nuclear magnetic resonance data it was found that *N*-mercaptoacetyl-L-histidine coordinated to Cu(II) through the sulfhydryl-sulfur and imidazole-nitrogen (Sugiura, 1977), the amide-nitrogen not being involved in the bonding. As in native copper proteins, the ligands must be adequate for both Cu(II) and Cu(I), and redox reactions not requiring great geometrical changes are favored. This is further evidence that this chelate is a good model for blue copper sites in proteins. As cysteine-sulfur is δ-donating, the histidine-imidazole would be required for π-back-donation: this is a prerequisite for the stabilization of the four-coordinated Cu(I) complex.

6.6. Polythiocompounds as Ligands

Macromolecular Cu(II) chelates with penicillamine and related compounds, $Cu_{14}[SC(CH_3)_2CHNH_2COO]_{12}Cl \cdot {\sim}18H_2O$ (Wright and Frieden, 1975) and $Cu_{14}[SC(CH_3)_2CH_2NH_2]_{12}Cl{\sim}3.5SO_4^{2-} \cdot {\sim}19H_2O$ (Schugar et al., 1976), which are deeply colored, were suggested as models for the copper-sulfur coordination in type 1 copper proteins. A broad absorption of the latter, at 518 nm ($\epsilon \simeq 3400\ M^{-1}\ cm^{-1}$ per Cu(II)), believed to contain eight Cu(I) and six Cu(II) on the basis of the crystallographic data, was assigned to the δ-component of S $\rightarrow$ Cu charge transfer. In fact, it was found that the actual charge density of the copper in the first case was that of Cu(I), merely suggesting that such a transition really occurs (Rupp and Weser, 1976a, 1976c; Younes and Weser, 1977).

Bereman et al. (1976) studied the properties of the bis(imidotetraphenyidithiophosphino-*S,S'*)-Cu(II) complex. A CuS_4 tetrahedral core is formed which is easily reduced to Cu(I) at room temperature in the solid state and in solution. Intense absorption bands were observed at 407 nm ($\epsilon = 4100\ M^{-1}\ cm^{-1}$) and 575 nm ($\epsilon = 3600\ M^{-1}\ cm^{-1}$). They were assigned, unlike the former examples, as Cu $\rightarrow$ S(δ) and Cu $\rightarrow$ S(π) charge-transfer bands. The electron paramagnetic resonance parameters were found to be consistent with a rather covalent pseudotetrahedral environment ($A_\parallel = 121$ G, $A_\perp = 20.2$ G, $g_\parallel = 2.107$, and $g_\perp = 2.030$).

6.7. Copper Chelates with Mercaptoimidazole Derivatives

Another approach to obtain model complexes of type 1 copper chromophores was to use mercaptoimidazole derivatives as ligands of the copper. Dobry-Duclaux and Périchon (1976) attempted to synthesize such chelates, using 2-

mercaptoimidazole, 4,5-dichloro-2-mercaptoimidazole, and ergothioneine. The complexes obtained, however, were unstable in air. Using 1-methyl-2-mercaptoimidazole, they synthesized a complex with Cu(II) salts which had a molar absorption coefficient at 620 nm of 630 M^{-1} cm^{-1} and which they described as complex A, having a structure of [Cu(II)(mercaptoimidazole)$_2$]$_2$. Another complex (complex B), obtained with $CuClO_4 \times 4CH_3CN$ and $Cu(CH_3COO)_2$, was described as [Cu(II)(Cu(I)mercaptoimidazole)$_5$]mercaptoimidazole $\times$ ClO_4 and had an $\epsilon_{620} = 3650\ M^{-1}$ cm^{-1}. The color of both complexes disappeared upon reduction and reappeared in the presence of oxygen. Complex A also showed a catalytic action on oxidation of ascorbate by molecular oxygen. In another study, Henry and Dobry-Duclaux (1976) examined the electron paramagnetic spectra of both complexes. Complex A had the following parameters: $g_{\parallel} = 2.278$, $g_{\perp} = 2.070$, and $A_{\parallel} = 6$ mK. Complex B showed a quasi-isotropic electron paramagnetic resonance signal at $g = 2.072$; no hyperfine structure due to copper or its ligands could be resolved.

If in these complexes, as in the blue copper centers, a S(π) → Cu charge transfer should occur, one would expect the charge density of copper to be altered. In an X-ray photoelectron spectrometric study carried out in our laboratory, the copper was found to be in oxidation state +1 because of the absence of any detectable copper $2p_{3/2}$ satellites in either complex (Figure 15) (Younes et al., 1979). In other words, the coordinating sulfur has been polarized and the

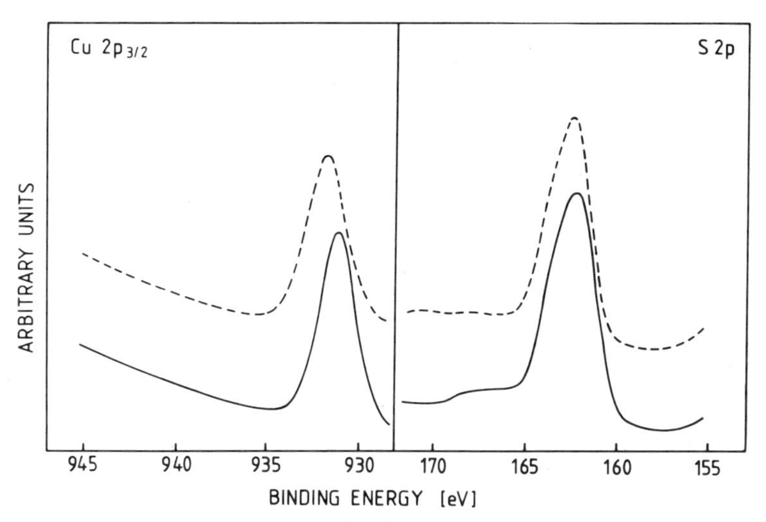

Figure 15. X-ray photoelectron spectra of both the copper $2p_{3/2}$ and the s2p levels of the two copper complexes with 1-methyl-2-mercaptoimidazole, described in Dobry-Duclaux and Périchon (1976). Recording conditions: work function, 6.0 eV; analyzer energy, 100 eV; sweep width, 20 eV; sweep time, 20 sec; number of channels, 200; pressure range, 1–2 μtorr; number of scans, 10 in all cases. —, Cu(1-methyl-2-mercaptoimidazole)$_2$; - - - -, Cu_7(1-methyl-2-mercaptoimidazole)$_7$(ClO_4)$_2$.

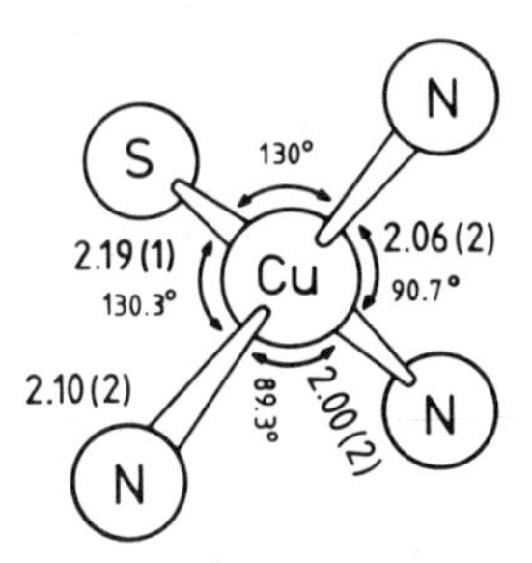

Figure 16. Coordination geometry about the copper atom in the K[Cu(HB(3,5-dimethyl-1-pyrazolyl)$_3$)(*p*-nitrobenzenethiolate)] · 2$(CH_3)_2CO$ complex as determined from X-ray crystallographic data.

electron resides at the copper site. Upon oxidation with hydrogen peroxide, Cu(II) was clearly detected by the shift of the main signal to 935.5 eV and the appearance of distinct satellites at 944.0 eV.

Recently the first copper redox pairs having a stoichiometry of Cu(I)N_3(SR) and Cu(II)N_3(SR), similar to that expected for the copper sites in proteins, were described. They were synthesized via the reaction of Cu(SR) or [Cu(SR)](ClO_4) (where SR corresponds to *p*-nitrobenzenethiolate or *o*-ethylcysteinate) with potassium hydrotris(3,5-dimethyl-1-pyrazolyl)borate (Thompson et al., 1977). The chelates were stable only at low temperatures, however, and had to be maintained at −78°C. The oxidized complexes showed intense optical absorption around 600 nm, which was assigned to a S → Cu charge-transfer transition. Resonance Raman lines were observed at 276, 339, 360, and 385 cm^{-1}, confirming the similarity of the complexes to native type 1 copper-containing proteins. Electron paramagnetic resonance studies revealed a similarity of *g* values to those of native blue proteins, although the synthetic system differed from the native protein in showing *no* unusually small hyperfine coupling constants ($A_{\parallel}$ = 17.0 to 17.1 mK). The coordination geometry around the copper was determined from X-ray data (Figure 16).

Not all small complexes are able to give a definite picture of the exact active sites of the proteins investigated. They do, however, mimic some of the characteristics of the proteins and thus facilitate the understanding of these features.

7. CONCLUSION

Many copper proteins are currently under investigation. In these proteins the metal is coordinated to a variable number of nitrogen, oxygen, and/or sulfur ligands. From a geochemical viewpoint, sulfur is one of the most exciting ligands in geological evolution. This moiety has to be on the list of the very early liganding species. Unfortunately, our knowledge of the copper-sulfur coordination in copper proteins is very limited, and further characterization of the copper-sulfur proteins remains a challenging task.

The sulfur-bearing protein ligands involved and the possible stabilization of a sulfur radical by Cu(I) reveal a new aspect of metal-sulfur redox reactions in bioenergetic systems. Geochemically viewed, the occurrence of sulfur radicals stabilized by polymeric ligands is a familiar phenomenon known in aquamarines and lapis lazuli. In these copper-free minerals, heteropolyacids, including polysilicates and polyaluminates, are the ligands for the successful formation of the sulfur radicals responsible for the deep blue color.

It may be concluded that one important reason for the dominant role of copper in biological systems is its capability to stabilize sulfur radicals.

ACKNOWLEDGMENTS

Parts of this work were supported by grants awarded to U.W. by the Deutsche Forschungsgemeinschaft (We 401/14) and the Fonds der Chemischen Industrie. L.M.S. is a predoctoral fellow of the Evangelisches Studienwerk e.V. Villigst. M.Y. is a recipient of a German Academic Exchange Service (DAAD) fellowship.

We are grateful to Dr. H.-J. Hartmann, Dr. H. Rupp, and Mrs. M. Kurth for valuable help and discussions.

REFERENCES

Achee, F. M., Chervenka, C. H., Smith, R. A., and Yasunobi, K. T. (1968). "Amine oxidase. XII: The Association and Dissociation and Number of Subunits of Beef Plasma Amine Oxidase," *Biochemistry,* **7,** 4329–4335.

Amano, D., Kagosaki, Y., Usui, T., Yamamoto, S., and Hayaishi, O. (1975). "Inhibitory Effects of Superoxide Dismutases and Various Other Proteins on the Nitroblue Tetrazolium Reduction by Phagocytozing Guinea Pig Polymorphonuclear Leukocytes," *Biochem. Biophys. Res. Commun.,* **66,** 272–279.

Amaral, D., Bernstein, L., Morse, D., and Horecker, B. L. (1963). "Galactose Oxidase of *Polyporus Circinatus:* a Copper Enzyme," *J. Biol. Chem.,* **238,** 2281–2284.

Beauchamp, C. and Fridovich, I. (1971). "Superoxide Dismutase: Improved Assays and Assays Applicable to Acrylamide Gels," *Anal. Biochem.,* **44,** 276–287.

Bereman, R. D., Wang, F. T., Najdzionek, J., and Braitsch, D. M. (1976). "Stereoelectronic Properties of Metalloenzymes. 4: Bis(imidotetraphenyldithiodiphosphino-*S,S′*)copper(II) as a Tetrahedral Model for Type I Copper(II)," *J. Am. Chem. Soc.,* **98,** 7266–7268.

Blumberg, W. E. and Peisach, J. (1966). "The Optical and Magnetic Properties of Copper in *Chenopodium album* Plastocyanin," *Biochim. Biophys. Acta,* **126,** 269–273.

Bors, W., Saran, M., Michel, C., Lengfelder, E., Fuchs, C., and Spöttl, R. (1975). "Pulse

Radiolytic Investigations of Catechols and Catecholamines. 1: Adrenaline and Adrenochrome," *Int. J. Radiat. Biol.,* **28,** 353–371.

Bouchilloux, S., McMahill, P., and Mason, H. S. (1963). "The Multiple Forms of Mushroom Tyrosinase," *J. Biol. Chem.,* **238,** 1699–1707.

Boyle, E., Freeman, P. C., Gondie, A. C., Mangen, F. R., and Thompson, M. (1976). "The Role of Copper in Preventing Gastrointestinal Damage by Acidic Antiinflammatory Drugs," *J. Pharm. Pharmacol.,* **28,** 865–868.

Brändén, C.-I., Jörnvall, H., Eklund, H., and Furugren, B. (1975). "Alcohol Dehydrogenase." In P. D. Boyer, Ed., *The Enzymes,* Vol. XI. Academic Press, New York, pp. 103–190.

Brigelius, R., Spöttl, R., Bors, W., Lengfelder, E., Saran, M., and Weser, U. (1974). "Superoxide Dismutase Activity of Low Molecular Weight Cu^{2+}-Chelates Studied by Pulse Radiolysis," *FEBS Lett.,* **47,** 72–75.

Brigelius, R., Hartmann, H.-J., Bors, W., Saran, M., Lengfelder, E., and Weser, U. (1975). "Superoxide Dismutase Activity of $Cu(tyr)_2$ and Cu,Co-Erythrocuprein," *Z. Physiol. Chem.,* **356,** 739–745.

Carrico, R. J. and Deutsch, H. F. (1969). "Isolation of Human Hepatocuprein and Cerebrocuprein," *J. Biol. Chem.,* **244,** 6087–6093.

Carrico, R. J. and Deutsch, H. F. (1970). "The Presence of Zinc in Human Cytocuprein and Some Properties of the Apoprotein," *J. Biol. Chem.,* **245,** 723–727.

Dawson, C. R., Strothkamp, K. G., and Krul, K. G. (1975). "Ascorbate Oxidase and Related Copper Proteins," *Ann. N.Y. Acad. Sci.,* **258,** 209–220.

De Alvare, L. R., Goda, K., and Kimura, T. (1976). "Mechanism of Superoxide Anion Scavenging Reaction by Bis(salicylato)-Copper(II) Complex," *Biochem. Biophys. Res. Commun.,* **69,** 687–694.

Deinum, J., Reinhammar, B., and Marchesini, A. (1974). "The Stoichiometry of the Three Different Types of Copper in Ascorbate Oxidase from Green Zucchini Squash," *FEBS Lett.,* **42,** 241–245.

Dobry-Duclaux, A. and Périchon, P. (1976). "Complexes de Cuivre et de Mercapto Imidazoles, Modèles des Groupes Chromophores des Protéines Bleues—1," *J. Chim. Phys.,* **73,** 1058–1067.

Dockal, E. R., Jones, T. E., Sokol, W. F., Engerer, R. J., Rorabacher, D. B., and Ochrymowycz, C. A. (1976). "Redox Properties of Copper-Thioether Complexes Comparison to Blue Copper Protein Behaviour," *J. Am. Chem. Soc.,* **98,** 4322–4324.

Falk, K. E. and Reinhammar, B. (1972). "Visible and Near-Infrared Circular Dichroism of Some Blue Copper Proteins," *Biochim. Biophys. Acta,* **285,** 84–90.

Faraggi, M. and Leopold, J. G. (1976). "Pulse Radiolysis Studies of Electron Transfer Reactions in Molecules of Biological Interest. II: The Reduction of Cu(II)-Peptide Complexes," *Radiat. Res.,* **65,** 238–249.

Fee, J. A. (1975). "Copper Proteins: Systems Containing the Blue Copper Center." In *Structure and Bonding,* Vol. 23. Springer-Verlag, Berlin, pp. 1–60.

Forman, H. J. and Fridovich, I. (1972). "Electrolytic Univalent Reduction of Oxygen in Aqueous Solution Demonstrated with Superoxide Dismutase," *Science,* **175,** 339.

Freeman, H. C. (1973). "Metal Complexes of Amino Acids and Peptides." In G. E. Eichhorn, Ed., *Inorganic Biochemistry.* Elsevier, Amsterdam, pp. 121–166.

Fridovich, I. (1974a). "Superoxide Dismutases." In A. Meister, Ed., *Advances in Enzymology,* Vol. 41. Wiley Interscience, New York, pp. 35–97.

Fridovich, I. (1974b). "Superoxide Dismutase." In O. Hayaishi, Ed., *Molecular Mechanisms of Oxygen Activation.* Academic Press, New York, pp. 453–477.

Fridovich, I. (1976). "Oxygen Radicals, Hydrogen Peroxide and Oxygen Toxicity." In W. Prior, Ed., *Free Radicals in Biology,* Vol. 1. Academic Press, New York, pp. 239–277.

Frieden, E. (1971). "Ceruloplasmin, a Link Between Copper and Iron Metabolism," *Adv. Chem. Series,* **100,** 292–321.

Giordano, R. S. and Bereman, R. D. (1974). "Stereoelectronic Properties of Metalloenzymes. I: A Comparison of the Coordination of Copper(II) in Galactose Oxidase and a Model System, *N,N′*-Ethylenebis(trifluoracetylacetoniminato)copper(II)," *J. Am. Chem. Soc.,* **96,** 1015–1023.

Goldstein, M. (1966). "Dopamine-β-hydroxylase: a Copper Enzyme." In J. Peisach, D. Aisen, and W. E. Blumberg, Eds., *The Biochemistry of Copper.* Academic Press, New York, pp. 443–453.

Gordon, S., Hart, E. J., Matheson, M. S., Rabani, J., and Thomas, J. K. (1963). "Reaction Constants of the Hydrated Electron," *J. Am. Chem. Soc.,* **85,** 1375–1377.

Halliwell, B. (1975). "The Superoxide Dismutase Activity of Iron Complexes," *FEBS Lett.,* **56,** 34–38.

Hamilton, G. H. (1974). "Chemical Models and Mechanisms for Oxygenases." In O. Hayaishi, Ed., *Molecular Mechanisms of Oxygen Activation.* Academic Press, New York, pp. 405–451.

Harless, E. (1847). "Ueber das blaue Blut einiger wirbellosen Thiere und dessen Kupfergehalt," *Arch. Anat. Physiol. Müllers,* 148–156.

Hartmann, H.-J. and Weser, U. (1977). "Copper-Thionein from Fetal Bovine Liver," *Biochim. Biophys. Acta,* **491,** 211–222.

Henry, Y. and Dobry-Duclaux, A. (1976). "Complexes de Cuivre de Mercaptoimidazoles, Modèles des Groupes Chromophores dans les Protéines Bleues. II: Étude par RPE des Complexes des Cuivre et du 1-Méthyl-2-mercaptoimidazole," *J. Chim. Phys.,* **73,** 1068–1070.

Hill, J. M. and Mann, P. J. G. (1964). "Further Properties of the Diamine Oxidase of Pea Seedlings," *Biochem. J.,* **91,** 171–182.

Holm, R. H. (1975). "Eisen-Schwefel-Cluster in natürlichen und synthetischen Systemen," *Endeavour,* **XXXIV,** ***121,*** 38–43.

Holm, R. H. and O'Connor, M. J. (1971). "The Stereochemistry of Bis-chelate Metal (II) Complexes," *Prog. Inorg. Chem.,* **14,** 241–401.

Hughes, M. N. (1972). "The Use of Model Compounds." In *The Inorganic Chemistry of Biological Processes.* Wiley-Interscience, New York, pp. 96–98.

Joester, K. E., Jung, G., Weber, U., and Weser, U. (1972). "Superoxide Dismutase Activity of Cu^{2+}-Amino Acid Chelates," *FEBS Lett.,* **25,** 25–28.

Jones, T. E., Zimmer, L. L., Diaddario, L. L., Rorabacher, D. B., and Ochrymowycz,

L. A. (1975a). "Macrocyclic Ligand Ring Size Effects on Complex Stabilities and Kinetics. Copper(II) Complexes of Cyclic Polythiaethers," *J. Am. Chem. Soc.,* **97,** 7163–7165.

Jones, T. E., Rorabacher, D. B., and Ochrymowycz, L. A. (1975b). "Simple Models for Blue Copper Proteins: The Copper-Thiaether Complexes," *J. Am. Chem. Soc.,* **97,** 7485–7486.

Jung, G., Ottnad, M., Bohnenkamp, W., and Weser, U. (1972). "X-ray Photoelectron Spectroscopy (XPS) of Bovine Erythrocuprein," *FEBS Lett.,* **25,** 346–352.

Jung, G., Ottnad, M., Bremser, W., Hartmann, H.-J., and Weser, U. (1973a). "Elektronenbindungsenergien von Kupfer, Zink und Kobalt in modifizierten Cupreinen," *Z. Physiol. Chem.,* **354,** 341–343.

Jung, G., Ottnad, M., Hartmann, H.-J., and Weser, U. (1973b). "Röntgenphotoelektronenspektroskopie von Metallen in Aminosäurekomplexen und Proteinen," *Z. Anal. Chem.,* **263,** 282–285.

Jung, G., Ottnad, M., Bohnenkamp, W., and Weser, U. (1973c). "X-ray Photoelectron Spectroscopic Studies of Copper and Zinc Amino Acid Complexes and Superoxide Dismutase," *Biochim. Biophys. Acta,* **295,** 77–86.

Kägi, J. H. R., Himmelhoch, S. R., Whanger, P. D., Bethune, J. L., and Vallee, B. L. (1974). "Equine Hepatic and Renal Metallothioneins," *J. Biol. Chem.,* **249,** 3537–3542.

Keilin, D. and Hartree, E. F. (1938). "Cytochrome *a* and Cytochrome Oxidase," *Nature,* **141,** 870–871.

Keilin, D. and Mann, T. (1938). "Polyphenoloxidase: Purification, Nature and Properties," *Proc. R. Soc.,* **B125,** 187–204.

Keilin, D. and Mann, T. (1939). "Laccase, a Blue Copper-Protein Oxidase from the Latex of *Rhus succedanea,*" *Nature,* **143,** 23–24.

Klug, H. P. (1966). "The Crystal Structure of Zinc Dimethyldithiocarbamate," *Acta Crystallogr.,* **21,** 536–546.

Klug-Roth, D. and Rabani, J. (1976). "Pulse Radiolytic Studies on Reactions of Aqueous Superoxide Radicals with Cu(II)-Complexes," *J. Phys. Chem.,* **80,** 588–591.

Kovács, K. and Matkovics, B. (1975). "Properties of Enzymes. IV: Nonspecific Superoxide Dismutase Activity of Iron-Porphyrin Systems," *Enzyme,* **20,** 1–5.

Lee, M. H. and Dawson, C. R. (1973). "Ascorbate Oxidase," *J. Biol. Chem.,* **248,** 6603–6609.

Lengfelder, E., Fuchs, C., Younes, M. and Weser, U. (1979). "Functional Aspects of the Superoxide Dismutating Action of Cu-Penicillamine," *Biochim. Biophys. Acta,* **567,** 492–502.

Levine, W. G. (1966). "Laccase, a Review." In J. Peisach, P. Aisen, and W. E. Blumberg, Eds., *The Biochemistry of Copper.* Academic Press, New York, pp. 371–387.

Levitzki, A., Pecht, I., and Anbar, M. (1965). "Oxidase-like Activity of the Copper(II)-Poly-L-histidine Complex," *Nature,* **207,** 1386–1387.

Lontie, R. and Witters, R. (1973). "Hemocyanin." In G. Eichhorn, Ed., *Inorganic Biochemistry.* Elsevier, Amsterdam, pp. 344–358.

Lovenberg, W., Ed. (1973). *Iron-Sulphur Proteins,* Vols. I and II. Academic Press, New York.

Lumbsden, J. and Hall, D. O. (1974). "Soluble and Membrane-Bound Superoxide-Dismutases in a Blue-Green Alga (*Spirulina*) and Spinach," *Biochem. Biophys. Res. Commun.,* **58,** 35–41.

Mahler, H. (1963). "Uricase." In P. D. Boyer, H. Lardy, and K. Myrbäck, Eds., *The Enzymes.* Academic Press, New York, pp. 285–296.

Malkin, R. (1973). "The Copper-Containing Oxidases." In G. L. Eichhorn, Ed., *Inorganic Biochemistry.* Elsevier, Amsterdam, pp. 689–709.

Malkin, R. and Malmström, B. G. (1970). "The State and Function of Copper in Biological Systems." In F. F. Nord, Ed., *Advances in Enzymology,* Vol. 33. Wiley-Interscience, New York, pp. 177–244.

Malmström, B. G. (1970). "Plenary Lecture." In *XIIth International Conference on Coordination Chemistry,* Sydney, 1969. Butterworths, London.

Malmström, B. G. and Vänngard, T. (1960). "Electron Spin Resonance of Copper Proteins and Some Model Complexes," *J. Mol. Biol.,* **2,** 118–124.

Malmström, B. G., Reinhammar, B., and Vänngard, T. (1970). "The State of Copper in Stellacyanin and Laccase from the Lacquer Tree *Rhus vernicifera," Biochem. Biophys. Acta,* **205,** 48–57.

Malmström, B. G., Andréasson, L. E., and Reinhammar, B. (1975). "Copper-Containing Oxidases and Superoxide Dismutase." In P. Boyer, Ed., *The Enzymes* Academic Press, New York, pp. 507–579.

Mann, T. and Keilin, D. (1938). "Haemocuprein and Hepatocuprein, Copper-Protein Compounds of Blood and Liver in Mammals," *Proc. R. Soc. London,* **B126,** 303–315.

Martell, A. E. and Taqui Khan, M. M. (1973). "Metal Ion Catalysis of Reactions of Molecular Oxygen." In G. L. Eichhorn, Ed., *Inorganic Biochemistry.* Elsevier, Amsterdam, pp. 654–688.

Massey, V., Palmer, G., and Ballou, D. (1971). "On the Reactions of Reduced Flavins and Flavoproteins with Molecular Oxygen." In H. Kamin, Ed., *Flavins and Flavoproteins.* University Park Press, Baltimore, and Butterworths, London, pp. 349–361.

McClune, G. J., Fee, J. A., McCluskey, G. A., and Groves, J. T. (1977). "Catalysis of Superoxide Dismutation by Fe-EDTA Complexes: Mechanism of Reaction and Evidence for the Direct Formation of an Iron(III)-EDTA Peroxo Complex from the Reaction of Superoxide with Fe(II)-EDTA," *J. Am. Chem. Soc.,* **99,** 5220–5222.

McCord, J. M. (1974). "Free Radicals and Inflammation: Protection of Synovial Fluid by Superoxide Dismutase," *Science,* **185,** 529–531.

McCord, J. M. and Fridovich, I. (1969). "Superoxide Dismutase," *J. Biol. Chem.,* **244,** 6049–6055.

McElroy, W. D. and Glass, B. Eds. (1950). *Copper Metabolism.* A Symposium on Animal, Plant, and Soil Relationships Sponsored by the McCollum-Pratt Institute of the Johns Hopkins University, Baltimore. Johns Hopkins Press, Baltimore, Md.

Michelson, A. M., McCord, J. M., and Fridovich, I. Eds. (1977). *Superoxide and Superoxide Dismutase.* Academic Press, London.

Miskowski, V., Tang, S.-P. W., Spiro, T. G., Shapiro, E., and Moss, T. H. (1975). "The

Copper Coordination Groups in the "Blue" Copper Proteins: Evidence from Raman Resonance Spectra," *Biochemistry,* **14,** 1244–1250.

Miskowski, V. M., Thich, J. A., Solomon, R., and Schugar, H. J. (1976). "Electronic Spectra of Substituted Copper(II)-Thioether Complexes," *J. Am. Chem. Soc.,* **98,** 8344–8350.

Misra, H. P. and Fridovich, I. (1972). "The Role of Superoxide Anion in the Autoxidation of Epinephrine and a Simple Assay for Superoxide Dismutase," *J. Biol. Chem.,* **247,** 3170–3175.

Morpurgo, L., Giovagnoli, C., and Rotilio, G. (1973). "Studies of the Metal Sites of Copper Proteins. V: A Model Compound for the Copper Site of Superoxide Dismutase," *Biochim. Biophys. Acta,* **322,** 204–210.

Morpurgo, L., Finazzi-Agró, A., Rotilio, G., and Mondovi, B. (1976). "Anion Complexes of Cu(II) and Co(II) Bovine Carbonic Anhydrase as Models for the Copper Site of Blue Copper Proteins," *Eur. J. Biochem.,* **64,** 453–457.

Mosbach, R. (1963). "Purification and Some Properties of Laccase from Polyporus Versicolor," *Biochim. Biophys. Acta,* **73,** 204–212.

Okaya, Y. and Knobler, C. B. (1964). "Refinement of the Crystal Structure of Tris(thiourea)copper(I) Chloride," *Acta Crystallogr.,* **17,** 928–930.

Oyanagui, Y. (1976). "Participation of Superoxide Anion at the Prostaglandine Phase Swelling of Carrageenan Foot-Odema, "Biochem Pharmacol., **25,** 1465–1472.

Paschen, W. and Weser, U. (1973). "Singlet Oxygen Decontaminating Activity of Erythrocuprein (Superoxide Dismutase)," *Biochim. Biophys. Acta,* **327,** 217–222.

Paschen, W. and Weser, U. (1975). "Problems Concerning the Biochemical Action of Superoxide Dismutase (Erythrocuprein)," *Z. Physiol. Chem.,* **356,** 727–737.

Patterson, G. S. and Holm, R. H. (1975). "Structural and Electronic Effects on the Polarographic Half-Wave Potentials of Copper(II)-Chelate Complexes," *Bioinorg. Chem.,* **4,** 257–275.

Peisach, J., Aisen, P., and Blumberg, W. E. Eds. (1966). *The Biochemistry of Copper.* Academic Press, New York.

Peisach, J., Levine, W. G., and Blumberg, W. E. (1967). "The Structural Properties of Stellacyanin, a Copper Mucoprotein from *Rhus vernicifera,* the Japanese Lac Tree," *J. Biol. Chem.,* **242,** 2847–2858.

Rabani, J., Klug-Roth, D., and Lilie, J. (1973). "Pulse Radiolytic Investigations of the Catalyzed Disproportionation of Peroxy Radicals: Aqueous Cupric Ions," *J. Phys. Chem.,* **77,** 1169–1175.

Rainsford, K. D. and Whitehouse, M. W. (1976). "Gastric Mucus Effusion Elicited by Oral Copper Compounds: Potential Anti-ulcer Activity," *Experientia,* **32,** 1172–1173.

Rapp, U., Adams, W. C., and Miller, R. W. (1973). "Purification of Superoxide Dismutase from Fungi and Characterization of the Reaction of the Enzyme with Catechols by Electron Spin Resonance Spectroscopy," *Can. J. Biochem.,* **51,** 158–171.

Richardson, J. S., Thomas, K. A., and Richardson, D. C. (1975). "Alpha-Carbon Coordinates for Bovine Cu,Zn-Superoxide Dismutase," *Biochem. Biophys. Res. Commun.,* **63,** 986–992.

Richardson, J. S., Richardson, D. C., Thomas, K. A., Silverton, E. W., and Davies, D. R. (1976). "Similarity of the Three Dimensional Structure between the Immunoglobulin Domain and the Cu,Zn-Superoxide Dismutase Subunit," *J. Mol. Biol.*, **102,** 221–235.

Richter, C., Azzi, A., and Wendel, A. (1976). "The Effect of Tyrosine-Divalent Copper Complex on the Hepatic Microsomal Demethylation," *FEBS Lett.*, **64,** 332–337.

Richter, C., Azzi, A., Weser, U., and Wendel, A. (1977). "Hepatic Microsomal Demethylation: Inhibition by a Tyrosine-Cu(II) Complex Provided with Superoxide Dismutase Activity," *J. Biol. Chem.*, **252,** 5061–5066.

Rist, G. H., Hyde, J. S., and Vänngard, T. (1970). "Electron-Nuclear Double Resonance of a Protein That Contains Copper: Evidence for Nitrogen Coordination to Cu(II) in Stellacyanin," *Proc. Natl. Acad. Sci.*, **67,** 79–86.

Rotilio, G., Morpurgo, L., Giovagnoli, C., Calabrese, L., and Mondovi, B. (1972). "Studies of the Metal Sites of Copper Proteins: Symmetry of Copper in Bovine Superoxide Dismutase and Its Functional Significance," *Biochemistry,* **11,** 2187–2192.

Rupp, H. and Weser, U. (1976a). "Copper(I) and Copper(II) in Complexes of Biochemical Significance Studied by X-ray Photoelectron Spectroscopy," *Biochim. Biophys. Acta,* **446,** 151–165.

Rupp, H. and Weser, U. (1976b). "X-ray Photoelectron Spectroscopy of Copper(II), Copper(I), and Mixed Valence Systems," *Bioinorg. Chem.*, **6,** 45–59.

Rupp, H. and Weser, U. (1976c). "Reactions of D-Penicillamine with Copper in Wilson's Disease," *Biochem. Biophys. Res. Commun.*, **72,** 223–229.

Rupp, H. and Weser, U. (1978). "Circular Dichroism of Metallothioneins: a Structural Approach," *Biochim. Biophys. Acta,* **533,** 209–226.

Schugar, H. J., Ou, C., Thich, J. A., Potenza, J. A., Lalancette, R. A., and Furey, W. J. (1976). "Molecular Structure and Copper(II)-Mercaptide Charge-Transfer Spectra of a Novel $Cu_{14}[SC(CH_3)_2CH_2NH_2]_{12}Cl$ Cluster," *J. Am. Chem. Soc.*, **98,** 3047–3048.

Sharma, V. S. and Schubert, J. (1969). "Catalytic Activity of Metal Chelates and Mixed Ligand Complexes in the Neutral pH-Region. I: Copper Imidazole," *J. Am. Chem. Soc.*, **91,** 6291–6296.

Sharma, V. S., Schubert, J., Brooks, H. B., and Sicilio, F. (1970). "Catalytic Activity of Metal Chelates and Mixed Ligand Complexes in the Neutral pH-Region. II: Copper Histidine," *J. Am. Chem. Soc.*, **92,** 822–826.

Shichi, H. and Hackett, D. P. (1963). "Purification and Properties of a Blue Protein from Etiolated Mung Bean Seedlings," *Arch Biochem. Biophys.*, **100,** 185–191.

Sigel, H. (1969). "Zur katalatischen und peroxidatischen Aktivität von Cu^{2+}-Komplexen," *Angew. Chem.*, **81,** 161–171.

Siiman, O., Young, N. M., and Caray, P. R. (1976). "Resonance Raman Spectra of "Blue" Copper Proteins and the Nature of Their Copper Sites," *J. Am. Chem. Soc.*, **98,** 744–748.

Solomon, E. J., Hare, J. W., and Gray, H. B. (1976). "Spectroscopic Studies and a Structural Model for Blue Copper Centers in Proteins," *Proc. Natl. Acad. Sci.*, **73,** 1389–1393.

Sorenson, J. R. J. (1976a). "Copper Chelates as Possible Active Forms of the Antiarthritic Agents," *J. Med. Chem.*, **19,** 135–148.

Sorenson, J. R. J. (1976b). "Some Copper Coordination Compounds and Their Anti-inflammatory and Antiulcer Activities," *Inflammation,* **1,** 317–331.

Stigbrand, T. (1971). "Structural Properties of Umecyanin, a Copper Protein from Horseradish Root," *Biochim. Biophys. Acta,* **236,** 246–252.

Sugiura, Y. (1977). "Proton Magnetic Resonance Study of Copper(I)-Complexes with Peptides Containing Sulfhydryl and Imidazole Groups as Possible Model Ligands for Copper Proteins," *Eur. J. Biochem.*, **78,** 431–435.

Sugiura, Y. and Hirayama, Y. (1977). "Copper(II)- and Nickel(II)-Complexes of Sulfhydryl and Imidazole Containing Peptides. Characterization and Model for "Blue" Copper Sites," *J. Am. Chem. Soc.,* **99,** 1581–1585.

Sugiura, Y., Hirayama, Y., Tanaka, H., and Ishizu, U. (1975). "Copper(II)-Complexes of Sulfur-Containing Peptides: Characterization and Similarity of Electron Spin Resonance Spectrum to the Chromophore in Blue Copper Proteins," *J. Am. Chem. Soc.,* **97,** 5577–5581.

Sutherland, I. W. and Wilkinson, J. F. (1963). "Azurin: a Copper-Protein Found in *Bordatella," J. Gen. Microbiol.,* **30,** 105–112.

Thompson, J. S., Marks, T. J., and Ibers, J. A. (1977). Blue Copper Proteins: Synthesis, Spectra, and Structures of $Cu(I)N_3(SR)$ and $Cu(II)N_3(SR)$ Active Site Analogues," *Proc. Natl. Acad. Sci.,* **74,** 3114–3118.

Valentine, J. S. and Curtis, A. B. (1975). "A Convenient Preparation of Solutions of Superoxide Anion and the Reaction of Superoxide Anions with a Copper(II)-Complex," *J. Am. Chem. Soc.,* **97,** 224–226.

Valerino, D. M. and McCormack, J. J. (1969). "Oxidation of Epinephrine by Milk Xanthine Oxidase," *Fed. Proc.,* **28,** 545.

Vallee, B. L. and Williams, R. J. P. (1968). "Metalloenzymes: the Entatic Nature of Their Active Sites," *Proc. Natl. Acad. Sci.,* **59,** 498–505.

Vänngard, T. (1972). "Copper Proteins." In H. M. Swartz, J. R. Bolton, and D. C. Borg, Eds., *Biological Applications of Electron Spin Resonance.* Wiley-Interscience, New York, pp. 411–447.

Weser, U. (1973). "Structural Aspects and Biochemical Function of Erythrocuprein." In *Structure and Bonding,* Vol. 17. Springer Verlag, Berlin, pp. 1–65.

Weser, U. and Paschen, W. (1972). "Mode of Singlet Oxygen Induced Chemiluminescence in the Presence of Erythrocuprein," *FEBS Lett.,* **27,** 248–250.

Weser, U. and Schubötz, L. M. (1978). "Conformational Aspects of Superoxide Dismutase Active Copper Chelates and Catechols," *Bioinorg. Chem.,* **9,** 505–519.

Weser, U., Bunnenberg, E., Cammack, R., Djerassi, C. Flohé, L., Thomas, G., and Voelter, W. (1971). "A Study on Purified Bovine Erythrocuprein," *Biochim. Biophys. Acta,* **243,** 203–213.

Weser, U., Paschen, W., and Younes, M. (1975). "Singlet Oxygen and Superoxide Dismutase," *Biochem. Biophys. Res. Commun.,* **66,** 769–777.

Weser, U., Hartmann, H.-J., Fretzdorff, A., and Strobel, G.-J. (1977a). "Homologous

Copper(I)(thiolate)$_2$-Chromophores in Yeast Copper-Thionein," *Biochim. Biophys. Acta,* **493,** 465–477.

Weser, U., Richter, C., Wendel, A., and Younes, M. (1978b). "Reactivity of Anti-inflammatory and Superoxide Dismutase-Active Cu(II)-Salicylates," *Bioinorg. Chem.,* **8,** 201–213.

Weser, U., Younes, M., Hartmann, H. J. and Zienau, S. (1979). "X-Ray Photoelectron Spectrometric Aspects of the Copper Chromophore in Plastocyanin," *FEBS Lett.,* **97,** 311–313.

Wharton, D. C. (1973). "Cytochrome Oxidase." In G. L. Eichhorn, Ed., *Inorganic Biochemistry.* Elsevier, Amsterdam, pp. 955–987.

Williams, R. J. P. (1973). "Fifth Keilin Memorial Lecture: Electron Transfer and Oxidative Energy," *Biochem. Soc. Trans.,* **1,** 1–26.

Wright, J. R. and Frieden, E. (1975). "Properties of the Red-Violet Complex of Copper and Penicillamine and Further Insight into Its Formation Reaction," *Bioinorg. Chem.,* **4,** 163–175.

Yamada, H., Kumagai, H., Kawasaki, H., Matsui, H., and Ogata, K. (1967). "Crystallization and Properties of Diamine Oxidase from Pig Kidney," *Biochem. Biophys. Res. Commun.,* **29,** 723–727.

Yost, F., Jr. and Fridovich, I. (1973). "An Iron-Containing Superoxide Dismutase from *Escherichia coli,*" *J. Biol. Chem.,* **248,** 4905–4908.

Younes, M. and Weser, U. (1976a). "Inhibition of Nitroblue Tetrazolium Reduction by Cuprein (Superoxide Dismutase), Cu(tyr)$_2$, and Cu(lys)$_2$," *FEBS Lett.,* **61,** 209–212.

Younes, M. and Weser, U. (1976b). "Reactivity of Superoxide Dismutase-Active Cu(II)-Complexes on the Rate of Adrenochrome Formation," *FEBS Lett.,* **71,** 87–90.

Younes, M. and Weser, U. (1977). "Superoxide Dismutase Activity of Copper-Penicillamine: Possible Involvement of Cu(I) Stabilized Sulphur Radical," *Biochem. Biophys. Res. Commun.,* **78,** 1247–1253.

Younes, M., Lengfelder, E., Zienau, S. and Weser, U. (1978). "Pulse Radiolytically Generated Superoxide and Cu(II)-Salicylates," *Biochem. Biophys. Res. Commun.,* **81,** 576–580.

Younes, M., Pilz, W. and Weser, U. (1979). "Models for Metal-Sulfur Coordination in Copper Proteins," *J. Inorg. Biochemistry,* **10,** 29–39.

9

CERULOPLASMIN: THE SERUM COPPER TRANSPORT PROTEIN WITH OXIDASE ACTIVITY

Earl Frieden

Department of Chemistry, Florida State University, Tallahassee, Florida

1. INTRODUCTION AND SOME EVOLUTIONARY IMPLICATIONS

A blue protein from the α_2-globulin fraction of human serum possessing oxidase activity was first reported by Holmberg in 1944. It was named "ceruloplasmin," meaning a blue substance from plasma, by Holmberg and Laurell (1948, 1951) in later papers describing its purification and reporting other basic observations on its chemical properties. Since ceruloplasmin accounts for over 95% of the circulating copper in a normal mammal and fluctuates greatly in several disease and hormonal states, the study of this protein has excited the imagination of biomedical scientists, who have generated several thousand papers on its chemistry and biology. Ceruloplasmin is an attractive protein for study because, like serum albumin, it appears to be multifunctional: as the principal copper transport protein, as a molecule directly involved in iron mobilization from the iron storage sites to the plasma by means of its ferroxidase activity, and, possibly, as a regulator of circulating biogenic amine levels through its oxidase activity. In this chapter we summarize the key chemical and physical characteristics of ceruloplasmin, describe its enzymatic activities, and try to relate these to its multifunctional role in copper and iron metabolism. In so doing, we attempt to avoid overlap and duplication of effort with recent reviews by Poulik and Weiss (1975), Fee (1975), Evans (1973), and Scheinberg and Morell (1973).

Ceruloplasmin may represent a current end point in the parallel development of copper and iron biochemistry in the natural selection of aerobic cells during the past 3 billion years (Figure 1). That copper and iron as metals have played a dominant part in the recent unfolding of human civilization, as well as in the lengthy evolution of essential metalloproteins and metalloenzymes, is a remarkable coincidence. In a review tracing the evolution of the essential metal ions, Frieden (1974) has pointed out the many close associations of copper and iron that have evolved in aerobic cells. A primary role was assigned to the development of enzymes to protect cells from unavoidable toxic oxygen by-products—superoxide ion, singlet oxygen, and hydrogen peroxide. This resulted in

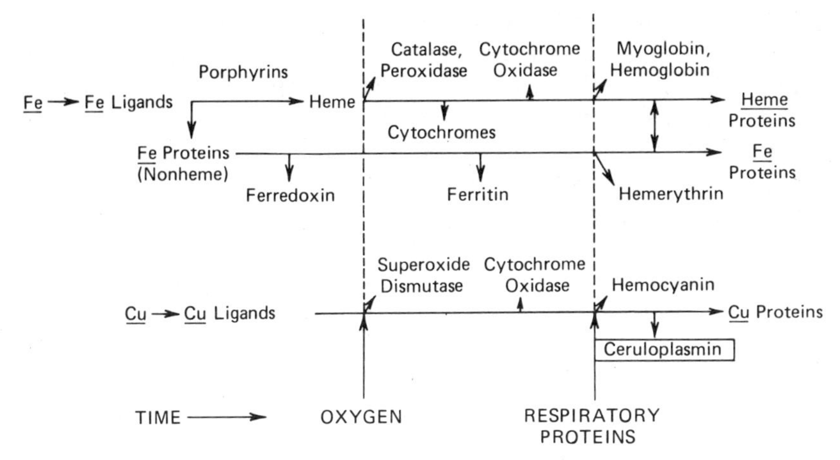

Figure 1. The role of ceruloplasmin in the evolutionary sequence and the development of the iron and copper proteins. The horizontal axis is essentially a time line with the advent of oxygen and respiratory proteins as biological markers.

the ubiquitous occurrence of the copper-zinc enzyme (superoxide dismutase) and the heme enzymes (catalase and peroxidase). The success of the aerobes was accompanied by the development of more sophisticated iron and copper enzymes, notably the cytochromes, cytochrome oxidase, and the numerous electron transferases in plants. With the increasing complexity of organisms, the cellular machinery utilizing iron and copper expanded greatly for the production of the oxygen-carrying proteins—hemoglobins, hemerythrins, and hemocyanins. This adaptation required the elaboration of storage and transport proteins exclusively for copper and iron: ceruloplasmin, ferritin, and transferrin. Later stages of evolution were accompanied by the appearance of crucial biosynthetic enzymes associated with connective tissue and other, more specific processes. A final example of the continuing close connection between iron and copper in the vertebrates is the ability of the copper protein of plasma, ceruloplasmin, to mobilize iron into transferrin for iron transport and distribution (Frieden, 1971).

Dawson et al. (1975) have discussed some interesting parallels between ceruloplasmin and two other blue copper oxidases, ascorbate oxidase and laccase, which occur despite their different origins and biological roles. They note that all three of these multicopper proteins have in common an oxidase activity toward ascorbate, although ascorbate oxidase has 10^3 to 10^4 times more enzymic activity than laccase or ceruloplasmin. It was pointed out also that there are numerous similarities in various categories of amino acids (aromatic, hydrophobic, acidic-basic, and bivalent), particularity with ascorbate oxidase and ceruloplasmin, as well as with their copper content (0.30%). More information is needed to establish close homology or common tertiary or quaternary structure between these three copper oxidases. On the basis of its more versatile substrate range, laccase is proposed as the primitive type of blue oxidase, perhaps involved in the protection of the early aerobes against excess oxygen. As the aerobes evolved the ability to utilize molecular oxygen and to protect against toxic oxygen by-product, the need for this primal laccase disappeared. Later, modified forms of the primal laccase emerged and were used for other, more specialized functions—ascrobate oxidase as a terminal oxidase in plants, and ceruloplasmin in copper transport and iron metabolization.

An interesting comparison has also been made by Kingston et al. (1977) of the amino acid sequence around a crucial invariant cysteine residue. This cysteine might be responsible for the blue color of type 1 Cu(II). A relatively good match between the cysteine, histidine, and methionine residues in sequences from ceruloplasmin, azurin (from *Pseudomonas aeruginosa*), and plastocyanin (from *Anabaena variabilis*) was noted, suggesting a common site for Cu(II)-polypeptide interaction in these widely diverse copper proteins.

2. THE CHEMISTRY OF CERULOPLASMIN

2.1. Molecular Properties and the State of Copper

Many of the fundamental chemical properties of ceruloplasmin are now well established (Fee, 1975; Frieden and Hsieh, 1976). Tables 1 to 3 summarize many of these key molecular properties. Uncertainties about the molecular weight of 134,000 and the six copper atoms per molecule (Table 1) have been resolved by the work of Rydén and Björk (1976), Kingston et al. (1977), and Huber and Frieden (1970). Amino acid and carbohydrate analyses (Table 2) suggest a large typical serum glycoprotein. Data to be presented later support the view that the 1065 residues are linked in a single chain with no subunits.

Details of any of the copper-binding sites are completely lacking. Kingston et al. (1977) have discussed a possible site for type 1 Cu(II), which is responsible for the intense blue color of several copper proteins, including ascorbate oxidase laccase, azurin, plastocyanin, stellacyanin, and umecyanin. The unique properties of type 1 Cu(II) suggest that it probably occurs in a similar environment in all of these proteins. The SH groups from cysteine and the nitrogens of histidine are the logical ligand groups in the protein. Sequences in three of these proteins, azurin, ceruloplasmin, and plastocyanin, show a reasonable match in regard to the location of cysteine, histidine, and methionine residues. This region, identified by Kingston et al. in the 20,000-molecular-weight fragment of ceruloplasmin, may be involved in binding one of the type 1 Cu(II) ions found in ceruloplasmin.

Table 3 summarizes a proposed stoichiometry of the prosthetic copper of ceruloplasmin (Rydén and Björk, 1976). The designation of at least three different types of prosthetic copper agrees with Deinum and Vanngard (1973). In the investigations, type 2 copper was found to an extent of 33% of total

Table 1. Key Molecular Parameters of Human Ceruloplasmin[a]

Property	Value
Molecular weight	132,000
Percent copper	0.30 ± 0.3
Copper atoms per mole	6
Amino acid composition	
N-terminus	1 valine
—SH groups	4[b]
Carbohydrate composition (%)	7–8
Carbohydrate chains	9–10
Sialic acid	9
$E^{1\%}_{1cm}$ at 610 nm	0.69 ± 0.01
$E^{1\%}_{1cm}$ at 280 nm	15.0 ± 0.4
Isoelectric point, pH	4.4
Sedimentation constant	7.1
Axial ratio (*a*/*b*)	3.6

[a] For further details and other hydrodynamic, optical, EPR, and electrolytic properties, see also Fee (1975) and Scheinberg and Morell (1973).
[b] One exposed and three buried —SH groups.

EPR-detectable copper; furthermore, two different type 1 copper atoms could be distinguished. It is thus clear that three EPR-detectable copper atoms exist. The total copper content of six atoms therefore requires that three EPR-nondetectable copper ions be present. Two of these may form a spin-coupled pair of cupric ions, as has been postulated for the laccases (Malkin and Malmström, 1970) and as is suggested by the many fundamental similarities among all the blue copper-containing oxidases. In contrast, the third EPR-nondetectable copper ion cannot be included in such a model, and Rydén and Björk propose a type 4 copper, either Cu(I) or Cu(II), to account for the sixth ion. It is agreed that only four of the six copper ions need to be part of an active site, with the slow type 1 copper and the type 4 copper the least essential of the six for participation in the catalytic activity of ceruloplasmin.

2.2. Ceruloplasmin as a Single-Chain Protein

There has been considerable controversy over the question of whether ceruloplasmin is a single-chain protein or is composed of subunits. Despite Rydén's evidence (Rydén, 1971, 1972) for a single-chain structure for human ceruloplasmin, several papers continue to support subunit structures (Freeman and Daniel, 1973; McCombs and Bowman, 1976). Rydén's position has now received strong support from Frank W. Putnam's laboratory. Kingston et al. (1977)

Table 2. Amino Acid and Carbohydrate Composition of Human Ceruloplasmin[a]

Amino Acid	Number of Residues per Mole
Aspartic acid	127
Threonine	77
Serine	62
Glutamic acid	120
Proline	46
Glycine	77
Alanine	49
Half-cysteine	14
Cysteine	3
Cystine	6
Valine	62
Methionine	27
Isoleucine	53
Leucine	70
Tyrosine	64
Phenylalanine	49
Histidine	39
Lysine	66
Ammonia	113
Tryptophan	22
Arginine	41
Sum of amino acid residues	1065
Monosaccharide	**Number of Residues per Mole**
Mannose	14
Galactose	12
Fucose	2
Sialic acid	9
Sum of monosaccharide residues	53

[a] Adapted from data on ceruloplasmin Form I, normalized to a molecular weight of 134,000 as reported by Rydén and Björk (1976). This paper should be consulted for other details concerning the analyses.

described chemical evidence that proteolytic cleavage causes the heterogeneity found in preparations of ceruloplasmin from human serum and the presence of what appeared to be discrete polypeptide subunit fragments, particularly a 20,000-molecular-weight fragment. Limited plasmin digestion of essentially undegraded ceruloplasmin could account for the variety of molecular forms reported for ceruloplasmin (Freeman and Daniel, 1973). Other fragments of

Table 3. Proposed Stoichiometry of Ceruloplasmin Prosthetic Copper[a]

Number of Copper	Designation	EPR Signal	Other Properties
1	Fast type 1 Cu^{2+}	Detectable	Blue, reoxidized fast
1	Slow type 1 Cu^{2+}	Detectable	Blue, reoxidized slowly
1	Permanent type 2 Cu^{2+}	Detectable	Nonblue, binds anions
2	Type 3 Cu	Nondetectable	Postulated to be spin-coupled pair of Cu^{2+} in analogy with other blue oxidases
1	Type 4 Cu	Nondetectable	Required by total copper content

[a] Based on a molecular weight of 134,000 and a copper content of six atoms per mole. Adapted from Rydén and Björk (1976).

29,000, 76,000, 94,000, and 116,000 have also been identified by gel electrophoresis from plasmin digests. The lack of similarity in the tryptic peptide maps suggested that these fragments were not similar polypeptide chains in different states of aggregation. The data on the fragmentation of ceruloplasmin by plasmin could be rationalized by assuming at least one labile peptide bond which produces one large fragment of 116,000 molecular weight and a frequently encountered smaller fragment of 20,000. The origin of the other fragments probably involves two or more cleavages of single-chain ceruloplasmin.

Kingston et al. (1977) concluded that the heterogeneity of ceruloplasmin can be attributed to limited proteolysis during isolation and/or storage. This proteolysis results in fragments that resemble subunits. These conclusions are consistent with the single-chain structure of ceruloplasmin proposed by Rydén (1971, 1972).

2.3. Clinical Significance of Serum Ceruloplasmin

The level of ceruloplasmin in the serum is highly sensitive to a variety of pathological states. However, appreciation of the clinical significance of ceruloplasmin has been limited to obvious disorders of copper metabolism, Wilson's and Menkes' diseases, and copper deficiency. In these pathologies the reductions in both serum copper and serum ceruloplasmin are widely accepted as an important diagnostic tool.

In Wilson's disease serum ceruloplasmin is 12 to 20% of the normal level of 31 mg %. However, hepatic copper levels may be elevated 20- to 30-fold. Table 4 lists the various conditions in which the serum ceruloplasmin levels are de-

Table 4. Conditions That Lead to Changes in Serum Ceruloplasmin Levels

Reduced Serum Ceruloplasmin

1. *Menkes' disease*—a genetic inadequacy in copper utilization.
2. *Wilson's disease*—accumulation of copper in several key tissues (liver, brain, pancreas) probably due to a genetic defect in the biosynthesis of ceruloplasmin.
3. *Copper deficiency*.
4. *Metal ion toxicity*—especially excess Zn, Cd, Mo, Ag.
5. Conditions that lead to *protein loss* or *liver hypofunction*—nephrotic syndrome, hepatic necrosis, malnutrition.
6. *Huntington's chorea*—a 30% reduction in plasma ceruloplasmin is claimed in this fatal disease leading to the degeneration of the basal ganglia.

Elevated Serum Ceruloplasmin

1. *Estrogens*—oral contraceptives and pregnancy produce a two-to-three fold increase in ceruloplasmin within several days.
2. *Malignancies*—including lymphoma, leukemia, Hodgkin's disease, multiple myeloma.
3. *Infections and inflammations*—including rheumatoid arthritis, rheumatic fever, lupus erythematosis, liver disease; also possibly due to a general factor (LEM: leukocytic endotoxin mediator).
4. *Myocardial infarction*—of several *normal* proteins of human plasma, ceruloplasmin showed the greatest and most prolonged elevation, even beyond 40 days, after an acute episode.

creased or increased. A more detailed discussion of the variation of serum ceruloplasmin levels in health and disease may be found in a review by Poulik and Weiss (1975).

A well-documented, striking elevation of serum ceruloplasmin occurs in women during pregnancy or during treatment with most oral contraceptives or other estrogenic steroids. Over 100% increases are frequently observed in human sera. This increase is attributed to the effect of estrogens in promoting the synthesis and secretion of ceruloplasmin by the liver (Evans, 1973). The impact of these changes on the liver cells and on local or general copper metabolism has yet to be carefully studied and evaluated.

A vigorous effort has been made by Shokeir (1975) to use the measurement of serum ceruloplasmin and serum dopamine β-hydroxylase, another copper enzyme, as a predictive test for Huntington's disease. It was reported that the ceruloplasmin level in several hundred Huntington's disease patients was 72% of the values for known unaffected relatives or matched unrelated controls.

Dopamine β-hydroxylase was 132 and 141%, respectively, of the levels for identical control groups. These relatively modest differences will require considerable ancillary reinforcement with other tests before any confidence can be placed in their predictive value.

3. THE CATALYTIC ACTIVITY OF CERULOPLASMIN

3.1. Background

In 1948 Holmberg and Laurell explored the oxidase activity of ceruloplasmin on numerous reducing substances. Their preparations of ceruloplasmin enhanced the oxidation of aryldiamines, diphenols, and other reducing substances, including ascorbate, hydroxylamine, and thioglycolate. Later Curzon (1960) found that the oxidation of aryldiamines in the presence of ceruloplasmin could be activated or inhibited by certain transition metal ions. Finally, Curzon and O'Reilly (1960) reported that Fe(II) could reduce ceruloplasmin and suggested a coupled iron-ceruloplasmin oxidation system (Curzon, 1961).

Our interest at Florida State University was originally stimulated by the possibility that ceruloplasmin was a mammalian ascorbate oxidase, an enzyme that had been clearly identified in plants but had eluded detection in animal tissues. First we showed that the ascorbate oxidase activity of ceruloplasmin was not due to traces of free Cu(II) (Osaki et al., 1964). However, it was found that, at low concentrations of ascorbate, oxidation was greatly stimulated by traces of iron ions, present in most ceruloplasmin preparations unless special precautions were taken to eliminate the iron impurity. The most useful reagent for the removal of Fe(III) proved to be apotransferrin (McDermott et al., 1968). At that time we proposed three major groups of substances to describe the oxidase action of ceruloplasmin (McDermott et al., 1968):

1. Fe(II), the substrate with the highest V_m and the lowest K_m.
2. An extensive group of bifunctional aromatic amines and phenols, which do not depend on traces of iron ions for their activity. This group includes the two classes of biogenic amines, the epinephrine and 5-hydroxylindole series, and the phenothiazine series.
3. A third group of pseudosubstrates comprising numerous reducing agents that can rapidly reduce Fe(II) or partially oxidized (free radical) intermediates of class 2. We consider these compounds to be secondary substrates by way of an iron cycle or an aromatic diamine acting as a shuttle.

These three classes of substrates are shown in Figure 2. In principle, any reductant can be a substrate if it can transfer an electron to oxidized ceruloplasmin without poisoning or blocking the autooxidizability of reduced ceruloplasmin. For example, in our laboratory D. J. McKee has shown that 10 mM $VOSO_4$ rapidly bleaches blue ceruloplasmin in its conversion to VO_2^+. Thus VO^{2+} may

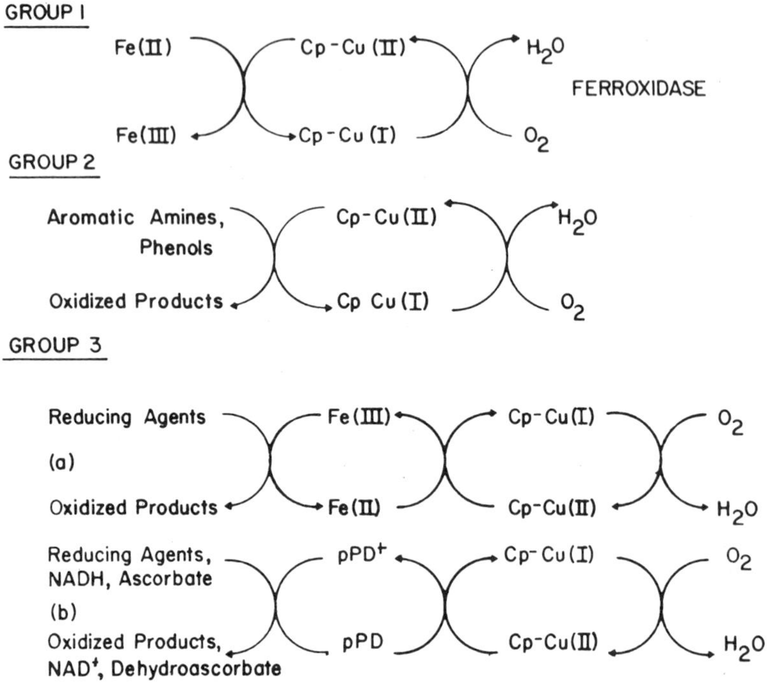

Figure 2. The various substrate groups of ceruloplasmin and the ways in which they react. Groups 1 and 2 are true substrates since they react directly with the oxidized form of ceruloplasmin (Cp). Groups 3a and 3b may be considered pseudosubstrates since their reactions are mediated by a group 1 or 2 substrate.

represent an additional ceruloplasmin substrate that fits the first group along with Fe(II).

There has been some uncertainty as to whether certain organic compounds are true substrates (group 2) or pseudosubstrates (group 3). The issue is complicated by the fact that the cyclical iron-catalyzed reactions are faster, in general, than direct electron transfer to ceruloplasmin. After considerable early controversy and uncertainty, Curzon and Young (1972) reported that ascorbate is a true ceruloplasmin substrate with a rather high K_m, 5.2 mM, and a typical V_m, 4.0 electrons/Cu atom·min. In their experiments the role of iron impurities was assumed to be eliminated by using 100 μM EDTA (Young et al., 1972). Young and Curzon (1972) also found that catechol was a true ceruloplasmin substrate with the largest K_m of all, 282 mM.

Similarly, Lovstad (1972) reported that D- or L-Dopa could also be catalytically oxidized, though very weakly, in the presence of the iron chelator Desferal. A summary of K_m and V_m data on organic ceruloplasmin substrates is given in Table 7 and is discussed later.

The reactions illustrated by group 3 pseudosubstrates have been used extensively to study the kinetics of ceruloplasmin. Ascorbate at concentrations well below its effective substrate range (100 μM) was used by Huber and Frieden

(1970a) to study Fe(II) oxidation, as in the reaction sequence in Figure 2. Young and Curzon (1972) also used ascorbate (50 μM) as the reducing agent in the reaction sequence in Figure 2 to study the oxidation of *N,N*-dimethyl-*p*-phenylenediamine. Walaas and Walaas (1961) introduced the use of NADH and NADPH to provide the electrons necessary to reduce partially oxidized or free radical intermediates resulting from the action of ceruloplasmin on aromatic diamines, phenols, or other oxidizable substrates. Since NADH does not react directly with ceruloplasmin, it has been widely used as an electron donor to study group 2 substrates.

Although the role of ceruloplasmin in iron mobilization is now widely documented, its catalytic activity toward any other class of substrates has not been related as directly to its biological function. We have suggested, therefore, that the name "ferroxidase" be used when describing the activity of ceruloplasmin as an enzyme (Osaki et al., 1966). It was further proposed that the enzyme be designated as a ferro-O_2-oxidoreductase and be assigned International Union of Biochemists No. E.C.1.12.3.1. It was realized, however, that the name "ceruloplasmin" might be retained when referring to the copper transport protein of the plasma because of its historical significance and widespread familiarity.

The presence of a *p*-phenylenediamine oxidase activity has been reported in a wide variety of vertebrate sera. Seal (1964) found no evidence of an oxidase activity in bullfrog sera. Inaba and Frieden (1967) were able to isolate from frog sera a blue oxidase resembling human ceruloplasmin very closely in several oxidase parameters. The ferroxidase activities of human, pig, and rat ceruloplasmin were compared by Williams et al. (1974), who estimated ratios of ferroxidase activity to plasma, copper, and *p*-phenylenediamine oxidase activity (Table 5). The pig and human enzymes compared more closely than did rat ceruloplasmin. Despite the low ferroxidase activity it was still possible to show that ceruloplasmin is essential for the flow of iron from reticuloendothelial cells to transferrin in rat plasma.

3.2. Ferroxidase Activity

The Ferroxidase Activity of Ceruloplasmin

The ferroxidase activity of human ceruloplasmin was first reported by Curzon and O'Reilly (1960, 1961), but its significance in iron metabolism and its substrate characteristics have been explored almost exclusively by the Florida State University group. In their first paper Osaki et al. (1966) compared the effect of oxygen and Fe(II) concentration on the ceruloplasmin and the nonenzymic reaction under normal serum conditions (see Figures 3 and 4). The nonenzymic rate of Fe(II) oxidation was first order with respect to both Fe(II) and oxygen concentrations. In contrast, Fe(II) oxidation catalyzed by ceruloplasmin showed typical saturation kinetics, reaching zero order at $>10\ \mu M\ O_2$ and $>50\ \mu M$

Table 5. Plasma Ferroxidase Activity in Pig, Man, and Rat, Assayed in Phosphate Buffer, pH 6.7, in the Presence of 300 μM Ascorbate[a]

Determination	Pig	Man	Rat
Number	6	41	27
Plasma copper (μg/100 ml)	170 ± 13.5	117 ± 3.2	129 ± 4.8
*p*PD oxidase (A_{530}/nm)	0.600 ± 0.0653	0.317 ± .00073	0.431 ± .00123
Ceruloplasmin ferroxidase (μmol/ml·hr)	118 ± 8.6	43 ± 1.3	6 ± 1.2
*p*PD-ox:Cu ratio (A_{530}:μg)	0.35	0.27	0.33
Cp ferroxidase:Cu ratio (μmol/hr:μg)	69	37	5
Cp ferroxidase:*p*PD-ox ratio (μmol/hr:A_{530})	197	136	14

[a] From Williams et al. (1974).

Fe(II). From the estimates of normal serum oxygen and Fe(II) levels it was estimated that the ceruloplasmin-catalyzed oxidation of Fe(II) was 10 to 100 times as fast as the nonenzymatic oxidation. This estimate does not include any correction for the presence of reducing metabolites, such as ascorbate at 40 μM, which did not affect the ceruloplasmin-catalyzed oxidation of Fe(II) but significantly reduced the net rate of the nonenzymatic oxidation of Fe(II).

Further studies of the kinetics of ferroxidase revealed biphasic curves in v versus v/Fe(II) plots (Osaki, 1966) with two K_m values, 0.6 and 50 μM, which differed by almost two orders of magnitude. While these data were originally interpreted in terms of two binding sites, Huber and Frieden (1970a) reported an excellent fit between experimental points and calculated values based on a rate-determining substrate-activation mechanism that resulted in a rate expression of the form

$$v = 4 \times 60[\mathrm{Cp}] \frac{k_{\mathrm{III}}(K_a[\mathrm{Fe(II)}]) + k_{13}/k_{\mathrm{XIII}}}{K_a[\mathrm{Fe(II)}] + 1}$$

The constants are identified in Figure 5. Curves calculated from this expression fit the experimental data within the allowed error at five temperatures, as shown in Figure 6. Consistent with the activation mechanism are the loss of blue color (at 610 nm) with low concentrations of Fe(II) and the activation observed by other divalent metal ions that are not substrates of ferroxidase (Huber and Frieden, 1970a; McKee and Frieden, 1971).

The mechanism of iron oxidation has been studied using rapid, stopped-flow methods to determine the kinetic parameters (Osaki et al., 1967). The following minimum reaction sequence, with rate constants indicated, was proposed:

$$\mathrm{Cu\text{-}Cu(II)} + \mathrm{Fe(II)} \xrightarrow{k_1} \mathrm{Cp\text{-}Cu(II)\text{-}Fe(II)}$$

$$k_1 = 1.2 \times 10^6\ M^{-1}\ \mathrm{sec}^{-1} \quad (1)$$

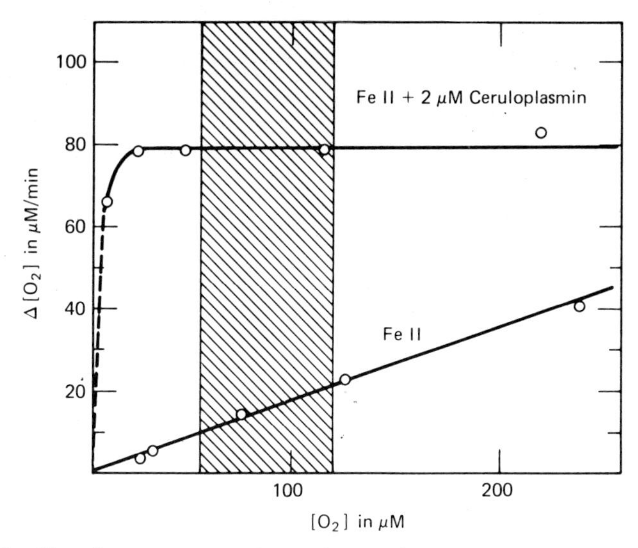

Figure 3. The effect of oxygen concentration on the rate of enzymatic and nonenzymatic oxidation of Fe(II) at 30°C. The oxygen concentration change per minute is plotted against various oxygen concentrations. The reaction mixture contained 70 μM ferrous ammonium sulfate in 0.0133 M phosphate buffer (pH 7.35) with and without ceruloplasmin. The shaded part of the figure indicates the oxygen concentration range in human vein (minimum) and artery (maximum). Data from Osaki et al. (1966).

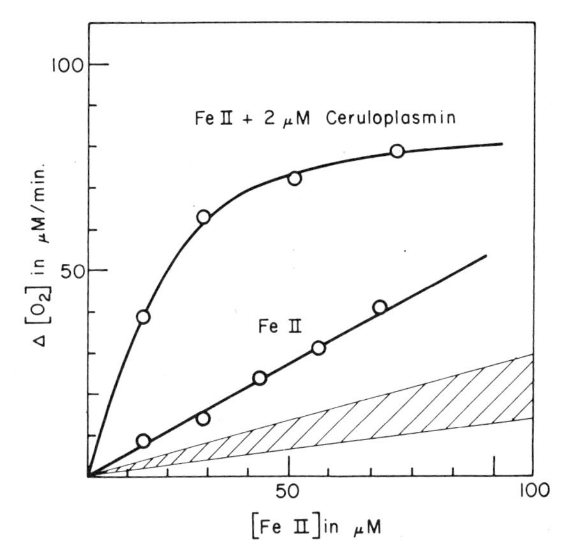

Figure 4. The effect of Fe(II) concentration on the rate of enzymatic and nonenzymatic oxidation of Fe(II) at 30°C. The oxygen concentration change per minute was measured with and without 2 μM ceruloplasmin. The reaction mixture contained 211 μM oxygen and variable amounts of ferrous ammonium sulfate in 0.0133 M phosphate buffer (pH 7.35). The estimated nonenzymatic oxidation rate at a lower oxygen concentration (55 to 120 μM; cf. Figure 3) is indicated by the shaded area. Data from Osaki et al. (1966).

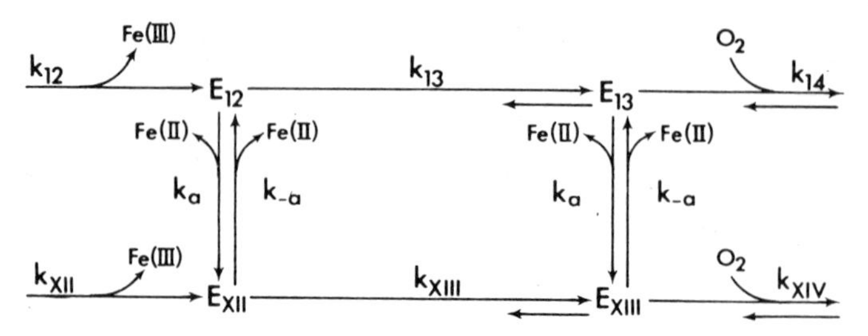

Figure 5. The essential steps in the proposed activation mechanism. The enzyme forms in the unactivated pathway (arabic numbers) are converted to the corresponding activated forms (roman numerals) by the binding of Fe(II) to the site of activation with the equilibrium constant K_a. The rate constants for the rate-limiting steps in the unactivated and activated pathways are k_{13} and k_{XIII}, respectively, with $k_{XIII} > k_{13}$. Data from Huber and Frieden (1970a).

$$\text{Cp-Cu(II)-Fe(II)} \xrightarrow{k_2} \text{Cp-Cu(I)-Fe(III)} \tag{2}$$

$$\text{Cp-Cu(I)-Fe(III)} \xrightarrow{k_3} \text{Cp-Cu(I)} + \text{Fe(III)} \tag{3}$$

$$\text{Cp-Cu(I)} \xrightarrow{k_4} [\text{Cp-Cu(I)}]^1 \qquad k_4 = 1.1 \text{ sec} \tag{4}$$

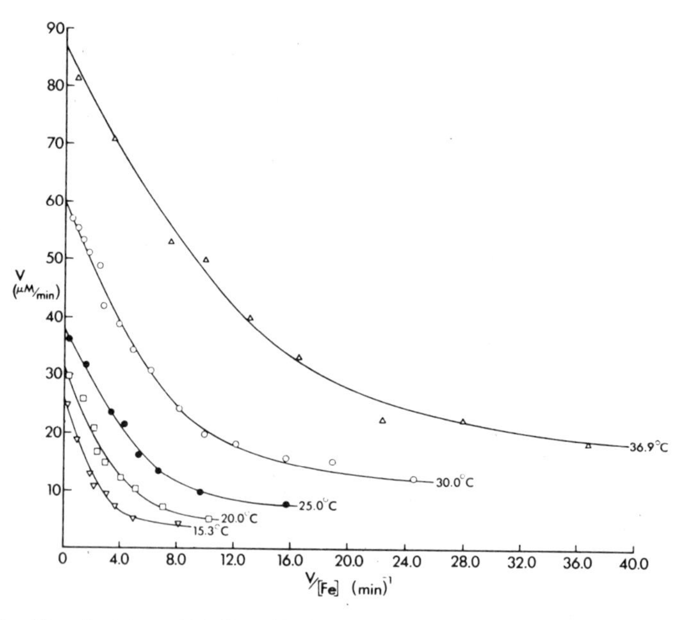

Figure 6. Plot of velocity of Fe(II) oxidation against velocity/[Fe(II)] at various temperatures. The reaction mixture contained 0.083 M Chelex-treated acetate buffer (pH 6.0), 100 μM Chelex-treated ascorbate, and 0.120 μM Chelex-treated ferroxidase. Every point is the average of at least three experiments. The drawn curve is calculated from the activation model. Data from Huber and Frieden (1970a).

$$[\text{Cp-Cu(I)}]^1 + O_2 + 4H^+ \xrightarrow{k_5} \text{Cp-Cu(II)} + 2H_2O$$
$$k_5 = 5.7 \times 10^5\ M^{-1}\ \text{sec}^{-1} \quad (5)$$

These data suggest that ceruloplasmin reacts with Fe(II) much faster than with other substrates and indicate the presence of a substrate-independent rate-determining step, for example, reaction 4 with the smallest rate constant, k_4. That the slowest step involved a conformational change, as depicted in reaction 4, was suggested by the large entropy change, $\Delta S^{\ddagger} = -23$ cal/mol·°C, estimated from the effect of temperature on the ferroxidase reaction (Osaki and Walaas, 1967). As emphasized in the discussion of other ceruloplasmin substrates, this rate-limiting step appears to be common to all substrates and to be relatively independent of the chemical nature of the substrate. Carrico et al. (1973) also have shown that the Fe(II)-mediated anaerobic reduction of ceruloplasmin by subequivalent ascorbate leads to a rapid reduction of the blue chromophore (reaction 2). Later Gunnarsson et al. (1973) proposed an even simpler mechanism for the oxidation of aromatic substrates, which is considered later.

The inhibitory effects of trivalent and other metal ions on ceruloplasmin activity were investigated by Huber and Frieden (1970b), and the results are summarized in Table 6. All trivalent cations tested inhibited ferroxidase activity, but the strongest trivalent inhibitors have ionic radii of 0.81 Å or less. The inhibition by Al(III) was mixed competitive and uncompetitive with respect to one of the substrates, Fe(II). The uncompetitive portion of the inhibition was not the result of competition by Al(III) with the other substrate, oxygen. A mechanism for the mixed inhibition by Al(III), consistent with these results, was proposed. Comparison of the strong cationic inhibitors provides the following series in order of decreasing effectiveness of inhibition: In(III) > ZrO(II) > Al(III) > Sc(III) > Ga(III). Insofar as they relate, these results are consistent with those from some earlier studies on the inhibition of aryldiamine oxidase activity of human ceruloplasmin reported by Curzon (1960), McDermott et al. (1968), and McKee and Frieden (1971).

Alternative Ferroxidase Activities

Since the recognition of the possible importance of the ferroxidase activity of ceruloplasmin in relation to its function in iron mobilization (Osaki et al., 1966), alternative ferroxidase activities have been proposed. These proposals fit into three categories:

1. Proteins, other than ceruloplasmin, that may have true ferroxidase activity. Attempts to distinguish these alternative ferroxidases have usually involved determining the effect of 1 mM azide, which inhibits ceruloplasmin over 98%.
2. Substances, including proteins, that have strong electron-accepting groups, thereby transforming Fe(II) to Fe(III).

Table 6. Effect of Cations on Ferroxidase Activity[a,b]

Inhibitory	
VO(II),ZrO(II)	Strong
In(III),Al(III),Sc(III),Ga(III)	
Y(III),La(III),Ce(III)	
Rh(III),Cr(III)	
Ni(II),Zn(II),Pb(II)(high conc.)	Weak
Activating	
Fe(II) > Co(II)> Mn(II) > Ni(II); Zn(II), Cd(III),Mg(II)	
No effect (at <1 m*M*)	
Li(I), Na(I), K(I), Sn(II), Ba(II), Ca(II)	

Cation	Linear or Nonlinear	I/Enzyme	$I_{50\%}$ (μM)	K_I (M)
Al(III)	L	1.9	2.1	6.2×10^{-12}
In(III)	L	1.0	0.45	4.7×10^{-7}
Sc(III)	L	1.2	13	—
Ga(III)	NL	—	~50	—
ZrO^{2+}	NL	—	0.5 – 2.0	—

[a] Data from Huben and Frieden (1970b).
[b] All reversible inhibition.

3. Compounds that strongly and preferentially chelate Fe(III), thereby enhancing the Fe(II) to Fe(III) reaction. We designate these activities as pseudoferroxidase activity.

The second and third categories can be distinguished from true ferroxidases because they are consumed in the reaction and exhibit a stoichiometric relationship between the Fe(II) oxidized and the Fe(III) chelated. Therefore they do not satisfy the basic prerequisite for catalysis, since they end up in a different chemical state after each reaction cycle. The true ferroxidases also show strong binding for Fe(II) and typical Michaelis-Menten saturation kinetics for both Fe(II) and O_2.

The presence of a nonceruloplasmin ferroxidase was suggested by a residual non-azide-sensitive iron-oxidizing activity in human serum, particularly in that from patients with Wilson's disease. Some of this activity may be due to the pseudoferroxidase activity of citrate, as proposed by Lee et al. (1969). However, most of the additional ferroxidase activity was associated with the protein fraction of the serum. The isolation and purification of a second ferroxidase, ferroxidase-II, was reported by Topham and Frieden (1970). This protein differed from ceruloplasmin in numerous respects, including not being inhibited by azide, exhibiting no *p*-phenylenediamine oxidase activity, being yellow, and containing a lipid moiety. The properties of ferroxidase-II as a cupro-lipoprotein have been explored more fully by Topham and co-workers (1974, 1975).

In further studies on the source of the non-azide-sensitive ferroxidase activity, Williams et al. (1974) and Sexton et al. (1974) found evidence for an induction of this activity in human serum. Williams et al. (1974) showed that dialysis of human serum at pH 5.5 in acetate buffer resulted in a tenfold increase in azide-resistant ferroxidase-II activity at pH 5.5 but no change at pH 7.7. In the presence of 15 μM $CuCl_2$, comparable activation occurred at pH 7.7. This activation was reduced at higher pH's and by low temperature, nitrogen, and a variety of protecting agents such as albumin, hemin, hemoproteins, EDTA, and butylated hydroxyanisole. Other evidence indicated a noncatalytic participation of this activity in the oxidation of Fe(II). Sexton (1974) correlated the increased ferroxidase activity with the degree of hydroperoxidation of a β-lipoprotein fraction of human serum. He found that during ferroxidase-II activation there was a change in the spectrum of this protein between 300 and 600 nm that was identical to changes reported by Gurd and others for the peroxidation of serum β-lipoprotein. It is also known that copper contamination during the dialysis of serum may lead to oxidative damage of serum lipoproteins. In the widely used Cohn fractionation, the β-lipoproteins are sensitive to oxidative damage in that the unsaturated fatty acids of the lipid moiety undergo hydroperoxidation reactions. Therefore Sexton's data raised the possibility that ferroxidase-II is a peroxidized lipoprotein with oxidative activity toward Fe(II) arising from the reduction of hydroperoxides by Fe(II) in the classical reaction between the metal ion and free radical reduction of lipohydroperoxides. Thus ferroxidase-II may be a peroxidized β-lipoprotein whose reducing power toward Fe(II) is induced during isolation and storage.

Topham (1975) contends that ferroxidase-II is a discrete serum protein and not an artifact. He believes that the association of protein, lipid, and copper components is indispensable for the catalytic activity of ferroxidase-II. When rats are fed a copper-deficient diet, ferroxidase-II activity falls to 10 to 20% of its original value (Topham et al., 1975). Topham's data differ, however, from those of Sexton (1974) and Williams et al. (1974) in that they deny that ferroxidase-II can be induced in fresh serum or can be found in Cohn IV-I fractions. Topham attributes the increase in ferroxidase-II activity observed upon acid treatment or dialysis of serum to dissociation of a low molecular weight inhibitor and not to lipid peroxidation. Finally, Topham states that "in normal human sera, ferroxidase-II contributes too small a percentage of the total ferroxidase activity to appear to play a major physiological role in iron metabolism. The actual primary role of human ferroxidase-II remains to be determined."

Pseudoferroxidase Activity

At least four biologically important substances have been reported to have pseudoferroxidase activity: apoferritin, transferrin, phosvitin, and citrate. The pseudoferroxidase activity of a dialyzable, heat-stable component of human serum, identified as citrate, was first pointed out by Lee et al. (1969), who proposed citrate as an alternative source of ferrous iron oxidizing activity in serum with low ceruloplasmin levels. Phosvitin, a highly phosphorylated protein in hens, was also shown to possess significant Fe(II) oxidizing activity (Taborsky, 1963).

The uptake of iron by apoferritin, the iron storage protein, requires this element in the ferrous form (Macara et al., 1972). Two groups (Marcara et al., 1972; Bryce et al., 1973) showed that apoferritin appeared to have a catalytic effect on the oxidation of Fe(II), which aided in the formation of Fe(III)-ferritin. Bates et al. (1973) showed that Fe(II) oxidation was increased by apotransferrin, the iron transport protein. They proposed that Fe(II) forms a weak complex with apotransferrin, and it is this complex that results in a faster oxidation of Fe(II) to Fe(III) with subsequent formation of $Fe(III)_2$-transferrin.

The facilitation of Fe(II) autooxidation by Fe(IIII)-complexing agents was studied by Harris and Aisen (1973), who proved that the rate of oxidation of Fe(II) by atmospheric oxygen at pH 7.0 was significantly enhanced by low molecular weight Fe(III)-complexing agents. The order of activity was EDTA $\simeq$ nitrilotriacetate > citrate > phosphate > oxalate. Under the conditions of Harris and Aisen, Fe(II) had a $t_{1/2}$ of 2700 sec; with EDTA (4×10^{-4} M) it has a $t_{1/2}$ of about 10 sec. The authors pointed out that there was nothing unique about the ability of apotransferrin to stimulate Fe(II) oxidation. Thus the effect of preferential Fe(III) binding accounts for the "ferroxidase" activity shown by apotransferrin, apoferritin, and phosvitin and may be termed pseudoferroxidase activity.

Concurrently, Frieden and Osaki (1974) proposed a kinetic scheme based on relatively simple assumptions that accounted for pseudoferroxidase activity. In the oxidation of Fe(II) and Fe(III) any effect on the reaction system that tends to reduce the free Fe(III) concentration has an enhancing effect on the rate of Fe(II) disappearance or Fe(III) complex formation. The derivation presented below is based on the proposed reaction sequence, the assumption of a steady state of free [Fe(III)], and the rate expression for the formation of the Fe(III) complex:

$$\mathrm{Fe(II)} + \mathrm{H}^+ + \mathrm{O}_2 \underset{k_{-1}}{\overset{k_1}{\rightleftharpoons}} \mathrm{Fe(III)} + \mathrm{O}_2^-$$

$$\mathrm{Fe(III)} + \mathrm{aTf} \xrightarrow{k_2} \mathrm{Fe(III)}\text{—}\mathrm{Tf}$$

$$\frac{d[\mathrm{Fe(III)}]}{dt} = k_1[\mathrm{Fe(II)}][\mathrm{H}^+][\mathrm{O}_2] - k_{-1}[\mathrm{Fe(III)}] - k_2[\mathrm{Fe(III)}][\mathrm{aTf}]$$

Assume steady state so that

$$\frac{d[\mathrm{Fe(III)}]}{dt} = \mathrm{OH}^+ \text{ and } p\hat{\mathrm{O}}_2 \text{ constant}$$

$$\mathrm{O} = k_1'[\mathrm{Fe(II)}] - k_{-1}[\mathrm{Fe(III)}][\mathrm{O}_2^-] = k_2[\mathrm{Fe(III)}][\mathrm{aTf}]$$

$$v_3 = k_2[\mathrm{Fe(III)}][\mathrm{aTf}] = k_1'[\mathrm{Fe(II)}] - k_{-1}[\mathrm{Fe(III)}][\mathrm{O}_2^-]$$

Assume that

$$v_3 = \frac{-d[\text{Fe(II)}]}{dt} = \frac{+d[\text{Fe(III)Tf}]}{dt}$$

Anytime that [Fe(III)] is reduced by [aTf] chelation, v_3 also increases.

3.3. Other Oxidase Activities of Ceruloplasmin

The relationship between ferroxidase activity and the iron mobilization properties of ceruloplasmin has focused much recent attention on Fe(II) as a substrate. However, stimulated by the fact that group 2 substrates include two important types of biogenic amines, that is, the epinephrine and 5-hydroxindole series and the phenothiazine series (tranquilizers), numerous investigators, particularly Barrass, Coult, Curzon, Lovstad, Pettersson, and their co-workers, have vigorously pursued the study of the ceruloplasmin-catalyzed oxidation of arylamines and phenols. Much of this work has been summarized by Young and Curzon (1972), Fee (1975), and Gunnarsson (1974).

Oxidation of Aromatic Amines and Phenols

A most comprehensive study of group 2 substrates was presented by Young and Curzon (1972). Table 7 adopts their method of presenting the data but also includes results from several other sources, particularly the data of Lovstad on the phenothiazines (Lovstad, 1975). The basic parameters, K_m and V_m, were estimated from reciprocal plots under standard conditions at 25°C and pH 5.5, in the presence of 10 mM EDTA and 50 mM ascrobate. The EDTA was used to eliminate trace iron effects, and the ascorbate to assure linear kinetics by preventing the accumulation of free radical intermediates. The data are remarkably consistent for an extensive series of p-phenylenediamines, aminophenols, catechols, and 5-hydroxyindoles. All the V_m values except the one for Fe(II) and a few others, fall into the range of 1 to 10 electrons/Cu atoms·min. This seems to emphasize the common substrate-independent, rate-determining step mentioned earlier. Young and Curzon (1972) pointed out a negative correlation between log K_m and log V_m. Replacement of a benzene ring by an indole group had little effect; the variation within a group of 20 p-amino compounds was only about twofold. The presence of a side chain reduced the V_m for both the catechol and hydroxyindole series. No significant relation was noted to pK or the size, position, or electronic character of substituents on any of the ring systems studied.

In contrast to the limited range of the V_m values, the K_m values varied over a 10^4-fold range. With the exception of the special role of Fe(II), a group of substituted p-phenylenediamines (pPD) appear to have the smallest K_m values and, presumably, the more favorable interaction with ceruloplasmin. The three preferred substrates with K_m values less than 10^{-4} M—all have an additional

Table 7. K_m and V_m Values of Ceruloplasmin Substrates

	K_m (μM)	V_m (electrons/ Cu atom·min)
Ferrous ion	0.650	22
Substituted *p*-phenylenediamines(*p*PD)		
N-(*p*-Methoxyphenyl-*p*PD)	21	6.5
N-Phenyl-*p*PD	48	4.8
N-Ethyl-*N*-2-(*S*-methylsulfon-amido)-ethyl-*p*PD	87	6.1
N-Ethyl-*N*-(2-hydroxyethyl)-*p*PD	110	7.4
2-Methoxy-*p*PD	161	6.2
N,N'-Dimethyl-*p*PD	164	4.3
Durenediamine	171	6.0
N,N,N,'N'-Tetramethyl-*p*PD	197	5.1
N,N-Dimethyl-*p*PD	203	5.1
2-Methyl-*p*PD	213	6.3
2-Chloro-*p*PD	241	5.4
*p*PD	292	4.4
N,N-Diethyl-*p*PD	556	3.2
N,N'-Di-*s*-butyl-*p*PD	620	6.1
2-Nitro-*p*PD	1,260	4.5
2-Sulfonic acid-*p*PD	2,620	3.8
o-Phenylenediamine	2,950	1.3
N,N-Dimethyl-*m*-phenylenedia-mine	3,050	4.0
N-Acetyl-*p*PD	12,300	3.4
m-Phenylenediamine	36,000	5.6
Catechols		
L-Epinephrine	2,550	2.3
L-Norepinephrine	2,810	2.7
3,4-Dihydroxyphenethylamine	2,850	7.5
Pyrogallol	57,900	10.8
4-Methylcatechol	60,300	6.8
Quinone	65,700	5.6
Catechol	282,000	9.0
Mixed aminophenols and others		
o-Dianisidine	180	15
p-Aminophenol	1,540	3.5
o-Aminophenol	2,880	3.6
Ascorbate	5,200	4.1
p-Anisidine	6,140	4.1
5-Hydroxyindoles		
5-Hydroxytryptamine	908	5.7

Table 7. (*continued*)

	K_m (μM)	V_m (electrons/ Cu atom·min)
5-Hydroxytryptophol	5,100	2.9
5-Hydroxyindol-3-ylacetic acid	8,340	1.5
5-Hydroxytryptophan	16,300	1.8
Phenothiazines		
Prochlorperazine	900	7.0
Perphenazine	1,300	8.0
Promazine	1,300	6.5
Thioridazine	1,400	8.8
Alimemazine	1,400	3.8
Periciazine	2,000	0.8
Diethazine	2,300	2.0
Promethazine	2,300	1.8
Trifluoperazine	2,800	3.8
Chlorpromazine	3,500	10.3
Fluphenazine	5,000	4.3
Triflupromazine	10,000	3.3

Active but with unfavorable constants: *p*-methoxyphenol, 3,4 dihydroxyphenylalanine, 3,4-dihydroxyphenylacetic acid, *N,N*-dimethylaniline, *N,N*-diethylaniline.

Undetectable activity: aniline, gentisic acid, mercaptoethanol, resorcinol.

benzene ring: —*N*-(*p*-methoxyphenyl)-*p*PD, *N*-phenyl-*p*PD, and *N*-ethyl-*N*-2-(*S*-methylsulfonamido)ethyl-*p*PD. The K_m values increase in the order para, ortho, meta, suggesting that electronic rather than steric factors are dominant. In fact, rigid steric restrictions for ceruloplasmin substrates are not indicated. For example, the tetra *N*-methyl-*p*PD has a K_m comparable to that of *p*PD or its partially substituted derivatives. However, ring-substituted *p*PD derivatives with strongly electron-withdrawing 2-nitro and 2-sulfonic acid groups have much greater K_m values than do other 2-substituted *p*PD derivatives. There is some evidence that negatively charged groups can increase K_m values, possibly because of repulsion of a negative charge on the enzyme near the active site.

Gunnarsson et al. (1970) reported that, within a limited comparison group, eight ceruloplasmin substrates had especially high energies of the highest occupied molecular orbital. The high orbital energies were correlated with lower K_m values. While most ceruloplasmin substrates have at least two electron-donating groups, Gunnarsson et al. (1971), from their orbital energy calculations, explored a special group of *N*-alkylated anilines as ceruloplasmin substrates. *N*-Methylaniline and *N,N*-dibutylaniline were oxidized first to yellow and then to blue pigments. The reaction sequence, including the ceruloplasmin steps, is shown in Figure 7. However, no quantitative data on these substrates were reported. A detailed study of the ceruloplasmin-catalyzed oxidation of a related oxidation product, *o*-dianisidine (4,4-diamino-3,3-dimethoxybiphenyl), has been published (Schosinsky et al., 1974). These data emphasize that electronic rather then steric characteristics are of prime importance for the activity of a ceruloplasmin substrate.

Newer Substrate Groups

Lovstad (1975) has discovered a new class of ceruloplasmin substrates, the phenothiazine derivatives. A highly suggestive clue was provided by Barrass and Coult (1972), who reported that the phenothiazines activated the ceruloplasmin-catalyzed oxidation of catecholamines. As mentioned previously, the data included in Table 7 on the phenothiazines were adapted from a recent paper by Lovstad (1975). As with other organic substrates, V_m does not vary greatly. In this series K_m also varies only over a tenfold range. At low concentrations, compounds with piperazinylpropyl side chains are more rapidly oxidized than those with an aliphatic side chain, suggesting greater enzyme affinity for the former compounds. Phenothiazines with three-carbon side chains (promazine, alimemazine) are more rapidly oxidized by ceruloplasmin than those with any two-carbon-atom side chains in the 10-position (promethazine, diethazine). Lovstad also confirmed that these phenothiazine derivatives activated the ceruloplasmin-catalyzed oxidation of catecholamines. The substrates most rapidly oxidized by ceruloplasmin also activated the oxidation of dopamine most effectively.

Barrass et al. (1974) made a comprehensive study of the phenylalkylamines and the indoles and their isosteres as substrates for ceruloplasmin. However, much of this work was qualitative, with no V_m values and a limited number of K_m values reported. Furthermore, the enzyme and other components used in the reaction mixture were not always carefully screened for traces of iron ions or other trace elements. Therefore it is possible that some of these substrates might be of the group 3, rather than the group 2, type. Barrass and co-workers found that all the 3,4-dihydroxyphenylalkylamines are substrates. Numerous compounds of the substituted amphetamines series were also oxidized. Only 3,4-dihydroxyphenylalanine and 3-aminotryosine were oxidized among eight substituted phenylalanines tested. The best substrates have a 3,4-dioxygen pattern with at least one free OH group. A primary amino group could replace one of the OH groups. The alkylamine side chain was essential for maximum substrate activity, but the length of the side chain was not critical. Higher homologues of dopamine were effective substrates, with the smallest K_m (40 μM) observed from the propylamine side chain. Monosubstitution of the α-carbon atom of the side chain had little effect, but disubstitution at this point greatly reduced substrate activity.

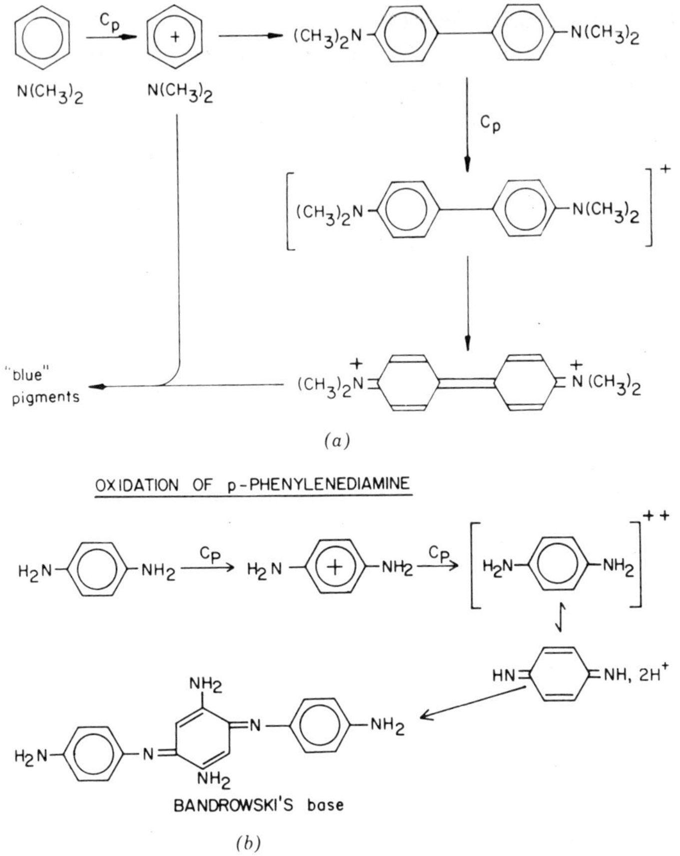

Figure 7. (*a*) Sequence of reactions in the oxidation of *N,N*-dimethylaniline to a blue benzidine derivative. At least two of the reactions involve ceruloplasmin; the other reactions are believed to be spontaneous. This is a modification of Figure 1 of Gunnarson et al. (1971). (*b*) Sequence of oxidative reactions by which ceruloplasmin and H_2O_2 convert *p*-phenylenediamine (*p*PD) to Bandrowski's base, identified by Rice (1962) as the ultimate oxidation product. Part of the pPD^{2+} is formed by a rapid disproportionation reaction: $2pPD^{+} \rightarrow pPD + pPD^{2+}$.

In a survey of the indole series, Barrass et al. (1973) confirmed the essentiality of a hydroxy group on the aromatic ring in the 4-, 5-, or 6-position. An aminoalkyl chain, free or substituted, at position 3 is necessary for high substrate activity, but, again, the distance separating the basic group of the side chain from the indole ring is not critical. Among indole isosteres, analogues of 5-hydroxytryptamine, only an imino group at C-1 of the bicyclic system is compatible with substrate activity.

The possibility that there may be other classes of substrates for ceruloplasmin has been suggested by Albergoni and Cassini (1975). They reported that bovine ceruloplasmin has a cysteine oxidase activity several-fold greater than the maximal activity reported for Fe(II) oxidation. A significant feature of this activity was a pH optimum of 7.4, suggesting that cysteine may be a significant substrate for ceruloplasmin in serum.

However, there are numerous differences between cysteine oxidation and the oxidation of other ceruloplasmin substrates, such as lack of sensitivity to 1.5 mM azide. Many features of cysteine oxidation are similar to those observed with free Cu(II). It remains to be confirmed that this oxidation is due to an intact ceruloplasmin molecule or to Cu(II) liberated by reduction by cysteine of one or more Cu(II) from this copper protein.

Mechanism of Oxidation of Aromatic Diamines

In a study of the pH dependence of the ceruloplasmin-catalyzed oxidation of *N,N*-dimethyl-*p*PD (DPD), Gunnarsson et al. (1972) were able to resolve some of the kinetic parameters of the four major substrate forms of DPD. They reported K_m values for these substrates as DPD· = 0.2 μM, DPD = 45 μM, DPD^+ = 70 μM, and DPD H^+ = 1100 μM. The extremely low K_m for DPD compares favorably with that of Fe(II). The argument of Gunnarsson et al. that this challenges the designation of ceruloplasmin as a ferroxidase is still not valid, since DPD is not a native substrate and no other potentially physiological substrate has been shown to have a V_m one-quarter as great or a K_m one-thousandth as small as that of Fe(II).

A simplified mechanism to describe the ceruloplasmin-catalyzed oxidation of organic substrates has been proposed by Gunnarsson et al. (1972) as follows:

$$E + S \xrightarrow{k_1} E' + P, \qquad E' \xrightarrow{k_2} E$$

where k_1 is identical with the second-order rate constant for reduction of the 610-nm chromophore, and k_2 is a pseudo-first-order constant corresponding to the rate-limiting reaction step. Thus $V_m = k_2$, $K_m = k_2/k_1$, and $V_m/K_m = k_1$. This mechanism assumes the substrate-independent rate-limiting step, the kinetic insignificance of the rate of formation of enzyme-substrate complexes in these reactions, and the saturation of the system by oxygen (a K_m value of 3.9 μM for O_2 had been reported earlier by Frieden et al., 1965). It was claimed that the correspondence of k_1 values determined directly by reduction of the 610-nm chromophore and k_1 values estimated from V_m/K_m ratios obtained from steady-state data supported this mechanism.

Little information is available on the identity of the oxidation products produced when aromatic amines and phenols are exposed to ceruloplasmin. Phenols are oxidized to the corresponding quinones, which frequently react further to form cyclization products, such as adrenochrome. The oxidation of dimethylaniline produces biphenyl derivatives that are oxidized further, probably spontaneously to a blue dye (Figure 7*a*). For *p*-phenylenediamine, Rice (1962) showed that the principal product for both ceruloplasmin and NH_3—H_2O_2 oxidations was Bandrowski's base (Figure 7*b*), known since 1889. It was proposed that the molar absorption of this compound at 540 nm (= 1910) could serve as a basis for defining the molecular activity of ceruloplasmin.

Inhibition of Oxidase Activity

The inhibition of the catalytic activity of ceruloplasmin toward aromatic diamines has been reviewed extensively by Curzon (1960, 1961), Curzon and O'Reilly (1960), Gunnarsson (1974), and Fee (1975). Curzon and Cumings (1966) identified seven categories of inhibitors, including inorganic anions, carboxylate anions, —SH compounds, chelating agents, hydrazines, 5-hydroxyindoles, and a miscellaneous group that included metal ions. Probably the most useful group is the inorganic anions, which include two of the strongest inhibitors—cyanide and azide. Holmberg and Laurell (1951) observed that the oxidase activity of ceruloplasmin was affected by virtually any anion. In Curzon's extensive studies (1960) at pH 5.5, 10 m*M* acetate buffer, and 4 μM EDTA, the order of inhibition of human ceruloplasmin was $CN^- > N_3^- > F^- > I^- > NO_3^- > Cl^- > Br^- > OCN^- > SCN^- > HPO_4^- > SO_4^{2-}$. A similarily ordered series was observed for rat ceruloplasmin by Lovstad and Frieden (1973). The metal-binding feature of these ions is the strongest but not the only factor determining the inhibitory impact. The two most powerful inhibitors are cyanide and azide with inhibitory constants (k_i) of about 2×10^{-6} *M*.

Azide is obviously the more convenient of the two and has been used frequently in attempts to distinguish between ceruloplasmin catalysis of Fe(II) oxidation and other ferroxidase activity in biological media such as plasma. To assess the importance of other ferroxidases in human serum, Sexton (1974) compared the effect of 1.0 m*M* azide on the ferroxidase activity of fresh human serum with crystalline ceruloplasmin with and without bovine serum albumin (7%). The percent of non-azide-inhibited activity was 1.5 ± 0.2% in all samples, suggesting that 98.5% of the ferroxidase activity was due to ceruloplasmin.

Azide has been the most thoroughly studied of all of the inhibitors of ceruloplasmin in regard to the mechanism of inhibition. Curzon and Cumings (1966) reported that azide was a reversible but virtually stoichiometric inhibitor of ceruloplasmin—an azide concentration not much greater than that of ceruloplasmin itself was required for inhibition at 25°C. They also noted that only one azide reacted with each ceruloplasmin molecule, regardless of the number of coppers at the active site. Reciprocal plots ($1/v$ vs. $1/s$) at various azide concentrations were linear and parallel, suggesting that azide binds to an intermediate form of the enzyme during catalysis. A variety of methods by different authors (Fee, 1975) support the view that azide inhibits primarily by impeding the breakdown of the reduced form of the enzyme in the oxidative reaction sequence.

4. THE BIOLOGICAL FUNCTION OF CERULOPLASMIN

4.1. The Mobilization of Plasma Iron

Copper-Iron Relationships

Almost 50 years ago Hart et al. (1928) noted that the copper-deficient animal became anemic. This defect in iron metabolism in the copper-deficient animal has been studied for several decades by Cartwright, Wintrobe, and Owens, and their associates. Lahey et al. (1952) compared blood variables of 5-day-old pigs fed an iron-supplemented milk diet with those of littermate controls on the same milk diet but supplemented with both iron and copper. The serum copper fell rapidly and was followed by a decrease in serum iron and erythrocyte copper, and finally a dramatic reduction in red cell volume due to both hypochromic and microcytic anemia. A similar sequence of events in copper-deficient rats was reported by Owen et al. (1956); there was a rapid fall in plasma ceruloplasmin and copper, followed by a slower but steady decline in liver copper and in hemoglobin.

Analysis of the role of copper in hemoglobin biosynthesis has followed three main lines: the biosynthesis of globin, the biosynthesis of protoporphyrin or heme, and the utilization of iron. There is no evidence for a general impairment of globin or protein biosynthesis in copper deficiency. Early efforts to find a copper-dependent step in heme biosynthesis were unsuccessful. Lee et al. (1968) observed that, as anemia developed in copper-deficient swine, there was a two- to threefold increase in the activities of several of the enzymes involved in heme biosynthesis, including δ-aminolevulinic acid synthetase, α-ketoglutaric acid-dependent glycine decarboxylase, and heme synthetase.

Meanwhile, our laboratory had focused attention on the role of copper in iron mobilization as the result of a study of the ferrous iron oxidase activity of ceruloplasmin, which was described in an earlier section. In 1966, in a paper entitled "The Possible Significance of the Ferrous Oxidase Activity of Ceruloplasmin in Normal Human Serum" (Osaki et al., 1966), it was proposed that ceruloplasmin might be a molecular link between copper and iron metabolism. This point of view has now been extensively confirmed by both *in vivo* and *in vitro* experiments. Much of the early work and documentation concerning this proposal has been discussed in earlier reviews (Frieden et al., 1971, 1974, 1976), and only new evidence and essential points are outlined here. It should also be emphasized that ceruloplasmin may not represent the only molecular involvement of copper in total red cell function, metabolism, and/or synthesis.

The link between ceruloplasmin and iron mobilization was suggested primarily by the study of the ferroxidase activity of ceruloplasmin and its effect on the rate of ferri-transferrin formation (Osaki et al., 1966). Indeed, the rate of Fe(II) uptake by apotransferrin can be used to quantitatively determine the oxidase activity of ceruloplasmin (Johnson et al., 1967). We also made a careful comparison of the rate of formation of ferri-transferrin from Fe(II) in human serum with and without ceruloplasmin at pH 7.4. The estimated amount of Fe(III)-transferrin generated is 3 to 5 mg/day without ceruloplasmin and over 60 mg/day with ceruloplasmin. Since the estimated daily utilization of iron is 35 to 40 mg/day, ceruloplasmin was necessary to restore iron oxidation to at least a normal level.

Ferroxidase and Iron Mobilization

It has now been amply confirmed both *in vitro* and *in vivo* that ceruloplasmin can mobilize plasma iron from the iron storage sites in the liver. The *in vitro* evidence was provided by liver perfusion studies in our laboratories at Florida State University (Osaki et al., 1971). Carefully excised livers were perfused, and the mobilized iron appearing in the perfusate was detected in a flow cell at 460 nm as the Fe(III)-transferrin complex or at 530 nm as the α,α-dipyridyl chelate. In a typical experiment, shown in Figure 8, the release of iron into the perfusate is shown in response to 0.57 μM ceruloplasmin. The rate of Fe(III)-transferrin formation was estimated to be 3 μm/100 sec (220 μg in 20 min) or enough to mobilize up to 192 mg Fe/day from the liver of a 70-kg man.

These experiments were extended to determine how sensitive iron efflux from perfused livers was to ceruloplasmin concentration. The rates of iron efflux were determined from the slopes of the recorded responses, as in Figure 8, and were plotted as relative rates versus ceruloplasmin concentration (shown in Figure 9). An effect of ferroxidase was observed at concentrations as low as 4×10^{-9} M, with a maximum effect at 2×10^{-7} M, which is 10% of the normal human serum level. These data correspond closely to the *in vivo* results of Roeser et al. (1970), mentioned earlier, in which only one tenth of the normal level of ceruloplasmin was required to produce maximum iron mobilization. Thus normal serum has a tenfold excess of ceruloplasmin as far as the apparent maximum requirement for iron mobilization is concerned. However, an appreciable response is noted at 1% of the normal ceruloplasmin level. This may account for the failure to observe defects in iron mobilization unless the plasma ceruloplasmin levels are extremely low, that is, less than 0.02 μM, 1% of the normal ceruloplasmin level.

The specificity of the iron mobilization response in the perfusate system was also studied. Only ceruloplasmin among the compounds tested proved to have any activity in the perfused liver. No iron-mobilizing activity was shown by 30 μM apotransferrin, 3.5 mM HCO_3^-, 10 μM $CuSO_4$, 5 mM glucose, 0.6 mM fructose, 120 μM citrate, or 36 μM bovine serum albumin ± 21 μM $CuSO_4$.

Convincing *in vivo* evidence (Lee et al., 1968, 1969) that ceruloplasmin (ferroxidase) plays a direct role in regulating plasma iron levels has been reported in a series of papers, mainly from the laboratories of Cartwright, Lee, and coworkers. Ragan et al. (1969) studied the effect of injected copper and ceruloplasmin on copper-deficient pigs supplemented with iron. In addition to the

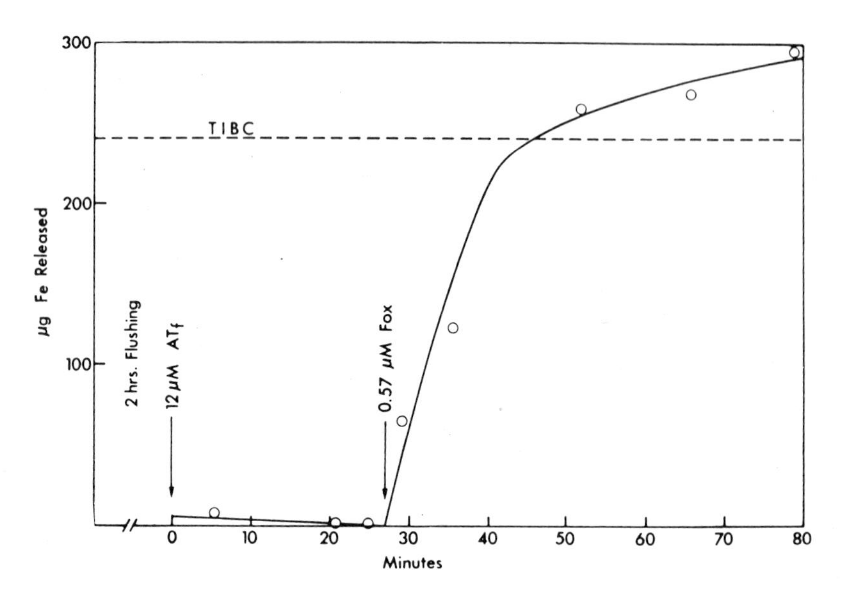

Figure 8. The effect of ferroxidase on a perfused liver prepared from a copper-deficient pig. The arrows in the figure indicate the times when the infusions were made. The total volume of the perfusate was 350 ml, and its total iron-binding capacity was 245 μg of Fe(III) per total perfusate. ATf and Fox represent apotransferrin and ceruloplasmin, respectively. Data from Osaki et al. (1971).

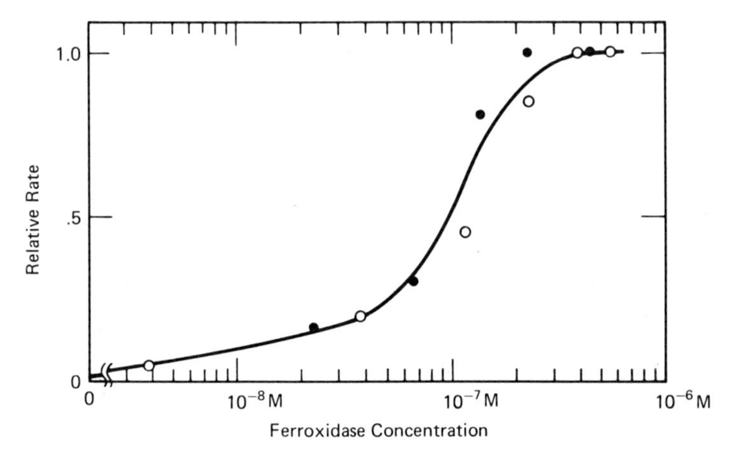

Figure 9. The effect of perfusate ferroxidase concentration on iron mobilization from dog livers. The open and filled circles represent two independent experiments. An additional observation at 1.2 *M* ferroxidase-I gave the same relative rate of 1.0. At zero enzyme concentration (×), the rate of iron inflow into the perfusion system was less than 0.15 μM of Fe(III)/100 sec (a relative rate of 0.030) or about one third of the value observed at 4-nm enzyme concentration. Data from Osaki et al. (1971).

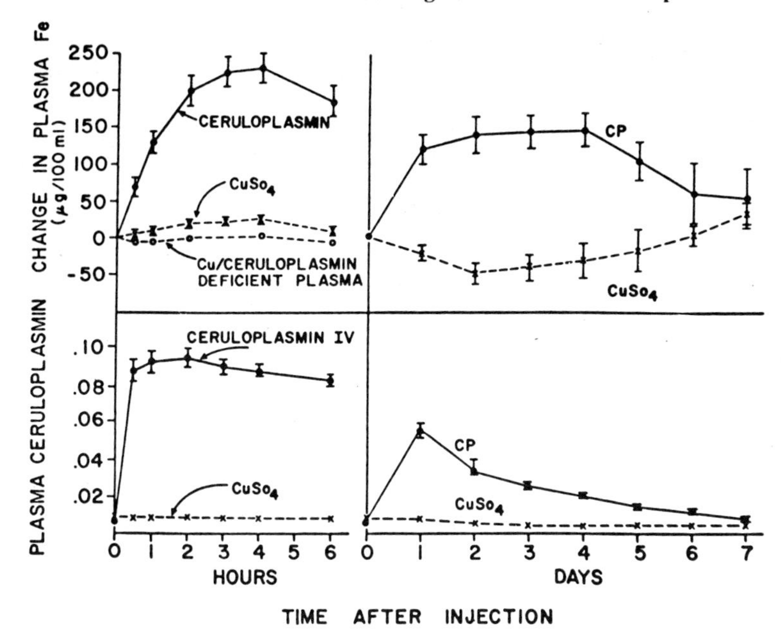

Figure 10. The effect of the intravenous administration of ceruloplasmin (●—●), copper sulfate (X- - -X), and copper-deficient (O- - -O) plasma on the plasma iron and plasma ceruloplasmin (expressed as *p*PD oxidase activity) in copper-deficient swine. Sufficient ceruloplasmin (Cp) or $CuSO_4$ was given to achieve 10% of the normal level (15 μg/100 ml plasma). Data from Ragan et al. (1969).

typical copper- and iron-deficient milk diet, each pig was given intramuscularly a total of 2.0 g Fe as iron dextrin (Pigdex) from 5 to 30 days of age. When a profound state of copper deficiency (80 days) was evident, the effects of injected pig ceruloplasmin, $CuSO_4$, or copper-deficient pig plasma were determined (Figure 10). The amount of ceruloplasmin injected was only enough to increase the plasma concentration 15 μg %, or 10% of the normal level. A remarkably rapid rise in plasma iron accompanied the injection of ceruloplasmin, with a peak in 3 to 4 hr. The increase in plasma iron was significant after 5 min. The maximum increase in plasma iron was about twice the normal level and persisted for 6 days. Neither $CuSO_4$ nor other pig plasma factors produced this iron increase. In fact, $CuSO_4$ actually reduced the iron levels after 2 days, presumably by stimulating red blood cell formation.

Roeser et al. (1970) performed many additional *in vivo* experiments that further support the role of ferroxidase in iron metabolism. In copper-deficient pigs the ceruloplasmin level fell to less than 1% of the normal, usually to about 0.5% ferroxidase activity. This deficiency of serum ferroxidase precedes the development of hypoferremia in the copper-deficient pig with the accumulation of iron in the liver. These authors demonstrated that a rise in serum ferroxidase activity precedes a rise in serum iron after copper injection. Serum iron does not increase until the serum ceruloplasmin reaches about 1% of normal. However, any hypoferremia can be corrected immediately by the administration of ceruloplasmin. When injected intravenously, Fe(II) disappeared rapidly from the circulation in the absence of ferroxidase activity and did not bind as readily to apotransferrin as did iron injected as Fe(III). In other words, in the absence of adequate ferroxidase, although there was no difference between control and copper-deficient pigs in the serum levels maintained when Fe(III) was injected, a 50% reduction in serum iron levels occurred when Fe(II) was injected. This was interpreted as a demonstration of the direct physiological role of ceruloplasmin in the control of serum iron. Finally, Roeser and co-workers studied asialoceruloplasmin, which was rapidly removed by the liver. This protein shows little or no iron-mobilizing activity when injected, although it does have ferroxidase activity in *in vitro* tests, supporting the idea that ceruloplasmin must function in the circulatory system. The comprehensive experiments by the Utah group lead to the conclusion that the defect in the release of iron in the copper-deficient animal can be reversed promptly by intravenous ceruloplasmin, an effect that cannot be due to copper alone. In no case did any of these workers' *in vivo* observations conflict with the hypothesis that the ferroxidase activity of ceruloplasmin is directly involved in iron mobilization.

Evans and Abraham (1973) showed that in the growing rat an early increase in liver iron accompanied the fall in ceruloplasmin activity during copper deprivation (Figure 11). Three to four weeks later, the hemoglobin reached a plateau level of about one half of normal, from 14 to 70 g %. These changes from normal metabolism were corrected several days after feeding the rats 35 ppm $CuCO_3$ in their diet. In a study of the effects of copper on iron metabolism and vice versa, Owen (1973) found that blood hemoglobin decreased most rapidly when rats were both copper and iron deficient. Copper-deficient rats accumulated large amounts of iron in their livers. These careful experiments by Owen provide further confirmation that copper is required for normal iron utilization.

Williams et al. (1974) extended their observations to the copper-deficient rat. A rapid fall in ceruloplasmin (oxidase activity) was followed closely by a reduction in plasma copper. The plasma iron remained unaffected until the 36th day, after which it declined from 235 to 133 μg % by the 50th day of copper depletion. Purified rat or pig ceruloplasmin infusions at 20% of normal raised the plasma iron to 303 and 238 μg %, respectively. However, neither $CuSO_4$ nor saline had any effect. These authors also confirmed the relatively low activity level of ferroxidase activity in the rat (about one-fourth that of human), but we consider this level adequate for iron mobilization.

How does ceruloplasmin mobilize iron? Is iron mobilization due to the ferroxidase activity of this copper protein, or is some specific receptor site(s) on the membrane of the iron storage cells activated by it? Ceruloplasmin has two

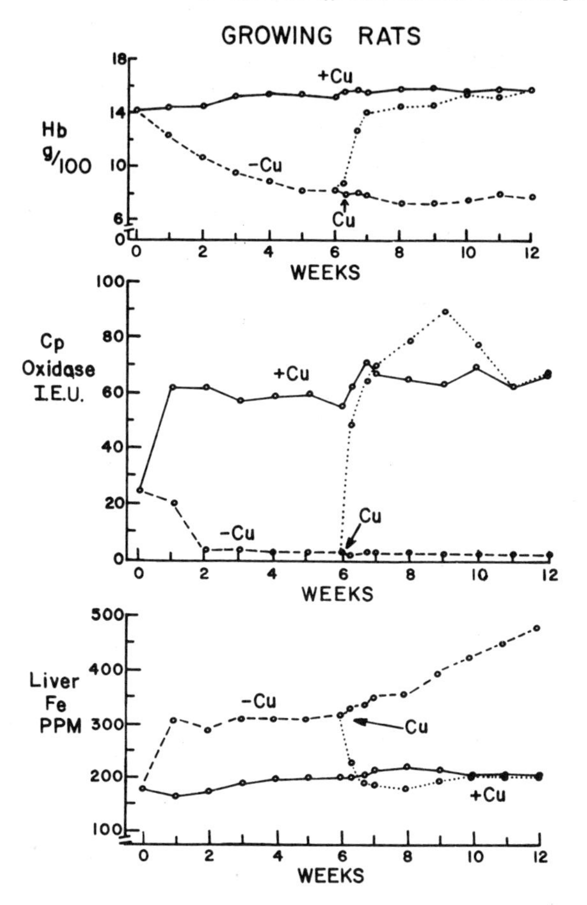

Figure 11. Changes in hemoglobin (Hb) levels, plasma ceruloplasmin (Cp), and liver iron associated with dietary copper. The recovery of copper-repleted rats is indicated by an arrow. —, Copper fed; - - -, copper depleted; ····, copper repleted. Data from Evans and Abraham (1973).

properties that may be related to this question: its ferroxidase activity and its relatively unique ability to complex strongly with Fe(II) in preference to Fe(III). As pointed out earlier, the formation of the Fe(II)-ceruloplasmin complex is the first reaction in the ferroxidase sequence and is extremely fast, $k = 10^7\ M^{-1}\ sec^{-1}$. This reaction may be a limiting factor in iron mobilization from the iron storage cells in the liver. The binding of Fe(II) could provide the impetus for the removal of Fe(II) from the liver iron stores as the iron is reductively released from ferritin (Figure 12). Williams et al. (1974) similarly describe the role of ceruloplasmin in the movement of iron from reticuloendothelial cells to transferrin as follows: Fe(II) occupies specific iron-binding sites on the membranes of reticuloendothelial cells; ceruloplasmin interacts with these iron-binding sites and then forms a Fe(II)-ceruloplasmin intermediate that transfers iron to apotransferrin by a specific ligand exchange reaction. Kinetic evidence for iron-

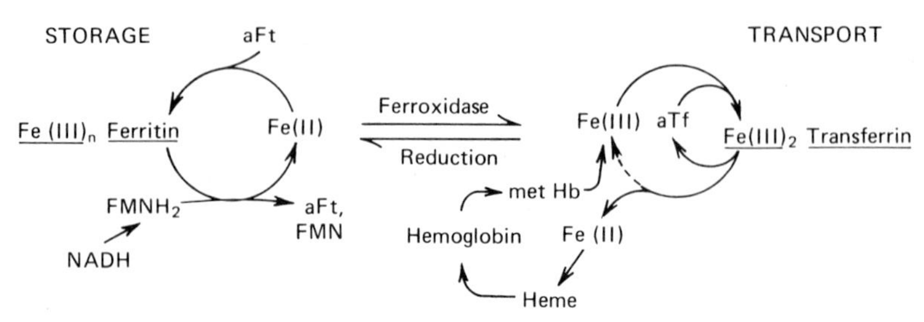

Figure 12. The central role that ceruloplasmin (ferroxidase, Fox) plays in regulating ferrous to ferric cycles, which in turn affect the storage, transport, biosynthesis, and catabolism of iron compounds.

ceruloplasmin complexes has been reported (Osaki et al., 1967, 1968) and was discussed earlier.

Additional Supporting Experiments

The ingestion of excessive amounts of other heavy metals, particularly zinc, eventually leads to anemia in the experimental animal. Following our proposal that ceruloplasmin might be a rate-limiting factor in hemoglobin biosynthesis, Lee and Matrone (1969) observed in rats a dramatic fall in serum ceruloplasmin to zero within 7 days after the start of zinc feeding. The reduction in ceruloplasmin preceded by several days the development of anemia in these rats. The ceruloplasmin level could be partially restored by injecting 100 μg Cu. Other metals also reduce ceruloplasmin levels. Whanger and Weswig (1970) found that several metals were copper antagonists to plasma ceruloplasmin levels, the order of effect being Ag(I) > Cd(II) > Mo(VI) > Zn(II). These provocative data suggest that these metals are interfering with the utilization of copper for the biosynthesis of ceruloplasmin in the liver.

In vertebrates the functional relationships among serum iron and ferroxidase activity, iron-binding proteins, and, ultimately, hemoglobin biosynthesis are well established. A study of the dramatic increase (five- to tenfold) in serum iron and ferroxidase activity and other blood parameters in normal, copper-deficient, and estrogenized roosters has been reported by Planas and Frieden (1973). Serum ferroxidase activity, iron, hemoglobin, and hematocrit values were greatly reduced when roosters were maintained on a copper- and iron-deficient diet for 17 days. Their anemic condition was exacerbated after 40 days, correlating with the total disappearance of ferroxidase activity. Dietary supplementation with copper, with or without iron, produced the largest increase in ferroxidase activity. The injection of copper salts also raised ferroxidase activity and, later, serum iron. The administration of estrogen, as either estradiol or diethylstilbestrol, also induced the appearance of ferroxidase in both normal and copper- or iron-deficient animals. The maximum ferroxidase activity enhancement was obtained on the second day, and the largest serum iron increase on the third day. There was no change in transferrin; apparently a more metabolically labile iron-binding protein, phosvitin, is produced to serve as a supplementary iron carrier. Diethylstilbestrol increased the ferroxidase activity almost to normal levels in copper- and iron-deficient roosters and those fed high doses of zinc and silver. Planas (1973) has reported a five- to tenfold increase in serum iron and copper and up to a 20-fold increase in ferroxidase activity in the serum of laying hens after estrogen treatment. He thus confirms the metabolic link between serum iron, copper, and ferroxidase, but his work also emphasizes the uniqueness of phosvitin, the special iron-binding protein that is produced in the livers of several different vertebrates, for example, birds and amphibians, in response to estrogen stimulation.

Remarkable changes in metalloprotein metabolism occur during the metamorphosis of the amphibian tadpole (Osaki et al., 1974), thus providing a useful model system for testing many functional relations of iron and copper proteins. There is a complete switch in the hemoglobin chains—the three tadpole hemoglobins have no chain in common with the three to four different frog hemoglobins. Extensive iron reutilization must be involved, since all new hemoglobin synthesis is believed to proceed *de novo*. The total resynthesis of hemoglobins coincides with the beginning of the metamorphic climax (Stage XX) and is preceded by significant increases in ferroxidase activity (Inaba and Frieden, 1967).

Since these metalloprotein systems seem to be related in the evolutionary history of the iron and copper metalloproteins (Frieden, 1974), further evidence is being sought for the sequence: liver iron → ferrireductase → ferroxidase (ceruloplasmin) → transferrin → hemoglobin.

Objections to the Role of Ceruloplasmin as a Physiological Ferroxidase

Despite the extensive evidence that ceruloplasmin is directly involved in iron mobilization, several objections to this idea have been advanced. Shokeir (1972) questioned the significance of ceruloplasmin as a normal ferroxidase on the basis of finding that blood from newborn infants had high transferrin saturations, but blood from the mothers had low transferrin saturations. It was also found that ceruloplasmin concentrations were low in the babies but high in their mothers. Shokeir conceded that only a small fraction (less than 1%) of normal ceruloplasmin activity is required to effect normal saturation of transferrin. This has been amply substantiated in several reports (Frieden, 1971; Roeser et al., 1970). It should be pointed out, also, that the control of transferrin saturation is a complex metabolic adjustment in which ceruloplasmin is only one factor with an effect on the iron mobilization parameter from the reticulo-endothelial (RE) cells of the liver. There are numerous examples of the mother providing crucial biomolecules, for example, calcium and iron, vitamins, and essential amino acids, to the fetus at the expense of her own metabolic economy. Thus it would not be surprising if unique and specialized mechanisms operate in governing iron transfer between mother and fetus.

As recently as 1973, Bates and Schlaback (1973) stated, "We do not believe, therefore, that the rate-limiting step of iron metabolism could be the oxidation of free Fe(II) to Fe(III) in blood and a subsequent rapid reaction to form Fe(III)-transferrin." This statement was based on the expected observation that apotransferrin, comparable to other Fe(III) chelators, has pseudoferroxidase activity. However, Bates and Schlaback presented no data showing that this pseudoferroxidase activity could account for the rate of Fe(III)-transferrin formation necessary for iron turnover, especially in the presence of plasma proteins and reducing agents such as 10^{-4} *M* ascorbate. Their rate determinations at 2×10^{-4} *M* Fe(II) are not relevant, since free Fe(II) concentration in plasma is considered to be much lower, $<10^{-6}$ *M*. However, no measurements with ceruloplasmin in this system were reported by Bates and Schlaback. Another factor they did not consider is that ceruloplasmin would easily out-compete the weak binding ability, if any, of any apotransferrin for Fe(II).

In another paper Bates et al. (1973) wrote that their results "emphasized that transferrin would react only partially with free Fe(II) in plasma and very poorly with iron which had been oxidized to Fe(III) by an agent such as ceruloplasmin." They used H_2O_2 as the oxidant and found that the rate of Fe(III)-transferrin formation was reduced. Granting that these data are valid for H_2O_2, they are not valid for ceruloplasmin. When we substituted 0.20 to 2.3 μM ceruloplasmin for 200 μM H_2O_2, Fe(II) was rapidly and stoichiometrically converted to Fe(III)-transferrin with the speed proportional to the concentration of ceruloplasmin. Bates et al. (1973) suggest that, if the iron is oxidized to Fe(III) before binding to transferrin, it becomes essentially unavailable. This is not the case when ceruloplasmin is the oxidizing agent. These authors presented no data showing that Fe(III)transferrin formation was adversely affected by the action of ceruloplasmin on Fe(II). Therefore neither the data of Bates et al. nor their arguments support the contention that the ferroxidase activity of ceruloplasmin has been shown to be unrelated to its function in mobilizing iron.

Brittin and Chee (1969) found no relationship between ceruloplasmin activity and iron absorption in iron-deficient and iron-loaded rats. No claims have been made that ceruloplasmin is involved in iron absorption, although evidence for a ferroxidase activity in intestinal mucosa has been reported (Manis, 1973). Manis (1973) proposed that Fe(II) is perferentially taken up by the mucosal cell into a portion of cellular iron that reacts chemically as a Fe(II) pool; this pool is a precursor of both the iron transported to the serosa and the sequestered Fe(III) pool. The ferroxidase activity is postulated to be an intracellular enzymatic mechanism for the formation of trivalent iron before its entry into the Fe(III) pool.

4.2. Ceruloplasmin as the Transport Form of Copper

Equally important is the role of ceruloplasmin in the transport of copper from the liver to the blood, where it provides the element in a stable form for distri-

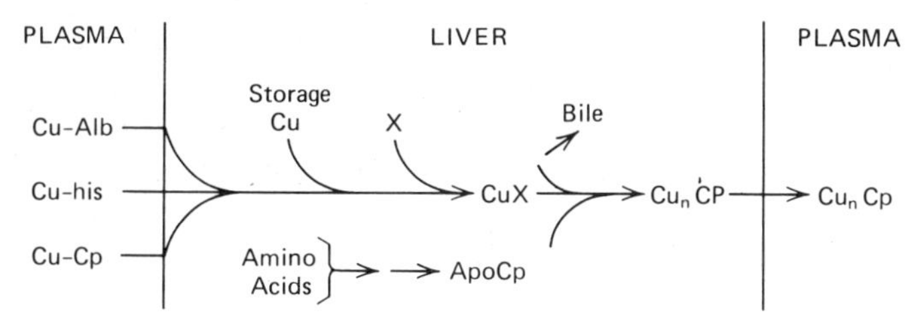

Figure 13. The proposed path of copper, after absorption, in transport to the liver for biosynthesis of the major copper protein of the plasma, ceruloplasmin. Cu-alb = copper-albumin, Cu-his = copper-histidine.

bution to the tissues to fulfill the need for this vital metal in copper enzymes and copper proteins. To understand the transport function of ceruloplasmin it is necessary first to review some essential facts about the circulating forms of copper and the biosynthesis and turnover of ceruloplasmin.

Plasma copper is only slightly more concentrated than the copper of the erythrocyte (115 ± 30 vs. 90 ± 20 μg %). However, the red cell copper, of which about 60% is associated with superoxide dismutase (formerly erythrocuprein), is relatively inert metabolically and is not known to be involved in transport. Plasma copper is composed of two major fractions when classified on the basis of the strength of copper binding (Figure 13). First, there is ceruloplasmin, in which the copper ion is an integral part of the molecule and, in the normal animal, accounts for over 95% of the plasma copper (Scheinberg and Morell, 1973). The other category is the dialyzable, or more dissociable, form of copper, which has been shown to include Cu(II)-histidine, and, perhaps, ternary complexes of copper with other amino acids. The latter fraction is believed to be a highly labile transport form, primarily involved in the preceruloplasmin movement of copper from the gut to the liver after absorption. However, a possible role for these labile complexes in subsequent copper transport has not been excluded.

Biosynthesis and Turnover

It has been shown that ingested copper disappears rapidly from plasma with a concomitant increase in hepatic copper, which is incorporated into ceruloplasmin and then released into the blood (Scheinberg and Sternlieb, 1960). The appearance of ceruloplasmin copper in the plasma reaches a maximum within 24 hr after copper intake (Holtzman and Gaumnitz, 1970). It is known that the plasma ceruloplasmin levels are under a wide range of humoral and hormonal controls (Evans, 1973). In the rat, dramatic increases in serum ceruloplasmin are observed within 3 weeks after birth at the expense of copper stores, after estrogen administration, and after stress and inflammation (Evans, 1973; Scheinberg and Morell, 1973; Linder and Munro, 1973). Virtually nothing, however, is known about the intracellular mechanism involved in ceruloplasmin biosynthesis, including both the apoceruloplasmin moiety and how and when the copper is inserted into the protein. We know that the vertebrate hepatocyte has an impressive capacity to synthesize ceruloplasmin, given sufficient available copper. The metabolic defect in Wilson's disease, wherein copper accumulates particularly in the liver and the brain, with low plasma copper and ceruloplasmin, could involve either site. The lower availability of copper might arise from stronger binding by intracellular carrier proteins, as reported by Evans et al. (1973), or there could be a block in the biosynthetic utilization of copper in the last step of ceruloplasmin production (Scheinberg and Morell, 1973). The probable sequence of the biosynthesis of ceruloplasmin involves the synthesis of the peptide chain, glycosylation, and finally, copper addition (Scheinberg and Morell, 1973). Although the amount of copper storage proteins may be affected, the synthesis of apoceruloplasmin appears to be independent of copper status (Holtzman and Gaumnitz, 1970). Apoceruloplasmin was found to be released into the plasma of copper-deficient rats at the same rate as was ceruloplasmin in rats kept on a diet adequate in copper (Holtzman and Gaumnitz, 1970). The question of whether the injection of copper will induce ceruloplasmin synthesis is still open, since the response seems to vary with the dose of copper (Fee, 1975). However, copper has been reported to induce the biosynthesis of ceruloplasmin in human and monkey liver slices (Neifakh et al., 1969).

Once it reaches the blood, ceruloplasmin has a survival $t_{1/2}$ of 54 hr in the rabbit (Hickman et al., 1970), shorter than that of other staple proteins of the plasma. A $t_{1/2}$ of about 12 hr in the rat has been estimated (Holtzman and Gaumnitz, 1970). These turnover times are adequate to account for the rate of utilization of copper by the tissues. In a series of ingenious papers, Ashwell and Morell (1974) have shown that the survival of ceruloplasmin in the plasma depends on an intact carbohydrate moiety, particularly sialic acid. Desialation of ceruloplasmin with neuraminidase reduces the $t_{1/2}$ of asialoceruloplasmin to less than 0.5 hr (Hickman et al., 1970).

The life cycle of the copper in ceruloplasmin is a one-time journey to the tissues or a return to the liver for resynthesis. The ability to add copper to the protein moiety seems to be an exclusive property of certain cells, particularly in the liver. Holtzman and Gaumnitz (1970) showed that, once apoceruloplasmin reaches the circulation, it is incapable of adding copper to form an active ceruloplasmin molecule. However, Owen (1975) more recently found that *in vitro* ceruloplasmin could exchange its copper for ionic copper at pH 7.4 in the presence of a reducing substrate, *p*-phenylenedianime. This observation is in accord with earlier findings that the copper ion in copper proteins exchanges much more readily in the cuprous state. In fact, reducing agents usually precede chelating agents in the preparation of apoproteins from copper proteins (Erickson et al., 1970).

Transport of Copper

The role of ceruloplasmin as a copper transport protein has been proposed for more than a decade. Broman (1967a, 1967b), particularly, advocated such a role because of the abundance of ceruloplasmin in the blood. Shokeir and

Shreffler (1969) found that leukocytes from Wilson's disease patients showed a great reduction of cytochrome oxidase activity but only a moderate decrease in heterozygous carriers, compared to normal controls. A similar trend was observed for plasma ceruloplasmin concentration. On the basis of limited data, Shokeir and Shreffler (1969) proposed that ceruloplasmin functioned as copper donor to copper-containing proteins. Along this line, Owen (1965) observed that, after intravenous injection of radioactive copper, rats did not accumulate radioactivity in extrahepatic organs until after the emergence of [^{64}Cu]ceruloplasmin. After intravenous injection of plasma containing [^{67}Cu]ceruloplasmin, which was prepared from donor rats pretreated with $^{67}CuCl_2$, Marceau and Aspin (1972) found that ^{67}Cu activity in rat plasma decreased with the increase of radioactivity in various organs. A similar result was obtained by Owen (1971). Following the same method of treatment, Marceau and Aspin (1973a) reported that radioactive copper was tightly bound to cytochrome *c* oxidase in the liver and brain after injection of ^{67}Cu-labeled plasma. In addition, the ^{67}Cu activity was also found to be incorporated into liver cytocuprein (superoxide dismutase) (Marceau and Aspin, 1973b). These results, however, do not exclude the possibility that the radioactive copper found in cytochrome *c* oxidase or superoxidase dismutase may come from sources other than ceruloplasmin. Although approximately 90% of ^{67}Cu in the plasma preparation Marceau and Aspin used was ceruloplasmin bound (Evans, 1973; Scheinberg et al., 1973), there was still 5% radioactive copper bound to albumin or amino acids (Evans, 1973). Thus the radioactive copper found in cytochrome *c* oxidase or superoxide dismutase after injection of isolated plasma might have been derived from other sources, since no quantitative data to the contrary were presented. Marceau and Aspin (1973b) did observe that copper derived from [^{67}Cu]plasma was tightly bound to cytochrome *c* oxidase, but copper from [^{64}Cu]plasma was only loosely bound to the oxidase. This, however, was not an appropriate comparison, since rats injected with [^{67}Cu]plasma were killed after 6 days, whereas rats injected with [^{64}Cu]albumin were sacrificed after 2 hr.

A critical test of the transport role of ceruloplasmin was provided by a series of experiments in the authors' laboratory (Figure 14) (Hsieh and Frieden, 1975). Rats were fed a copper-deficient diet for about 2 months to reduce the cytochrome *c* oxidase activity. Groups of these rats were then injected intravenously with ceruloplasmin, albumin-bound copper, histidine-bound copper, and $CuCl_2$. The copper compounds were injected at three times the level of copper found in the plasma of rats fed a copper-supplemented diet. Cytochrome *c* oxidase activity in the liver, spleen, and heart greatly increased in 6 days in rats receiving ceruloplasmin (Figure 14), but the enzyme activity in these tissues from other groups increased only moderately. In the latter case the increase of cytochrome *c* oxidase activity occurred after ceruloplasmin activity in the blood reached a maximal level, again indicating that ceruloplasmin probably is the source of copper in cytochrome *c* oxidase.

The transport of copper by ceruloplasmin probably requires specific receptor

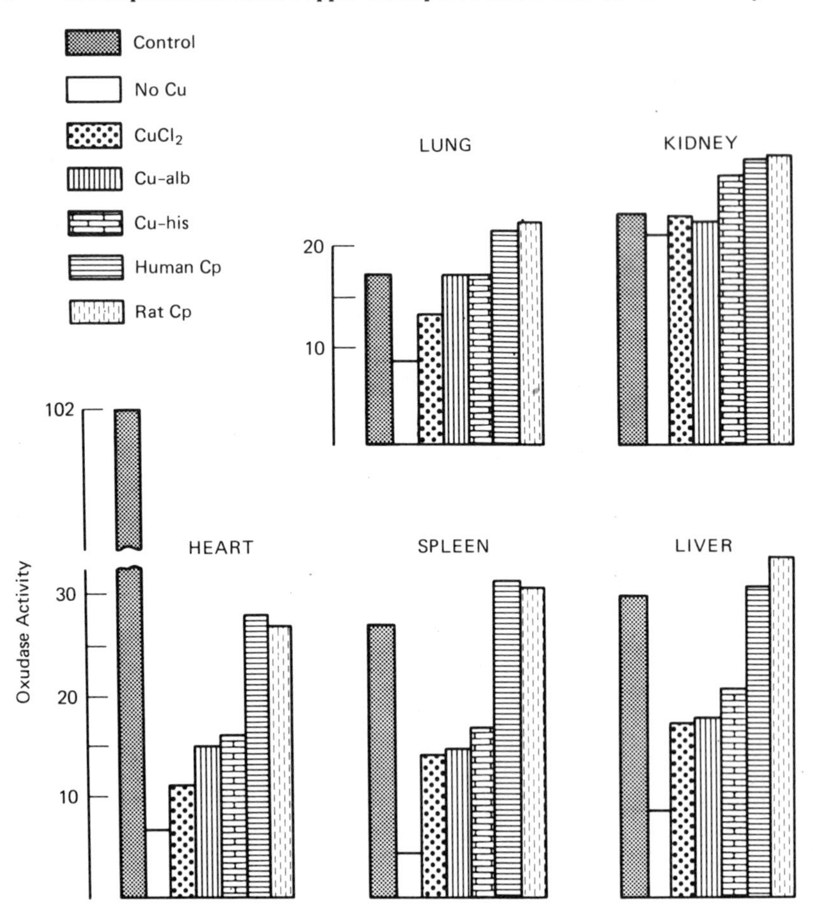

Figure 14. Preferential restoration of cytochrome *c* oxidase activity in tissues of copper-deficient rats. Rats were injected intravenously with the indicated compounds corresponding to 4.1 μg Cu/100 g body weight on days 1, 3, and 5. On day 6 all rats were sacrificed, and 10% homogenates were prepared from each tissue for measurement of enzyme activity. Cu-alb = copper-albumin, Cu-his = copper histidine. Oxidase activity is defined as nanomoles cytochrome *c* oxidized per minute per milligram protein. Each value represents the average of data from three rats. Data from Hsieh and Frieden (1975).

mechanisms in the various target tissues. The greater lability of Cu(I), mentioned earlier, strongly suggests a reductive step in copper release. Ample reductive mechanisms are available once ceruloplasmin is within the reactive sphere of the cell. First, there are the numerous endogenous substrates described earlier. Second, ceruloplasmin has been shown to be able to tap the electron transport machinery of the cell. Brown and White (1961) have reported that in the presence of cytochrome *c* and typical oxidative substrates, for example, succinate and NADH, heart muscle particles can reduce ceruloplasmin. The reaction occurs under anaerobic conditions and is reversed by oxygen; it is sensitive to

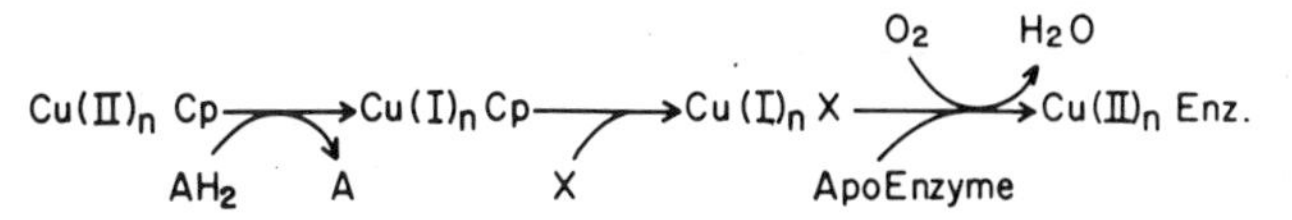

Figure 15. Proposed mechanism for the transfer of copper from ceruloplasmin (Cp) to an intracellular copper enzyme. AH_2 is a reducing substrate, and X is a hypothetical intracellular Cu(I) acceptor and/or ligand exchanger.

cyanide, carbon monoxide, and antimycin A. Under aerobic conditions, ceruloplasmin inhibits the electron transport system, possibly by a reaction with essential —SH groups.

On the basis of these ideas we conclude by proposing a simplified mechanism for the incorporation of the copper from ceruloplasmin into an intracellular copper enzyme or copper protein (Figure 15). The first step is the reduction of the Cu(II) of ceruloplasmin by any of the substrates or reaction sequences described earlier. If this occurs at the cell membrane, the Cu(I) is likely to be inside the cell, and the hypothetical acceptor X might not be as necessary for intracellular transport. In its Cu(I) form the copper is added to an apoenzyme, where it is fixed into the holoenzyme in the Cu(II) state with the aid of oxygen. This mechanism takes into account the primary role of ceruloplasmin as the copper transport and donor molecule, the high exchangeability of Cu(I) ligands, and the greater stability of copper ion in Cu(II) proteins.

4.3. Regulation of Plasma or Tissue Levels of Biogenic Amines

Barrass and Coult (1972) have summarized the effects of drugs used in the treatment of mental illness, for example, tranquilizers and antidepressants, on the ceruloplasmin-catalyzed oxidation of the biogenic amines—noradenaline and 5-hydroxytryptamine. The suggestion here is that ceruloplasmin or an enzyme with similar properties may be of importance in affecting the relative concentrations of nonadrenaline and 5-hydroxytryptamine in the serum, and, eventually, in those areas of the brain where these compounds act as neurotransmitters. Thus a ceruloplasmin-like enzyme, by its effect on the lifetime of biogenic amines, could play an important role in the regulation of brain chemistry necessary for mental function, and interference with this enzyme may lead to the appearance of abnormal mental states.

A varied spectrum of drug effects has been observed. Hallucinogens such as LSD accelerate the ceruloplasmin-catalyzed oxidation of noradrenaline but inhibit the enzymatic oxidation of 5-hydroxytryptamine. Tranquilizers of the phenothiazine type accelerate the enzymatic oxidation of both substrates. As discussed earlier, and as shown in Table 7, Lovstad (1975) found that many phenothiazines were effective substrates for ceruloplasmin. Antidepressant drugs, for example, imipramine, inhibited the enzymatic oxidation of both substrates but only at relatively high concentrations (10^{-2} M). However, some phenylethylamines and anticholinergics with central nervous system activity showed no effect on the enzymatic oxidation of the two biogenic amines.

The mode of action of LSD was of particular interest since this drug elevates brain 5-hydroxytryptamine levels and depresses brain catecholamines. When tested with ceruloplasmin, LSD inhibited the oxidation of 5-hydroxytryptamine by 50% at a concentration one-tenth that of the substrate and enhanced the oxidation of noradrenaline fourfold when treated at one tenth of the substrate. The K_m values for noradrenaline and 5-hydroxytryptamine are similar, 3 mM and 1 mM, respectively. Ceruloplasmin could exert close control over the relative concentrations of these two compounds in key parts of the brain. Thus LSD may produce its central effects through ceruloplasmin by perturbing the balance between these two groups of biogenic amines.

Barrass et al. (1972) also proposed a possible involvement of ceruloplasmin in Parkinson's disease. The compound 3-hydroxy-4-methoxyphenethylamine is one of the endogenous toxins that accumulate in Parkinsonians, producing tremors and hypokinesia. *In vitro* it has been shown that this compound enhances dopamine oxidation catalyzed by ceruloplasmin. Since Parkinson's disease may be associated with decreased catecholamine and increased 5-hydroxytryptamine levels and an elevated serum ceruloplasmin, the latter has been suggested as a systemic basis for the etiology of this disease.

Most recently Shokeir (1975) has reported reduced ceruloplasmin levels in patients with Huntington's disease. This has long been suspected, since many of the early symptoms of Huntington's chorea resemble those of another genetic disorder, Wilson's disease. This hepatolenticular disease is accompanied by copper accumulation and toxicity in the brain and liver and by decreased ceruloplasmin biosynthesis with, almost invariably, low serum.

5. SUMMARY

Ceruloplasmin, the blue copper protein of vertebrate plasma, has been surveyed mainly from a functional point of view. A few salient comments about its chemistry, molecular properties, evolutionary background, and clinical significance have been included. The relationships between ceruloplasmin and other blue copper oxidases and copper proteins have been described. Current evidence suggests that, despite its molecular weight of 134,000, ceruloplasmin is a single-chain glycoprotein, easily fragmented into large polypeptides by plasmin and other proteases. The amino acid and carbohydrate compositions are known, and a few sequences have been determined. The six copper ions present comprise three or, possibly, four different types of copper with several functional differences in catalysis. Finally, it was predicted that the use of ceruloplasmin levels in human sera for the diagnosis of disease states will be greatly extended.

Ceruloplasmin possesses significant oxidase activity toward Fe(II) and numerous aromatic amines and phenols. Its ferroxidase activity has led to the discovery that it is a molecular link between copper and iron metabolism. Ceruloplasmin mobilizes iron into the plasma from iron storage cells in the liver. An equally important function is that ceruloplasmin, after its rapid biosynthesis in the liver, serves as a major copper transport vehicle, comparable to transferrin. Evidence is accumulating that the copper atoms of ceruloplasmin are a prerequisite for copper utilization in the biosynthesis of cytochrome oxidase and other copper proteins. The ability of ceruloplasmin to release copper at specific cellular sites may be related to its broad substrate spectrum of biological reducing agents. A possible third role of ceruloplasmin is as a contributor to regulation of the balance of biogenic amines through its oxidase action on the epinephrine and the hydroxyindole series. Thus ceruloplasmin is a copper protein with several important functions, all of which are directly related to its oxidase activity.

ACKNOWLEDGMENT

This work was supported by National Science Foundation Grant PCM-7622171.

REFERENCES

Albergoni, and Cassini, (1975). *FEBS Lett.,* **55,** 261.

Ashwell, G. and Morell, A. G. (1974). *Adv. Enzymol.,* **41,** 99.

Barrass, B. C. and Coult, D. B. (1972a). *Biochem. Pharmacol.,* **21,** 677.

Barrass, B. C. and Coult, D. B. (1972b). In, P. B. Bradley and R. W. Brimble, Eds., *Progress in Brain Research,* Vol. 36, Elsevier, Amsterdam, pp. 97–104.

Barrass, B. C., Coult, D. B., and Pinder, R. M. (1972). *J. Pharm. Pharmacol.,* **24,** 499–501.

Barrass, B. C., Coult, D. B., Pinder, R. M., and Skeels, M. (1973). *Biochem. Pharmacol.,* **22,** 2891.

Barrass, B. C., Coult, D. B., Rich, P., and Tutt, K. J. (1974). *Biochem. Pharmacol.,* **23,** 47.

Bates, G. W. and Schlaback, M. R. (1973). *J. Biol. Chem.,* **248,** 3228–3232.

Bates, G. W., Workman, E. F., and Schlaback, M. R. (1973). *Biochem. Biophys. Res. Commun.,* **50,** 84.

Britten G. M. and Chee, Q. T. (1969). *J. Lab. Clin. Med.,* **74,** 53.

Broman, L. (1958). *Nature,* **182,** 1655.

Broman, L. (1967a). In O. Walaas, Ed., *Molecular Basis of Some Aspects of Mental Activity,* Vol. 2. Academic Press, New York, p. 131.

Broman, L. (1967b). In J. Peisach, P. Aisen, and W. R. Blumberg, Eds., *The Biochemistry of Copper.* Academic Press, New York, p. 131.

Brown, F. C. and White, Jr. (1961). *J. Biol. Chem.,* **236,** 911.

Bryce, C. F. A., Crichton, R. R., and Harrison, P. M. (1973). *Biochem. J.*, **133,** 301.

Carrico, R. J., Malmström, B. G., and Vanngard, T. (1973). *Eur. J. Biochem.*, **37,** 41.

Curzon, G. (1960). *Biochem. J.*, **77,** 66.

Curzon, G. (1961). *Biochem. J.*, **79,** 656.

Curzon, G. and Cumings, J. N. (1966). In J. Peisach, P. Aisen, W. E. Blumberg, Eds., *The Biochemistry of Copper.* Academic Press, New York, pp. 545–558.

Curzon, G. and O'Reilly, S. (1960). *Biochem. Biophys. Res. Commun.*, **2,** 284.

Curzon, G. and Young, S. N. (1972). *Biochim. Biophys. Acta,* **268,** 41.

Dawson, C. R., Strothkamp, K. G., and Krul, K. G. (1975). *Ann. N. Y. Acad. Sci.*, **258,** 209–220.

Deinum, J. and Vanngard, T. (1973). *Biochim. Biophys. Acta,* **310,** 321.

Erickson, J., Gray, R., and Frieden, E. (1970). *Proc. Soc. Exp. Biol. Med.*, **134,** 117.

Evans, G. (1973). *Physiol. Rev.*, **53,** 535–570.

Evans, J. L. and Abraham, P. A. (1973). *J. Nutr.*, **103,** 196.

Evans, G. W., Dubois, R. S., and Hambidge, K. M. (1973). *Science,* **181,** 1175.

Fee, J. A. (1975). *Struct. Bonding,* **23,** 1–60.

Freeman, S. and Daniel, E. (1973). *Biochemistry,* **12,** 4806.

Frieden, E. (1971). *Bioinorg. Chem.*, **100,** 292–321.

Frieden, E. (1974). In M. Friedman, Ed., *Protein-Metal Interactions,* Plenum Press, New York, pp. 1–31.

Frieden, E. (1976). *Trends. Biochem. Sci.*, **1,** 273.

Frieden, E. and Hsieh, H. S. (1975). *Adv. Enzymol.*, **187,** 236.

Frieden, E. and Osaki, S. (1974). In M. Friedman, Ed., *Protein-Metal Interactions,* Plenum Press, New York, pp. 235–265.

Frieden, E., Osaki, S., and Kobayashi, H. (1965). *J. Gen. Physiol.*, **49,** 213.

Gunnarsson, P. O. (1974). Thesis, Union, Lund, Sweden.

Gunnarsson, P. O., Pettersson, G., and Pettersson, I. (1970). *Eur. J. Biochem.*, **17,** 586.

Gunnarsson, P. O., Lindstrom, A., and Pettersson, G. (1971). *Acta Chem. Scand.*, **25,** 770.

Gunnarsson, P. O., Nylen, V., and Pettersson, G. (1972). *Eur. J. Biochem.*, **27,** 572.

Gunnarsson, P. O., Nylen, V., and Pettersson, G. (1973). *Eur. J. Biochem.*, **37,** 41.

Harris, D. C., and Asien, P. (1973). *Biochim. Biophys. Acta,* **329,** 156.

Hart, E. B., Steenbock, H., Waddell, J., and Elvehjem, C. A., (1928). *J. Biol. Chem.*, **77,** 797.

Hickman, J., Ashwell, G., Morell, A. G., Van den Hamer, C. J. A., and Scheinberg, I. H. (1970). *J. Biol. Chem.*, **245,** 759.

Holmberg, C. G. (1944). *Acta Physiol. Scand.*, **8,** 227.

Holmberg, C. G. and Laurell, C. B. (1951a). *Acta Chem. Scand.*, **2,** 476.

Holmberg, C. G. and Laurell, C. B. (1951b). *Acta Chem. Scand.*, **5,** 476.

Holtzman, N. A. and Gaumnitz, B. M. (1970). *J. Biol. Chem.*, **245,** 2354.

Hsieh, H. S. and Frieden, E. (1975). *Biochem. Biophys. Res. Commun.*, **67,** 1326.

Huber, C. T. and Frieden, E. (1970a). *J. Biol. Chem.*, **245,** 3973.

Huber, C. T. and Frieden, E. (1970b). *J. Biol. Chem.*, **245,** 3979.

Inaba, T. and Frieden, E. (1967). *J. Biol. Chem.*, **242,** 4789.

Johnson, D. A., Osaki, S., and Frieden, E. (1967). *J. Clin. Chem.*, **13,** 142.

Kingston, I. B., Kingston, B. L., and Putnam, F. W. (1977). *Proc. Natl. Acad. Sci.*, **74,** 5377–5381.

Lahey, M. E., Guber, C. J., Chase, M. S., Cartwright, G. E., and Wintrobe, M. M. (1952). *Blood,* **7,** 1053.

Lee, D., Jr., and Matrone, G. (1969). *Proc. Soc. Exp. Biol. Med.*, **130,** 1190.

Lee, G. R., Cartwright, G. E., and Wintrobe, M. M. (1968). *Proc. Soc. Exp. Biol. Med.*, **127,** 977.

Lee, G. R., Nacht, S., Christensen, D., Hansen, S. P., and Cartwright, G. E. (1969). *Proc. Soc. Exp. Biol. Med.*, **131,** 918–923.

Linder, M. C. and Munro, H. N. (1973). *Enzyme,* **15,** 111.

Lovstad, R. A. (1972). *Acta Chem. Scand.*, **26,** 2832.

Lovstad, R. A. (1975). *Biochem. Pharmacol.*, **24,** 475.

Lovstad, R. A. and Frieden, E. (1973). *Acta Chem. Scand.*, **27,** 121.

Lowenstein, H. (1975). *Int. J. Peptide Protein Res.*, **7,** 1.

Macara, T. G., Hoy, T. G., and Harrison, P. M. (1972). *Biochem. J.*, **126,** 151.

Malkin, R., and Malström, B. G. (1970). *Adv. Enzymol.*, **33,** 177.

Manis, J. (1973). *Proc. Soc. Exp. Biol. Med.*, **144,** 2025–2032.

Marceau, N. and Aspin, N. (1972). *Am. J. Physiol.*, **222,** 106.

Marceau, N. and Aspin, N. (1973a). *Biochem. Biophys. Acta,* **328,** 338.

Marceau, N. and Aspin, N. (1973b). *Biochem. Biophys. Acta,* **328,** 351.

McCombs, M. L. and Bowman, B. H. (1976). **434,** 452–461.

McDermott, J. A., Huber, C. T., Osaki, S., and Frieden, E. (1969). *Biochim. Biophys. Acta,* **151,** 541.

McKee, D. and Frieden, E. (1971). *Biochemistry,* **10,** 3880.

Morell, A. G., Van den Hamer, C. J. A., Scheinberg, I. T., and Ashwell, G. J. (1966). *J. Biol. Chem.*, **241,** 3745.

Neifakh, S. A., Monkhov, N. K., Shaposhnikov, A. M., and Zubzhitski, Y. N. (1969). *Experimentia,* **25,** 337.

Osaki, S. (1966). *J. Biol. Chem.*, **241,** 5053.

Osaki, S. and Walaas, O. (1967). *J. Biol. Chem.*, **242,** 2653.

Osaki, S., McDermott, J. A., and Frieden, E. (1964). *J. Biol. Chem.*, **239,** 3570.

Osaki, S., Johnson, D. A., and Frieden, E. (1966). *J. Biol. Chem.*, **241,** 2746.

Osaki, S., Johnson, D., and Frieden, E. (1971). *J. Biol. Chem.*, **246,** 3018.

Osaki, S., James, G. T., and Frieden, E. (1974). *Dev. Biol.*, **39,** 158–163.

Owen, C. A. J. (1965). *Am. J. Physiol.*, **209,** 900.

Owen, C. A. J. (1971). *Am. J. Physiol.*, **221,** 1722.

Owen, C. (1973). *Am. J. Physiol.*, **224,** 514–518.

Owen, C. A. J. (1975). *Proc. Soc. Exp. Med.,* **144,** 681.

Owen, C. A. J. Hazelrig, J. B., Bush, J. A., and Wintrobe, M. M. (1956). *Blood,* **11,** 143.

Planas, J. (1973). *Rev. Esp. Fisiol.,* **29,** 293.

Planas, J. and Frieden, E. (1973). *Am. J. Physiol.,* **225,** 423.

Poulik, M. D. and Weiss, M. L. (1975). In F. W. Putnam, Ed., *The Plasma Proteins,* Vol. II, 2nd ed. Academic Press, New York, pp. 51–108.

Ragan, H. A., Nacht, S., Lee, G. R., Bishop, C. R., and Cartwright, G. E. (1969). *Am. J. Physiol.,* **217,** 1320.

Rice, E. W. (1962). *Anal. Biochem.,* **3,** 542.

Roeser, H. P., Lee, G. R., Nacht, S., and Cartwright, G. E. (1970). *J. Clin. Invest.,* **49,** 2408.

Rydén, L. (1971a). *FEBS Lett.,* **18,** 321.

Rydén, L. (1971b). *J. Protein Res.,* **3,** 121.

Rydén, L. (1972a). *Eur. J. Biochem.,* **26,** 380.

Rydén, L. (1972b). *Eur. J. Biochem.,* **28,** 46.

Rydén, L. and Björk, I. (1976). *Biochemistry,* **15,** 3411–3417.

Scheinberg, I. H. and Morell, A. G. (1957). *J. Clin. Invest.,* **36,** 1193.

Scheinberg, I. H. and Morell, A. G. (1973). In G. I. Eichorn, Ed., *Inorganic Biochemistry.* Elsevier, New York, pp. 306–319.

Scheinberg, I. H. and Sternlieb, I. (1960). *Pharmacol. Rev.,* **12,** 355.

Schosinsky, K. H., Lehmann, H. P., and Beeler, M. G. (1974). *Clin. Chem.,* **20,** 1556.

Seal, U. S. (1961). *Comp. Biochem. Biophys.,* **95,** 151.

Sexton, R. C. (1974). Ph.D. dissertation, Florida State University.

Sexton, R. C., Osaki, S., and Frieden, E. (1974). *Fed. Proc.,* **33,** 1569.

Shokeir, M. H. K. (1972). *Clin. Biochem.,* **5,** 115–120.

Shokeir, M. H. K. (1975). *Clin. Genet.,* **7,** 354.

Shokeir, M. H. K. and Schreffler, D. C. (1969). *Proc. Natl. Acad. Sci.,* **62,** 867.

Taborsky, G. (1963). *Biochemistry,* **2,** 266.

Topham, R. W. (1975). Personal communication.

Topham, R. W. and Frieden, E. (1970). *J. Biol. Chem.,* **245,** 6698.

Topham, R. W. and Johnson, D. A. (1974). *Arch. Biochem. Biophys.,* **160,** 647.

Topham, R. W., Sung, C., Morgan, F. G., Prince, W. D., and Jones, S. H. (1975). *Arch. Biochem., Biophys.,* **167,** 129.

Walaas, O. and Walaas, E. (1961). *Arch. Biochem. Biophys.,* **95,** 151.

Whanger, P. D. and Weswig, P. H. (1970). *J. Nutr.,* **100,** 341–348.

Williams, D. M., Christensen, D. D., Lee, G. R., and Cartwright, G. E. (1974a). *Biochim. Biophys. Acta,* **350,** 129.

Williams, D. M., Lee, G. R., and Cartwright, G. E. (1974b). *Am. J. Physiol.,* **227,** 1094.

Young, C. and Curzon, G. (1972). *Biochem. J.,* **129,** 273.

10

COPPER TOXICITY STUDIES USING DOMESTIC AND LABORATORY ANIMALS

I. Bremner

Rowett Research Institute, Bucksburn, Aberdeen, Scotland

1. INTRODUCTION

Although copper is an essential element required for a wide range of metabolic processes in the body, excessive intakes of this metal can have serious effects in animals. Depending on the species involved, growth and food intake may be reduced, anemia can develop, and considerable damage may occur to the liver, kidneys, brain, and muscle, often resulting in death. Fortunately the tolerance of most domestic and laboratory animals to copper is relatively high, and copper poisoning presents a much less serious problem in animal husbandry than does copper deficiency. It is often necessary to increase dietary intakes of copper by 20- to 50-fold over normal levels before any lesions of copper toxicity develop. However, this is not invariably the case, and with some species copper poisoning may occur quite readily, even under natural conditions.

2. OCCURRENCE OF COPPER TOXICOSIS

2.1. Ruminant Animals

It has been known for many years that sheep are particularly susceptible to copper toxicosis, and much of the early work on the occurrence of this disease in ruminant animals has been reviewed previously (Eden, 1940; Todd, 1969; Bremner, 1974). Uncomplicated copper poisoning has been reported in sheep under natural grazing conditions in some parts of Australia, where copper-bearing parent rocks have resulted in high copper contents in soils and plants (Bull, 1951). It has also occurred in other areas of the same continent at lower levels of copper intake on acid soils where the predominant pasture species was *Trifolium subterraneum.* The development of the disease was attributed in this case to the low molybdenum content of the plant. Molybdenum is a potent antagonist of copper metabolism in ruminants (Dick, 1954), and it was suggested that the low molybdenum intake of the sheep promoted excessive hepatic accumulation of copper. A third instance of naturally occurring copper toxicosis is provided by the disease known as toxemic jaundice, which has developed in sheep eating the plant *Heliotropium europaeum* (Bull, 1951). This contains hepatotoxic alkaloids which apparently induce changes in the size and life span of liver cells and cause substantial increases in liver copper concentration and susceptibility to copper poisoning.

At present most cases of copper toxicosis in sheep, certainly within the United Kingdom, are associated with intensive rearing of the animals indoors on cereal-based diets (Todd, 1969). However, isolated cases of this disease have also occurred when sheep have grazed in orchards or on pastures that were treated with copper-containing fungicides (Biejers, 1932) and molluscicides (Gracey and Todd, 1960).

Copper toxicosis has also developed when sheep have inadvertently received diets specially formulated for pigs and containing large amounts of copper as a growth stimulant (Suveges et al., 1971). The inclusion of copper in these diets may also constitute a less direct hazard to sheep, as it has been claimed that the disposal onto pasture of the copper-rich slurry from pig units may lead to an increased incidence of copper toxicosis in grazing sheep (Batey et al., 1972). High concentrations of copper have been reported in treated pastures, but most of this is present as surface contamination and relatively little uptake of copper by the plant occurs (Batey et al., 1972). Nevertheless, this copper is still biologically available to animals, as has been shown by studies in which sheep received either dried slurry (Suttle and Price, 1976) or the solid residues of an aerobic digest of slurry (Dalgarno and Mills, 1975). There have been reports of deaths from copper poisoning (Ulsen, 1972; Feenstra and Ulsen, 1973) and of liver damage (Kneale and Smith, 1977) in sheep receiving contaminated grass or hay. In other trials, however, where sheep grazed severely contaminated pastures for 3 successive years no cases of copper poisoning were recorded (Gracey et al., 1976).

The magnitude of this environmental problem is difficult to assess at present, as it is affected by a range of factors. Dalgarno and Mills (1975) have pointed out that the likelihood of copper toxicity problems developing depends not only on the rate of application of the slurry to the pasture but also on the climate and on the conditions, anaerobic or aerobic, under which the slurry has been processed. This has been confirmed by the studies of Kneale and Smith (1977). It is noteworthy that, despite the widespread use of copper-supplemented pig rations in the United Kingdom for several years, there is still no evidence that this practice has caused copper poisoning problems in grazing sheep.

Young calves may also develop copper toxicosis at relatively low copper intakes, especially when they are receiving milk-based diets (Shand and Lewis, 1957; Weiss and Baur, 1968). Older calves and adult cattle, however, are much more tolerant of copper, and the few cases of copper poisoning which have been reported have involved the administration of excessively large amounts of this element (Kidder, 1949; Todd and Gribben, 1965). There are no reports of copper toxicosis occurring naturally in goats, although a recent study (Adam et al., 1977) has shown that this disease can be induced at relatively high levels of exposure.

2.2. Pigs

Pigs are much more tolerant of copper than are ruminant animals and under normal circumstances would be considered unlikely to suffer from copper poisoning. However, the discovery of the growth-promoting properties of copper has resulted in the supplementation of pig rations in some countries with around

250 mg Cu/kg (Braude, 1967). This practice has been adopted in most pig-rearing units in the United Kingdom for over a decade with no reports of deleterious effects. Nevertheless, concern has been expressed about the possible danger of copper poisoning developing in animals receiving these diets, as this has occurred in a few experimental studies where copper-supplemented practical and semipurified diets have been fed (Hoefer et al., 1960; Wallace et al., 1960; Ritchie et al., 1963; Gipp et al., 1973, 1974). The incidence and the severity of the disease are significantly increased at greater dietary copper concentrations. Intakes of 750 to 1000 mg Cu/kg have been found to produce a wide range of symptoms of copper toxicosis and in many cases to be fatal (Allen and Harding, 1962; Suttle and Mills, 1966a).

2.3. Other Species

Rabbits are also reasonably tolerant of high copper intakes, and increases in growth rate and food consumption have been reported in animals receiving 100 to 200 mg Cu/kg diet (King, 1975; Omole, 1977). The only adverse effect noted in the rabbit was a slight thinning of the cecum. Dietary supplementation with copper has also been found to influence growth rate and food consumption in chicks (Smith, 1969). Intakes of around 100 mg Cu/kg had a beneficial effect, but at concentrations over 350 mg/kg both growth rate and food consumption were significantly reduced. No overt signs of copper toxicity were found in these studies (Smith, 1969), but others have reported damage to gizzard linings (Fisher et al., 1973; Poupoulis and Jensen, 1976a) and changes in fatty acid composition (Poupoulis and Jensen, 1976b) at dietary copper concentrations of 400 to 600 mg/kg.

It has been suggested that dogs may be particularly susceptible to acute copper poisoning (Goresky et al., 1968), and particular breeds seem to suffer from an inherited form of chronic copper toxicity (Sternlieb et al., 1977).

Ponies are extremely tolerant of copper, and no symptoms of copper toxicity were detected when yearling animals received diets containing 800 mg/kg for about 6 months (Smith et al., 1975).

3. SYMPTOMS OF COPPER TOXICOSIS

3.1. Ruminant Animals

The development of chronic copper toxicosis in sheep is generally regarded as occurring in two distinct phases (Todd et al., 1962; Ishmael et al., 1971, 1972). In the first of these there are no overt signs of toxicosis, and both growth and food intake are generally normal, although reduced weight gains have been reported on occasion (Hill and Williams, 1965; Tait et al., 1971). Blood copper concentrations may be slightly elevated (Eden, 1940; McCosker, 1968; Bremner et al., 1976) or unchanged at this time (Barden and Robertson, 1962; Todd and Thompson, 1963; Ishmael et al., 1972). The main feature of this phase is the substantial accumulation of copper in the liver with concentrations increasing to perhaps 1000 μg Cu/g dry matter. Plasma activities of enzymes such as aspartate transaminase, sorbitol dehydrogenase, lactic dehydrogenase, and arginase are also increased (Todd and Thompson, 1963; Ross, 1966; Ishmael et al., 1972; Bremner et al., 1976), indicating that considerable tissue or, more specifically, liver damage has occurred. This is consistent with the necrosis of isolated parenchymal cells and of swollen copper-containing Kupffer cells, rich in acid phosphatase, detected in liver samples collected some weeks before the onset of the second, critical phase of the disease, commonly described as the hemolytic crisis (Ishmael et al., 1971).

The main clinical features of this second phase include jaundice, anorexia, excessive thirst, and hemoglobinuria (Todd, 1969). Dramatic reductions in blood concentrations of hemoglobin and glutathione occur within 1 to 2 days, and there is a transient increase in blood methemoglobin concentrations. Death frequently results within a few days, although some animals survive (Ishmael et al., 1972). The onset of these symptoms is associated with liberation of stored copper from the liver and a massive increase in blood copper concentrations. Substantial further increases also occur in the activities in plasma of liver-specific enzymes, confirming that extensive liver degeneration has occurred and consistent with the histological findings of focal necrosis, inflammatory cells, bile plugs, and large PAS-positive, diastase-resistant Kupffer cells in liver samples (Ishamel et al., 1971, 1972).

Considerable kidney damage also occurs at the time of the hemolytic crisis, with significant functional impairment and degeneration, necrosis, and loss of mitochondrial enzyme activity from the proximal convoluted tubules (Gopinath et al., 1974). Todd (1969) has suggested that the appearance of the kidney, which is black and gorged with hemoglobin degradation products, is one of the most strongly characteristic features of copper toxicosis in sheep. It is thought that the renal failure and associated uremia play an important part in the death of the sheep from copper poisoning (Gopinath et al., 1974).

Changes may also occur in other tissues, especially at the time of the hemolytic crisis. For example, there have been reports of spongy transformation and vacuolation of white matter of the brain (Doherty et al., 1969; Morgan, 1973). Muscle cell membranes may also be damaged, as increased levels of creatine phosphokinase have been detected in serum in the terminal stages of the disease (Thompson and Todd, 1974).

The symptoms of copper toxicosis in cattle and goats have been less extensively studied, but the main features of jaundice and rapid onset of hemoglobinuria and methemoglobinemia are similar to those in the sheep (Shand and Lewis, 1957; Todd and Gracey, 1959; Adam et al., 1977).

3.2. Pigs

There are, however, some differences in the symptoms of chronic copper toxicosis in other species. For example, dullness, weakness, respiratory distress, anemia, and jaundice, with pulmonary odema and ulceration of the esophageal region of the stomach, have been reported in pigs receiving 1000 mg Cu/kg diet (Allen and Harding, 1962). Suttle and Mills (1966a) suggested that the jaundice which developed was of hepatic origin, as histological examination of the liver revealed centrilobular necrosis and some disruption of the bile canaliculi. Further evidence of tissue damage was provided by the increase in the activity of serum aspartate aminotransferase which occurred when serum copper concentrations had increased but before other signs of toxicosis were evident. The damage to liver tissue was associated with considerable hepatic accumulation of copper, but there was no correlation between the degree of liver damage and tissue copper concentration.

Although anemia typically occurs in both copper-poisoned sheep and pigs, the characteristics of this condition differ between the species. No methemoglobinemia or hemoglobinuria occurs in the pig, and there is no acute hemolytic episode. Instead the anemia develops gradually and is of a hypochromic, microcytic type. As iron concentrations in liver and plasma are also reduced, it seems likely that the anemia arises from the onset of an iron-deficiency state, perhaps as a consequence of impaired iron absorption (Wallace et al., 1960; Bunch et al., 1961; Gipp et al., 1973, 1974).

Changes also occur in the fatty acid composition of pigs receiving 250 mg Cu/kg, and the adipose tissue of these animals is often softer than normal. This is associated with an increase in the proportion of 16:1 and 18:1 fatty acids relative to the corresponding saturated acids in tissue lipids and also with changes in the stereospecific distribution of fatty acids within the triacylglycerol molecule (Elliot and Bowland, 1968; Moore et al., 1969; Christie and Moore, 1969). It has been suggested that these changes could have resulted from increased activity of the Δ-9-desaturase enzyme in both hepatic and adipose tissue microsomes of the copper-supplemented pigs (Ho and Elliot, 1973, 1974; Thompson et al., 1973).

Further insight into the involvement of copper in this reaction was provided by Wahle and Davies (1975), who showed that Δ-9-desaturase activity in the hepatic microsomes of copper-deficient rats was only about half of that in copper-adequate animals. They noted, however, that there were apparent differences between rats and pigs in their responses to copper supplementation, as parenteral injection of copper did not increase the Δ-9-desaturase activity in normal rats.

3.3. Poultry

Similar changes in fatty acid composition have also been reported in chicks receiving 500 mg Cu/kg diet, but these were less severe than those observed in pigs and depended on the type of diet used (Poupoulis and Jensen, 1976b). In general, the concentrations of monoenoic fatty acids in adipose tissue were increased at the expense of the saturated acids (Husbands, 1972). The most marked effect of high copper intakes in chicks is the development of damage to the lining of the gizzard (Fisher et al., 1973; Poupoulis and Jensen, 1976a). This has been attributed to the shedding of gizzard glandular cells into the koilin layer and to disruption and cessation of koilin production. Liver copper concentrations in chicks receiving copper-supplemented diets are not greatly increased, and no liver damage occurs in these birds.

3.4. Dogs

In contrast, mean liver copper concentrations of 6600 mg/kg dry matter have been reported in some Bedlington terriers showing symptoms not unlike those reported in human beings with Wilson's disease (Sternlieb et al., 1977) and in sheep with copper toxicosis. The dogs suffered from anorexia, vomiting, depression, weight loss, diarrhea, jaundice, ascites, and anemia. In some cases increased activities of serum alanine and/or aspartate aminotransferases were observed. Histochemical staining of liver sections revealed the existence of necrosis and other structural injuries and the presence of lipofuscin granules, rich in both copper and acid phosphatase.

4. FACTORS INFLUENCING THE DEVELOPMENT OF COPPER TOXICOSIS

4.1. Dietary and Hepatic Copper Concentrations

Although the most important event leading to the development of symptoms of copper poisoning in sheep is the accumulation of large amounts of copper in the liver, it is not possible to define a critical hepatic concentration of copper at which the hemolytic crisis will invariably occur. The onset of the hemolytic episode has taken place, for example, at liver copper concentrations of 1000 (Tait et al., 1971) and 6000 mg/kg dry matter (MacPherson and Hemingway, 1969). Furthermore, only a limited proportion of copper-loaded animals on a particular diet usually succumbs to the crisis (Tait et al., 1971; Bremner et al., 1976). It is therefore not surprising that the incidence of copper toxicosis in sheep cannot be related to the dietary copper intake of the animals. Buck (1970) has claimed that a dietary concentration of only 8 mg Cu/kg is sufficient to induce toxicosis, but problems have arisen more commonly at intakes of 25 to 30 mg/kg (MacPherson and Hemingway, 1969; Tait et al., 1971; Van Adrichem, 1965; Bremner et al., 1976). However, Mills (1974) found no symptoms of copper toxicosis in lambs receiving up to 60 mg Cu/kg of a semisynthetic ration. Continued intakes of copper are not essential for development of the disease, and the hemolytic crisis can occur several weeks after removal of the copper-supplemented diet (Suveges et al., 1971; Gopinath and Howell, 1975).

One reason why copper toxicosis can develop over such a wide range of dietary copper contents is that the availability of the element is influenced by dietary composition and by the age, breed, and physiological state of the animal. For example, the hepatic retention of dietary copper in young calves receiving liquid milk substitutes can be as great as 50% (Bremner and Dalgarno, 1973), with liver copper concentrations increasing to 550 mg/kg dry matter at dietary copper concentrations of only 5 mg/kg dry matter. Suttle (1975) has similarly reported that the availability of dietary copper in suckling lambs may be as high as 70% but decreases to only 10% in fully weaned animals. It has been suggested that ewes are particularly susceptible to copper poisoning in late pregnancy (Suveges et al., 1971). This could arise from increased retention of copper during pregnancy, as has been demonstrated in the rat (Williams et al., 1977). There may also be sex differences in susceptibility to copper toxicosis, as Suttle (1977) found that male lambs showed higher mortality and higher plasma aspartate aminotransferase levels than did female lambs receiving the same copper-supplemented diet, although liver copper concentrations were similar in the two sexes.

4.2. Genetic Factors

There is also some evidence of a genetic influence on susceptibility to copper toxicity (see Hurley and Keen, this volume). Sheep of the Texel breed were more affected than those of the German Blackhead or the East Friesian breed (Luke and Wiemann, 1970). Scottish Blackface lambs retained less copper in their livers and showed fewer signs, both biochemical and histological, of liver damage than did Finnish Landrace lambs (Suttle, 1977). These findings are consistent with other observations on breed differences in liver and blood copper concentrations in sheep (Wiener and Field, 1971; Luke and Marquering, 1972; Wiener et al., 1976).

A further striking example of the increased susceptibility of some breeds to copper toxicosis is provided by sheep of the Orkney breed from North Ronaldsay (Wiener et al., 1977). Although these animals survive in their natural habitat on a diet consisting largely of seaweed, it was found that they succumbed very readily to copper poisoning when grazing on an upland farm regarded as marginally copper deficient. The elevated liver copper concentrations indicated that the absorption of copper was much greater in these animals than in other breeds. It was suggested that the capacity for highly efficient absorption and retention of copper had developed in these sheep by natural selection over many generations as a consequence of the low availability of copper in their natural feed of seaweed.

The incidence of a fatal liver disease associated with the excessive hepatic accumulation of copper in only certain strains of Bedlington terriers is also indicative of some ill-defined genetic influence on susceptibility to copper toxicosis in dogs (Sternlieb et al., 1977).

4.3. Protein

The reason for the apparently low availability of the copper in the seaweed diet of the Orkney sheep has not been determined, but a wide range of dietary factors can modify the absorption and retention of copper in this and other species. For example, MacPherson and Hemingway (1965) reported that both liver copper retention and the incidence of copper poisoning could be reduced in sheep by increasing the dietary protein concentration from 8.5 to 19%. Todd (1969) also found that increasing the protein content of a diet by inclusion of soyabean meal decreased liver copper retention, but he suggested that this could have arisen from the increased molybdenum content of the ration.

Similar effects of dietary protein have been observed in pigs receiving copper-supplemented diets. Thus Suttle and Mills (1966b) found that copper toxicosis developed in one experiment where fish meal was used as protein source but not when soybean meal or skim milk was used. Similarly, Parris and McDonald (1969) reported that copper accumulation in liver was greater with fish meal than with soybean meal diets. Moreover, signs of copper toxicosis, in the form of reduced weight gains, were evident only with the fish meal diets. The reasons for these differences in response have not been firmly established, but they need not reflect differences in the amino acid compositions of the proteins. It is possible that other dietary components could have influenced copper availability. For example, the high phytate content of the soybean diets could have restricted the availability of copper, as has been shown in the rat (Davies and Nightingale, 1975). Alternatively, the high calcium content of the fish meal could have reduced the availability of dietary zinc and so exacerbated the toxicity of copper (Suttle and Mills, 1966a).

4.4. Zinc and Iron

The existence of a close relationship between zinc status and susceptibility to copper toxicosis has been demonstrated in both pigs and sheep. Thus dietary supplementation with 100 or 500 mg Zn/kg reduced the incidence of copper toxicosis in pigs receiving 250 or 750 mg Cu/kg (Ritchie et al., 1963; Suttle and Mills, 1966a). Similarly, increasing the dietary zinc content to 220 or 420 mg/kg reduced the mortality rate and the severity of tissue damage in sheep receiving a high copper diet (Bremner et al., 1976). As the only adverse effect of zinc supplementation was the development of a slight anemia in the sheep receiving the greatest amount of zinc, it would seem that the dangers of copper toxicosis developing in sheep, as in pigs, could be substantially reduced by increasing the dietary zinc intake.

The mechanisms whereby zinc exerts its protective effect against copper toxicosis have not been firmly established (see Kirchgessner et al., this volume). The intestinal absorption of copper in the rat is inhibited by zinc (Van Campen and Scaife, 1967), and it is therefore not unexpected that decreased liver copper concentrations have been reported in both pigs (Ritchie et al., 1963) and sheep (Bremner et al., 1976) receiving zinc supplements. However, this is not invariably the case, as Suttle and Mills (1966a) did not find any reduction in liver copper concentrations in their zinc-supplemented pigs.

An alternative explanation is that the protective effect of zinc is related to alterations in the distribution of copper in the liver and perhaps other tissues. Considerable interest has been shown in this regard in the binding of copper to metallothionein, which is synthesized in liver in response to copper administration and is thought to have a role in the temporary storage or cellular detoxication of copper (Bremner and Davies, 1976a; Bremner and Young, 1976b). About 40% of the copper in porcine liver is generally in this form (Bremner, 1976; Bremner and Young, 1976a; Froslie and Norheim, 1977), but the proportion is much less in the livers of sheep and calves (Bremner and Marshall, 1974a, 1974b; Norheim and Søli, 1977), which are much more susceptible to copper poisoning. It is of interest therefore that the protective effect of zinc supplementation is associated with increased incorporation of copper into hepatic metallothionein (Bremner and Marshall, 1974b; Bremner et al., 1976).

Another explanation of the beneficial action of zinc supplementation is that it alleviates the zinc-deficiency syndrome which may develop in some copper-treated pigs and is manifested as parakeratosis (Suttle and Mills, 1966a; O'Hara et al., 1960). However, it has also been reported in some studies that copper alleviates rather than induces parakeratotic lesions in pigs (Wallace et al., 1960; Ritchie et al., 1963), a finding which suggests that additional factors may be involved in this complex interaction. One surprising feature is that, even though the copper-supplemented pigs are in a zinc-responsive state, liver zinc concentrations are often significantly increased by the copper treatment (Ritchie et al., 1963; Suttle and Mills, 1966a).

Although increasing dietary zinc intake can eliminate most symptoms of copper toxicity in pigs, it does not prevent the development of anemia (Suttle and Mills, 1966a), which is thought to result from the induction of an iron-deficiency state (Gipp et al., 1973, 1974). However, dietary supplementation with 100 to 140 mg Fe/kg does eliminate or decrease the severity of the anemia in copper-poisoned pigs and also prevents the development of other gross and histopathological lesions (Bunch et al., 1961; Gipp et al., 1973). Suttle and Mills (1966a) also found that an iron intake of 750 mg/kg eliminated jaundice and anemia and prevented any increase in serum copper and aspartate transaminase levels in pigs receiving 750 mg Cu/kg.

4.5. Molybdenum

Several investigations have been made on the protective effect of molybdenum against copper poisoning in sheep, as this metal has a marked inhibitory effect on copper availability, especially in the presence of sulfur (Dick, 1954; Kirchgessner et al., this volume). This is believed to be mediated through formation of thiomolybdates in the rumen (Dick et al., 1975; Mills et al., 1978). Although dietary supplementation with molybdenum and sulfate was found to be of limited value as a curative measure after the hemolytic crisis had developed (Todd et al., 1962), it is effective at reducing liver copper accumulation, liver damage, and mortality if given throughout the period during which higher copper diets are fed (Ross, 1964, 1966). The amounts of molybdenum and sulfate given daily in the studies cited were very high, and there was reluctance to adopt this measure for the control of copper toxicosis for fear of inducing a copper-deficiency state. However, it has been shown that much smaller amounts of molybdenum are also effective (Harker, 1976), and Suttle (1977) has suggested that the addition of only 4 mg Mo and 2 g S/kg feed should be sufficient under most circumstances to limit the hepatic accumulation of copper and prevent the development of clinical copper toxicosis. It is interesting that Todd (1972) has attributed the prevalence of copper poisoning in housed sheep receiving cereal-based rations to the naturally low levels of molybdenum and sulfur in these diets. Dietary supplementation with molybdenum and sulfate has no effect on the metabolism of copper in pigs (Gipp et al., 1967; Standish et al., 1975).

5. BIOCHEMICAL ASPECTS OF COPPER TOXICOSIS

It was shown in the preceding sections that a wide range of lesions affecting blood, liver, kidneys, and other tissues can develop in copper-poisoned animals. However, our understanding of the biochemical sequence of events leading to the onset of these lesions is still far from complete, as is exemplified by the investigations on the development of hemolytic anemia in copper-poisoned sheep and also in human beings with Wilson's disease.

5.1. Hemolytic Anemia

When human or ovine erythrocytes are incubated in the presence of relatively small concentrations of copper, the osmotic fragility of the erythrocyte membrane is increased and hemolysis occurs, suggesting a possible direct effect of copper on the membrane (Goldberg et al., 1956; Moroff et al., 1974; Thompson and Todd, 1976). However, no hemolysis occurs at the same copper concentration when albumin or plasma proteins are also present in the incubation medium; moreover, there is no increase in the osmotic fragility of erythrocytes in copper-poisoned sheep before the hemolytic crisis (Søli and Frøslie, 1977). It seems more likely therefore that changes which occur within the erythrocyte are responsible for the development of hemolytic anemia in copper poisoning.

The onset of hemolysis is preceded by a reduction in blood glutathione concentrations to near-zero values, by increases in methemoglobin concentration, and by the formation of Heinz bodies, indicating that marked oxidative changes must occur within the erythrocyte at this time (Todd and Thompson, 1963; Ishmael et al., 1972; Søli and Frøslie, 1977). The decrease in reduced glutathione concentrations in human erythrocytes has been attributed to inhibition of glutathione reductase (Flikweert et al., 1974) and to a catalytic autooxidation of glutathione by copper (Metz and Sagone, 1972). It has also been claimed, however (Agar and Smith, 1973; Thompson and Todd, 1976), that there is no direct action of copper on the regeneration of reduced glutathione in sheep erythrocytes. Furthermore, there is no significant inhibition of enzymes in the pentose phosphate pathway which would affect reduced glutathione regeneration, even though the activity of glucose-6-phosphate dehydrogenase in sheep erythrocytes is much lower than in erythrocytes of other species (Thompson and Todd, 1976) and an inhibitory effect of copper on the activity of this enzyme has been reported in human erythrocytes (Fairbanks, 1967).

The oxidative changes in the hemoglobin molecule probably also result from the decreased levels of reduced glutathione, although direct effects of copper in hemoglobin autooxidation have been reported (Salvati et al., 1969; Rifkind, 1974).

It is interesting that a predisposition to Heinz body formation and glutathione instability also exists in the erythrocytes of human beings with active liver disease and has been attributed to an acquired metabolic abnormality in the erythrocyte hexose monophosphate shunt (Smith et al., 1975). It may well be that this is an additional factor in the development of the hemolytic crisis in copper-poisoned sheep, especially as a massive increase in the severity of liver damage occurs immediately before periods of hemolysis (Ishmael et al., 1972).

5.2. Liver Damage

The damage to the liver in copper-poisoned animals is clearly associated with the accumulation of the metal in that organ, but the specific reason for the hepatotoxicity of copper has not been unequivocally established. Considerable interest has been shown in the hepatic distribution of copper in an attempt to explain this aspect of copper toxicosis, and evidence has been obtained suggesting that lysosomal accumulation of copper is important in this regard. For example, subcellular fractionation (Verity et al., 1967) and electron microscopical examination (Barka et al., 1964; Lindquist, 1967, 1968) of the livers of copper-loaded rats have shown that appreciable amounts of copper accumulate in this organelle. Similar results have been obtained from the livers of patients with Wilson's disease (Goldfischer and Sternlieb, 1968), of copper-loaded toads (Goldfischer et al., 1970), and of copper-poisoned sheep (T. P. King and I. Bremner, unpublished results).

There are conflicting views, however, as to the relationship between lysosomal accumulation of copper and the liver damage that occurs in copper poisoning. Lindquist (1968) suggested that the accumulation of copper in the lysosome would lead to rupture of the lysosomal membrane and leakage of acid hydrolases, perhaps as a result of lipid peroxidation or of interaction between copper and thiol groups in the membrane (Chvapil et al., 1972). However, this conclusion was based partly on the finding of increased activities of acid phosphatase in the liver of copper-treated animals, particularly in the cytoplasm. Later studies (McNatt et al., 1971) indicated that the acid phosphatase detected in the cytoplasm of these animals differed from the stabilized enzyme found in the lysosomes and therefore may not have originated within that organelle. An opposite view regarding the importance of lysosomal accumulation of copper was taken in studies on patients with Wilson's disease, as it was suggested that this accumulation constituted a protective measure against copper toxicosis (Goldfischer and Sternlieb, 1968).

One difficulty in interpreting the results of studies on rats and extrapolating to other animals is that there are considerable differences in liver copper concentrations and in the degrees of liver damage which occur in different species. For example, cellular necrosis has been reported in copper-poisoned sheep (Ishmael et al., 1971) and dogs (Sternlieb et al., 1977) but is not normally found in copper-loaded rats (Barka et al., 1964; Goldfischer, 1967).

Recent studies on the ultrastructure of the liver of sheep before and after the development of copper toxicosis have provided further information on the changes which occur as copper accumulates in the liver (T. P. King and I. Bremner, unpublished observations). As liver copper concentrations increased, there was a corresponding increase in the number of acid phosphatase-rich, copper-containing lysosomes in hepatocytes. These lysosomes consisted of autolysosomes and telolysosomes at copper concentrations <200 mg/kg wet weight, but at higher concentrations lipofuscin-like residual bodies were present, suggesting that some lipid peroxidation had occurred. In addition, there was a loss of rough endoplasmic reticulum, changes in smooth endoplasmic reticulum, and swelling and degeneration of mitochondria.

At later stages of copper accumulation there was evidence of cell death, and the relationship between copper concentration and number of lysosomes ceased to hold. An increasing number of necrotic cells, packed with copper-rich residual bodies, occurred around the portal tracts and a fine cirrhosis ensued. Examination of postmortem material indicated that a number of degenerative and regenerative changes were superimposed upon each other and that the considerable transfer of copper which occurred with cell death might have led to a form of acute copper poisoning.

Although the reason for the excessive hepatic accumulation of copper by the sheep has not been established, it may be associated with a relatively low excretion rate of the metal. For example, elevated copper concentrations are maintained in the liver of copper-loaded sheep long after the dietary intake of copper has been reduced (Gopinath and Howell, 1975), whereas copper is rapidly eliminated from porcine liver (Castell et al., 1975). Furthermore, biliary excretion of copper is not greatly increased in copper-poisoned sheep (Norheim and Søli, 1977). It is possible that there is some association between low biliary excretion and lysosomal accumulation of copper, as biliary obstruction in the rat appears to increase the deposition of this element in hepatic lysosomes (Worwood and Taylor, 1969). It has also been suggested that lysosomes are the source of biliary copper and that a lysosomal defect may account for the diminution in biliary copper excretion in patients with Wilson's disease (Sternlieb et al., 1973).

Little attention has been paid to the forms in which copper accumulates in the lysosomes of copper-poisoned animals, but it is possible that the copper is present as a polymeric form of metallothionein, as has been reported in neonatal calf liver (Porter, 1974). Rupp and Weser (1974) have suggested that this polymeric copper-thionein is formed as a result of oxidation of the monomeric copper protein. It may be that intracellular oxidation of copper-thionein and incorporation of the product into lysosomes occur very readily in ovine liver. This could explain the relatively small proportion of copper present in ovine liver as the monomeric protein (Bremner and Marshall, 1974a, 1974b; Norheim and Søli, 1977). It is perhaps significant that the copper-thionein in copper-poisoned sheep contains very little zinc, as the tendency for the copper protein to form aggregates *in vitro* is much greater when no zinc is present (Rupp and Weser, 1974; Bremner and Young, 1976a, 1976b).

5.3. Kidney Damage

There is also some evidence that the impaired renal function and degradation of proximal tubules in copper-poisoned sheep (Gopinath et al., 1974) could be associated with the extensive accumulation of copper-thionein in the kidneys (Bremner and Young, 1977). It has been pointed out that the kidney damage and renal distribution of the metal in these animals are similar to those found in chronic cadmium toxicity (Bremner and Young, 1977). In view of the claim (Nordberg et al., 1975) that the renal damage in cadmium-poisoned animals results from excessive tubular reabsorption of circulating cadmium-thionein, it seemed possible that the sudden increase in renal copper-thionein concentrations in copper-poisoned sheep could arise from reabsorption of the protein after its liberation from the liver into the bloodstream. A study was therefore made of the metabolism of intravenously administered copper-thionein in rats (Bremner et al., 1978). It was found that the metabolism of the injected copper protein was indeed similar to that of cadmium-thionein and that about 20% of the copper-thionein accumulated in intact form in the kidneys of the rats. Furthermore, there were signs of minor kidney damage in these animals (I. Bremner and N. T. Davies, unpublished observations), supporting the view that leakage of copper-thionein from the liver of copper-poisoned sheep at the time of the hemolytic crisis and its subsequent uptake by the kidneys may be responsible for the renal failure in these animals. However, it has not been possible to confirm this hypothesis by the detection of copper-thionein in the blood of these sheep (Bremner and Young, 1977; Norheim and Søli, 1977), although it may be that the rapid clearance of the protein from plasma, as was found in rats (Bremner et al., 1978), prevents its accumulation in significant amounts.

REFERENCES

Adam, S. E. I., Wasfi, I. A., and Magzoub, M. (1977). "Chronic Copper Toxicity in Nubian Goats," *J. Comp. Pathol.*, **87**, 623–628.

Agar, N. S. and Smith, J. E. (1973). "Effect of Copper on Red Cell Glutathione and Enzyme Levels in High and Low Glutathione-Sheep," *Proc. Soc. Exp. Biol. Med.*, **142**, 502–505.

Allen, M. M. and Harding, J. D. J. (1962). "Experimental Copper Poisoning in Pigs," *Vet. Rec.*, **74**, 173–179.

Barden, P. J. and Robertson, A. (1962). "Experimental Copper Poisoning in Sheep," *Vet. Rec.*, **74**, 252–256.

Barka, T., Scheuer, P. J., Schaffner, F., and Popper, H. (1964). "Structural Changes of Liver Cells in Copper Intoxication," *Arch. Pathol.*, **78**, 331–349.

Batey, T., Berryman, C., and Line, C. (1972). "The Disposal of Copper-Enriched Pig Manure Slurry on Grassland," *J. Br. Grassl. Soc.*, **27**, 139–143.

Biejers, J. A. (1932). "Kopervergiftiging bij Schapen," *Tijdschr. Diergeneeskd*, **59**, 1317–1324.

Braude, R. (1967). "Copper as a Stimulant in Pig Feeding," *World Rev. Anim. Prod.*, **3**, 69–78.

Bremner, I. (1974). "Heavy Metal Toxicities," *Quart. Rev. Biophys.*, **7**, 75–124.

Bremner, I. (1976). "The Relationship between the Zinc Status of Pigs and the Occurrence of Hepatic Copper- and Zinc-Binding Proteins," *Br. J. Nutr.*, **35**, 245–252.

Bremner, I. and Dalgarno, A. C. (1973). "Iron Metabolism in the Veal Calf. 2: Iron Requirements and the Effect of Copper Supplementation," *Br. J. Nutr.*, **30,** 61–76.

Bremner, I. and Davies, N. T. (1976). "Studies on the Appearance of a Hepatic Copper-Binding Protein in Normal and Zinc-Deficient Rats," *Br. J. Nutr.*, **36,** 101–112.

Bremner, I. and Marshall, R. B. (1974a). "Hepatic Copper- and Zinc-Binding Proteins in Ruminants. 1: Distribution of Cu and Zn among Soluble Proteins of Livers of Varying Cu and Zn Content," *Br. J. Nutr.*, **32,** 283–291.

Bremner, I. and Marshall, R. B. (1974b). "Hepatic Copper- and Zinc-Binding Proteins in Ruminants. 2: Relationship between Copper and Zinc Concentrations and the Occurrence of a Metallothionein-like Fraction," *Br. J. Nutr.*, **32,** 293–300.

Bremner, I. and Young, B. W. (1976a). "Isolation of Copper,Zinc-Thioneins from Pig Liver," *Biochem. J.*, **155,** 631–635.

Bremner, I. and Young, B. W. (1976b). "Isolation of (Copper,Zinc)-Thioneins from the Livers of Copper-Injected Rats," *Biochem. J.*, **157,** 517–520.

Bremner, I. and Young, B. W. (1977). "Copper-Thionein in the Kidneys of Copper-Poisoned Sheep," *Chem.-Biol. Interactions,* **19,** 13–23.

Bremner, I., Young, B. W., and Mills, C. F. (1976). "Protective Effect of Zinc Supplementation against Copper Toxicosis in Sheep," *Br. J. Nutr.*, **36,** 551–561.

Bremner, I., Hoekstra, W. G., Davies, N. T., and Williams, R. B. (1978). "Renal Accumulation of Copper,Zinc-Thioneins in Physiological and Pathological States." In M. Kirchgessner, Ed., *Trace Element Metabolism in Man and Animals*—3. Arbeitsgemeinschaft für Tierernahrungsforschung Weihenstephan, Freising-Weihenstephan, Germany, pp. 44–51.

Buck, W. B. (1970). "Diagnosis of Feed-Related Toxicosis," *J. Am. Vet. Med. Assoc.*, **156,** 1434–1443.

Bull, L. B. (1951). "The Occurrence of Chronic Copper Poisoning in Grazing Sheep in Australia." In *Plant and Animal Nutrition in Relation to Soil and Climatic Factors.* H. M. Stationery Office, London, pp. 300–310.

Bunch, R. J., Speer, V. C., Hays, V. W., Hawbaker, J. H., and Catron, D. V. (1961). "Effects of Copper Sulfate, Copper Oxide and Chlortetracycline on Baby Pig Performance," *J. Anim. Sci.*, **20,** 723–726.

Castell, A. G., Allen, R. D., Beames, R. M., Bell, J. M., Belzile, R., Bowland, J. P., Elliot, J. I., Ihnat, M., Larmond, E., Mallard, T. M., Spurr, D. T., Stothers, S. C., Wilton, S. B., and Young, L. G. (1975). "Copper Supplementation of Canadian Diets for Growing-Finishing Pigs," *Can. J. Anim. Sci.*, **55,** 113–134.

Christie, W. W. and Moore, J. H. (1969). "The Effect of Dietary Copper on the Structure and Physical Properties of Adipose Tissue Triglycerides in Pigs," *Lipids,* **4,** 345–349.

Chvapil, M., Ryan, J. N., and Brada, Z. (1972). "Effects of Selected Chelating Agents and Metals on the Stability of Liver Lysosomes," *Biochem. Pharmacol.*, **21,** 1097–1105.

Dalgarno, A. C. and Mills, C. F. (1975). "Retention by Sheep of Copper from Aerobic Digests of Pig Faecal Slurry," *J. Agric. Sci. (England)*, **85,** 11–18.

Davies, N. T. and Nightingale, R. (1975). "The Effects of Phytate on Intestinal Absorption and Secretion of Zinc, and Whole Body Retention of Zinc, Copper, Iron and Manganese in Rats," *Br. J. Nutr.*, **34,** 243–258.

Dick, A. T. (1954). "Studies on the Assimilation and Storage of Copper in Crossbred Sheep," *Aust. J. Agric. Res.*, **5,** 511–544.

Dick, A. T., Dewey, D. W., and Gawthorne, J. M. (1975). "Thiomolybdates and the Copper-Molybdenum-Sulphur Interaction in Ruminant Nutrition," *J. Agric. Sci. (England)*, **85,** 567–568.

Doherty, P. C., Barlow, R. M. and Angus, K. W. (1969). "Spongy Changes in the Brains of Sheep Poisoned by Excess Copper," *Res. Vet. Sci.*, **10,** 303–304.

Eden, A. (1940). "Observations on Copper Poisoning," *J. Comp. Pathol.*, **53,** 90–111.

Elliot, J. I. and Bowland, J. P. (1968). "Effects of Dietary Copper Sulfate on the Fatty Acid Composition of Porcine Depot Fats," *J. Anim. Sci.*, **27,** 956–960.

Fairbanks, V. F. (1967). "Copper Sulfate-Induced Hemolytic Anaemia," *Ann. Intern. Med.*, **120,** 428–432.

Feenstra, P. and Ulsen, F. W. van (1973). "Hay as a Cause of Copper Poisoning in Sheep," *Tijdschr. Diergeneeskd.*, **98,** 632–633.

Fisher, C., Laursen-Jones, A. P., Hill, K. J., and Hardy, W. S. (1973). "The Effect of Copper Sulphate on Performance and the Structure of the Gizzard in Broilers," *Br. Poultry Sci.*, **14,** 55–68.

Flikweert, J. P., Hoorn, R. K. J., and Staal, G. E. J. (1974). "The Effect of Copper on Human Erythrocyte Glutathione Reductase," *Int. J. Biochem.*, **5,** 649–653.

Froslie, A. and Norheim, G. (1977). "The Concentrations of Copper, Zinc and Molybdenum in Swine Liver and the Relationship to the Distribution of Soluble Copper- and Zinc-Binding Proteins," *Acta Vet. Scand.*, **18,** 471–479.

Gipp, W. F., Pond, W. G., and Smith, S. E. (1967). "Effects of Level of Dietary Copper, Molybdenum, Sulfate and Zinc on Body Weight Gain, Hemoglobin and Liver Copper Storage of Growing Pigs," *J. Anim. Sci.*, **26,** 727–730.

Gipp, W. F., Pond, W. G., Tasker, J., Van Campen, D., Krook, L., and Visek, W. J. (1973). "Influence of Level of Dietary Copper on Weight Gain, Hematology and Liver Copper and Iron Storage of Young Pigs," *J. Nutr.*, **103,** 713–719.

Gipp, W. F., Pond, W. G., Kallfelz, F. A., Tasker, J. B., Van Campen, D. R., Krook, L., and Visek, W. J. (1974). "Effect of Dietary Copper, Iron and Ascorbic Acid Levels on Hematology, Blood and Tissue Copper, Iron and Zinc Concentrations and ^{64}Cu and ^{59}Fe Metabolism in Young Pigs," *J. Nutr.*, **104,** 532–541.

Goldberg, A., Williams, C. B., Jones, R. S., Yanagita, M., Cartwright, G. E., and Wintrobe, M. M. (1956). "Copper Metabolism. XXII: Hemolytic Anaemia in Chickens Induced by Administration of Copper," *J. Lab. Clin. Med.*, **48,** 422–453.

Goldfischer, S. (1967). "Demonstration of Copper and Acid Phosphatase Activity in Hepatocyte Lysosomes in Experimental Copper Toxicity," *Nature (London)*, **215,** 74–75.

Goldfischer, S. and Sternlieb, I. (1968). "Changes in the Distribution of Hepatic Copper in Relation to the Progression of Wilson's Disease (Hepatolenticular Degeneration)," *Am. J. Pathol.*, **53**, 883–899.

Goldfischer, S., Schiller, B., and Sternlieb, I. (1970). "Copper in Hepatocyte Lysosomes of the toad, *Bufo marinus L.*," *Nature* (*London*), **228**, 172–173.

Gopinath, C. and Howell, J. McC. (1975). "Experimental Chronic Copper Toxicity in Sheep: Changes That Follow the Cessation of Dosing at the Onset of Haemolysis," *Res. Vet. Sci.*, **19**, 35–43.

Gopinath, C., Hall, G. A., and Howell, J. McC. (1974). "The Effect of Chronic Copper Poisoning on the Kidneys of Sheep," *Res. Vet. Sci.*, **16**, 57–69.

Goresky, C. A., Holmes, T. H., and Sass-Kortsak, A. (1968). "The Initial Uptake of Copper by the Liver in the Dog," *Can. J. Physiol. Pharmacol.*, **46**, 771–784.

Gracey, J. F. and Todd, J. R. (1960). "Chronic Copper Poisoning in Sheep Following the Use of Copper Sulphate as Molluscicide," *Br. Vet. J.*, **116**, 405–408.

Gracey, H. I., Stewart, T. A., Woodside, J. D., and Thompson, R. H. (1976). "The Effects of Disposing High Rates of Copper-Rich Pig Slurry on Grassland on the Health of Grazing Sheep," *J. Agric. Sci.* (*England*), **87**, 617–623.

Harker, D. B. (1976). "The Use of Molybdenum for the Prevention of Nutritional Copper Poisoning in Housed Sheep," *Vet. Rec.*, **99**, 78–81.

Hill, R. and Williams, H. L. (1965). "The Effects on Intensively Reared Lambs of Diets Containing Excess Copper," *Vet. Rec.*, **77**, 1043–1045.

Ho, S. K. and Elliot, J. I. (1973). "Supplementary Dietary Copper and the Desaturation of 1-^{14}C-Stearoyl-coenzyme A by Porcine Hepatic and Adipose Microsomes," *Can. J. Anim. Sci.*, **53**, 537–545.

Ho, S. K. and Elliot, J. I. (1974). "Fatty Acid Composition of Porcine Depot Fat as Related to the Effect of Supplementary Dietary Copper on the Specific Activities of Fatty Acyl Desaturase Systems," *Can. J. Anim. Sci.*, **54**, 23–28.

Hoefer, J. A., Miller, E. R., Ullrey, D. E., Ritchie, H. D., and Luecke, R. W. (1960). "Interrelationships between Calcium, Zinc, Iron and Copper in Swine Feeding," *J. Anim. Sci.*, **19**, 249–259.

Husbands, D. R. (1972). "The Effect of Dietary Copper on the Composition of Adipose Tissue Triglycerides in the Broiler Chicken," *Br. Poultry Sci.*, **13**, 201–205.

Ishmael, J., Gopinath, C., and Howell, J. McC. (1971). "Experimental Chronic Toxicity in Sheep: Histological and Histochemical Changes during Development of the Changes in the Liver," *Res. Vet. Sci.*, **12**, 358–366.

Ishmael, J., Gopinath, C., and Howell, J. McC. (1972). "Experimental Chronic Toxicity in Sheep: Biochemical and Haematological Studies during the Development of Changes in the Liver," *Res. Vet. Sci.*, **13**, 22–29.

Kidder, R. W. (1949). "Symptoms of Induced Copper Toxicity in a Steer," *J. Anim. Sci.*, **8**, 623–624.

King, J. O. L. (1975). "The Feeding of Copper Sulphate to Growing Rabbits," *Br. Vet. J.*, **131**, 70–75.

Kneale, W. A. and Smith, P. (1977). "The Effects of Applying Pig Slurry Containing High Levels of Copper to Sheep Pastures," *Exp. Husb.*, **32**, 1–7.

Lindquist, R. R. (1967). "Studies on the Pathogenesis of Hepatolenticular Degeneration.

I: Acid Phosphatase Activity in Copper-Loaded Livers," *Am. J. Pathol.*, **51**, 471–481.

Lindquist, R. R. (1968). "Studies on the Pathogenesis of Hepatolenticular Degeneration. III: The Effect of Copper on Rat Liver Lysosomes," *Am. J. Pathol.*, **53**, 903–922.

Luke, F. and Marquering, B. (1972). "The Mineral Content of the Liver in Sheep. 1: Nutritional and Genetic Effects on the Copper Content," *Zuchtungskunde*, **44**, 56–65.

Luke, F. and Wiemann, H. (1970). "Chronic Copper Poisoning and Copper Retention in the Liver in Sheep of Different Breeds," *Berlin Munch Tierarztl. Wochenschr.*, **83**, 253–255.

McCosker, P. J. (1968). "Observations on Blood Copper in Sheep," *Res. Vet. Sci.*, **9**, 103–116.

McNatt, E. N., Campbell, W. G., and Callahan, B. C. (1971). "Effects of Dietary Copper Loading on Livers of Rats. I: Changes in Subcellular Acid Phosphatases and Detection of an Additional *p*-Nitrophenylphosphatase in the Cellular Supernatant during Copper Loading," *Am. J. Pathol.*, **64**, 123–144.

MacPherson, A. and Hemingway, R. G. (1965). "Effects of Protein Intake on the Storage of Copper in the Liver of Sheep," *J. Sci. Food Agric.*, **16**, 220–227.

MacPherson, A. and Hemingway, R. G. (1969). "The Relative Merit of Various Blood Analyses and Liver Function Tests in Giving Early Diagnosis of Chronic Copper Poisoning in Sheep," *Br. Vet. J.*, **125**, 213–221.

Metz, E. N. and Sagone, A. L. (1972). "The Effect of Copper on the Erythrocyte Hexose Monophosphate Shunt Pathway," *J. Lab. Clin. Med.*, **80**, 405–413.

Mills, C. F. (1974). "Trace Element Interactions: Effects of Dietary Composition on the Development of Imbalance and Toxicity." In W. G. Hoekstra, J. W. Suttie, H. E. Ganther and W. Mertz, Eds., *Trace Element Metabolism in Animals*—2. University Park Press, Baltimore, Md., pp. 79–90.

Mills, C. F., Bremner, I., El-Gallad, T. T., Dalgarno, A. C., and Young, B. W. (1978). "Mechanism of the Molybdenum/Sulphur Antagonism of Copper Utilisation by Ruminants." In M. Kirchgessner, Ed., *Trace Element Metabolism in Man and Animals*—3. Arbeitsgemeinschaft für Tierernahrungsforschung Wiehenstephan, Freising-Weihenstephan, Germany.

Moore, J. H., Christie, W. W., Braude, R., and Mitchell, K. G. (1969). "The Effect of Dietary Copper on the Fatty Acid Composition and Physical Properties of Pig Adipose Tissue," *Br. J. Nutr.*, **23**, 281–287.

Morgan, K. T. (1973). "Chronic Copper Toxicity of Sheep: An Ultrastructural Study of Spongiform Leucoencephalopathy," *Res. Vet. Sci.*, **15**, 88–95.

Moroff, G., Poster, J., and Oster, G. (1974). "Enhancement of the Copper-Induced Hemolysis by Progesterone," *J. Steroid Biochem.*, **5**, 601–606.

Nordberg, G. F., Goyer, R. A., and Nordberg, M. (1975). "Comparative Toxicity of Cadmium-Metallothionein and Cadmium Chloride on Mouse Kidney," *Arch. Pathol.*, **99**, 192–197.

Norheim, G. and Søli, N. E. (1977). "Chronic Copper Poisoning in Sheep. II: The Distribution of Soluble Copper-, Molybdenum-, and Zinc-Binding Proteins from Liver and Kidney," *Acta Pharmacol. Toxicol.*, **40,** 178–187.

O'Hara, P. J., Newman, A. P., and Jackson, R. (1960). "Parakeratosis and Copper Poisoning in Pigs Fed a Copper Supplement," *Aust. Vet. J.*, **36,** 225–229.

Omole, T. A. (1977). "Influence of Levels of Dietary Protein and Supplementary Copper on the Performance of Growing Rabbits," *Br. Vet. J.*, **133,** 593–599.

Parris, E. C. C. and McDonald, B. E. (1969). "Effect of Dietary Protein Source on Copper Toxicity in Early-Weaned Pigs," *Can. J. Anim. Sci.*, **49,** 215–222.

Porter, H. (1974). "The Particulate Half-Cystine-Rich Copper Protein of Newborn Liver. Relationship to Metallothionein and Subcellular Localization in Nonmitochondrial Particles Possibly Representing Heavy Lysosomes," *Biochem. Biophys. Res. Commun.*, **56,** 661–668.

Poupoulis, C. and Jensen, L. S. (1976a). "Effect of High Dietary Copper on Gizzard Integrity of the Chick," *Poultry Sci.*, **55,** 113–121.

Poupoulis, C. and Jensen, L. S. (1976b). "Effect of High Dietary Copper on Fatty Acid Composition of the Chick," *Poultry Sci.*, **55,** 122–129.

Rifkind, J. M. (1974). "Copper and the Autoxidation of Haemoglobin," *Biochemistry*, **13,** 2475–2481.

Ritchie, H. D., Luecke, R. W., Baltzer, B. V., Miller, E. R., Ullrey, D. E., and Hoefer, J. A. (1963). "Copper and Zinc Interrelationships in the Pig," *J. Nutr.*, **79,** 117–123.

Ross, D. B. (1964). "Chronic Copper Poisoning in Lambs," *Vet. Rec.*, **76,** 875–876.

Ross, D. B. (1966). "The Diagnosis, Prevention and Treatment of Chronic Copper Poisoning in Housed Lambs," *Br. Vet. J.*, **122,** 279–284.

Rupp, H. and Weser, U. (1974). "Conversion of Metallothionein into Copper-Thionein, the Possible Low Molecular Weight Form of Neonatal Hepatic Mitochondrocuprein," *Biochem. Biophys. Res. Commun.*, **44,** 293–297.

Salvati, A. M., Ambrogioni, M. T., and Tentoni, L. (1969). "The Autoxidation of Hemoglobin: Effect of Copper," *Ital. J. Biochem.*, **18,** 1–18.

Shand, A. and Lewis, G. (1957). "Chronic Copper Poisoning in Calves," *Vet. Rec.*, **69,** 618–621.

Smith, J. D., Jordan, R. M., and Nelson, M. L. (1975). "Tolerance of Ponies to High Levels of Dietary Copper," *J. Anim. Sci.*, **41,** 1645–1649.

Smith, J. R., Kay, N. E., Gottlieb, A. J., and Oski, F. A. (1975). "Abnormal Erythrocyte Metabolism in Hepatic Disease," *Blood*, **46,** 955–964.

Smith, M. S. (1969). "Responses of Chicks to Dietary Supplements of Copper Sulphate," *Br. Poultry Sci.*, **10,** 97–108.

Søli, N. E. and Froslie, A. (1977). "Chronic Copper Poisoning in Sheep. I: The Relationship of Methaemoglobinemia to Heinz Body Formation and Haemolysis during the Terminal Crisis," *Acta Pharmacol. Toxicol.*, **40,** 169–177.

Standish, J. F., Ammerman, C. B., Wallace, H. D., and Combs, G. E. (1975). "Effect of High Dietary Molybdenum and Sulphate on Plasma Copper Clearance and Tissue Minerals in Growing Swine," *J. Anim. Sci.*, **40,** 509–513.

Sternlieb, I., Van der Hamer, C. J. A., Morell, A. G., Alpert, S., Gregoriadis, G., and

Scheinberg, I. H. (1973). "Lysosomal Defect of Hepatic Copper Excretion in Wilson's Disease (Hepatolenticular Degeneration)," *Gastroenterology,* **64,** 99-105.

Sternlieb, I., Twedt, D. C., Johnson, G. F., Gilbertson, S., Korotkin, E., Quintana, N., and Scheinberg, I. H. (1977). "Inherited Copper Toxicity of the Liver in Bedlington Terriers," *Proc. R. Soc. Med.,* **70,** Suppl. 3, 8-9.

Suttle, N. F. (1975). "Changes in the Availability of Dietary Copper to Young Lambs Associated with Age and Weaning," *J. Agric. Sci. (England).,* **84,** 255-261.

Suttle, N. F. (1977). "Reducing the Potential Copper Toxicity of Concentrates to Sheep by the Use of Molybdenum and Sulphur Supplements," *Anim. Feed Sci. Technol.,* **2,** 235-246.

Suttle, N. F. and Mills, C. F. (1966a). "Studies of the Toxicity of Copper to Pigs. 1: Effects of Oral Supplements of Zinc and Iron Salts on the Development of Copper Toxicosis," *Br. J. Nutr.,* **20,** 135-148.

Suttle, N. F. and Mills, C. F. (1966b). "Studies of the Toxicity of Copper to Pigs. 2: Effect of Protein Source and Other Dietary Components on the Response to High and Moderate Intakes of Copper," *Br. J. Nutr.,* **20,** 149-161.

Suttle, N. F. and Price, J. (1976). "The Potential Toxicity of Copper-Rich Animal Excreta to Sheep," *Anim. Prod.,* **23,** 233-242.

Suveges, T., Ratz, F., and Salyi, G. (1971). "Pathogenesis of Chronic Copper Poisoning in Lambs," *Acta Vet. Acad. Sci. Hung.,* **21,** 383-391.

Tait, R. M., Krishnamurti, C. R., Gilchrist, E. W., and MacDonald, K. (1971). "Chronic Copper Poisoning in Feeder Lambs," *Can. Vet. J.,* **12,** 73-75.

Thompson, R. H. and Todd, J. R. (1974). "Muscle Damage in Chronic Copper Poisoning of Sheep," *Res. Vet. Sci.,* **16,** 97-99.

Thompson, R. H. and Todd, J. R. (1976). "Role of Pentose-Phosphate Pathway in Haemolytic Crisis of Chronic Copper Toxicity of Sheep," *Res. Vet. Sci.,* **20,** 257-260.

Thompson, E. H., Allen, C. E., and Meade, R. J. (1973). "Influence of Copper on Stearic Acid Desaturation and Fatty Acid Composition in the Pig," *J. Anim. Sci.,* **36,** 868-873.

Todd, J. R. (1969). "Chronic Copper Toxicity of Ruminants," *Proc. Nutr. Soc.,* **28,** 189-198.

Todd, J. R. (1972). "Copper, Molybdenum and Sulphur Contents of Oats and Barley in Relation to Chronic Copper Poisoning in Housed Sheep," *J. Agric. Sci. (England),* **79,** 191-195.

Todd, J. R. and Gracey, J. F. (1959). "Chronic Copper Poisoning of a Young Heifer," *Vet. Rec.,* **71,** 145-146.

Todd, J. R. and Gribben, H. J. (1965). "Suspected Chronic Copper Poisoning in a Cow," *Vet. Rec.,* **77,** 498-499.

Todd, J. R. and Thompson, R. H. (1963). "Studies of Chronic Copper Poisoning. II: Biochemical Studies of the Blood of Sheep during the Haemolytic Crisis," *Br. Vet. J.,* **119,** 189-198.

Todd, J. R., Gracey, J. F., and Thompson, R. H. (1962). "Studies on Chronic Copper

Poisoning. 1: Toxicity of Copper Sulphate and Copper Acetate," *Br. Vet. J.*, **118,** 482–491.

Ulsen, F. W. (1972). "Sheep, Swine and Copper," *Tijdschr. Diergeneeskd.*, **97,** 735–738.

Van Adrichem, P. W. M. (1965). "Changes in Activity of Serum Enzymes and in LDH Isoenzyme Pattern in Chronic Copper Poisoning of Sheep," *Tijdschr. Diergeneeskd.*, **90,** 1371–1381.

Van Campen, D. R. and Scaife, P. U. (1967). "Zinc Interference with Copper Absorption in Rats," *J. Nutr.*, **91,** 473–476.

Verity, M. A., Gambell, J. K., Reith, A. R., and Brown, W. J. (1967). "Subcellular Distribution and Enzyme Changes Following Subacute Copper Intoxication," *Lab. Invest.*, **16,** 580–590.

Wahle, K. W. J. and Davies, N. T. (1975). "Effect of Dietary Copper Deficiency in the Rat on Fatty Acid Compoisition of Adipose Tissue and Desaturase Activity of Liver Microsomes," *Br. J. Nutr.*, **34,** 105–112.

Wallace, H. D., McCall, J. T., Bass, B., and Combs, G. E. (1960). "High Levels of Copper for Growing-Finishing Swine," *J. Anim. Sci.*, **19,** 1153–1163.

Weiss, E. and Baur, P. (1968). "Experimental Studies on Chronic Copper Poisoning in Calves," *Zentralbl. Vet. Med.*, **15A,** 156–184.

Wiener, G. and Field, A. C. (1971). "The Concentrations of Minerals in the Blood of Genetically Diverse Groups of Sheep," *J. Agric. Sci.* (*England*), **76,** 513–520.

Wiener, G., Herbert, J. G., and Field, A. C. (1976). "Variation in Liver and Plasma Copper Concentrations of Sheep in Relation to Breed and Haemoglobin Type," *J. Comp. Pathol.*, **86,** 101–110.

Wiener, G., Field, A. C., and Smith, C. (1977). "Deaths from Copper Toxicity of Sheep at Pasture and the Use of Fresh Seaweed," *Vet. Rec.*, **101,** 424–425.

Williams, R. B., Davies, N. T., and McDonald, I. (1977). "The Effects of Pregnancy and Lactation on Copper and Zinc Retention in the Rat," *Br. J. Nutr.*, **38,** 407–416.

Worwood, M. and Taylor, D. M. (1969). "Subcellular Distribution of Copper in Rat Liver after Biliary Obstruction," *Biochem. Med.*, **3,** 105–116.

11

TOXICITY OF COPPER TO AQUATIC BIOTA

Peter V. Hodson

Uwe Borgmann

Harvey Shear

Great Lakes Biolimnology Laboratory, Fisheries and Oceans Canada, Canada Centre for Inland Waters, Burlington, Ontario

1. INTRODUCTION

In a discussion of the toxic effects of copper on aquatic biota, concepts such as synergism, antagonism, metal complexation, and bioaccumulation in the food chain need to be included. Study of the toxicity of copper to aquatic biota is not merely an academic exercise. It is also of considerable economic interest in terms of the management strategies employed in an aquatic ecosystem where diverse and often conflicting uses must be accommodated.

Copper in water is exceedingly toxic to aquatic biota, in contrast to low toxicity to mammalian consumers of water. Whereas concentrations as low as 5 to 25 $\mu g/l$ are lethal to some invertebrate and fish species within 4 days, the recommended standard for public water supplies, based on palatability, is 1000 $\mu g/l$ (U.S. Environmental Protection Agency, 1976). The great sensitivity of most aquatic biota is a result of high surface–volume ratios of algae and, in invertebrates and fish, high respiratory water flows plus an extensive, highly permeable gill surface area that facilitates rapid uptake of large amounts of copper. Consequently, sensitivity to waterborne metals is analogous to mammalian sensitivity to airborne metals as a result of rapid respiratory uptake. Ingested copper is available only in limited amounts since freshwater biota generally do not drink water and fecal material competes with the stomach and intestinal tract for absorption of ingested copper. Since no published evidence for toxicity to aquatic biota of ingested copper was found, this chapter pertains only to waterborne copper.

As waterborne copper concentrations increase from background levels (generally less than 5 $\mu g/l$), the incidence of mortality of exposed aquatic biota increases and survival times decrease. Since the amount of copper taken up per organism is unknown, a median lethal dose (LD_{50}) cannot be estimated. However, the probit of percent mortality is linearly related to $\log_{10}$ (copper concentration), allowing calculation of the median lethal concentration (LC_{50}) by probit analysis (Finney, 1971). The LC_{50}, by referring to the average animal, provides a convenient basis for comparing toxicities but must also refer to exposure time (e.g., 96-hr LC_{50}) (Sprague, 1969) since the total exposure is a function of (concentration $\times$ time). There is usually a threshold concentration below which no mortality occurs and survival is indefinite. This may be termed the asymptotic LC_{50} (Brown, 1973).

2. COPPER TOXICITY TO ALGAE

2.1. Introduction

Many studies, some of which date back to the early part of this century (Moore and Kellerman, 1904), have been carried out on the effects of copper on algae. The early studies and many recent ones (e.g., Fitzgerald, 1964a, 1964b), were concerned with the control of algal blooms in lakes and reservoirs by the application of compounds containing copper. Most of the recent studies that will be discussed here relate to the effects of copper on the physiology and morphology of algac and to the factors governing toxicity.

2.2. Copper Levels in Natural Waters

Copper Concentrations in the Marine Environment

Boyle (this volume, Part I) has reviewed the available data on copper in seawater and showed that the concentrations range from an average of about 0.15 $\mu g/l$ in the open sea to around 1.0 $\mu g/l$ in polluted near-shore waters. Copper levels in marine sediments were found to range from a high of 740,000 $\mu g/kg$ (dry weight basis) (Goldberg and Arrhenius, 1958) to a low of 2000 $\mu g/kg$ (dry weight basis) (Segar and Pellenberg, 1973).

On the basis of the data of many investigators, Bernhard and Zattera (1973) gave a generalized value for the copper concentrations in marine organisms of about 1000 $\mu g/kg$ fresh weight. There are, of course, exceptions to this general value, particularly in seaweeds, where some species (*Padina* sp., *Thallassia testudinum*) are able to accumulate 3 to 7 times this level of copper (Bernhard and Zattera, 1973).

Copper Concentrations in Freshwater Environments

Concentrations of copper in the Laurentian Great Lakes have been found to range from about 1.0 to 2.5 μg/l in the open waters of the lakes (Pluarg, 1978). In lakes near Sudbury, Ontario (an active nickel and copper mining area), Stokes et al. (1973) found that total copper concentrations generally decreased with distance from the site of mining activity. Trollope and Evans (1976) in their survey of freshwater ponds near zinc smelting waste areas of south Wales found copper concentrations of 20 μg/l. In the south Wales ponds, algae comprised three divisions (Chlorophyta, Xanthophyta, and Cyanophyta) and were found to have accumulated from 12,000 to 83,000 times the ambient copper concentration in the water. The algae investigated by Stokes et al. (1973) belonged to the Chlorophyta (*Chlorella* sp. and *Scenedesmus* sp.). These algae were able to bioaccumulate up to 2400 mg Cu/kg on a dry weight basis, representing a 3400 bioaccumulation factor.

In the Laurentian Great Lakes, considerable work has been carried out on the chemical composition of the sediments. Kemp and Thomas (1976) indicate concentrations of copper in the range of 40 to 120 mg/kg for Lakes Huron, Erie and Ontario. This work also shows rather clearly the recent (post-1850) anthropogenic inputs of copper to the lakes, as recorded in sediment profiles.

2.3. Empirical Copper Toxicity

There is a substantial body of evidence on the toxicity of copper as observed under nonlaboratory conditions. Whitton (1971) reviewed the work of several authors regarding *in situ* copper toxicity. In the English Trent River system it was found that two species of green algae and one diatom were the first to colonize tributaries downstream of a copper-processing plant discharging copper to the stream. No microflora were observed within 6.4 km (4 miles) of the plant outfall; then colonies of *Chlorococcum* sp., *Achnanthes affinis,* and *Stigeoclonium tenue* were observed. The normal flora consists of two pennate diatoms, namely, *Cocconeis* sp. and *Nitzschia palea.* From numerous other empirical observations, Whitton indicated a range of copper concentrations that could be tolerated by algae. These values were 750 μg/l for *Synura* sp. and *Dinobryon* sp; 800 μg/l for *S. tenue;* 1500 μg/l for *N. palea, Navicula viridula, Cymbella ventricosa,* and *Gomphonema parvalum;* and 2000 μg/l for *A. affinis.*

A significant amount of empirical information exists on commercial algicides and on the concentrations needed to control algal blooms in reservoirs, ponds, and even the ocean. Fitzgerald (1975) investigated the toxic effects of several commercially available algicides, including copper sulfate, on several species of green and blue-green algae. He concluded that cell density (number of algal cells per milliliter), contact time, and composition of the test medium were all factors influencing copper toxicity.

Fitzgerald (1975) also found that two commercially available algicides containing copper were more effective at similar dose rates and contact times than was copper sulfate. As little as 50 μg Cu/l in a commercial preparation proved algicidal to *Microcystis aeruginosa,* whereas 200 μg Cu/l as $CuSO_4$ was needed to achieve the same effect. Niemi (1972) found that the effects of copper sulfate additions to *in situ* assimilation and dark fixation of carbon by phytoplankton communities were reflected in the differences in susceptibilities of spring and summer communities, and these differences could be related directly to the susceptibility of the dominant species.

In the marine environment a bloom of the "red tide" alga, *Gymnodinium breve,* was controlled effectively for 2 weeks by the application of 180 μg Cu/l (Rounsfell and Evans, 1958). The high concentration of copper was necessitated because of dilution effects in the near-shore zone and the precipitation of copper at the pH (8.2) of seawater. A very much lower lethal concentration of copper (0.5 μg/l) was found to be effective against this same alga (Marvin et al., 1961) in laboratory experiments. In a seawater lagoon system, however, toxic levels of copper could never be achieved by the addition of copper-bearing ore. It was concluded that there was too much chelation capacity in the lagoon to permit toxic levels of copper to build up.

2.4. Experimental Studies

Effects on Algal Physiology

By far the largest portion of the literature surveyed was concerned with the sublethal effects of copper on algae. Physiological processes such as photosynthesis, growth, and nitrogen fixation have been studied in some detail.

Effects on Photosynthesis and Growth. Steeman Nielsen et al. (1969) began studying the effects of copper on the growth and photosynthesis of *Chlorella pyrenoidosa.* Copper toxicity was maximal during the first 24 hr of the experiments. Toxicity to copper was found to decrease with increasing cell density. Much of this work was carried out at high copper concentrations (≥17 μg/l); however, some depression of photosynthesis was seen at copper levels of 1 μg/l.

Steeman Nielsen and Wium-Andersen (1970) worked with *Chlorella pyrenoidosa* and *Nitzschia palea,* studying growth and photosynthesis. They found that ionic copper was extremely toxic to both species when no metal complexing was present. Copper concentrations of 1 to 2 μg/l had a noticeable effect on the rate of photosynthesis of *N. palea,* which was more sensitive to copper than was *C. pyrenoidosa.* The suppression of photosynthesis was found to increase with increasing time of contact with copper. Effects on growth were noted at 5 μg/l for both *C. pyrenoidosa* and *N. palea.* These authors also pointed out the possible errors that might occur in ^{14}C experiments using commercially prepared $NaH[^{14}CO_3]$ diluted with ordinary distilled water, which, they found, could contain about 300 μg Cu/l as a contaminant. It is apparent that, if ^{14}C ampoules were made using such distilled water, and if a 1-ml ampoule was used for a 100-ml experimental flask, a concentration of about 3 μg Cu/l would result. This concentration of copper, as noted above, would suppress photosynthesis. Much information reported in the literature, therefore, could have been the result of an inherent toxic effect of the tracer used.

In further work carried out on the mechanism of copper toxicity, Steeman Nielsen and Wium-Andersen (1972) found that 6 μg Cu/l caused increased photosynthetic inhibition in the light phase of a culture of *N. palea* grown in a régime of 12 hr light and 12 hr dark. Copper also reduced the rate of photosynthesis during the dark phase to about 5% of a control culture. These workers found that 3 μg Cu/l gave similar positive results, although the absolute amount of photosynthetic inhibition was not as great.

Steeman Nielsen and Wium-Andersen (1972) used their results to explain the so-called afternoon depression of photosynthesis in marine systems reported by Doty and Oguri (1957). They postulated that the afternoon depression could be caused by copper contamination of the ^{14}C ampoules used in the experiments. Open ocean waters, with little organic matter to bind copper, exhibited this depression, whereas Newhouse et al. (1967) did not observe this phenomenon in later work in near-shore waters, where there is abundant organic matter for complexing copper.

Overnell (1975, 1976) examined the oxygen evolution of several marine and freshwater algae in response to copper. He found that O_2 evolution in whole cells and in chloroplasts (the Hill reaction) of *Chlamydomonas reinhardii* was suppressed at copper concentrations of 635 to 6350 μg/l. In work with marine phytoplankton, Overnell observed a 50% reduction in O_2 evolution in five of the seven test species at copper concentrations ranging from 1270 to 12700 μg/l. He also found that copper at a concentration of 63,500 μg/l was not toxic to *Phaeodactylum tricornutum.* All these results were obtained after 15-min exposure to copper. An exposure of 4.5 or 24 hr produced a toxic effect in all the test organisms.

In their work with *Chlorella sorokiniana,* Cedeno-Maldonado and Swader (1974) indicated that autotrophic growth, photosynthesis, and respiration were all inhibited by copper and that photosynthesis was more sensitive to copper than was respiration. After 2-min exposure to 63,500 μg Cu/l, photosynthesis was inhibited. Long-term growth experiments demonstrated sensitivity to much lower copper levels; 63.5 μg Cu^{2+} ion/l resulted in a significant decline in growth of *C. sorokiniana.* Cedeno-Maldonado and Swader further found that exposure for a short time to inhibitory copper levels permanently destroyed the photosynthetic apparatus of the cells and rendered them incapable of growth in fresh, copper-free medium.

Much additional work has been carried out on the effects of copper on the growth of phytoplankton. Bartlett et al. (1974) found that copper at 300 μg/l proved toxic to *Selenastrum capricornutum.* Sublethal effects were found at 50 to 60 μg/l, where the lag phase of growth of this alga was extended. Above 90 μg Cu/l all growth ceased. Hutchinson (1973) studied the effects of several heavy metals on four species of algae. His results indicated that copper was toxic to the growth of three algae (*Chlorella vulgaris, Scenedesmus acuminata,* and *Chlamydomonas eugametas*) at concentrations of 50 μg/l and greater. It is interesting that growth of *Haematococcus capensis* was initially stimulated by additions of copper up to 100 μg/l. Above this concentration, however, copper proved inhibitory to growth. *Chlorella vulgaris* was most sensitive to copper, being totally eliminated above a concentration of 100 μg/l.

Effects on Nitrogen Fixation. Nitrogen fixation by heterocystous blue-green algae can account for a major portion of the nitrogen budget of a lake (Horne and Goldman, 1972). Many suggestions have been advanced about the control of eutrophication through nitrogen removal at source. Unfortunately, these programs are generally costly and not particularly efficient. The work of Horne and Goldman (1974) suggests that trace additions of copper to a lake dominated by blue-green algae could substantially reduce N_2 fixation. These workers found that the addition of 5 μg Cu/l reduced *in situ* N_2 fixation by 76% in 2 days and by 90 to 95% in 8 days. The addition of 10 μg Cu/l caused a rapid, permanent drop in N_2 fixation. In contrast, rates of photosynthesis were reduced by only 7 to 32%. Horne and Goldman could not conclude, however, that the low indigenous copper concentrations in Clear Lake, California, are in fact inhibiting either N_2 fixation or photosynthesis at present. However, they concluded that an external addition of copper could reduce the total nitrogen budget of the lake by up to 50%.

Effects on Algal Morphology and Reproduction

Several workers have examined other aspects of the effects of copper on algal physiology. Khobot'yev et al. (1975) observed distinct morphological changes and alterations in pigment composition in *Scenedesmus quadricauda* and *Anabaena variabilis* exposed to copper. In *S. quadricauda,* after exposure to 7.8×10^{-5} *M* $CuCl_2$, the cells were found to be enlarged. First the outer cells of the coenobium and then the inner cells showed this effect. Eventually the cells became large and round, and the typical shape of the coenobium disappeared. It is interesting that this effect was found to be reversible when copper was withdrawn from the medium. The chlorophyll content of the cells was found to decrease with increasing copper concentrations, and the phaeophytin levels were seen to increase. Saward et al. (1975), in work carried out on marine phytoplankton populations, found that chlorophyll *a* levels and total pigment content declined with increasing copper concentrations of 10 to 100 μg/l.

Betzer and Kott (1969) found that 5.0 mg $CuSO_4$/l did not result in any morphological change in *Cladophora* sp. after 24-hr exposure. If, however, 20.0 mg $CuSO_4$/l was added to the culture medium, the cell contents leaked through the cell wall at 3- and 24-hr exposure. Severe leakage and contraction of the cytoplasm occurred when the alga was exposed to 50.0 mg $CuSO_4$/l. At 100.0-mg/l exposure there was a definite change in cell morphology after 3 hr. These effects were reversible, depending on exposure time and concentration of copper used. At 5 mg $CuSO_4$/l or less, an exposure time of up to 4 days could be used with subsequent regrowth of the alga. At higher concentrations, exposure times of as little as 0.5 hr were sufficient to prevent regrowth.

The effect of copper on the morphology of *Thalassiosira pseudonana* was examined by Erickson (1972). With a range of copper concentrations of 5 to 30 μg/l mean cell volumes increased with increasing copper concentrations. In work with *Chlorella vulgaris,* Foster (1977) found that increasing copper concentrations resulted in the production of giant cells, and that cell separation, and hence reproduction, were severely impaired.

Steeman Nielsen and Kamp-Nielsen (1970) found, in working with *Chlorella pyrenoidosa,* that cell division was also affected by copper. If the cells initially divided after exposure to copper, they would continue to divide; but if the initial division was prevented, subsequent divisions would not occur. This information was determined from specialized experiments involving light/dark cycles of 16 hr/8 hr. A synchronized culture was produced, and it was possible for the authors to find the stage of cell growth at which cell division was inhibited by copper. In related work on *C. pyrenoidosa* Steeman Nielsen et al. (1969) found that copper is bound to the cytoplasmic membrane of the cell, and it is this that prevents cell division.

2.5. Factors Affecting Copper Toxicity

Considerable work has been carried out on the factors that control copper toxicity to algae. It is the general conclusion of the work reviewed that chelation is the single most important factor in reducing copper toxicity.

Fogg and Westlake (1955) found that the addition of ethylenediaminetetraacetic acid (EDTA) reduced copper toxicity to *Anabaena cylindrica,* and that the addition of algal extracellular products had a similar effect in mitigating toxicity.

Gächter et al. (1973), also in work with *A. cylindrica,* deduced that ionic copper was the toxic form (at 6.4 ng/l). They found that nitrilotriacetic acid (NTA), EDTA, extracellular polypeptides, a protein digest of *A. cylindrica,* and an extract of zooplankton all reduced the toxicity of copper. Erickson et al. (1970) reported that NTA significantly reduced the effects of copper in laboratory studies of marine algae. In further investigations Erickson (1972) found that copper toxicity to *Thalassiosira pseudonana* was dependent upon the organic content of the culture medium. The higher the content of detritus and dead cells in the culture, the greater was the reduction in copper toxicity.

Much work has been carried out in the field on the assessment of natural regulators of copper toxicity. Marvin et al. (1961) attempted to control a bloom of *Gymnodinium breve* in a lagoon by means of copper ore additions, but did not succeed. They concluded that the seawater had a very high content of some chelator that was effectively binding the copper. In a survey of 169 British rivers McLean (1974) noted that *Stigeoclonium tenue* tolerated high copper levels in rivers with high contents of organic matter. Ramamoorthy and Kushner (1975) indicated, in their work with heavy metal binding in the Ottawa river, that 95% of such binding was associated with large molecular weight (>45,000) substances. They concluded that humic substances were acting as the site of metal binding. Horne and Goldman's (1974) work on the suppression of nitrogen fixation by additions of copper to Clear Lake, California, concluded that indigenous copper levels could not be shown to be having any effect. Horne and Goldman postulated that natural chelators in the lake were binding the copper and rendering it nontoxic.

Morris and Russell (1973) stated that copper tolerance in a marine environment might be very rare because of the presence of chelators in seawater. Davey et al. (1973) indicated that, in the absence of all chelators, copper levels of 1 μg/l caused inhibition of growth. Further work in a series of freshwater lakes near Sudbury, Ontario, Canada, confirms this conclusion. Stokes and Hutchinson (1976) sampled a series of lakes around the mining and smelting operations at Sudbury and performed analyses for copper content, pH, and organic carbon content. Bioassays were also carried out to determine the effects of the various lake waters on algal growth. One conclusion drawn was that the complexing capacity of a lake had a marked effect on the toxicity of copper to algae, although other chemical variables sometimes complicated the interpretation of results.

The synergistic, additive, or antagonistic effects of combinations of metals have also been examined to some extent. Hutchinson (1973) found that copper and nickel act synergistically to reduce the growth of *Haematococcus capensis* by 82% at 50 μg total Cu + Ni/l. All combinations of copper and nickel concentrations tested showed synergistic effects. Young and Lisk (1972) investigated the combined effects of silver and copper on *Anacystis nidulans* and *Chlorella vulgaris* and obtained two different results. With *C. vulgaris* the metal combination proved synergistic at 200 to 300 μg/l total metal concentration, whereas with *A. nidulans* there was an antagonistic effect. Steeman Nielsen and Kamp-Nielsen (1970), in work on *Chlorella pyrenoidosa* and *Nitzschia palea,* found that Fe(III) counteracted the toxicity of copper, probably as a result of the metal being adsorbed onto particles of ferric hydroxide generated in the culture medium. Antagonistic effects of other metals were noted by Overnell (1976), who found that calcium, magnesium, and potassium all acted to mitigate copper toxicity to *Phaeodactylum tricornutum* in natural seawater. Braek et al. (1976) investigated the synergistic and antagonistic effects of copper and zinc on four species of marine algae; these elements acted antagonistically toward *P. tricornutum* and synergistically toward *Skeletonema costatum, Thalassiosira pseudonana,* and *Amphidinium carteri.* Copper alone was found to be more toxic to all species than zinc alone.

Other factors have been investigated with respect to their effects on copper toxicity to algae. The pH of the algal test medium (either artificial or natural) is critical in influencing copper toxicity. Steeman Nielsen et al. (1969) and Steeman Nielsen and Kamp-Nielsen (1970) found that low pH levels depressed copper toxicity. Hargreaves and Whitton (1976) obtained similar results in studies on *Hornidium rivulare.* Copper toxicity was seen to be greatly enhanced at pH 6.0, whereas copper was only slightly toxic at pH 2.8 and 3.5.

The density of the cell suspension being tested, and presumably the density of a natural population, also affect toxicity. Gibson (1972) and Steeman Nielsen and Kamp-Nielsen (1970) indicate that dense populations of algae are more resistant to copper than less dense ones, and that in terms of copper strategies it is more effective to treat a less dense population. Several mechanisms may be at work in this situation. It is possible that the copper is being bound to algal cell walls, being incorporated into the cells, or being chelated by algal extracellular products or breakdown products.

One additional factor has been found to influence copper toxicity, and that is the O_2 concentration in water. Hassall (1962) reported that copper toxicity to *Chlorella vulgaris* was reduced if the O_2 content of the test medium was near saturation.

2.6. Phytoplankton as Indicators of Copper Pollution

Several authors, in studying natural algal populations and their tolerances of copper, have attempted to identify certain "indicator" species, that is, those presumed to be indicative of copper pollution in rivers and lakes. Whitton (1971) concluded that abundant growths of *Cladophora* sp. in fresh water indicated no long-term copper pollution, whereas growths of *Stigeoclonium tenue* could be indicative of copper pollution. Aside from these two species, Whitton could not draw any generalizations about algal indicator species. In other work, Whitton (1970) concluded that *Cladophora* was too sensitive a species to be useful as an indicator of metal pollution. *Stigeoclonium tenue,* however, proved to be a valuable indicator species.

These conclusions about the suitability of *Cladophora* as a monitor are in contrast with those of Keeney et al. (1976), who found that *C. glomerata* could act as a biological monitor. What was monitored in this case was heavy metal accumulation, rather than toxic effects and the presence or absence of correlation with copper levels discussed by Whitton (1970, 1971).

Bryan and Hummerstone (1973) concluded, in their work with the brown seaweed *Fucus vesiculosis,* that this organism could serve as a useful biological monitor of heavy metal pollution. Analysis of the weed gave a reasonable indication of the average conditions to be found in water at various points along the estuaries studied. A method was then available to compare heavy metal inputs (including copper) from year to year, or from estuary to estuary. These workers also noted that the alga was unlikely to show a response to short-term variations in heavy metal concentrations.

2.7. Summary and Research Needs

The work reviewed here indicates that copper is toxic to algae in varying degrees, and that the toxic response is highly variable, depending on the organism. Many factors mitigate copper toxicity, and further research is essential on the active species of copper that causes toxicity. The usefulness of biological monitors has been generally restricted to fish or invertebrates, and work is needed on suitable algal indicators of copper pollution.

3. COPPER TOXICITY TO AQUATIC INVERTEBRATES

3.1. Acute Toxicity

Some concentrations of copper found to be acutely toxic to invertebrates are listed in Tables 1 and 2 and shown diagramatically in Figure 1. Sensitivity to copper varies greatly from species to species. For example, 48-hr LC_{50} values range from 5 to over 100,000 μg Cu/l (Tables 1 and 2). *Daphnia* appears to be the most sensitive organism and LC_{50}'s for this genus are consistently below 100 μg/l. Euphausids, ctenophores, and medusae are also very sensitive, although these organisms have rarely been tested. The sensitivity of *Daphnia* is probably at least partially due to the fact that standard tests with this genus usually employ 12 ± 12 hr old young (e.g., Biesinger and Christensen, 1972; Winner and Farrell, 1976). Anderson (1948) stressed the need for using immature individuals of arthropods in toxicity studies instead of, for example, adult copepods, which no longer molt after reaching maturity. Nevertheless, adult *Daphnia hyalella* were found to be much more sensitive than copepods in equivalent tests (Baudouin and Scoppa, 1974). In general, larvae or younger stages are more sensitive to copper than are adults (see Birge and Black, this volume). This is seen in crayfish (Hubschman, 1967a), several marine crustaceans (Conner, 1972), *Artemia* (Brown and Ahsanullah, 1971), and polychaetes (Reish et al., 1976). Similarly,

Table 1. Acute Toxicity of Copper to Freshwater Invertebrates

Species	Toxicity Unit	Copper ($\mu g/l$)	Comments	Reference
Rotifers				
Philodina acuticornis	96-hr LC_{50}	700	Soft water	Buikema et al. (1974)
	96-hr LC_{50}	1100	Hard water	Buikema et al. (1974)
Ectoprocts				
Plumatella casmiana	96-hr LC_{50}	500–1000		Bushnell (1974)
	Threshold LC_{50}	<500		Bushnell (1974)
Arthropods—crustaceans				
Daphnia magna (water flea)	48-hr LC_{50}	9.8	No food	Biesinger and Christensen (1972)
	48-hr LC_{50}	60	With food	Biesinger and Christensen (1972)
	64-hr LC_{50}	13		Anderson (1948)
	72-hr LC_{50}	86	With food	Winner and Farrell (1976)
D. hyalella	48-hr LC_{50}	5	No food	Baudouin and Scoppa (1974)
D. ambigua	72-hr LC_{50}	68	With food	Winner and Farrell (1976)
D. parvula	72-hr LC_{50}	72	With food	Winner and Farrell (1976)
D. pulex	72-hr LC_{50}	86	With food	Winner and Farrell (1976)

Eudiaptomus padanus (copepod)	48-hr LC_{50}	500		Baudouin and Scoppa (1974)
Cyclops abyssorum (copepod)	48-hr LC_{50}	2,500		Baudouin and Scoppa (1974)
Gammarus pseudolimnaeus	96-hr LC_{50}	20	No food	Arthur and Leonard (1970)
G. pulex (amphipod)		250–500	Mortality threshold 1 week	Liepolt and Weber (1958)
Gammarus sp.	96-hr LC_{50}	910		Rehwoldt et al. (1973)
Orconectes rusticus (crayfish)	Toxicity threshold	60–125	Newly hatched young	Hubschman (1967a)
		125	Increase in egg mortality	Hubschman (1967a)
	Toxicity threshold	500–1,000	Adults	Hubschman (1967a)
Ascellus meridianus (isopod)	48-hr LC_{50}	1,200–2,500		Brown (1976)
Arthropods—insects				
Chironomous sp. (midge)	96-hr LC_{50}	30		Rehwoldt et al. (1973)
C. thummi		500	Mortality threshold 1 week	Liepolt and Weber (1958)

Table 1. Continued

Species	Toxicity Unit	Copper (μg/l)	Comments	Reference
Ephmerella grandis (mayfly)	14-day LC_{50}	180–200		Nehring (1976)
E. subvaria (mayfly)	48-hr LC_{50}	320		Warnick and Bell (1969)
Heptagenia lateralis (mayfly)		500	Mortality threshold 1 week	Liepolt and Weber (1958)
Damsel fly	96-hr LC_{50}	4,600		Rehwoldt et al. (1973)
Caddis fly	96-hr LC_{50}	6,200		Rehwoldt et al. (1973)
Hydropsyche betteni (Caddis fly)	96-hr LC_{50}	>64,000		Warnick and Bell (1969)
	14-day LC_{50}	32,000		Warnick and Bell (1969)
Acroneuria lycorias (stone fly)	96 h LC_{50}	8,300		Warnick and Bell (1969)
Pteronarcys californica (stone fly)	14-day LC_{50}	10,000–13,900		Nehring (1976)

Annelids				
Nais sp.	96-hr LC_{50}	90		Rehwoldt et al. (1973)
Tubifex rivulorum		250–500	Mortality threshold 1 week	Liepolt and Weber (1958)
T. tubifex	48-hr LC_{50}	6.4	Distilled water	Brković-Popović and Popović (1977)
	48-hr LC_{50}	210	34 mg/L hardness	Brković-Popović and Popović (1977)
	48-hr LC_{50}	890	261 mg/l hardness	Brković-Popović and Popović (1977)
Molluscs				
Physa integra (snail)	96-hr LC_{50}	39		Arthur and Leonard (1970)
Campeloma decisum (snail)	96-hr LC_{50}	1700		Arthur and Leonard (1970)
Amnicola sp. (snail)	96-hr LC_{50}	900	Adults	Rehwoldt et al. (1973)
	96-hr LC_{50}	9,300	Eggs	Rehwoldt et al. (1973)

Table 2. Acute Toxicity of Copper to Marine Invertebrates

Species	Toxicity Unit	Copper ($\mu g/l$)	Comments	Reference
Rotifers				
Branchionus plicatilis	24-hr LC_{50}	100		Reeve et al. (1976)
Ectoprocts				
Watersipora cucullata	2-hr LC_{50}	570	Larvae	Wisely and Blick (1967)
Bugula neritina	2-hr LC_{50}	3,800	Larvae	Wisely and Blick (1967)
Ctenophores				
Pleurobranchia pilens	24-hr LC_{50}	33		Reeve et al. (1976)
Mnemiopsis mecradyi	24-hr LC_{50}	17–29		Reeve et al. (1976)
Medusa				
Phialidium sp.	24-hr LC_{50}	36		Reeve et al. (1976)
Chaetognaths				
Sagitta hispida	24-hr LC_{50}	43–460		Reeve et al. (1976)
Crustaceans				
Acartia clausi (copepod)	Toxicity threshold	≈300		Corner and Sparrow (1956)
Nitocra spinipes	24-hr LC_{50}	>26,000		Barnes and Stanbury (1948)

(copepod)				
Calanoid copepods (several spp.)	24-hr LC_{50}	105–2778		Reeve et al. (1976)
Euphausia pacifica	24-hr LC_{50}	14–30		Reeve et al. (1976)
Balanus eburneus (barnacle)	29-hr LC_{50}	60	Nauplius larvae	Clarke (1947)
B. balanoids and *B. crenatus*	6-hr LC_{50}	270–460	Nauplius larvae	Pyefinch and Mott (1948)
Elminius modestus (barnacle)	Toxicity threshold	≈10,000	Larvae	Corner and Sparrow (1956)
Crangon crangon (shrimp)	48-hr LC_{50}	330	Larvae	Conner (1972)
	48-hr LC_{50}	29,500	Adults	Conner (1972)
Carcinus maenus (crab)	48-hr LC_{50}	600	Larvae	Conner (1972)
	48-hr LC_{50}	109,000	Adults	Conner (1972)
Homarus gammarus (lobster)	48-hr LC_{50}	100–330	Larvae	Conner (1972)
H. americanus	48-hr LC_{50}	560	13°C (adults)	McLeese (1974)
	48-hr LC_{50}	1000	5°C	McLeese (1974)
	Lethal threshold	56	13°C and 5°C	McLeese (1974)

Table 2. Continued

Species	Toxicity Unit	Copper ($\mu g/l$)	Comments	Reference
Leander squilla (prawns)	Toxicity threshold	<500		Raymont and Shields (1964)
Carcinus sp.	Toxicity threshold	1,000–2,000		Raymont and Shields (1964)
Artemia salina (brine shrimp)		1,000	LT_{50} = 110 hr	Brown and Ahsanullah (1971)
		1,000	LT_{50} = 168 hr (adults)	Brown and Ahsanullah (1971)
	24-hr LC_{50}	2,050–2,554		Reeve et al. (1976)
Annelids				
Spirorbis lamellosa	2-hr LC_{50}	510	Larvae	Wiseley and Blick (1967)
Galeolaria caespitosa	2-hr LC_{50}	2,900	Larvae	Wiseley and Blick (1967)
Phyllodoce maculata	Toxicity threshold	70–80		McLusky and Phillips (1975)
Nereis virens	Toxicity threshold	100		Raymont and Shields (1964)
N. diversicolor	96-hr LC_{50}	200	5% Salinity	Jones et al. (1976)
	96-hr LC_{50}	370–480	10–34% Salinity	Jones et al. (1976)
Capitella	96-hr LC_{50}	180	Trochophore	Reish et al. (1976)

capitata			larvae	
	96-hr LC_{50}	200	Adult	Reish et al. (1976)
Neanthes arenaceo-dentata	96-hr LC_{50}	300	Juvenile and adult	Reish et al. (1976)
	28-day LC_{50}	140	Juvenile	Reish et al. (1976)
	28-day LC_{50}	250	Adult	Reish et al. (1976)
Ctenodrilus serratus	96-hr LC_{50}	250–500		Reish and Carr (1978)
Ophryotrocha diadoma	96-hr LC_{50}	100–250		Reish and Carr (1978)
O. labronica		1,000	4.5-hr LT_{50}	Brown and Ahsanullah (1971)
Molluscs				
Mytilus edulis (mussel)	2-hr LC_{50}	22,000	Larvae	Wiseley and Blick (1967)
	Lethal threshold	200–300		Delhaye and Cornet (1975)
	Lethal threshold	100–200		Scott and Major (1972)
Several mollusc spp.	Lethal threshold	100–200		Marks (1938)
Crassostrea gigas (oyster)	96-hr LC_{50}	560		Okazaki (1976)
	14-day LC_{50}	>100		Okazaki (1976)
Japanese oyster	96-hr LC_{50}	1,900		Fujiya (1960)

Table 2. Continued

Species	Toxicity Unit	Copper ($\mu g/l$)	Comments	Reference
Rangia cupeata (clam)	96-hr LC_{50}	210	<1% salinity	Olson and Harrel (1973)
	96-hr LC_{50}	8,000	5.5% salinity	Olson and Harrel (1973)
	96-hr LC_{50}	7,000	22% salinity	Olson and Harrel (1973)
Tellina tenuis (bivalve)	96-hr LC_{50}	1,000		Stirling (1975)
Busycon canaliculatum	Toxicity threshold	200–500		Betzer and Yevich (1975)

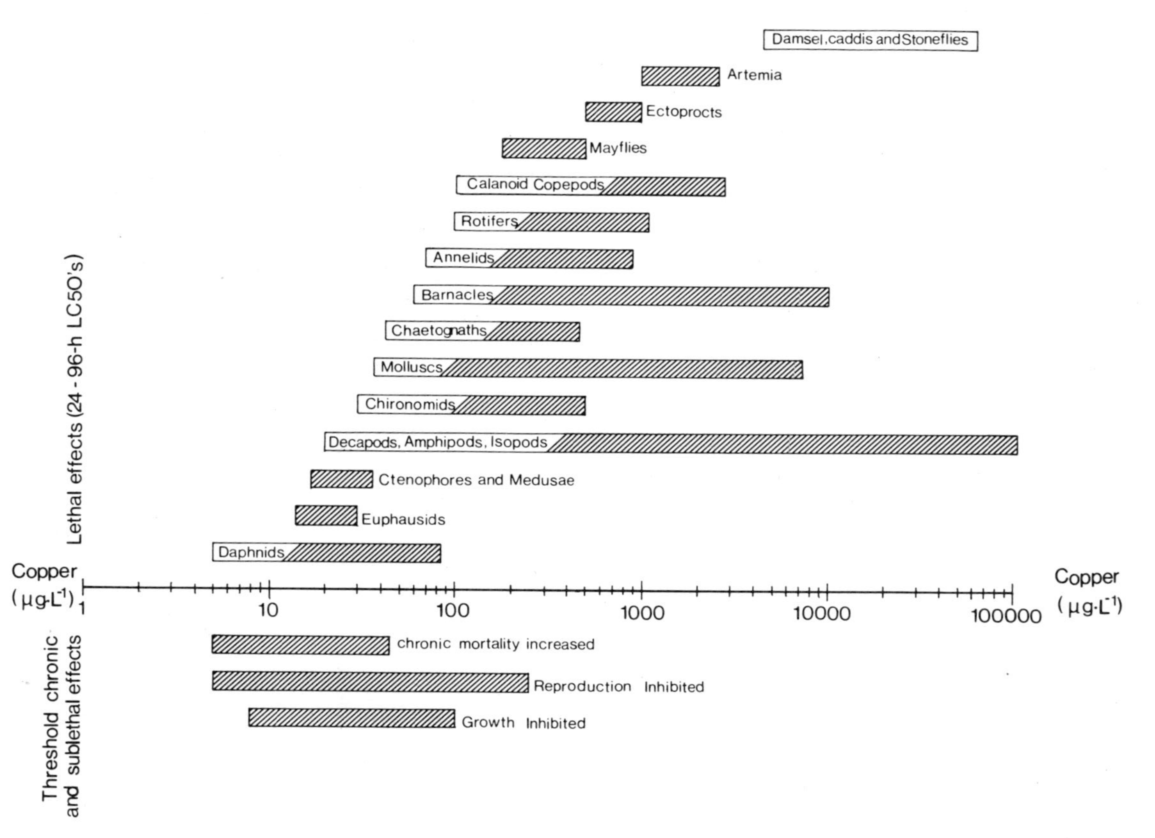

Figure 1. Summary diagram of the lethal, chronic, and sublethal effects of copper on aquatic invertebrates (tests done in distilled water are omitted). The chronic and sublethal values represent the lower concentrations observed to have a particular effect on any group (e.g., copepods). More severe effects will obviously occur at higher concentratons. Data are derived from Tables 1 to 3 and the text.

the nauplius stage of barnacles is more sensitive than the later cyprid stage (Pyefinch and Mott, 1948). Eggs, however, may be more resistant than adults or young. Crayfish eggs exposed to 125 μg Cu/l for 2 weeks suffered 10% mortality, whereas recently hatched young suffered over 50% mortality in the same time period (Hubschman, 1967a). Also, snail (*Amnicola*) eggs are more resistant than adults (Rehwoldt et al., 1973).

Some of the variation in toxicity between species is related to the nature of the body covering. For example, in 2-hr tests, young bivalves, which immediately close their shells when placed in high copper concentrations, are more resistant than the less well protected bryozoans and tubeworms. In fact, young oysters are resistant to all concentrations of copper for 2 hr unless the copper concentration is sufficiently high to reduce the pH to the point where the shell dissolves (Wisely and Blick, 1967). Similarly, Scott and Major (1972) suggested that the length of time of survival of mussels is dependent on their ability to keep their shells closed. Among freshwater snails, *Physa integra,* which does not have an operculum, is far more sensitive than *Campeloma decisum,* which keeps its operculum closed during exposure to high copper concentrations (Arthur and Leonard, 1970). Such high resistance to copper would, of course, not be expected in long-term sublethal tests, since molluscs cannot remain closed indefinitely (see Cheng, this volume). Like the mollusc shell, the exoskeleton of arthropods probably provides some protection against acute copper toxicity. In tests lasting longer than a few hours, several marine crustaceans can tolerate several thousand micrograms of copper per liter, whereas the soft-bodied marine polychaetes cannot survive at 1000 μg Cu/l (see Table 2 and Figure 1). Similarly, several insect species are quite resistant to high copper concentrations (Table 1 and Figure 1).

Reeve et al. (1976) observed that 24-hr LC_{50} values of copper and mercury were greater for larger zooplankton within the same species (*Euphausia pacifica* being the only exception), and also often between species. This is consistent with the observations mentioned above, that the young are usually more sensitive than adults. Since, as already noted, it appears that acute toxicity of copper is related to the nature of the body covering, the higher sensitivity of the smaller or younger individuals may be related to a higher surface:volume ratio.

Within-species variations in acute copper toxicity may be due to several factors, one of which is hardness. Rotifers are more resistant to copper in hard than in soft water (Buikema et al., 1974). The LC_{50} for *Daphnia magna* reported by Winner and Farrell (1976), using water of hardness 130 to 160 mg/l (as $CaCO_3$), is higher than that found by Biesinger and Christensen (1972) in water of hardness 45 mg/l, even though the latter workers' exposure time was longer. Andrew et al. (1977) obtained lower mortality rates for *Daphnia* upon addition of bicarbonate, and on the basis of various copper ion equilibria concluded that the toxicity of copper was proportional to cupric and copper hydroxyl ion concentrations and inversely proportional to soluble copper carbonate complexes.

A strong relationship between toxicity and hardness also exists for *Tubifex* (Brković-Popović and Popović, 1977a; Table 1 and Figure 2). (Also see Section 4 on copper toxicity to fish.)

Although temperature affected the rate of mortality during exposure to copper, it did not influence the final lethal threshold in lobsters. Furthermore, copper toxicity to lobsters was not affected by salinity changes from 20 to 30 parts per thousand (ppt) (McLeese, 1974). Copper toxicity to clams was similar at 5.5 and 22 ppt, but toxicity increased if clams were exposed in fresh water (Olson and Harrel, 1973). Toxicity to polychaetes was not appreciably affected by salinities between 10 and 34 ppt (although slightly higher in worms from a low salinity area exposed at 34 ppt), but was higher at 5 ppt (Jones et al., 1976). *Marinogammarus* exposed to 5000 μg Cu/l survived over 100 hr if salinity was greater than 70% seawater, but toxicity increased rapidly at lower salinities, with a mean activity time of only 3 hr in distilled water (Russell-Hunter, 1949). Hence for several marine species salinity does not appear to have a marked effect on copper toxicity unless the salt concentration drops below a critical level (dependent on the species), with toxicity being greater at lower salinities.

Caution is needed in interpreting the data in Tables 1 and 2, since some authors have observed delayed mortality after animals were placed in clean water following the exposure period. Although the 96-hr LC_{50} for crayfish is 3000 μg/l, delayed mortality resulted in large numbers of deaths after 2 to 3 weeks among crayfish exposed to 2500 μg Cu/l for 24 hr (Hubschman, 1967a). Similarly, Stephenson and Taylor (1975) observed continuing mortality in clams for 16 days after exposure to copper was terminated.

3.2. Chronic Toxicity and Sublethal Effects

Some concentrations of copper found to produce chronic toxicity or sublethal effects are shown in Table 3 and Figure 1. Since copper is a component of various biological systems, such as the blood respiratory pigment hemocyanin, found in molluscs and decapods, and cytochrome oxidase, found in most animals, it is a nutrient and is beneficial to organisms in small concentrations. Wilson and Armstrong (1961) observed that addition of sufficient copper to seawater to raise the concentration from 0.6 to 1.5 μg/l improved larval development of sea urchins, but copper had adverse effects above about 20 μg/l. Similarly, Bougis (1959) observed great sensitivity to copper in developing sea urchin larvae. Whereas 24 μg/l inhibited larval settlement of barnacles (Pyefinch and Mott, 1948), copper concentrations of 50 to 600 μg/l stimulated settlement of oyster larvae (Prytherch, 1934). Hence copper may be either beneficial or harmful at concentrations found naturally in seawater, and has been implicated as an important component in the ecology of marine invertebrates (e.g., Bougis, 1962). All stimulatory effects of copper, however, cannot be regarded as due to a nu-

Table 3. Chronic Toxicity or Sublethal Effects of Copper to Invertebrates

Environment and Species	Maximum Concentration, No Effect (μg/l)	Minimum Concentration + Effect (μg/l)	Effect	Reference
Freshwater—crustaceans				
Daphnia magna		22	16% Reproduction impairment	Biesinger and Christensen (1972)
		44	3-week LC_{50}	Biesinger and Christensen (1972)
D. ambigua	40	60	Significant drop in instantaneous rate of population growth	Winner and Farrell (1976)
D. parvula	40	60	Significant drop in instantaneous rate of population growth	Winner and Farrell (1976)
D. pulex	40	60	Significant drop in instantaneous rate of population growth	Winner and Farrell (1976)
D. magna	60	80	Significant drop in instantaneous rate of population growth	Winner and Farrell (1976)
Gammarus pseudolimnaeus (amphipod)	4.6	8	Second-generation growth affected	Arthur and Leonard (1970)
	8	14.8	Growth and survival affected	Arthur and Leonard (1970)

Orconectes rusticus (crayfish)		15	Growth retarded	Hubschman (1976a)
Ascellus meridianus (isopod)		100	Growth reduced except in animals from highly polluted regions	Brown (1976)
Freshwater—molluscs				
Physa integra (snail)	8	14.8	Growth and survival affected	Arthur and Leonard (1970)
Marine—hydroids				
Eirene viridula	30	60	Colony growth affected	Karbe (1972)
Campanularia flexuosa	10	13	Colony growth affected	Stebbing (1976)
Marine—ectoprocts				
Bugula neritina		50	Growth retarded	Miller (1946)
Marine—echinoderms				
Arbacia lixula		60	Plutens larvae development affected	Bernhard (1955) (from Bernhard and Zattera, 1973)
Paracentrotus lividus		≈10	Development of pluteus affected	Bougis (1959)
		50	No growth	Bougis (1959)

Table 3. Continued

Environment and Species	Maximum Concentration, No Effect ($\mu g/l$)	Minimum Concentration + Effect ($\mu g/l$)	Effect	Reference
Echinus esculentus		1.5	Improved development over 0.6 $\mu g/l$	Wilson and Armstrong (1961)
	<10	20	Adverse effect on development	Wilson and Armstrong (1961)
Marine—crustaceans				
Euchaeta japonica (copepod)		≈6	Increased mortality of prefeeding stages	Lewis et al. (1972, 1973)
Tigriopus japonicus (copepod)	6.4	64	Egg production inhibited	D'Agostino and Finney (1974)
Calanus, Metridia (copepods)		5	Decreased feeding	Reeve et al. (1976)
Acartia (copepod)		20	Decreased egg production	Reeve et al. (1976)
Pseudocalanus, Calanus, Euphausia (copepods, euphausid)		5	Decreased feeding and/or egg production and increased mortality	Reeve et al. (1977)
Balanus balanoides (barnacle)		24	Larval settlement inhibited	Pyefinch and Mott (1948)

Artemia salina	25	50	Significant decrease in growth and no reproduction	Saliba and Ahsannullah (1973)
Marine—annelids				
Capitella capitata		10	Abnormal larvae	Reish et al. (1974)
Ophryotrocha labronica		25	Significant decrease in growth	Saliba and Ahsannullah (1973)
		100	No reproduction	Saliba and Ahsannullah (1973)
O. diadema	50	250	Reproduction reduced	Reish and Carr (1978)
Ctenodrilus serratus	50	100	Reproduction reduced	Reish and Carr (1978)
Marine—molluscs				
Ostrea virginica (oyster)		50–600	Stimulates setting of larvae	Prytherch (1934)
Crassostrea virginica (oyster)	10	32.8	Decrease in growth of larvae	Calabrese et al. (1977)
		19–43	Increased mortality in adults and juveniles and inhibition of spawning	Mandelli (1975)
Mercenaria mercenaria (clam)	4.9	16.4	Decrease in growth of larvae	Calabrese et al. (1977)
Tellina tenuis (bivalve)	10	30	Decrease in siphon dry weight	Saward et al. (1975)

tritive effect. For example, low copper concentrations (≈1 μg/l) stimulated growth of hydroids, but this stimulation was only transitory and might have been a normal response to stress (Stebbing, 1976).

The total range of copper concentrations found to have toxic effects on growth and development (≈5 to 100 μg/l; Table 3) is far less than the range of values found to be acutely toxic (Tables 1 and 2). For example, although acute copper toxicity is far greater to *Daphnia* than to *Artemia* (Tables 1 and 2 and Figure 1), inhibition of growth occurs at similar concentrations in both organisms (Table 3). Hence there is no good correlation between acute and chronic toxicity, and one cannot be used to estimate the other. This suggests that the mode of action of copper may be quite different in the two types of test.

The toxicity of copper is affected by complexation and chelation. As previously mentioned, hardness affects acute toxicity to *Daphnia,* presumably since copper carbonate complexes are less toxic than cupric or copper hydroxyl ions (Andrew et al., 1977). Chronic toxicity may be similarly affected, since growth effects on *Daphnia magna* are greater at a hardness of 45 mg/l (Biesinger and Christensen, 1972) than at 130 to 160 mg/l (Winner and Farrell, 1976). It is interesting to note, however, that chronic toxicity in seawater appears to be roughly similar to that in fresh water, in spite of the much higher salt concentrations in the former (Table 2). An important factor affecting copper toxicity is the degree of complexation with organic substances. The complexing agent EDTA can reverse the toxic effects of added copper on copepods (Lewis et al., 1972) and clams (Stephensen and Taylor, 1975). Conversely, EDTA can be detrimental to developing sea urchin larvae (Wilson and Armstrong, 1961) and sea urchin spermatozoa (Young and Nelson, 1974), possibly because of the removal of copper or other trace metals required in minute amounts. Another chelator, NTA, can greatly reduce mortality and reproductive impairment in *D. magna* exposed to copper (Biesinger et al., 1974). Naturally occurring substances, such as those extracted from marine sediments, may also reduce copper toxicity (Lewis et al., 1973). The difference in acute toxicity to *Daphnia* with and without food, and the higher acute toxicity of copper in the absence of food, as compared to chronic toxicity (when food must be added) (Tables 1, 2 and 3; Biesinger and Christensen, 1972), are probably due to the complexation of copper with the food, since these filter-feeding animals require food in the form of a fine suspension. It is not surprising, therefore, that toxicity is reported to be greater in organisms such as amphipods, snails, and crayfish, which are not filter feeders and which can be maintained in continuous-flow systems (Table 3; Arthur and Leonard, 1970; Hubschman, 1967a).

Behavior

In addition to its effects on survival, growth, and reproduction, copper is also known to influence invertebrate behavior. As already mentioned, when exposed to high copper concentrations, oysters and mussels close their valves (Wisely and Blick, 1967) and the snail *Campeloma* closes its operculum (Arthur and Leonard, 1970). Copper may also act as a repellent to oyster and mussel larvae (Wisely and Blick, 1967). In addition, the production of byssal threads (used for attachment) by mussels is reduced by copper at 250 μg/l (Martin et al., 1975). The Asiatic freshwater clam does not actively pump at above 500 μg Cu/l, and some clams can detect copper as low as 12 μg/l (Martin, 1971). Stephensen and Taylor (1975) observed that the burrowing activity of clams (*Venerupis decussata*) was impaired at 10 μg Cu/l, but addition of 1 mg EDTA/l abolished this effect, even at 100 μg Cu/l. Hence copper chelation is important in its effect on behavior, as well as on survival. The bivalve *Tellina tenuis* also displays impaired burrowing activity in the presence of copper (Stirling, 1975).

Harry and Aldrich (1963) observed that the snail *Taphius glabratus* is retracted within its shell at high concentrations of metal ions, but displays a distress syndrome at lower values (e.g., 50 to 100 μg Cu/l), during which the body is extended but the snail is unable to attach its foot, and muscular activity is impaired. MacInnes and Thurberg (1973) observed the same distressed reaction in the mud snail at 100 μg Cu/l. Polychaete worms demonstrate distress by wriggling and mucus production at 120 μg Cu/l (McLusky and Phillips, 1975).

The attraction to wheat germ is inhibited by copper in the snail *Australorbis glabratus,* which can detect copper salts below 1 μg/l (Etges, 1963). Similarly, lobsters demonstrate less attraction to herring muscle extract after exposure to 40 μg Cu/l, possibly because of impairment of their chemosensory ability (McLeese, 1975).

An interesting response in the behavior of sea urchin spermatozoa is their intense activity, accompanied by increased oxygen consumption, immediately upon dilution in seawater. This response may be initiated by trace metals (including copper) in natural seawater, since small amounts of copper stimulate motility and respiration whereas EDTA inhibits motility. Higher copper concentrations, however, are inhibitory (Rothschild and Tuft, 1950; Young and Nelson, 1974).

Histopathology

Copper can produce deleterious effects on invertebrate tissues. At high concentrations it can result in the deterioration and fragmentation of entire polychaetes (McLusky and Phillips, 1975; Saliba and Ahsanullah, 1973). Similarly, 3000 μg Cu/l causes tissue disintegration in a few hours in hydroids, with morphological changes occurring at concentrations as low as 60 μg/l (Karbe, 1972). Exposure of channeled whelk to 1000 μg Cu/l resulted in swelling of the leaflets of the gill, followed by necrosis and sloughing of the epithelial cells. Microscopic changes in other tissues were not observed (Betzer and Yevich, 1975). Histopathological changes in the digestive tubules, gills, and mantle epithelium of the Asiatic freshwater clam were caused by copper in amounts as low as 12 to 50 μg/l (Martin, 1971). Copper also induced damage to the digestive diverticula and stomach epithelia in oysters (Fujiya, 1960). However, Hubschman (1967b) observed no evidence of tissue damage in crayfish at high copper concentrations (up to 10,000 μg/l), although degeneration of the cells of the antennal gland was evident after exposure to 1000 or 500 μg Cu/l for 30 days.

Oxygen Consumption

Respiration has been observed to decrease in many invertebrates exposed to copper. Copper lowered the oxygen consumption of tubificids (Brković-Popović and Popović, 1977b), amphipods (Russell-Hunter, 1949), mud snails (MacInnes and Thurberg, 1973), and mussels (Delhaye and Cornet, 1975; Scott and Major, 1972). The heart rate of mussels also drops after exposure to copper (Scott and Major, 1972). Copper inhibited respiration but not motility in *Artemia* (Corner and Sparrow, 1956). Jones (1942) observed that copper inhibited oxygen consumption in *Gammarus* and *Polycelis* but suggested that this effect was a symptom of the toxic process and not the cause of death. Brown and Newell (1972) reported that copper inhibited oxygen consumption of whole *Mytilus* and gill tissue, but not digestive gland tissue or tissue homogenates. They suggested that the drop in metabolism was due to inhibition of an energy-consuming process such as ciliary activity, and not a direct effect on respiratory enzymes. Copper inhibited succinate-dependent oxygen consumption of crayfish hepatopancreas *in vitro* and also after *in vivo* exposure of crayfish to 5000 μg/l. However, oxygen consumption of hepatopancreas was not affected by exposing crayfish to 1000 μg Cu/l or less (Hubschman, 1967b).

3.3. Acclimation to Copper

Few attempts have been made to test the ability of invertebrates to adapt to chronic copper exposure. After exposure to 25 to 100 μg Cu/l for 2 weeks, *Artemia* larvae survived longer at 1000 μg/l than did control larvae. However, acclimation did not significantly affect survival time at 10,000 to 100,000 μg/l. After exposure to 100 μg Cu/l for 3 weeks, adults also survived longer at 1000 μg/l than did controls, although again no difference in survival time was observed at 10,000 to 100,000 μg Cu/l. Second- and third-generation *Artemia* raised in copper appear to have lost some of this increased resistance to the element (Saliba and Ahsanullah, 1973). *Ophryotrocha labronica* does not appear to be able to acquire increased tolerance to copper after exposure to sublethal levels (Saliba and Ahsanullah, 1973). Brown (1976) observed that isopods (*Asellus meridianus*) collected from an area with high copper in sediments and animals had higher 48-hr LC_{50} values and grew better in water containing 100 μg Cu/l than did animals collected from areas with lower copper contamination. Similarly,

Bryan and Hummerstone (1971) observed that polychaetes (*Nereis diversicolor*) collected from an area with copper-rich sediment were more resistant to copper than those collected from an area with relatively little copper in the sediment. However, exposing worms from one area for 76 days in the laboratory in sediment collected from the other site resulted in very little change in copper tolerance, suggesting that such tolerance may be genetically controlled and not affected by acclimation.

3.4. Toxicity of Metal Mixtures

Most studies on the toxicity of copper to invertebrates in the presence of additional metals suggest synergistic effects. For example, neither 4.4 μg Cu/l nor 6.4 μg Cu/l had any effect on production of a second generation of the harpacticoid copepod *Tigriopus japonicus,* but if both metals were present together, no eggs were produced (D'Agostino and Finney, 1974). In another harpacticoid, *Nitrocera spinipes,* 17% mortality occurred in 24 hr in the presence of 400 μg Hg/l, but this was raised to 94% if 260 μg Cu/l was also present, even though this concentration of copper by itself caused only 11% mortality. Other copper and mercury concentrations also indicated synergistic action in this copepod (Barnes and Stanbury, 1948). Low concentrations of copper, which were virtually nontoxic by themselves, appreciably increased the toxicity of mercury to the amphipod *Marinogammarus marinus* (Russell-Hunter, 1949). Copper and mercury, if present together, also act synergistically on the survival of *Artemia* and *Acartia,* and pretreatment of *Artemia* with copper or mercury increased the subsequent toxicity of the other metal (Corner and Sparrow, 1956). Interaction between metals was also observed in the oxygen consumption of the mud snail, where 250 μg Cu/l depressed respiration and 1000 μg Cu/l increased respiration, yet the presence of both metals together decreased oxygen consumption to below the level observed in the presence of copper alone (MacInnes and Thurberg, 1973).

3.5. Mode of Action of Copper

Much more is known about the levels of copper that cause toxic effects in invertebrates than about the means by which this toxicity is produced, but some comments can be made. At moderate to high concentrations, damage to the gill tissues of molluscs has been mentioned (e.g., Betzer and Yevich, 1975; Martin, 1971). Similar effects have been observed with fish (see Section 4) and may be related to the decreased oxygen consumption observed in several invertebrates, suggesting a possible suffocation, at least in some organisms and at some concentrations. Direct inhibition of respiratory enzymes *in vivo* seems unlikely, at least in some species (Brown and Newell, 1972) and at lower concentrations (1000 μg Cu/l or less) (Hubschman, 1967b). The increased toxicity of copper at low salinity in marine organisms (Jones et al., 1976; Olson and Harrel, 1973; Russell-Hunter, 1949) suggests a possible effect of this metal on osmoregulation. In fact, the osmoregulatory ability of estuarine crabs (Thurberg et al., 1973) and isopods (Jones, 1975) is impaired by copper at low salinity. It is interesting to compare these data with the observation in crayfish that copper histologically damages the antennal gland, which is involved in osmotic and ionic regulation (Hubschman, 1967b).

In *Artemia* it has been suggested that copper acts synergistically with mercury by increasing the permeability of the organism to mercury. It was further suggested that the lower degree of synergism between copper and mercury in the copepod *Acartia* was consistent with the lower resistance of this animal to mercury (presumably due to its greater permeability to this element), since addition of copper could not potentially increase permeability as much in an organism that is already more permeable to mercury (Corner and Sparrow, 1956). The possibility of one metal increasing the permeability of another was also considered as one possible method of metal interaction in studies with copper and mercury on the copepod *Nitocera* (Barnes and Stanbury, 1948). It has already been mentioned, however, that, although *Artemia* is among the most resistant of invertebrates to acute copper poisoning (Table 2), its sensitivity to chronic copper exposure is roughly similar to that of other species (Table 3). Hence the mode of action in chronic tests may be different from that in acute tests. It is possible, for example, that permeability and surface effects may be relatively less important in long-term toxicity, when copper may have sufficient time to accumulate and produce toxic effects inside the organism.

The ultimate effects of copper and other metal ions must occur at the molecular level. Commonly, attempts have been made to explain the action of metals in relation to sulfhydryl (—SH) groups in enzymes and proteins. For example, the stimulatory effect of metals on the motility and respiration of sea urchin spermatozoa has been attributed to metal binding to essential regulatory sites (possibly involving soluble —SH groups), while inhibitory effects at higher concentrations may be due to metal binding to fixed —SH groups, resulting in inhibition of enzymatic activity (Young and Nelson, 1974). Biesinger and Christensen (1972) observed a significant correlation between the chronic toxicity of several metals to *Daphnia* and the solubility of the respective metal sulfides, and a model relating toxicity to metal sulfide solubility has been proposed by Shaw and Grushkin (1957). However, toxicity may also be related to other factors as well, such as the electronegativity or equilibrium constant for the metal-ATP complex (Biesinger and Christensen, 1972). It is interesting to note, however, that chronic toxicity of copper is observed at roughly similar concentrations for a wide variety of invertebrates (Table 3), suggesting the possibility of a similar mode of action among diverse groups of organisms.

3.6. Summary

Some copper concentrations toxic to invertebrates are shown in Figure 1. Only LC_{50} values for 24 hr or longer are included. When comparing lethal concentrations, it must be remembered that, whereas a 96-hr exposure for an adult lobster, for example, is definitely an acute test, a 96-hr exposure using invertebrates that have life cycles as short as a week or two (e.g., *Daphnia*) must almost be considered as a chronic test. Apart from species differences, the factor most important in affecting acute copper toxicity appears to be the binding capacity of the water; hence toxicity decreases as hardness or organic content increases. Chronic toxicity to copper varies much less from species to species than does acute toxicity. Adaptation to high copper concentrations in water or sediment appears to occur but is probably due to genetic selection of hardy individuals rather than acclimation.

In addition to effects on survival, growth, and reproduction, copper also can cause histopathologically observable tissue damage, decreased oxygen consumption, or distress behavior in invertebrates. The mode of action of copper on invertebrates is not well known, although impairment of osmotic and ionic regulation may be one possible cause of death. The mode of action in acute tests may be different from that in chronic studies, permeability to copper being more important during acute exposure.

4. COPPER TOXICITY TO FISH

4.1. Acute Lethal Toxicity

Many factors affect lethal copper toxicity to fish through changes in (*a*) the availability of copper to the fish (e.g., the degree of precipitation or complexation of copper); (*b*) the permeability of the fish to copper (the ability of the fish to remove copper from the water); and (*c*) the sensitivity of the fish to a given amount of copper taken up.

Environmental Factors

Many dissolved constituents of water control copper lethality by controlling the availability of copper to the fish. Increased hardness dramatically reduces copper lethality. Figures 2*a* and 2*b* (see also Table 4) illustrate that (*a*) the slope relating LC_{50}'s to hardness for the Centrarchidae and Cyprinidae approximates 1.0, while that for the Salmonidae is 0.7; (*b*) when allowance is made for hardness, species within the same genera have similar sensitivities to copper (with the exception of carp); and (*c*) the Centrarchidae, Percicthyidae, and Anguillidae are much more resistant to copper than most other species, killifish and carp are intermediate in sensitivity, and zebrafish are much more sensitive than all other species.

Table 4. Fish Species Used in Figure 2 (After Scott and Crossman, 1973)

Family	Species	Symbol	References
Salmonidae			
Rainbow trout	(*Salmo gairdneri*)	A	Brown et al. (1974); Folmar (1976); Fogels and Sprague (1977); Goettl et al. (1976); Hale (1977); Hodson, unpublished data; Howarth (in Fogels and Sprague, 1977); Herbert and Van Dyke (1964); Lett et al. (1976)
Atlantic salmon	(*S. salar*)	B	Sprague and Ramsay (1965); Sprague, (1964)
Brook trout	(*Salvelinus fontinalis*)	C	McKim and Benoit (1971)
Coho salmon	(*Oncorhynchus kisutch*)	D	Chapman (in Lorz and McPherson, 1976); Lorz and McPherson (1976)
Cyprinidae			
Fathead minnow	(*Pimephales promelas*)	E	Brungs et al. (1976); Geckler et al. (1976); Mount (1968); Mount and Stephan (1969); Pickering and Henderson (1966)
Bluntnose minnow	(*P. notatus*)	F	Geckler et al. (1976)
Creek chub	(*Semotilus atromaculatus*)	G	Geckler et al. (1976)
Blacknose dace	(*Rhinichthys atratulus*)	H	Geckler et al. (1976)
Stone roller	(*Campestoma anomolum*)	I	Geckler et al. (1976)
Striped shiner	(*Notropis chrysocephalus*)	J	Geckler et al. (1976)
Goldfish	(*Carassius auratus*)	K	Pickering and Henderson (1966)
Carp	(*Cyprinus carpio*)	L	Rehwoldt et al. (1971)

Ictaluridae			
Brown bullhead	(*Ictalurus nebulosus*)	M	Brungs et al. (1973); Geckler et al. (1976)
Anguillidae			
American eel	(*Anguilla rostrata*)	N	Rehwoldt et al. (1971)
Cyprinodontidae			
Banded killifish	(*Fundulus diaphanus*)	O	Rehwoldt et al. (1971)
Percichthyidae			
Striped bass	(*Roccus saxatalis*)	P	Rehwoldt et al. (1971)
White perch	(*R. americanus*)	Q	Rehwoldt et al. (1971)
Centrarchidae			
Bluegills	(*Lepomis macrochirus*)	R	Benoit (1975); Pickering and Henderson (1966); O'Hara (1971)
Pumpkinseed sunfish	(*L. gibbosus*)	S	Rehwoldt et al. (1971)
Percidae			
Rainbow darter	(*Etheostoma caeruleum*)	T	Geckler et al. (1976)
Orangethroat darter	(*E. spectabile*)	U	Geckler et al. (1976)
Johnny darter	(*E. nigrum*)	V	Geckler et al. (1976)
Miscellaneous			
Guppies	(*Poecelia reticulata*)	W	Pickering and Henderson (1966)
Stoneloach	(*Noemecheilus barbatulus*)	X	Solbe and Cooper (1976)
Zebrafish	(*Brachydanio rerio*)	Y	Fogels and Sprague (1977)
Flagfish	(*Jordanella floridae*)	Z	Fogels and Sprague (1977)
Invertebrate			
polychaete	(*Tubifex tubifex*)	●	Brkovič-Popovič and Popovič (1977)

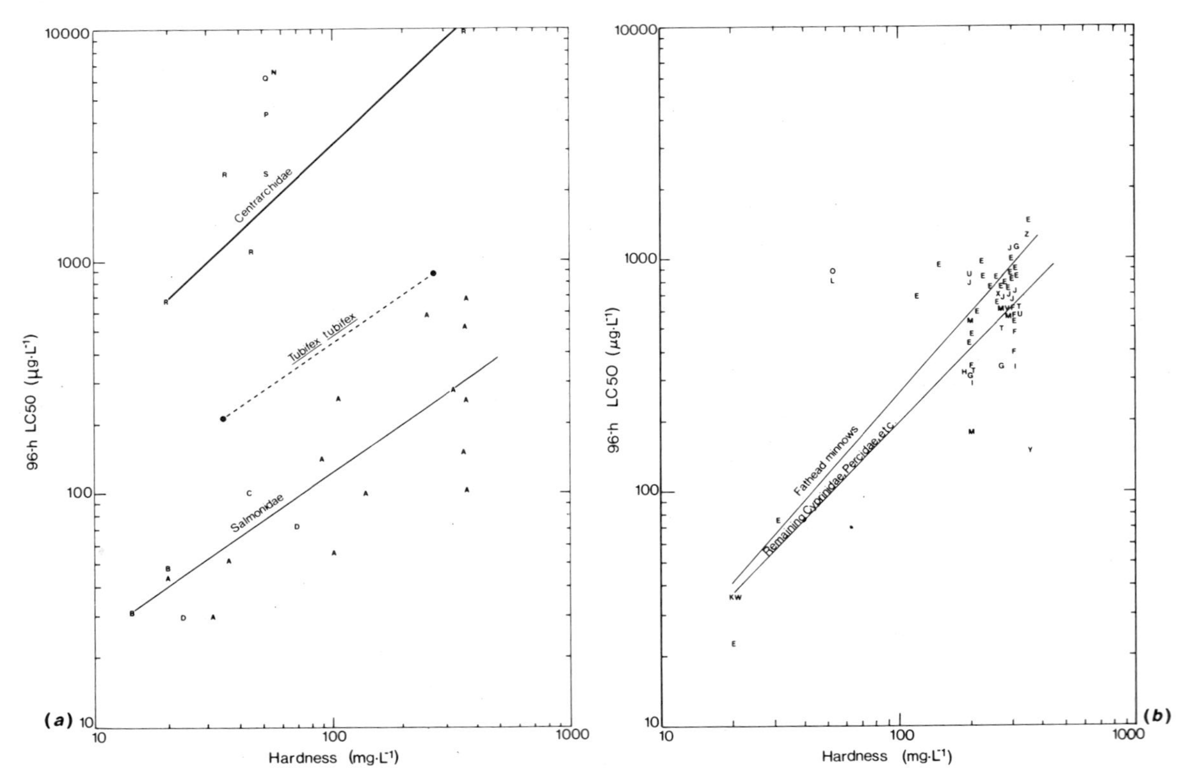

Figure 2. Variation of 96-hr LC_{50}'s of fish in fresh water with water hardness. Species represented in Figure 2 are listed in Table 4. (*a*) The LC_{50}-water hardness relationship for the Salmonidae [$\log_{10} LC_{50}\ (\mu g/l) = 0.6942 + 0.6955 \log_{10}$ hardness (mg/l); $s^2y{\cdot}x = 0.061$; $s_b^2 = 0.014$; $r = .829$] and the Centrarchidae ($\log_{10} LC_{50} = 1.591 + 0.9518 \log_{10}$ hardness; $s^2y{\cdot}x = 0.019$; $s_b^2 = 0.022$; $r = .977$). Data points for the American eel, striped bass, and white perch are included for comparison, as are data for one invertebrate species. (*b*) The LC_{50}-water hardness relationship for fathead minnows, for which there is a large data base ($\log_{10} LC_{50} = 0.1266 + 1.143 \log_{10}$ hardness; $s^2y{\cdot}x = 0.024$; $s_b^2 = 0.011$; $r = .923$), and for a variety of other cyprinids plus brown bullheads, darters, guppies, stoneloach, and flagfish ($\log_{10} LC_{50} = 0.2112 + 1.041 \log_{10}$ hardness; $s^2y{\cdot}x = 0.022$; $s_b^2 = 0.008$; $r = .900$). Data for zebrafish, carp, and the banded killifish were not included in the calculation of the line because of their marked deviations from the other data points. Because of the similarity of the fathead minnow and "other fish" regressions, the data were combined for the following regression (not plotted): $\log_{10} LC_{50} = 0.2399 + 1.0567 \log_{10}$ hardness; $s^2y{\cdot}x = 0.028$; $s_b^2 = 0.005$; $r = .908$.

The probable reason for the hardness effect is complexation or precipitation of copper by dissolved anions such as CO_3^{2-}, PO_4^{2-}, and OH^-, which form the basis of alkalinity. Hardness, generally a measure of cations (Ca^{2+}, Mg^{2+}), usually provides a close estimate of alkalinity, and most toxicology papers report hardness in place of alkalinity. Since calcium may control the permeability of biological membranes, hardness may also interact with the ability of the fish to take up copper and hence may be the basis for changes in copper toxicity. However, the binding constant of copper with organic substrates is much greater than that of calcium or magnesium (Zitko and Carson, 1976). Consequently, increasing the hardness by adding calcium or magnesium to a natural surface water while maintaining a constant alkalinity resulted in no change in short-term copper lethality to salmon (Zitko and Carson, 1976) or fathead minnows (Andrew, 1976). Increasing the bicarbonate alkalinity, however, reduced acute lethality to fathead minnows, and lethality was closely correlated to cupric ion activity (as measured by a cupric ion electrode) (Andrew, 1976). Complexed or precipitated copper did not appear to contribute to toxicity, so that variance between observations of copper toxicity based on total copper would be considerably reduced by basing toxicity on cupric ion activity. Using a specific ion electrode, Shaw and Brown (1974) concluded that the toxicity of copper to rainbow trout varied as the sum of cupric ion and $CuCO_3$ concentrations, rather than as the concentration of cupric ions alone. Zitko et al. (1973) assumed that lethality of copper to Atlantic salmon was constant at a specific cupric ion activity. Cupric ion activity was measured in a series of total copper concentrations at several levels of fulvic acid, a copper-binding humic acid. The total copper concentrations corresponding to the cupric ion activity assumed to be lethal for salmon was then taken as the predicted LC_{50}. Measured LC_{50} values at the same fulvic acid concentrations gave ratios of predicted to measured LC_{50}'s that varied from 1.65 to 6.6. The predictability was greatest at low fulvic acid concentrations and high hardnesses. These results suggest that either the methodology varies between authors or that cupric ion activity may not totally explain toxicity, that is, Cu^{2+} may not be the sole toxic species or cupric ion activity may not be a true measure of Cu^{2+}.

Pagenkopf et al. (1974) used the equilibrium equations of Perrin and Sayce (1967) to predict the concentrations of free Cu^{2+} ions and copper complexed by carbonate and hydroxide ions in water from published studies of copper toxicity to fish. Toxicity was found to correlate more closely with predicted concentrations of free Cu^{2+} than with any other predicted complexed species.

Increasing salinity should theoretically reduce copper toxicity by increasing the concentration of inorganic complexing ions that reduce available copper. However, reported 96-hr LC_{50}'s for larval pinfish (*Lagodon rhomboides*), spot (*Leiostomus xanthurus*), Atlantic croaker (*Micropogon undulatus*), and Atlantic menhaden (*Brevoortia tyrannus*) were 150, 160, 210, and 610 $\mu g/l$, respectively (Engel et al., 1976). These values indicate that marine fish are equally as sensitive as freshwater fish to the lethal effects of copper.

The preceding review indicates that any factor increasing the complexing capacity of water should reduce the apparent toxicity of copper in water, that is, the toxicity of cupric ions will remain unchanged but a greater concentration of total copper will be required to achieve a toxic concentration of cupric ion. This is confirmed by observations that pH, inorganic phosphate, sewage treatment plant effluents, pulp mill effluents, amino acids, sodium nitrilotriacetic acid (NTA), suspended organic materials, and humic substances affected the lethality of total copper (Andrew, 1976; Brown et al., 1974; Chynoweth et al., 1976; Cook and Côte, 1972; Grande, 1967; Howarth and Sprague, 1978; Sprague, 1968; Wilson, 1972; Zitko et al., 1973). Wilson (1972) demonstrated that the reduction of copper toxicity by spent sulfite liquor (SSL) from a pulp mill was predictable from the SSL concentration. Prediction of toxicity in the presence of humic acids was partially successful based on a cupric ion electrode, as outlined above (Zitko et al., 1973). The ability of humic acids to reduce copper lethality is also a function of hardness. Water hardnesses of 45 mg/l or greater prevent reduction of copper toxicity by humic acids (Cook and Côte, 1972). These authors hypothesized that calcium displaces copper from the binding sites of humic acids and makes the cupric ion available for toxicity. However, the binding constants of copper, magnesium, and calcium reported by Zitko and Carson (1976) indicate that copper should not be displaced by these cations, so that hardness and humic acids must interact in some other way.

Grande (1967) reported that fish were present in Norwegian waters only when the copper concentrations were less than 60 μg/l, a value at the threshold for lethal effects for juvenile salmonids in his laboratory. Since other authors report lethality at much lower copper concentrations in water of similar hardness (e.g., Sprague, 1964), Grande suggested that humic acids typical of Norwegian waters were protecting the fish. Van Loon and Beamish (1977) found fish in all lakes except ones that were heavily contaminated with metals from nickel smelter operations. In lakes with fish populations, copper was 23 μg/l or less, whereas in the barren lake it was 450 μg/l. However, the situation was confounded by correspondingly high levels of zinc (8000 μg/l), cadmium (50 μg/l), and iron (1600 μg/l) and a pH of 4.0 in the barren lake.

Howarth and Sprague (1978) tested the effects of pH values ranging from 5 to 9 on copper lethality to rainbow trout at water hardnesses of 30, 100, and 360 mg/l. Water at the two lower hardnesses was obtained by reverse osmosis deionization of the high-hardness water. Their studies indicated that increasing hardness decreased copper toxicity, pH had a bimodal effect on toxicity that interacted with hardness, and copper lethality was complete within 72 hr at pH 8 and 9, but times to thresholds often exceeded 5 days at lower pH values. Little pH effect on copper toxicity was observed at 30 mg/l hardness. At the two higher hardnesses, toxicity was greatest at pH 6 and 7, decreased at pH 5 and 8, and increased again at pH 9. In other words, the pH effect was not uniform at high hardness. Since the pH effect was greatest at the high hardness and decreased to zero at the low hardness, it is apparent that some factor interacting with copper and pH, for example, the bicarbonate buffering system, was removed during reverse osmosis. Alternatively, some factor opposing the pH effect might have assumed greater importance. Chloride, for example, was not removed by reverse osmosis (Howarth and Sprague, 1978) and constituted a greater proportion of the total dissolved solids at low hardnesses. Howarth and Sprague also calculated the concentrations of various ionic copper species, using the equilibrium equations of Perrin and Sayce (1967), to predict copper toxicity as described by Pagenkopf et al. (1974). When the lethal concentrations of the sum of (Cu^{2+} + $CuOH^+$ + $Cu_2OH_2^{2+}$) were compared to pH and hardness, a very regular decrease in copper toxicity was observed with increasing hardness and decreasing pH. In other words, copper was less toxic at low pH values, and the pH effect on copper toxicity was greatest at high hardnesses. Decreased toxicity at low pH's seems unusual in view of low-pH effects on gill permeability, ionoregulation, and respiration of fish (Packer and Dunson, 1970, 1972).

Dissolved oxygen concentrations influence copper lethality through fish respiration rates. Lloyd (1961b) showed that the amount of copper (and zinc, lead, or phenol) required to cause 100% mortality in a fixed time decreased as percent saturation of oxygen decreased. At 40% saturation, the toxic concentration of copper decreased by one third from that at 100% saturation. Lloyd postulated that this decrease was due to increased availability and uptake of copper as the fish increased ventilation frequency and respiratory water flow to compensate for reduced dissolved oxygen. He also postulated that any factor increasing respiratory water flow and oxygen uptake (e.g., exercise, high temperature) would increase the uptake and toxicity of metals.

Other toxic substances in a mixture with copper may interact to increase or decrease expected toxicity. Interactions may be assessed by expressing the lethal concentration (e.g., 96-hr LC_{50} or asymptotic LC_{50}) as 1.0 toxic unit. Twice this concentration is 2.0 toxic units, while one-half this concentration is 0.5 toxic unit. Sprague (1970) has outlined five basic interactions: (*a*) additive: total observed toxic effects are equivalent to the sum of the expected toxic effects of the individual components of the mixture acting alone [e.g., 0.5 toxic unit of A in a mixture with 0.5 unit of B (or 0.2 of A plus 0.8 of B) produces 50% mortality in 96 hr]; (*b*) less than additive: total observed effects are less than the sum of the expected individual effects because one or more of the components have become less toxic through some biological or chemical interactions (e.g., 0.8 toxic unit of A plus 0.8 of B is required for 50% mortality in 96 hr); (*c*) more than additive: total observed effects are greater than the sum of the expected individual toxic effects because the toxicity of one or more components of the mixture has increased (e.g., only 0.2 toxic unit of A plus 0.2 of B is required for 50% mortality in 96 hr); (*d*) no interaction: the presence of one toxicant has no effect on the response to a second (e.g., 1.0 unit of A is still required to produce 50% mortality in 96 hr despite the presence of 0.9 unit of B); and (*e*) antagonism:

the presence of one toxicant reduces the response to a second (e.g., 1.5 units of A is required to produce 50% mortality in 96 hr when 0.5 unit of B is present).

Table 5 demonstrates that the first three interactions are observed in tests of copper lethality in mixtures. Total toxicity is less than expected when copper is mixed with nonionic detergents and an organophosphate pesticide, malathion. However, anionic detergents and two other organophosphates, Sevin and parathion, caused greater than expected toxicity when mixed with copper. The different effects of anionic and nonionic detergents may result from their dif-

Table 5. Relative Toxicities of Lethal Toxicant Mixtures Containing Copper

Less than Additive Toxicity	Additive Toxicity	Greater than Additive Toxicity
Copper-nonionic detergent (Calamari and Marchetti, 1973)	Copper-phenol Brown and Dalton, 1970)	Copper-anionic detergent (Calamari and Marchetti, 1973)
Copper-malathion (Macek, 1975)	Copper-zinc-phenol (Brown and Dalton, 1970)	Copper-cadmium (Eisler and Gardner, 1973)
	Copper-zinc-nickel (Brown and Dalton, 1970)	Copper-zinc-cadmium (Eisler and Gardner, 1973)
		Copper-zinc (high concentrations) (Lloyd, 1961a)
	Copper-zinc (low concentrations) (Lloyd, 1961a)	Copper-parathion (Macek, 1975)
		Copper-Sevin (Macek, 1975)
	Copper-zinc (Sprague and Ramsay, 1965)	
	Copper-ammonia (Herbert and Van Dyke, 1964)	

ferent interactions with calcium and magnesium ions, which may influence gill membrane permeability characteristics in different ways (Calamari and Marchetti, 1973). Consequently, the ability of the fish to take up copper is affected by changes in gill membrane permeability to copper. No explanation was given for the variable effects of organophosphates.

No consistent effects were observed for copper interactions with other metals. Although most authors reported that zinc was simply additive with copper (Table 5), Lloyd (1961) observed that copper and zinc together in high concentrations had a greater than additive toxicity. Nickel, phenol, and ammonia toxicity were additive with copper toxicity, whereas cadmium and cadmium plus zinc had a greater than additive toxicity with copper. For toxicants demonstrating additivity with copper, prediction of the toxicities of complex mixtures in the laboratory and in rivers receiving industrial effluents has been successful (Brown, 1968; Brown and Dalton, 1970; Herbert and Van Dyke, 1964).

Biological Factors

As indicated above, most freshwater fish species within a single genus appear equally sensitive to copper when allowance is made for hardness (Figure 2). "Lumping" these data may result in oversimplification since many papers comparing toxicity of copper to several species under the same conditions show a consistent order of sensitivity (Geckler et al., 1976). However, LC_{50} data within a species may vary by 2 to 6 times (Fogels and Sprague, 1977), as illustrated in Figure 2. The highest LC_{50} of rainbow trout is about 7 times the lowest at a common hardness of 365 mg/l (Figure 2*a*), while at a hardness of 200 mg/l one LC_{50} value for bullheads is 3 times another (Figure 2*b*). If only the extreme high or low LC_{50}'s were available, there could be a false indication of high or low sensitivity. Consequently, the high tolerance of carp and killifish and the high sensitivity of zebrafish should be confirmed.

A possible explanation for the within-species variation in LC_{50} values is the observed effect of fish weight on copper toxicity. Howarth and Sprague (1978) demonstrated that copper toxicity to rainbow trout weighing from 0.7 to 10 g each could be expressed by the equation

$$\log_{10}(LC_{50}) = 1.582 + 0.3477 \log_{10}(\text{weight})$$

The result is that a tenfold increase in fish weight (e.g., from 1 to 10 g) would increase the 96-hr LC_{50} by 2.2 times. Whether this equation would apply for very large fish was not demonstrated. The probable cause of the weight effect is a reduction in metabolic rate per gram of fish with increasing weight. The net effect is a reduced concentration of copper in fish tissue for a given exposure (concentration × time), that is, a greater exposure is required to achieve a toxic concentration in the target tissue.

The slope of the line relating 96-hr LC_{50}'s of Salmonidae (principally rainbow trout) to hardness was 0.7, which is the same as the slope of 48-hr LC_{50}'s shown by Brown (1968) for rainbow trout tested at the same time in the same laboratory. The predicted variation in 96-hr LC_{50}'s of rainbow trout with hardness at pH 8.0, using equation 2 of Howarth and Sprague (1978), provides a slope of about 1.08. Use of their observed 96-hr LC_{50}'s of 30 and 561 μg Cu/l at pH 8.0 and hardnesses of 31 and 371 mg/l, respectively, yields a slope of 1.14. These results would place the response and the sensitivity of Salmonidae closer to those of Cyprinidae.

Although the variation of bluegill LC_{50}'s with hardness parallels that for other fish species, bluegill LC_{50} values were about 10 times higher than those for other species at the same hardness. Since the data for bluegills originate from three separate authors, the results represent a true species difference rather than a difference due to experimental design. The reason for bluegill resistance is unexplained but may be due to a greater ability to detoxify or excrete copper.

A criticism of all of the above studies relating copper toxicity to various biological and environmental factors is that all of them were conducted within fixed, relatively short ($\leq$96 hr) time frames. If the treatments affected the rate of mortality (as shown for pH: Howarth and Sprague, 1978), then an apparent but erroneous change in toxicity may be observed. For example, while less mortality and hence toxicity may be observed in 96 hr because of a treatment, prolongation of the exposure might result in the same mortality as previously observed. This could occur if the treatment affected the permeability of the fish to the copper or affected the rate of response of the fish to the copper taken up. Consequently, the concept of additivity (or nonadditivity) and the effects of changing water quality on copper lethality may not apply to longer term exposures. This problem is exemplified by the effects of temperature on zinc toxicity to salmon (Hodson and Sprague, 1975). Examination of the short-term effects of zinc indicate greatest toxicity at high temperatures within 4 days. However, the ultimate threshold responses indicate a complete reversal with greatest toxicity at low temperatures because of a much slower rate of response.

4.2. Chronic Toxicity and Sublethal Effects

Sublethal effects may be considered on a whole-organism basis (e.g., reproductive success, resistance to disease, lingering mortality, behavior, swimming endurance, growth) or on a physiological basis (e.g., respiratory, hematological, neurological, enzymatic, histological, responses).

Reproduction, Growth, and Mortality

Effects of prolonged exposure to very low copper concentrations on fish reproduction and growth are summarized in Table 6. These data indicate that, in fresh water, various aspects of spawning, growth, and survival are affected at copper concentrations between 5 and 40 μg/l in laboratory water low in organic material, and between 66 and 120 $\mu g/l$ in laboratory water drawn from a river enriched with organic materials. The most sensitive parameter listed in Table 3 is not consistent between species. Spawning frequency, and hence overall egg production, is reduced in fathead minnows, but not in brook trout or bluegills, the only other species in which spawning was studied. Egg production per spawning was the most sensitive parameter for brook trout and in one study with fathead minnows. For Atlantic salmon and rainbow trout, egg mortality was the most sensitive aspect of reproduction, in contrast to two other salmonids, brook trout and Chinook salmon, where growth and long-term survival were most sensitive. Survival was also the most sensitive parameter for walleyes, channel catfish, bluegills, mummichogs, and Atlantic silversides. The major problem with these data is a lack of comparability, since not all parameters were studied by each author and the efficiency with which each was studied was not constant.

A species sensitivity comparison indicates that most salmonids tested responded to copper at concentrations between 10 and 20 $\mu g/l$, although growth of brook trout was affected at values as low as 5 $\mu g/l$. Channel catfish and walleyes were as sensitive as the salmonids, but fathead minnows were only as sensitive in soft water. In hard water, fathead minnows were about as sensitive as bluegills, the most resistant freshwater species tested. A comparison of mummichog and Atlantic silversides responses is confounded by the salinity of the test water. The high resistance of bluegills to sublethal copper concentrations parallels their great resistance to the lethal effects of copper.

Pickering et al. (1977) demonstrated that length of exposure of fathead minnows did not change the effects of copper on reproduction. Exposure of fish for 6 or 3 months before spawning had the same effect as exposure only during spawning. In other words, the sexually mature adults appear to be the sensitive stage in the fathead minnow life cycle. Effects on spawning were statistically significant at all concentrations of copper greater than or equal to 37 $\mu g/l$.

Water hardness does not seem to have a consistent or great effect on sublethal toxicity (Table 6). The studies by Mount (1968), Mount and Stephan (1969), and Pickering et al. (1977) indicate a twofold decrease in copper toxicity for a sixfold increase in hardness and a consistent toxicity between authors at the same hardness. Concentrations of copper affecting reproduction of brook trout vary with hardness, but not concentrations affecting growth (McKim and Benoit, 1971; Sauter et al., 1976). Consequently, copper effects on overall production of brook trout will probably be the same at all hardnesses. There was no observable hardness effect on copper toxicity to rainbow trout (Goettl et al., 1976; Grande, 1967) or walleye (Sauter et al., 1976).

Since the hardness effect was not consistent, and since it was of relatively minor magnitude, hardness may not have influenced copper availability to the fish. Rather, hardness may reduce the permeability of fish to copper, perhaps through calcium interactions with membrane permeability. At the low copper

Table 6. Lowest Copper Concentrations (μg/l) Having an Effect (and No Effect) on Fish Reproduction, Growth, and Survival

Species	Spawning	Overall Egg Production	Percent Hatch	Fry, Juvenile, or Adult Mortality	Growth	Most Sensitive Parameter	Lowest Effective Concentration	Comments	Reference
Salmonidae									
Rainbow trout (*Salmo gairdneri*)			19 (12)	19 (12)		Percent hatch, mortality	19	Hardness = 100 mg/l	Goettl et al. (1976)
			20	40		Percent hatch	20	Hardness = 7.8 mg/l	Grande (1967)
Atlantic salmon (*Salmo salar*)			10	40		Percent hatch	10	Hardness = 7.8 mg/l	Grande (1967)
Chinook salmon (*Oncorhynchus tschawytscha*)			>80	20	20	Mortality, growth	20		Hazel and Meith (1970)
Brook trout (*Salvelinus fontinalis*)	>33	17 (10)	33 (17)	17 (10)	17 (10)	Egg production, mortality, growth	17	Hardness = 45 mg/l	McKim and Benoit (1971)
			13 (7)	27 (13)	5	Growth	5	Hardness = 38 mg/l	Sauter et al. (1976)
			74 (49)	49 (27)	8	Growth	8	Hardness = 187 mg/l	Sauter et al. (1976)
	>9.4	>9.4	>9.4	>9.4	>9.4			Hardness = 45 mg/l	McKim and Benoit (1974)
Cyprinidae									
Fathead minnow (*Pimephales promelas*)	33 (15)	33 (15)	>33	95 (33)	95 (33)	Spawning, egg production	33	Hardness = 198 mg/l	Mount (1968)
	18 (11)	18 (11)	>11	18 (11)	>11	Spawning, egg	18	Hardness = 31	Mount and

						production, mortality		mg/l	Stephan (1969)
	61 (38)	37 (24)	>100	>100	>100	Egg production	37	Hardness = 200 mg/l	Pickering et al. (1977)
	120 (66)	120,66 (34)	120	180 (120)		Spawning, egg production	66–120	Effects marginal at 66 μg/l; hardness fluctuated; high organic content in water	Brungs et al. (1976)
Ictaluridae									
Channel catfish (*Ictalurus nebulosus*)			>66	19 (13)	19 (13)	Mortality, growth	19	Hardness = 38 mg/l	Sauter et al. (1976)
			>66	19 (13)	19 (13)	Mortality, growth	19	Hardness = 187 mg/l	Sauter et al. (1976)
Cyprinodontidae									
Mummichog (*Fundulus heteroclitus*)			500 (250)	<250		Mortality	<250	20% Salinity	Gardner and LaRoche (1973)
Atherinidae									
Atlantic silversides (*Menidia menidia*)			500 (250)	<250		Mortality	<250	20% Salinity	Gardner and LaRoche (1973)
Centrarchidae									
Bluegill (*Lepomis macrochirus*)	162 (77)	>77	162 (7.7)	40 (21)	162 (77)	Mortality	40	Hardness = 45 mg/l	Benoit (1975)
Percidae									
Walleye (*Stizostedion vitreum*)			47	13		Mortality	13	Hardness = 13 mg/l	Sauter et al. (1976)

concentrations characteristic of sublethal toxicity, the majority of the metal would be available for uptake and the effect of hardness on permeability would predominate, whereas at the high copper concentrations characteristic of lethal effects, the hardness effect on copper availability would predominate and would mask the effect on permeability.

Salinity does not seem to affect chronic copper toxicity since eggs of spot, a marine fish, showed 50% mortality at a cupric ion activity of 10^{-9} *M* (Engel et al., 1976). Organic material has a more dramatic effect on sublethal copper toxicity. In water drawn from a stream receiving a large volume of effluent from a secondary sewage treatment plant, the spawning success of fathead minnows was significantly reduced at copper concentrations of 120 μg/l or greater (Brungs et al., 1976). A slight effect was observed at 66 μg/l. Hardness in this experiment fluctuated from 88 to 352 mg/l with monthly means ranging from 226 to 321 mg/l. Since only a small hardness effect was evident in the studies described above, the decreased toxicity of copper was probably due to complexation by organic material from the sewage effluent, that is, less copper was available to the fish. Andrew (1976) has also suggested that inorganic phosphate may have reduced copper availability in this study.

The reduction in embryo production of fathead minnows by a mixture of copper, cadmium, and zinc was no greater than that due to zinc alone (Eaton, 1973). The number of spawnings per female was reduced from control levels when copper in the mixture was greater than or equal to 6.7 μg/l, a concentration lower than the ones observed in the studies reported above. However, the effects of zinc on spawning success, at the concentrations tested, accounted for the observed inhibition. Therefore there appears to be little interaction of these three metals at sublethal concentrations, in contrast to the more than additive effects observed in lethal studies.

The effects of copper on growth, cited above, were measured as part of life cycle bioassays. Since energy consumption and utilization were not standardized through forced swimming and measured, weight-adjusted rations, no conclusion can be drawn as to the causes of observed growth effects, for example, changes in appetite, conversion efficiency, energy utilization, or differential mortality of large or small individuals.

In a highly controlled growth experiment using rainbow trout held at a constant caloric content in exercising chambers, Lett et al. (1976) studied the effects of copper on appetite, growth, and proximate body composition. The initial copper effect was a cessation of feeding, with a gradual return to control levels. The higher the copper concentration, the slower was the return of appetite, so that fish exposed to 300 μg Cu/l required 15 days at a 2% ration before feeding was normal. Consequently, growth rates were depressed by copper but recovered with appetite to approach those of control fish after 40 days. Assimilation efficiency was unchanged, indicating that depressed growth was a response to appetite suppression rather than to a decreased ability to digest. Waiwood and

Beamish (in press b) observed that appetite suppression by copper was enhanced by low pH and low hardness. Rainbow trout were exposed to copper in swimming chambers at a fixed ration and activity regime. Water hardness and pH equaled 365, 100, and 30 mg/l (EDTA) and 6.0, 7.5, 7.8, and 8.0, respectively. The appetite suppression was greatest at low pH, low hardness, and high copper concentrations but occurred only at 30 μg Cu/l at the highest hardness. The result was decreased growth correlated to appetite suppression similar to that observed by Lett et al. (1976). Growth was also suppressed when appetite was normal, because of a lower gross conversion efficiency [(change in wet weight per day ÷ dry weight of food eaten per day) × 100] of copper-treated fish. The copper effect on conversion efficiency and growth was also enhanced by decreasing pH and hardness. Low pH and hardness alone did not affect appetite, conversion efficiency, or growth. Growth and conversion efficiency could be predicted by these equations:

$$G = -0.272 - 0.98(C \times H) - 4.25(C \times H \times T) + 1.29(C \times H \times \mathrm{pH}) + 0.159(W)$$

$$E = 0.36 - 1.64(C \times H) + 1.48(C \times H \times \mathrm{pH}) + 0.0135(C \times T) + .0555(C \times \mathrm{pH} \times T) + 0.0085(C) - 0.0073(C \times \mathrm{pH})$$

where G = growth (% wet weight/day)
E = conversion efficiency (%)
C = total copper (μg/l)
H = hardness (mg/l)
pH = pH + (pH − 6.88)
T = exposure time (days)
W = wet body weight (g)

The growth equation estimates a 25% growth reduction at copper concentrations of 206 μg/l at pH 8.0 and 27 μg/l at pH 6.0 at a hardness of 36.5 mg/l, 32 μg/l at pH 7.75 and 8 μg/l at pH 6.0 at a hardness of 100 mg/l, and 6 μg/l at pH 7.5 and 2 μg/l at pH 6.0 at a hardness of 30 mg/l.

Waiwood and Beamish (in press b) also evaluated copper speciation according to Pagenkopf et al. (1974) and concluded that only Cu^{2+} and $CuOH^+$ were significantly correlated to the observed growth response. However, hardness and pH did not cause significant variation in the proportions of these ions. Consequently, their results support the hypothesis that hardness and pH effects on sublethal toxicity are a result of interactions with biological membranes and the ability of the fish to take up available copper.

Waiwood and Beamish (in press b) speculated that the reduced growth rate was a result of increased maintenance energy requirement and decreased efficiency of energy utilization. This was supported by their previous observation (Waiwood and Beamish, in press a) of decreased swimming performance with increased copper. This effect was also enhanced by low pH values and hardness.

These results correspond to the observations of appetite suppression in brook trout at copper concentrations as low as 9 μg/l (Drummond et al., 1973), in Atlantic salmon at 20 μg/l (Grande, 1967), and in coho salmon at 5 to 20 μg/l (Lorz and McPherson, 1976). The reduced appetite of coho salmon may have caused an observed reduction in the seaward migration of copper-exposed fish. Since migration is size dependent, and since the copper-exposed fish were released into streams before reaching migration size, they may have been unable within the time period of the experiment to attain a size adequate to enable them to migrate (Lorz and McPherson, 1976). Since timing of migration may affect the carrying capacity of the stream, the coincidence of migrants and abundant food in estuaries and the sea, and water levels and temperature during the migration, these results may have serious implications for fish production.

Resistance to Disease

Pippy and Hare (1969) reported mass mortalities of Atlantic salmon grilse, white suckers, and shiners (*Notropis cornutus*) in the Miramichi River, New Brunswick, Canada, that were associated with heavy infestations of a pseudomonad bacterium (*Aeromonas liquifaciens*), high water temperatures (>20°C), and high concentrations of copper and zinc (approaching or exceeding 1.0 toxic unit). Although high temperatures, metal pollution, or the presence of bacteria in fish did not alone kill the salmon, the coincidence of all three was postulated as the case of the epizootic. Rödsaether et al. (1977) observed that eels held in fresh water developed the symptoms of "red pest" disease or vibrio (*Vibrio anguillarum*) when the water was contaminated with 30 to 60 μg Cu/l. No disease occurred in eels held under identical conditions without added copper. *Vibrio anguillarum* does not live in fresh water but was isolated from the blood of eels exposed to copper. This suggests that the eels normally carry the bacterium and that the copper initiated the disease.

Behavior

Warner et al. (1966) stated that behavior represents "the final integrated result of a diversity of biochemical and physiological processes." In this context, causes of behavioral changes are difficult to ascertain—rather it is the consequences of these changes that are important. Sprague (1964) demonstrated that Atlantic salmon would avoid copper solutions when given a choice between clean and "contaminated" water in an avoidance "tube." The response was extremely sensitive with significant avoidance at concentrations as low as 4.3 μg/l ($\simeq$0.02 toxic unit). Although the reason for this avoidance was unknown, the consequence was a strong correlation between the downstream migration of adult salmon and the occurrence of high copper and zinc concentrations in the Miramichi River system in New Brunswick, Canada (Sprague et al., 1965; Saunders and Sprague, 1967). When mining in the upper watershed was active, metal levels were high (often greater than 1.0 toxic unit), and a high percentage of unspawned adults returned downstream from spawning areas. About one third reascended, and a few were observed in adjacent unpolluted streams (2%), but most (62%) were not seen again. During years of no mining, more fish ascended the river, fewer returned before spawning, and overall spawning was more successful (Saunders and Sprague, 1967). Downstream migration coincided with metal levels exceeding 0.4 toxic unit of combined copper and zinc. Laboratory avoidance of copper by rainbow trout in a Y-maze has been observed at concentrations of 0.1, 1.0, and 10 μg/l (Folmar, 1976). Although these results confirm the high sensitivity of fish to the presence of copper, they may have been influenced by some other factor, since background copper concentrations, even in very soft waters, are often greater than 1.0 μg/l (Sprague, 1964; International Joint Commission, 1976).

The migration of coho salmon was also affected, but in a different way. Yearling coho salmon undergo a parr-smolt transformation in fresh water that induces saltwater tolerance and allows migration to the sea. Exposure to 5 to 20 μg Cu/l for as little as 144 hr reduced the numbers of successful migrants in a release experiment by 30 to 70% (Lorz and McPherson, 1976). The suspected cause was a marked *in vivo* inhibition of Na^+,K^+-activated ATPase, an enzyme required for adaptation to seawater. The saltwater tolerance of these animals was also markedly reduced by these concentrations of copper.

Drummond et al. (1973) observed that brook trout exposed to as little as 6 μg Cu/l became more active for a period of up to 6 to 8 hr. Kleerekoper et al. (1972, 1973) also observed changes in the locomotor behavior of goldfish exposed to as little as 11 to 17 μg/l. Upon entering a shallow gradient of copper in an "open-field" situation, they avoided 10 μg Cu/l at 21.1°C but were attracted at 21.5°C (Kleerekoper et al., 1973). At higher concentrations (up to 50 μg/l) the fish spent more time in the copper-treated areas, and turning behavior changed (increased average size of turns). These results represent a more mathematical description of the increased activity of brook trout.

Respiration and Osmoregulation

Drummond et al. (1973) found a short-term coughing response in brook trout to copper concentrations as low as 9.5 μg/l. This response was analogous to mammalian coughing due to air pollution, perhaps as a result of olfactory or taste bud irritation. The response disappeared within a day, probably because of short-term receptor adaptation. Bluegills demonstrated an increasing inhibition of oxygen uptake over 14 days as copper was raised above a threshold concentration of 300 μg/l; at lethal concentrations oxygen uptake was increased, but activity remained constant (O'Hara, 1971). At near-lethal concentrations rainbow trout showed increased coughing frequency, ventilation frequency, and amplitude of opercular and buccal respiratory pressure changes after 6 hr (Sellers et al., 1975). These changes were greatest at 40 to 60 μg Cu/l and less at higher and lower concentrations. There was little variation in blood P_{O_2} and blood pH. These results indicated less respiratory response to copper than to zinc, a metal causing death through gill destruction and hypoxia. Consequently, the acute lethal action of copper is probably not respiratory failure. Osmoregulatory failure is more likely because channel catfish and common shiners (*Notemigonus crysoleucas*) had reduced serum osmolarity when exposed in salt water (235 mOsm NaCl) (Lewis and Lewis, 1971). Isoosmotic salt concentrations also prolonged the survival of copper-exposed fish. Gill ion exchange or kidney salt regulation may be the target of acute copper toxicity. The free alternative is supported by copper inhibition of gill Na^+,K^+-activated ATPase and reduced saltwater tolerance of coho salmon (Lorz and McPherson, 1976). Schreck and Lorz (1978) showed that exposure of coho salmon to 140 μg Cu/l caused little change in serum osmolarity, but serum chloride decreased by 30% after 78-hr exposure. Consequently, death of copper-treated coho salmon may be the result of failure to regulate chloride. If this were the case with rainbow trout, the hardness and pH effects observed by Howarth and Sprague (1978) might be confounded by the high chloride content of their experimental soft water.

Histopathology

Histological examination of fish exposed to acutely toxic copper concentrations demonstrated pathological changes in kidneys, liver, hematopoietic tissue, and gills of winter flounder (*Pseudopleuronectes americanus*) (Baker, 1969); mechanoreceptors of lateral line canals, olfactory organs including chemoreceptors, kidney, and brain of adult mummichog and Atlantic silversides (Gardner and LaRoche, 1973; Eisler and Gardner, 1973); and taste buds of the palatal organs of goldfish (Vijayamadhavan and Iwai, 1975). Copper rapidly permeated the taste buds of goldfish before the onset of cell destruction (Vijayamadhavan and Iwai, 1975), and changes in fish appetite (Lett et al., 1977; Drummond et al., 1973) may be a response to the taste of copper or to the destruction of the sensory system necessary for feeding. The recovery of appetite after 15 days (Lett et al., 1976) may represent regeneration of copper-tolerant taste buds. Olfactory lesions occurred in adult mummichogs and silversides exposed to copper, but not in young fish raised in copper (Gardner and LaRoche, 1973). Consequently, sensory organs (mechano- and chemoreceptors) appear very sensitive to copper but may adapt to it in fish surviving copper exposure. The olfactory organ response to copper may also explain changes in fish behavior and migration. High concentrations causing destruction of olfactory cells may prevent avoidance of copper as observed by Kleerekoper et al. (1973), while exposures that do not destroy the cells may stimulate avoidance. Hara et al. (1976) studied the neural response of rainbow trout to nasal perfusion by 10^{-5} *M* L-serine, a "standard" olfactory stimulus. When copper was included in the perfusate, concentrations as low as 8 μg/l inhibited the olfactory response. Consequently, olfactory-dependent migration and feeding could be inhibited by copper.

Gill pathology of flounder exposed to low copper concentrations included changes in the epithelial cell layers covering lamellae, an increased number of chloride cells, and a decreased number of mucous cells (Baker, 1969). In mummichogs exposed to copper alone or to copper in a mixture with zinc and cadmium, copper-specific lesions were observed in the epithelium and tubule cells of the proximal and collecting tubules. These gill and kidney lesions suggest changes in the capacity to ionoregulate and osmoregulate and may explain the observed osmoregulatory responses to copper of common shiners and catfish (Lewis and Lewis, 1971).

Hematology

Prolonged exposure of fish to sublethal copper concentrations causes a variety of hematological responses. After 6 days of exposure to 50 μg Cu/l or more, blood glucose of brown bullheads increased 2 to 8 times the control levels and the increase persisted for at least 30 days (Christensen et al., 1972). No change was seen at 600 days, suggesting long-term adaptation to copper. Hematocrit and hemoglobin values were also elevated at 6 and 30 days, but the total red blood cell count was constant, indicating an increase in average cell size due to either cellular swelling or mortality of small immature cells and replacement by larger cells from the spleen. To a certain extent, these symptoms resemble those of diabetes mellitus (Schalm et al., 1975). Plasma lactate dehydrogenase (LDH) was unaffected by copper exposure, while plasma glutamic oxalacetic transaminase (GOT) was reduced at copper concentrations above 27 μg/l after 600-day exposure. Increased levels of plasma GOT usually indicate liver damage and release of the enzyme into the blood. Therefore the observed decreased activity of GOT may represent a direct inhibition of the enzyme by copper as observed *in vitro* by Christensen (1971–72).

Exposure of brook trout to copper for long periods produced a reaction similar to that observed in bullheads (McKim et al., 1970). Transient increases in hematocrit, hemoglobin, and numbers of red blood cells disappeared after 21 days, as did decreases in plasma chloride and osmolarity. In this case the increased hematocrit, hemoglobin, and red blood cell counts might be offset by hemodilution, as shown by reduced osmolarity. After 337-day exposure, plasma GOT was again inhibited at copper concentrations above 9.5 μg/l.

Enzyme Inhibition

Christensen (1971–72) measured the *in vitro* plasma activities of GOT and LDH from white suckers (*Catastomus commersoni*). When the plasma was incubated in the presence of copper, concentrations of 65 and 130 mg/l *in vitro* were required to produce 20% inhibition (I_{20}) of LDH and GOT, respectively. The I_{20} values of 22 metals were found to be strongly correlated to the toxicities of the metals to the stickleback and to the equilibrium constants of metal sulfides, suggesting sulfhydryl enzyme inhibition as a mode of enzyme and whole-fish toxicity.

Enzyme inhibition due to metals *in vitro* is different from that observed on exposure *in vivo*. Copper *in vitro* inhibited alkaline phosphatase, acid phosphatase, xanthine oxidase, and catalase from killifish, but only acid phosphatase and catalase were inhibited by *in vivo* exposure (Jackim et al., 1970). The difference in responses probably reflects the inability of copper to penetrate various liver cell compartments at a toxic concentration during *in vivo* exposures. Jackim et al. (1970) also found that RNAase of killifish livers showed an increasing degree of inhibition with increasing time of exposure to 3.2 mg Cu/l. The time-dependent inhibition was probably due to the slow buildup of copper in the tissues as an equilibrium between uptake and excretion was established. The opposite result was seen when killifish δ-aminolevulinic acid dehydrase (ALA-D) was assayed after exposure of the fish to 1 mg Cu/l. Four days of exposure resulted in 11.7% inhibition of ALA-D, whereas 14 days resulted in only 7% inhibition (Jackim, 1973). If these results represent a true change in inhibition, they may result from an adaptation to the toxicant or induction of some copper excretion mechanism. Since the copper concentration in the exposure system decreased from 1 to 0.6 mg/l, less copper may also have been available for toxic effects. Hodson et al. (1977) observed no inhibition of erythrocyte ALA-D in rainbow trout after a 7-day exposure to 86 μg Cu/l.

Bilinski and Jonas (1973) found no correlation between inhibition of gill lactate oxidation and copper lethality to rainbow trout. Lactate oxidation was inhibited by 50% at 64 μg Cu/l, but no inhibition was observed at lower copper concentrations that were nevertheless lethal.

Endocrinology

Sockeye salmon showed a strong corticosteroid response to copper exposure (Donaldson and Dye, 1975). Total corticosteroids and cortisone concentrations in plasma of exposed salmon were significantly elevated from control within 2-hr exposure to 6.5 μg Cu/l. At 65 μg/l or higher, cortisol, cortisone, and total corticosteroids were significantly elevated within 1 hr. Fish exposed to 650 μg/l died within 24 hr, and corticosteroids remained elevated at this concentration for the full 24 hr. At lower concentrations, corticosteroids decreased with time toward control levels. The results indicate rapid hormonal response to stress at concentrations that ranged from lethal to sublethal and an adaptation to stress at the sublethal concentrations. These results were confirmed with coho salmon by Lorz and Schreck (1978), who also showed that acute cadmium exposure did not elicit the stress response and that copper exposure reduced the resistance of the fish to subsequent handling or salinity stress. The technique described could be a rapid method for identifying potentially harmful concentrations of copper.

4.3. Summary

Figure 3 summarizes the various responses of fish to copper. Lethal effects may

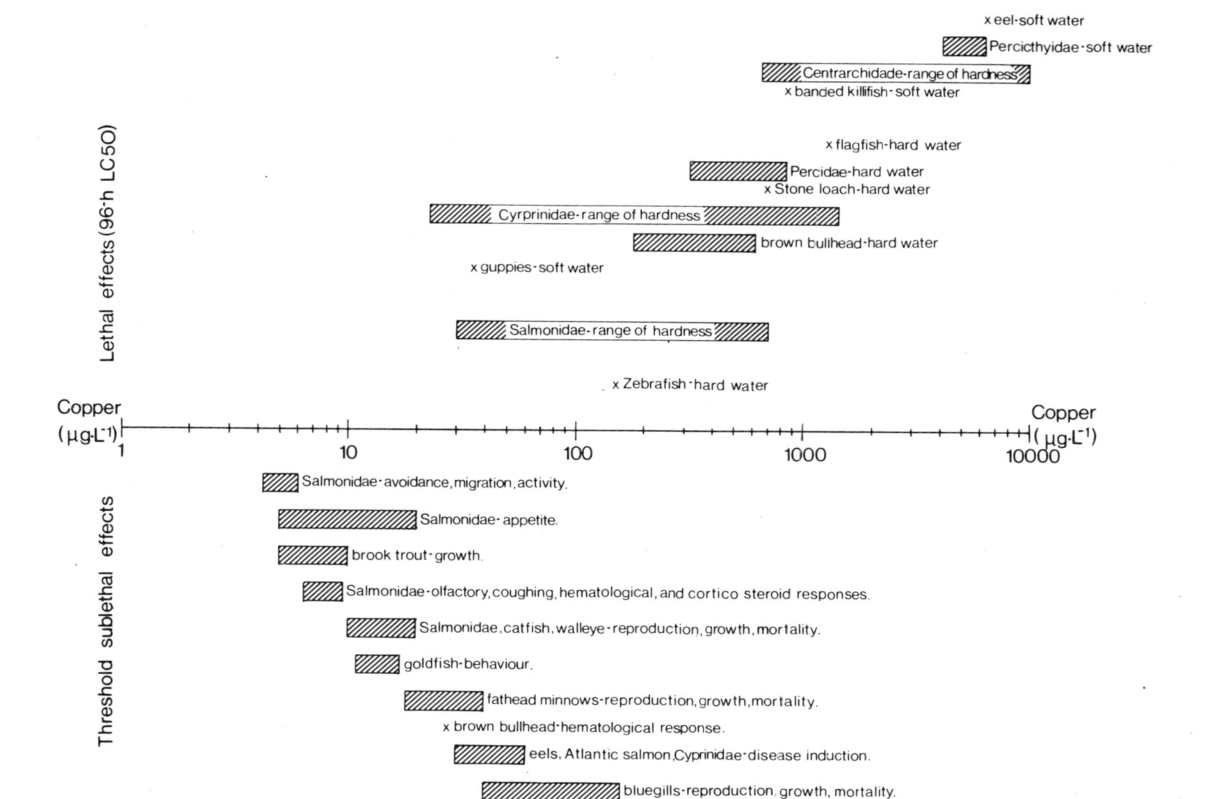

Figure 3. Summary diagram of the lethal, chronic, and sublethal effects of copper on fish. Data are derived from Figure 2, Table 6, and the text.

be observed over a wide range of copper concentrations (23 to 10,200 μg/l); the variations were primarily a result of the effects of water hardness, organic complexing capacity, and species sensitivity. Numerous histological, physiological, and enzymatic responses of fish to near-lethal copper exposures suggest that osmoregulatory failure during exposure to the metal is the probable cause of death.

Sublethal effects of copper on fish occur at concentrations up to the lethal level. The thresholds of effect, however, are observed at concentrations less than 160 μg Cu/l. The behavior and growth of salmonids represent the most sensitive parameters measured (Figure 2). Significant effects on avoidance, activity, coughing, appetite, growth, and migration occurred at copper concentrations between 4 and 10 μg/l. Physiological responses, including inhibition of gill ATPase and plasma GOT activities, plus short-term corticosteroid elevation, were also observed between 5 and 10 μg/l. Threshold effects on reproduction, growth, and mortality, and hence fish production, were observed between 10 and 20 μg/l for salmonids, catfish, and walleyes, between 18 and 40 μg/l for fathead minnows, and between 40 and 160 μg/l for bluegills. Finally, disease was induced in eels, salmonids, suckers, and cyprinids at 30 to 60 μg Cu/l. These data indicate that fish populations are adversely affected at copper concentrations well below the lethal level, so that some species could disappear without direct observable mortality.

REFERENCES

Anderson, B. G. (1948). "The Apparent Thresholds of Toxicity to *Daphnia magna* for Chlorides of Various Metals When Added to Lake Erie Water," *Trans. Am. Fish. Soc.*, **78,** 96–113.

Andrew, R. W. (1976). "Toxicity Relationships to Copper Forms in Natural Waters." Chapter 6 in R. W. Andrew, P. V. Hodson, and D. E. Konasewich, Eds., *Workshop on Toxicity to Biota of Metal Forms in Natural Water.* International Joint Commission, Windsor, Canada, 327 pp.

Andrew, R. W., Biesinger, K. E., and Glass, G. E. (1977). "Effects of Inorganic Complexing on the Toxicity of Copper to *Daphnia magna,*" *Water Res.*, **11,** 309–315.

Arthur, J. W. and Leonard, E. N. (1970). "Effects of Copper on *Gammarus pseudolimnaeus, Physa integra,* and *Campeloma decisum* in Soft Water," *J. Fish Res. Board Can.*, **27,** 1277–1283.

Baker, J. T. P. (1969). "Histological and Electron Microscopical Observations on Copper Poisoning in the Winter Flounder (*Pseudopleuronectes americanus*)," *J. Fish. Res. Board Can.*, **26,** 2785–2793.

Barnes, H. and Stanbury, F. A. (1948). "The Toxic Action of Copper and Mercury Salts Both Separately and When Mixed on the Harpacticoid Copepod, *Nitocera spinipes* (Boeck)," *J. Exp. Biol.*, **25,** 270–275.

Bartlett, L., Rabe, F. W., and Funk, W. H. (1974). "Effects of Copper, Zinc and Cadmium on *Selenastrum capricornutum*," *Water Res.*, **8,** 179–185.

Baudouin, M. F. and Scoppa, P. (1974). "Acute Toxicity of Various Metals to Freshwater Zooplankton," *Bull. Environ. Contam. Toxicol.*, **12,** 745–751.

Benoit, D. A. (1975). "Chronic Effects of Copper on Survival, Growth and Reproduction of the Bluegill (*Lepomis macrochirus*)," *Trans. Am. Fish. Soc.*, **104,** 353.

Bernhard, M. (1955). "Die Kultur von Seeigellarven (*Arbacia lixula* L.) in künstichem and natürlichen Meerwasser mit Hilfe von Ionenaustauschsubstanzen and Komplexbildnern," *Pubbl. Staz. Zool. Napoli,* **XXIV,** 80–95.

Bernhard, M. and Zattera, A. (1973). "Major Pollutants in the Marine Environment." In E. A. Pearson and E. de Fraja Frangipare, Eds., *Marine Pollution and Marine Waste Disposal.* Pergamon Press, Toronto, pp. 195–300.

Betzer, N. and Kott, Y. (1969). "Effects of Halogens on Algae. II: *Cladophora* sp.," *Water Res.*, **3,** 257–264.

Betzer, S. B. and P. P. Yevich (1975). "Copper Toxicity in *Busycon canaliculatum* L.," *Biol. Bull.*, **148,** 16–25.

Biesinger, K. E. and Christensen, G. M. (1972). "Effects of Various Metals on Survival, Growth, Reproduction, and Metabolism of *Daphnia magna*," *J. Fish. Res. Board Can.*, **29,** 1671–1700.

Biesinger, K. E., Andrew, R. W., and Arthur, J. W. (1974). "Chronic Toxicity of NTA (Nitrilotriacetate) and Metal-NTA Complexes to *Daphnia magna*," *J. Fish. Res. Board Can.*, **31,** 486–490.

Bilinski, E. and Jonas, R. E. E. (1973). "Effects of Cadmium and Copper on the Oxidation of Lactate by Rainbow Trout (*Salmo gairdneri*) Gills," *J. Fish. Res. Board Can.*, **30,** 1553–1558.

Bougis, P. (1959). "Sur l'Effet Biologique du Cuivre en Eau de Mer," *C. R. Acad. Sci. Paris,* **249,** 326–328.

Bougis, P. (1962). "Le Cuivre en Ecologie Marine," *Pubbl. Staz. Zool. Napoli,* Suppl. 32, 497–514.

Braek, G. S., Jensen, A., and Mohus, A. (1976). "Heavy Metal Tolerance of Marine Phytoplankton. III: Combined Effects of Copper and Zinc Ions on Cultures of Four Common Species," *J. Exp. Mar. Biol. Ecol.*, **25,** 37–50.

Brković-Popović, I. and Popović, M. (1977a). "Effects of Heavy Metals on Survival and Respiration Rate of Tubificid Worms. I: Effects on Survival," *Environ. Pollut.*, **13,** 65–72.

Brković-Popović, I. and Popović, M. (1977b). "Effects of Heavy Metals on Survival and Respiration Rate of Tubificid Worms: II: Effects on Respiration Rate," *Environ. Pollut.*, **13,** 93–98.

Brown, B. E. (1976). "Observations on the Tolerance of the Isopod *Asellus meridianus* Rac. to Copper and Lead," *Water Res.*, **10**(6), 555–559.

Brown, B. and Ahsanullah, M. (1971) "Effects of Heavy Metals on Mortality and Growth," *Mar. Pollut. Bull.*, **2,** 182–187.

Brown, B. E. and Newell, R. C. (1972). "The Effect of Copper and Zinc on the Metabolism of the Mussel *Mytilus edulis*," *Mar. Biol.*, **16,** 108–118.

Brown, V. M. (1968). "The Calculation of the Acute Toxicity of Mixtures of Poisons to Rainbow Trout," *Water Res.,* **2,** 723–733.

Brown, V. M. (1973). "Concepts and Outlook in Testing the Toxicity of Substances to Fish." In G. E. Glass, Ed., *Bioassay Techniques and Environmental Chemistry.* Ann Arbor Science Publ., Ann Arbor, Mich., pp. 73–95.

Brown, V. M. and Dalton, R. A. (1970). "The Acute Lethal Toxicity to Rainbow Trout of Mixtures of Copper, Phenol, Zinc, and Nickel," *J. Fish. Biol.,* **2,** 211–216.

Brown, V. M., Shaw, T. L., and Shurben, D. G. (1974). "Aspects of Water Quality and the Toxicity of Copper to Rainbow Trout," *Water Res.,* **8,** 797–803.

Brungs, W. A., Geckler, J. R. and Gast, M. (1976). "Acute and Chronic Toxicity of Copper to the Fathead Minnow in a Surface Water of Variable Quality," *Water Res.,* **10,** 37–43.

Bryan, G. W. and Hummerstone, L. G. (1971). "Adaptations of the Polychaete *Nereis diversicolor* to Estuarine Sediments Containing High Concentrations of Heavy Metals. I: General Observations and Adaptation to Copper," *J. Mar. Biol. Assoc. U.K.,* **51,** 845–863.

Bryan, G. W. and Hummerstone, L. G. (1973). "Brown Seaweed as an Indicator of Heavy Metals in Estuaries in Southwest England," *J. Mar. Biol. Assoc. U.K.,* **53**(3), 705–720.

Buikema, A. L., Jr., Cairns, J., Jr., and Sullivan, G. W. (1974) "Evaluation of *Philodina acuticornis* (Rotifera) as a Bioassay Organism for Heavy Metals," *Water Res. Bull.,* **10,** 648–661.

Bushnell, J. H. (1974). "Bryozoans (Ectaprocta)." Chapter 6 in C. W. Hart, Jr., and S. L. H. Fuller, Eds., *Pollution Ecology of Freshwater Invertebrates.* Academic Press, New York, 389 pp.

Calabrese, A., MacInnes, J. R., Nelson, D. A., and Miller, J. E. (1977). "Survival and Growth of Bivalve Larvae under Heavy-Metal Stress," *Mar. Biol.,* **41,** 179–184.

Calamari, D. and Marchetti, R. (1973). "The Toxicity of Mixtures of Metals and Surfactants to Rainbow Trout (*Salmo gairdneri* Rich.)," *Water Res.,* **7,** 1453.

Cedeno-Maldonado, A. and Swader, J. A. (1974). "Studies on the Mechanism of Copper Toxicity in *Chlorella,*" *Weed Sci.,* **22,** 443–449.

Christensen, G. M. (1971–72). "Effects of Metal Cations and Other Chemicals upon the *in vitro* Activity of Two Enzymes in the Blood Plasma of the White Sucker, *Catastomus commersoni* (Lacépède)," *Chem.-Biol. Interactions,* **4,** 351–361.

Christensen, G. M., McKim, J. M., Brungs, W. A., and Hunt, E. P. (1972). "Changes in the Blood of the Brown Bullhead [*Ictalurus nebulosus* (Lesueur)] following Short and Long Term Exposure to Copper(II)," *Toxicol. Appl. Pharmacol.,* **23,** 417.

Chynoweth, D. P., Black, J. A., and Mancy, K. H. (1976). "Effects of Organic Pollutants on Copper Toxicity to Fish." Chapter 7 in R. W. Andrew, P. V. Hodson, and D. E. Konasewich, Eds., *Workshop on Toxicity to Biota of Metal Forms in Natural Water.* International Joint Commission, Windsor, Canada, 327 pp.

Clarke, G. L. (1947). "Poisoning and Recovery in Barnacles and Mussels," *Biol. Bull.,* **92,** 73–91.

Conner, P. M. (1972). "Acute Toxicity of Heavy Metals to Some Marine Larvae," *Mar. Pollut. Bull.,* **3,** 190–192.

Cook, R. H. and Côte, R. P. (1972). *The Influence of Humic Acids on the Toxicity of Copper and Zinc to Juvenile Atlantic Salmon as Derived by the Toxic Unit Concept.* MS 72-5, Environmental Protection Service, Environment Canada.

Corner, E. D. S. and Sparrow, B. W. (1956). "The Modes of Action of Toxic Agents. I: Observations on the Poisoning of Certain Crustaceans by Copper and Mercury," *J. Mar. Biol. Assoc. U.K.,* **35,** 531–548.

D'Agostino, A. and Finney, C. (1974). "The Effect of Copper and Cadmium on the Development of *Tigriopus japonicus.*" In F. J. Vernberg and W. B. Vernberg, Eds., *Pollution and Physiology of Marine Organisms,* Symposium XIII. Academic Press, New York, pp. 445–463.

Davey, E. N., Morgan, N. J., and Erickson, S. J. (1973). "A Biological Measurement of the Copper Complexation Capacity of Seawater," *Limnol. Oceanogr.,* **18,** 993–997.

Delhaye, W. and Cornet, D. (1975). "Contribution to the Study of the Effect of Copper on *Mytilus edulis* during Reproductive Period," Comp. Biochem. Physiol., **50A,** 511–518.

Donaldson, E. M. and Dye, H. M. (1975). "Corticosteroid Concentrations in Sockeye Salmon (*Oncorhynchus nerka*) Exposed to Low Concentrations of Copper," *J. Fish. Res. Board Can.,* **32,** 533–539.

Doty, M. S. and Oguri, M. (1957). "Evidence for a Photosynthetic Daily Periodicity," *Limnol. Oceanogr.,* **2,** 37–40.

Drummond, R. A., Spoor, W. A., and Olson, G. F. (1973). "Some Short-Term Indicators of Sublethal Effects of Copper on Brook Trout, *Salvelinus fontinalis.*" *J. Fish. Res. Board Can.,* **30,** 698–701.

Eaton, J. G. (1973). "Chronic Toxicity of a Copper, Cadmium and Zinc Mixture to the Fathead Minnow (*Pimephales promelas* Rafinesque)," *Water Res.,* **7,** 1723–1736.

Eisler, R. and Gardner, G. R. (1973). "Acute Toxicology to an Estuarine Teleost of Mixtures of Cadmium, Copper, and Zinc Salts," *J. Fish. Biol.,* **5,** 131–142.

Engel, D. W., Sunda, W. G., and Thuotte, R. M. (1976). "Effects of Copper on Marine Fish Eggs and Larvae," *Environ. Health Perspect.,* **17,** 288–289.

Erickson, S. J. (1972). "Toxicity of Copper to a Marine Diatom in Unenriched Inshore Seawater," *J. Phycol.,* **8**(4), 318–323.

Erickson, S. J., Lackie, N., and Maloney, T. E. (1970). "A Screening Technique for Estimating Copper Toxicity to Estuarine Phytoplankton," *J. Water Pollut. Control Fed.,* **42,** 270–278.

Etges, F. J. (1963). "The Effects of Some Molluscicidal Chemicals on Chemokinesis in *Australorbis glabratus,*" *Am. J. Trop. Med. Hyg.,* **12,** 701–704.

Finney, D. J. (1971). *Probit Analysis,* 3rd ed. Cambridge University Press, Cambridge, England, 1971, 333 pp.

Fitzgerald, G. P. (1964a). "Evaluation of Potassium Permanganate as an Algicide for Water Cooling Towers," *IEC Prod. Res. Dev.,* **3,** 82–85.

Fitzgerald, G. P. (1964b). "Laboratory Evaluation of Potassium Permanganate as an Algicide for Water Reservoirs," *Southwest Water Works Assoc. J.,* **45**(10), 16–17.

Fitzgerald, G. P. (1975). "Are Chemicals Used in Algae Control Biodegradable?" *Water Sewage Works,* May, pp. 82–85.

Fogels, A. and Sprague, J. B. (1977). "Comparative Short-Term Tolerance of Zebra Fish, Flagfish and Rainbow Trout to Five Poisons Including Potential Reference Toxicants," *Water Res.,* **11,** 811–817.

Fogg, G. E. and Westlake, D. F. (1955). "The Importance of Extracellular Products of Algae in Freshwater," *Verh. Int. Verein. Limnol.,* **12,** 219–232.

Folmar, L. C. (1976). "Overt Avoidance Reaction of Rainbow Trout Fry to 9 Herbicides," *Bull. Environ. Contam. Toxicol.,* **15**(5), 509–514.

Foster, P. L. (1977). "Copper Exclusion as a Mechanism of Heavy Metal Tolerance in a Green Alga," *Nature,* **269,** 322–323.

Fujiya, M. (1960). "Studies on the Effects of Copper Dissolved in Sea Water on the Oyster," *Bull. Jap. Soc. Sci. Fish.,* **26,** 462–468.

Gächter, R., Lum-Shue-Chan, K., and Chau, Y. K. (1973). "Complexing Capacity of the Nutrient Medium and Its Relation to Inhibition of Algal Photosynthesis by Copper," *Schweiz. Z. Hydrol.,* **35,** 253–261.

Gardner, G. R. and LaRoche, G. (1973). "Copper-Induced Lesions in Estuarine Teleosts," *J. Fish. Res. Board Can.,* **30,** 363–368.

Geckler, J. R., Horning, W. B., Neiheisel, T. M., Pickering, Q. H., Robinson, E. L., and Stephan, C. E. (1976). *Validity of Laboratory Tests for Predicting Copper Toxicity in Streams.* U.S. Environmental Protection Agency, EPA 600/3-76-116, Duluth, Minn., 192 pp.

Gibson, C. E. (1972). "The Algicidal Effect of Copper on a Green and a Blue-Green Alga and Some Ecological Implications," *J. Appl. Ecol.,* **9,** 513–519.

Goettl, J. P. Jr., Davies, P. H., and Sinley, J. R. (1976). "Water Pollution Studies." In *Colorado Fisheries Research Review 1972–1975.* State Publication Code DOW-R-R-F72-75, pp. 68–74.

Goldberg, E. D. and Arrhenius, G. O. S. (1958). "Chemistry of Pacific Pelagic Sediments," *Geochim. Cosmochim. Acta,* **13,** 153–212.

Grande, M. (1967). "Effect of Copper and Zinc on Salmonid Fishes." In *Advances in Water Pollution Research, Proceedings of the 3rd International Conference,* Munich, Germany, September, 1966, Vol. 1 . Water Pollution Control Federation, Washington, D.C., pp. 97–111.

Hale, J. G. (1977). "Toxicity of Metal Mining Wastes," *Bull. Environ. Contam. Toxicol.,* **17,** 66–73.

Hara, T. J., Law, Y. M. C., and Macdonald, S. (1976). "Effects of Mercury and Copper on the Olfactory Response in Rainbow Trout, *Salmo gairdneri,*" *J. Fish. Res. Board Can.,* **33,** 1568–1573.

Hargreaves, J. W. and Whitton, B. A. (1976). "Effect of pH on Tolerance of *Hormidium rivulare* to Zinc and Copper," *Oecologia,* **26**(3), 235–243.

Harry, H. W. and Aldrich, D. V. (1963). "The Distress Syndrome in *Taphius glabratus* (Say) as a Reaction to Toxic Concentrations of Inorganic Ions," *Malacologia,* **1,** 283–289.

Hassall, K. A. (1962). "A Specific Effect of Copper on the Respiration of *Chlorella vulgaris,*" *Nature,* **193,** 90.

Hazel, G. R. and Meith, S. J. (1970). "Bioassay of King Salmon Eggs and Sac Fry in Copper Solutions," *Calif. Fish. Game,* **56,** 121–124.

Herbert, D. W. M. and Van Dyke, J. M. (1964). "The Toxicity to Fish of Mixtures of Poisons. II: Copper-Ammonia and Zinc-Phenol Mixtures," *Ann. Appl. Biol.,* **53,** 415–421.

Hodson, P. V. and Sprague, J. B. (1975). "Temperature-Induced Changes in Acute Toxicity of Zinc to Atlantic Salmon (*Salmo salar*)," *J. Fish. Res. Board Can.,* **32,** 1–10.

Hodson, P. V., Blunt, B. R., Spry, D. J., and Austen, K. (1977). "Evaluation of Erythrocyte δ-Aminolevulinic Acid Dehydratase Activity as a Short-Term Indicator in Fish of a Harmful Exposure to Lead," *J. Fish. Res. Board Can.,* **34,** 501–508.

Horne, A. J. and Goldman, C. R. (1972). "Nitrogen Fixation in Clear Lake, California. I: Seasonal Variation and the Role of Heterocysts," *Limnol. Oceanogr.,* **17,** 678–692.

Horne, A. J. and Goldman, C. R. (1974). "Suppression of Nitrogen Fixation by Blue-Green Algae in a Eutrophic Lake with Trace Additions of Copper," *Science,* **183,** 409–411.

Howarth, R. S. and Sprague, J. B. (1978). "Copper Lethality to Rainbow Trout in Waters of Various Hardness and pH," *Water Res.,* **12,** 455–462.

Hubschman, J. H. (1967a). "Effects of Copper on the Crayfish *Orconectes rusticus* (Girard). I: Acute Toxicity," *Crustaceana,* **12,** 33–42.

Hubschman, J. H. (1967b). "Effects of Copper on the Crayfish *Orconectes rusticus* (Girard). II: Mode of Toxic Action," *Crustaceana,* **12,** 141–150.

Hutchinson, T. C. (1973). "Comparative Studies of the Toxicity of Heavy Metals to Phytoplankton and Their Synergistic Interactions," *Water Pollut. Res. Canada,* **8,** 68–75.

International Joint Commission (1976). *Annual Report of the Water Quality Objectives Subcommittee.* Appendix A to the 1975 Water Quality Board Report. International Joint Commission, Windsor, Ontario, Canada.

Jackim, E. (1973). Influence of Lead and Other Metals on Fish δ-Aminolevulinate Dehydrase Activity," *J. Fish Res. Board Can.,* **30,** 560–562.

Jackim, E., Hamlin, J. M., and Sonis, S. (1970). "Effect of Metal Poisoning on Five Liver Enzymes in the Killifish (*Fundulus heteroclitus*)," *J. Fish. Res. Board Can.,* **27,** 383–390.

Jones, J. R. E. (1942). "The Effect of Ionic Copper on the Oxygen Consumption of *Gammarus pulex* and *Polycelis nigra," J. Exp. Biol.,* **18,** 153–161.

Jones, L. H., Jones N. V., and Radlett, A. J. (1976). "Some Effects of Salinity on the Toxicity of Copper to the Polychaete *Nereis diversicolor," Estuarine Coastal Mar. Sci.,* **4,** 107–111.

Jones, M. B. (1975). "Effects of Copper on Survival and Osmoregulation in Marine and Brackish Water Isopods (Crustacea)." In M. Barnes, Ed., *Proceedings of the 9th European Marine Biology Symposium.* University of Aberdeen Press, Aberdeen, Scotland, pp. 419–431.

Karbe, L. (1972). "Marine Hydroiden als Testorganismen zur Prüfung der Toxizität von Abwasserstoffen: Die Wirkung von Schwermetallen auf Kolonien von *Eirene viridula,*" *Mar. Biol.,* **12,** 316–328.

Keeney, W. L., Breck, W. G., Vanloon, G. W., and Page, J. A. (1976). "The Determination of Trace Metals in *Cladophora glomerata—C. glomerata* as a potential biological monitor," *Water Res.,* **10,** 981–984.

Kemp, A. L. W. and Thomas, R. L. (1976). "Cultural Impact on the Geochemistry of the Sediments of Lakes Ontario, Erie and Huron," *Geosci. Can.,* **3**(3), 191–207.

Khobot'yev, V. G., Kapkov, V. I., Rukhadze, Ye. G., Turrenina, N. V., and Shidlovskaya, N. A. (1975). "The Toxic Effect of Copper Complexes on Algae," *Hydrobiol. J.,* **11,** 33–38.

Kleerekoper, H., Westlake, G. F., Matis, J. H., and Gensler, P. J. (1972). "Orientation of Goldfish (*Carassius auratus*) in Response to a Shallow Gradient of Copper in an Open Field," *J. Fish. Res. Board Can.,* **29,** 45–54.

Kleerekoper, H., Waxman, J. B., and Matis, J. H. (1973). "Interaction of Temperature and Copper Ions as Orienting Stimuli in the Locomotor Behavior of the Goldfish (*Carassius auratus*)," *J. Fish. Res. Board Can.,* **30,** 725–728.

Lett, P. F., Farmer, G. J., and Beamish, F. W. H. (1976). "Effect of Copper on Some Aspects of the Bioenergetics of Rainbow Trout (*Salmo gairdneri*)," *J. Fish. Res. Board Can.,* **33,** 1335–1342.

Lewis, A. G., Whitfield, P. H., and Ramnarine, A. (1972). "Some Particulate and Soluble Agents Affecting the Relationship between Metal Toxicity and Organism Survival in the Calanoid Copepod *Euchaeta japonica,*" *Mar. Biol.,* **17,** 215–221.

Lewis, A. G., Whitfield, P., and Ramnarine, A. (1973). "The Reduction of Copper Toxicity in a Marine Copepod by Sediment Extract," *Limnol. Oceanogr.,* **18,** 324–326.

Lewis, S. D. and Lewis, W. M. (1971). "The Effect of Zinc and Copper on the Osmolarity of Blood Serum of the Channel Catfish, *Ictalurus punctatus* Rafinesque, and Golden Shiner, *Notemigonus crysoleucas* Mitchill," *Trans. Am. Fish. Soc.,* **100,** 639–643.

Liepolt, R. and Weber, E. (1958). "Die Giftwirkung von Kupfersulfat auf Wasserorganismen," *Wass. Abwasser,* 1958, pp. 335–353.

Lloyd, R. (1961a). "The Toxicity of Mixtures of Zinc and Copper Sulphates to Rainbow Trout (*Salmo gairdneri* Richardson)," *Ann. Appl. Biol.,* **49,** 535–538.

Lloyd, R. (1961b). "Effect of Dissolved Oxygen Concentrations on the Toxicity of Several Poisons to Rainbow Trout (*Salmo gairdneri* Richardson)," *J. Exp. Biol.,* **38,** 447–455.

Lorz, H. W. and McPherson, B. P. (1976). "Effects of Copper on Downstream Migration, Gill ATPase, and Adaption to Sea Water of Juvenile Coho Salmon (*Oncorhynchus kisutch*)," *J. Fish. Res. Board Can.,* **33,** 2023–2030.

Macek, K. J. (1975). "Acute Toxicity of Pesticide Mixtures to Bluegills," *Bull. Environ. Contam. Toxicol.,* **14,** 648–652.

MacInnes, J. R. and Thurberg, F. P. (1973). "Effects of Metals on the Behaviour and Oxygen Consumption of the Mud Snail," *Mar. Pollut. Bull.,* **4,** 185–186.

Mandelli, E. F. (1975). "The Effects of Desalination Brines on *Crassostrea virginica* (Gmelin)," *Water Res.,* **9,** 287–295.

Marks, G. W. (1938). "The Copper Content and Copper Tolerance of Some Species of Molluscs of the Southern California Coast," *Biol. Bull.*, **75**(2), 224–237.

Martin, M. J., Piltz, F. M., and Reish, D. J. (1975). "Studies on the *Mytilus edulis* Community in Alamitos Bay, California. 5: The Effects of Heavy Metals on Byssal Thread Production," *Veliger*, **18**(2), 183–188.

Martin, S. G. (1971). "An Analysis of the Histopathologic Effects of Copper Sulfate on the Asiatic Freshwater clam, *Corbicula fluminen* (Muller)." Ph.D. Thesis, University of Washington.

Marvin, K. T., Lansford, L. M., and Wheeler, R. S. (1961). "Effects of Copper Ore on the Ecology of a Lagoon," *U.S. Fish. Wildlife Serv. Fish. Bull.*, **184**, 153–160.

McKim, J. M. and Benoit, D. A. (1971). "Effects of Long-Term Exposures to Copper on Survival, Growth, and Reproduction of Brook Trout (*Salvelinus fontinalis*)," *J. Fish. Res. Board Can.*, **28**, 655–662.

McKim, J. M. and Benoit, D. A. (1974). "Duration of Toxicity Tests for Establishing 'No Effect' Concentrations for Copper with Brook Trout (*Salvelinus fontinalis*)," *J. Fish. Res. Board Can.*, **31**, 449–452.

McKim, J. M., Christensen, G. M., and Hunt, E. P. (1970). "Changes in the Blood of Brook Trout (*Salvelinus fontinalis*) after Short-Term and Long-Term Exposure to Copper," *J. Fish Res. Board Can.*, **27**, 1883–1889.

McLean, R. O. (1974). "The Tolerance of *Stigeoclonium tenue* Kütz. to Heavy Metals in South Wales," *Br. Phycol. J.*, **9**, 91–95.

McLeese, D. W. (1974). "Toxicity of Copper at Two Temperatures and Three Salinities to the American Lobster (*Homarus americanus*)," *J. Fish. Res. Board Can.*, **31**, 1949–1952.

McLeese, D. W. (1975). "Chemosensory Response of American Lobsters (*Homarus americanus*) in the Presence of Copper and Phosphamidon," *J. Fish. Res. Board Can.*, **32**(11), 2055–2060.

McLusky, D. S. and Phillips, C. N. K. (1975). "Some Effects of Copper on the Polychaete *Phyllodoce maculata*," *Estuarine Coastal Mar. Sci.*, **3**, 103–108.

Miller, M. A. (1946). "Toxic Effects of Copper on Attachment and Growth of *Bugula neritina*," *Biol. Bull.*, **90**, 122–140.

Moore, G. E. and Kellerman, K. F. (1904). *A Method of Destroying or Preventing the Growth of Algae and Certain Pathogenic Bacteria in Water Supplies*. Bull. Bur. Plant Ind. U.S. Dept. Agric. 64.

Morris, O. P. and Russell, G. (1973). "Effect of Chelation on Toxicity of Copper," *Mar. Pollut. Bull.*, **4**, 159–160.

Mount, D. I. (1968). "Chronic Toxicity of Copper to Fathead Minnows (*Pimephales promelas* Rafinesque)," *Water Res.*, **2**, 215–224.

Mount, D. I. and Stephan, C. E. (1969). "Chronic Toxicity of Copper to the Fathead Minnow (*Pimephales promelas*) in Soft Water," *J. Fish. Res. Board Can.*, **26**, 2449–2457.

Nehring, R. B. (1976). "Aquatic Insects as Biological Monitors of Heavy Metal Pollution," *Bull. Environ. Contam. Toxicol.*, **15**, 147–154.

Newhouse, J., Doty, M. S., and Tsuda, R. T. (1967). "Some Diurnal Features of a Neritic Surface Population," *Limnol. Oceanogr.*, **12**, 207–212.

Niemi, A. (1972). "Effects of Toxicants on Brackish Water Phytoplankton Assimilation," *Commentat. Biol. Soc. Sci. Fenn.*, **55**, 1–19.

O'Hara, J. (1971). "Alterations in Oxygen Consumption by Bluegills Exposed to Sublethal Treatment with Copper," *Water Res.*, **5**, 321–327.

Okazaki, R. K. (1976). "Copper Toxicity in the Pacific Oyster *Crassostrea gigas*," *Bull. Environ. Contam. Toxicol.*, **16**, 658–664.

Olson, K. R. and Harrel, R. C. (1973). "Effects of Salinity on Acute Toxicity of Mercury, Copper, and Chromium for *Rangia cuneata* (Pelecypoda, Mactridae)," *Contrib. Mar. Sci.*, **17**, 9.

Overnell, J. (1975). "The Effect of Some Heavy Metal Ions on Photosynthesis in a Freshwater Alga," *Pestic. Biochem. Physiol.*, **5**, 19–26.

Overnell, J. (1976). "Inhibition of Marine Algal Photosynthesis by Heavy Metals," *Mar. Biol.*, **38**, 335–342.

Packer, R. K. and Dunson, W. A. (1970). "Effects of Low Environmental pH on Blood pH and Sodium Balance of Brook Trout," *J. Exp. Zool.*, **174**, 65–72.

Packer, R. K. and Dunson, W. A. (1972). "Anoxia and Sodium Loss Associated with the Death of Brook Trout at low pH," *Comp. Biochem. Physiol.*, **41A**, 17–26.

Pagenkopf, G. K., Russo, R. C., and Thurston, R. V. (1974). "Effect of Complexation on Toxicity of Copper to Fishes," *J. Fish. Res. Board Can.*, **31**, 462–465.

Perrin, D. D. and Sayce, I. G. (1967). "Computer Calculations of Equilibrium Concentrations in Mixtures of Metal Ions and Complexing Species," *Talanta*, **14**, 833–842.

Pickering, Q. H. and Henderson, C. (1966). "The Acute Toxicity of Some Heavy Metals to Different Species of Warmwater Fish," *Air Water Pollut. Int. J.*, **10**, 453–463.

Pickering, Q., Brungs, W., and Gast, M. (1977). "Effects of Exposure Time and Copper Concentration on Reproduction of the Fathead Minnow (*Pimephales promelas*)," *Water Res.*, **11**, 1079–1084.

Pippy, J. H. C. and Hare, G. M. (1969). "Relationship of River Pollution to Bacterial Infection in Salmon (*Salmo salar*) and Suckers (*Catastomus commersoni*)," *Trans. Am. Fish. Soc.*, **98**, 685–690.

PLUARG (1978). *International Reference Group on Great Lakes Pollution from Land Use Activities.* Final Report, Environmental Management Strategy for the Great Lakes System. International Joint Commission, Windsor, Ontario, Canada, 173 pp.

Prytherch, H. F. (1934). "The Role of Copper in the Setting, Metamorphosis and Distribution of the American Oyster, *Ostrea virginica*," *Ecol. Monogr.*, **4**(1), 47–107.

Pyefinch, K. A. and Mott, J. C. (1948), "The Sensitivity of Barnacles and Their Larvae to Copper and Mercury," *J. Exp. Biol.*, **25**, 276–298.

Ramamoorthy, S. and Kushner, D. J. (1975). "Heavy Metal Binding Components of River Water," *J. Fish. Res. Board Can.*, **32**, 1755–1766.

Raymont, J. E. and Shields, J. (1964). "Toxicity of Copper and Chromium in the Marine Environment." *Adv. Water Pollut. Res. Proc. 1st Int. Conf. 1962*, **3**, 275–383.

Reeve, M. R., Grice, G. D., Gibson, V. R., Walter, M. A., Darcy, K., and Ikeda, T. (1976). "A Controlled Environmental Pollution Experiment (CEPEX) and Its Usefulness in the Study of Larger Marine Zooplankton under Toxic Stress." In A. P. M. Lockwood, Ed., *Effects of Pollutants on Aquatic Organisms.* Cambridge University Press, Cambridge, England, 193 pp.

Reeve, M. R., Gamble, J. C., and Walter, M. A. (1977). "Experimental Observations on the Effects of Copper on Copepods and Other Zooplankton: Controlled Ecosystem Pollution Experiment," *Bull. Mar. Sci.,* **27**(1), 92–104.

Rehwoldt, R., Bida, G., and Nerrie, B. (1971). "Acute Toxicity of Copper, Nickel and Zinc Ions to Some Hudson River Fish Species," *Bull. Environ. Contam. Toxicol.,* **6**, 445.

Rehwoldt, R., Lasko, L., Shaw, C., and Wirhowski, E. (1973). "The Acute Toxicity of Some Heavy Metal Ions toward Benthic Organisms," *Bull. Environ. Contam. Toxicol.,* **10,** 291–294.

Reish, D. J. and Carr, R. S. (1978). "The Effect of Heavy Metals on the Survival, Reproduction, Development, and Life Cycles for Two Species of Polychaetous Annelids," *Mar. Pollut. Bull.,* **9**(1), 24–27.

Reish, D. J., Piltz, F., Martin, J. M., and Word, J. Q. (1974). "Induction of Abnormal Polychaete Larvae by Heavy Metals," *Mar. Pollut. Bull.,* **5**(8), 125–126.

Reish, D. J., Martin, J. M., Piltz, F. M., and Word, J. Q. (1976). "The Effect of Heavy Metals on Laboratory Populations of Two Polychaetes with Comparisons to the Water Quality Conditions and Standards in Southern California Marine Waters," *Water Res.,* **10,** 299–302.

Rödsaether, M. C., Olafsen, J., Raa, J., Myhre, K., and Steen, J. B. (1977). "Copper as an Initiating Factor of Vibriosis (*Vibrio anguillarum*) in Eel (*Anguilla anguilla*)," *J. Fish. Biol.,* **10,** 17–21.

Rothschild, L. and Tuft, P. H. (1950). "The Physiology of Sea-Urchin Spermatozoa. The Dilution Effect in Relation to Copper and Zinc," *J. Exp. Biol.,* **27,** 59–72.

Rounsfell, G. A. and Evans, J. E. (1958). *Large-Scale Experimental Test of Copper Sulphate as a Control for the Florida Red Tide.* U.S. Dept. Interior, Fish. Wildlife Serv. Spec Sci. Rep. Fish. 270, 57 pp.

Russell-Hunter, W. (1949). "The Poisoning of *Marinogammarus marinas* by Cupric Sulphate and Mercuric Chloride," *J. Exp. Biol.,* **26,** 113–124.

Saliba, L. J. and Ahsanullah, M. (1973). "Acclimation and Tolerance of *Artemia salina* and *Ophryotrocha labronica* to Copper Sulphate," *Mar. Biol.,* **23,** 297–303.

Saunders, R. L. and Sprague, J. B. (1967). "Effects of Copper-Zinc Mining Pollution on a Spawning Migration of Atlantic Salmon," *Water Res.,* **1**, 419–432.

Sauter, S., Buxton, K. S., Macek, K. J., and Petrocelli, S. R. (1976). *Effects of Exposure to Heavy Metals on Selected Freshwater Fish.* U.S. Environmental Protection Agency, Office of Research and Development, EPA 600/3-76-088, Duluth, Minn.

Saward, D., Stirling, A., and Topping, G. (1975). "Experimental Studies on the Effects of Copper on a Marine Food Chain," *Mar. Biol.,* **29,** 351–361.

Schalm, O. W., Jain, N. C., and Carroll, E. J. (1975). *Veterinary Hematology,* 3rd ed. Lea and Febiger, Philadelphia, 807 pp.

Schreck, C. B. and Lorz, H. W. (1978). "Stress Response of Coho Salmon (*Oncorhynchus kisutch*) Elicited by Cadmium and Copper and Potential Use of Cortisol as an Indicator of Stress," *J. Fish. Res. Board Can.*, **35,** 1124–1129.

Scott, D. M. and Major, C. W. (1972). "The Effect of Copper(II) on Survival, Respiration, and Heart Rate in the Common Blue Mussel, *Mytilus edulis*," *Biol. Bull.*, **143,** 679–688.

Scott, W. B. and Crossman, E. J. (1973). *Freshwater Fishes of Canada.* Fish. Res. Board Can. Bull. 184, Ottawa, 966 pp.

Segar, D. A. and Pellenberg, R. E. (1973). "Trace Metals in Carbonate and Organic Rich Sediments," *Mar. Pollut. Bull.*, **4,** 138–142.

Sellers, Jr., C. M., Heath, A. G., and Bass, M. L. (1975). "The Effect of Sublethal Concentrations of Copper and Zinc on Ventilatory Activity, Blood Oxygen and pH in Rainbow Trout (*Salmo gairdneri*)," *Water Res.*, **9,** 401–408.

Shaw, T. L. and Brown, V. M. (1974). "The Toxicity of Some Forms of Copper to Rainbow Trout," *Water Res.*, **8,** 377–382.

Shaw, W. H. R. and Grushkin, B. (1957). "The Toxicity of Metal Ions to Aquatic Organisms," *Arch. Biochem. Biophys.*, **67,** 447–452.

Solbe, J. F. de L. G. and Cooper, V. A. (1976). "Studies on the Toxicity of Copper Sulphate to Stone Loach *Noemacheilus barbatulus* (L) in Hard Water," *Water Res.*, **10,** 523–527.

Sprague, J. B. (1964a). "Avoidance of Copper-Zinc Solutions by Young Salmon in the Laboratory," *J. Water Pollut. Control Fed.*, **36,** 990–1004.

Sprague, J. B. (1964b). "Lethal Concentrations of Copper and Zinc for Young Atlantic Salmon," *J. Fish. Res. Board Can.*, **21,** 17–26.

Sprague, J. B. (1968). "Promising Anti-pollutant: Chelating Agent NTA Protects Fish from Copper and Zinc," *Nature*, **220,** 1345–1346.

Sprague, J. B. (1969). "Measurement of Pollutant Toxicity to Fish. I: Bioassay Methods for Acute Toxicity," *Water Res.*, **3,** 793–821.

Sprague, J. B. (1970). "Measurement of Pollutant Toxicity to Fish. II: Utilizing and Applying Bioassay Results," *Water Res.*, **4,** 3–32.

Sprague, J. B. and Ramsay, B. (1965). "Lethal Levels of Mixed Copper-Zinc Solutions for Juvenile Salmon," *J. Fish. Res. Board Can.*, **22,** 425–432.

Sprague, J. B., Elson, P. F., and Saunders, R. L. (1965). "Sublethal Copper-Zinc Pollution in a Salmon River—a Field and Laboratory Study," *Air Water Pollut. Int. J.*, **9,** 531–543.

Stebbing, A. R. D. (1976). "The Effects of Low Metal Levels on a Clonal Hydroid," *J. Mar. Biol. Assoc. U.K.*, **56**(4), 977–994.

Steeman Nielsen, E. and Kamp-Nielsen, L. (1970). "Influence of Deleterious Concentrations of Copper on the Growth of *Chlorella pyrenoidosa*," *Physiol. Plant.*, **23,** 828–840.

Steeman Nielsen, E. and Wium-Andersen, S. (1970). "Copper Ions as Poison in the Sea and Freshwater," *Mar. Biol.*, **6,** 93–97.

Steeman Nielsen, E. and Wium-Andersen, S. (1972). "Influence of Copper on Photosynthesis of Diatoms, with Special Reference to an Afternoon Depression," *Verh. Int. Verein. Limnol.*, **18,** 78–83.

Steeman Nielsen, E., Kamp-Nielsen, L., and Wium-Andersen S., (1969). "The Effect of Deleterious Concentrations of Copper on the Photosynthesis of *Chlorella pyrenoidosa*," *Physiol. Plant.*, **22,** 1121–1133.

Stephenson, R. R. and Taylor, D. (1975). "The Influence of EDTA on the Mortality and Burrowing Activity of the Clam (*Venerupis decussata*) Exposed to Sublethal Concentrations of Copper," *Bull. Environ. Contam. Toxicol.*, **14,** 304–308.

Stirling, E. A. (1975). "Some Effects of Pollutants on the Behaviour of the Bivalve *Tellina tenuis*," *Mar. Pollut. Bull.*, **6,** 122–124.

Stokes, P. M. and Hutchinson, T. C. (1976). "Copper Toxicity to Phytoplankton as Affected by Organic Ligands, Other Cations and Inherent Tolerance of Algae to Copper." In R. W. Andrews, P. V. Hodson, and D. E. Konasewich, Eds., *Toxicity to Biota of Metal Forms in Natural Water.* Proceedings of a workshop in Duluth, Minn., Oct. 7–8, 1975, pp. 159–185.

Stokes, P. M., Hutchinson, T. C., and Krauter, K. (1973). "Heavy Metal Tolerance in Algae Isolated from Polluted Lakes near the Sudbury, Ontario Smelters," *Water Pollut. Res. Can.*, **8,** 178–201.

Thurberg, F. P., Dawson, M. A., and Collier, R. S. (1973). "Effects of Copper and Cadmium on Osmoregulation and Oxygen Consumption in Two Species of Estuarine Crabs," *Mar. Biol.*, **23,** 171–175.

Trollope, D. R. and Evans, B. (1976). "Concentrations of Copper, Iron, Lead, Nickel and Zinc in Freshwater Algal Blooms," *Environ. Pollut.*, **11,** 109–116.

U.S. Environmental Protection Agency (1976). *Quality Criteria for Water.* EPA-440/9-76-023. Washington, D.C. 20460, 501 pp.

Van Loon, J. C. and Beamish, R. J. (1977). "Heavy Metal Contamination by Atmospheric Fallout of Several Flin Flon Area Lakes and the Relation to Fish Populations," *J. Fish. Res. Board Can.*, **34,** 899–906.

Vijayamadhavan, K. T. and Iwai, T. (1975). "Histochemical Observations on the Permeation of Heavy Metals into Taste Buds of Goldfish," *Bull. Jap. Soc. Sci. Fish.*, **41,** 631–639.

Waiwood, K. G. and Beamish, F. W. H. (In press a). "Effect of copper, pH and Hardness on the Critical Swimming Performance of Rainbow Trout (*Salmo gairdneri* Richardson)," *Water Res.*, **12,** 611–620.

Waiwood, K. G. and Beamish, F. W. H., (In press b). "The Effect of Copper, Hardness and pH on Growth of Rainbow Trout (*Salmo gairdneri*)" (submitted to *J. Fish. Biol.*).

Warner, R. E., Patterson, K. K., and Borgman, R. (1966). "Behavioural Pathology in Fish: A Quantitative Study of Sublethal Pesticide Toxication," *J. Appl. Ecol.*, **3** (Suppl.), 223–247.

Warnick, S. L. and Bell, H. L. (1969). "The Acute Toxicity of Some Heavy Metals to Different Species of Aquatic Insects," *J. Water Pollut. Control Fed.*, **41,** 280–284.

Whitton, B. A. (1970). "Toxicity of Zinc, Copper and Lead to *Chlorophyta* from Flowing Waters," *Arch. Mikrobiol.*, **72,** 353–360.

Whitton, B. A. (1971). "Toxicity of Heavy Metals to Freshwater Algae: A Review," *Phykos,* **9,** 116–125.

Wilson, D. P. and Armstrong, F. A. J. (1961). "Biological Differences between Sea Waters: Experiments in 1960," *J. Mar. Biol. Assoc. U.K.*, **41,** 663–681.

Wilson, R. C. H. (1972). "Prediction of Copper Toxicity in Receiving Waters," *J. Fish. Res. Board Can.*, **29,** 1500–1502.

Winner, R. W. and Farrell, M. P. (1976). "Acute and Chronic Toxicity of Copper to Four Species of *Daphnia*," *J. Fish. Res. Board Can.*, **33,** 1685–1691.

Wisely, B. and Blick, R. A. P. (1967). "Mortality of Marine Invertebrate Larvae in Mercury, Copper, and Zinc Solutions," *Aust. J. Mar. Freshwater Res.*, **18,** 63–72.

Young, L. G. and Nelson, L. (1974). "The Effects of Heavy Metal Ions on the Motility of Sea Urchin Spermatozoa," *Biol. Bull.*, **147,** 236–246.

Young, R. G. and Lisk, D. J. (1972). "Effect of Copper and Silver Ions on Algae," *J. Water Pollut. Control Fed.*, **44,** 1643–1647.

Zitko, V. and Carson, W. G. (1976). "A Mechanism of the Effects of Water Hardness on the Lethality of Heavy Metals to Fish," *Chemosphere*, **5**, 299–303.

Zitko, V., Carson, W. V., and Carson, W. G. (1973). "Prediction of Incipient Lethal Levels of Copper to Juvenile Atlantic Salmon in the Presence of Humic Acid by Cupric Electrode," *Bull. Environ. Control Toxicol.*, **10,** 265.

12

EFFECTS OF COPPER ON EMBRYONIC AND JUVENILE STAGES OF AQUATIC ANIMALS

Wesley J. Birge

Jeffrey A. Black

T. H. Morgan School of Biological Sciences, University of Kentucky, Lexington, Kentucky

1. INTRODUCTION

Copper is widely distributed throughout the biosphere, reaching appreciable concentrations in aquatic and terrestrial ecosystems. Kopp and Kroner (1967) reported soluble copper in 74% of 1500 water samples taken from various regions of the United States. They observed a maximum concentration of 280 μg Cu/l and a mean level of 15 μg/l. Common sources of environmental contamination include mining, smelting, and metallurgy, as well as production and use of chemicals, pharmaceuticals, and numerous other products (McDermott et al., 1963; Environmental Protection Agency, 1976). Copper frequently is used in algicides, fungicides, and pesticides and is a common treatment for red leg disease in amphibians (Kaplan and Licht, 1955). Copper pollution also results from the mining and use of coal. Numerous investigations show coal to contain mean concentrations of 14 to 50 μg Cu/g (Ruch et al., 1974; Bolton et al., 1975; Sheibley, 1975), and this element has been detected in ash pond discharges, coal conversion effluents, and mine waters (Ahmad, 1973; Chu et al., 1975; Cherry et al., 1976; Hildebrand et al., 1976; Birge et al., 1978).

Copper is an essential element for most plants and animals (Frieden, 1968; EPA, 1976). It is required in the synthesis of chlorophyll and hemoglobin, and serves as the oxygen coupling site in hemocyanin, the respiratory blood pigment in many molluscs and certain other invertebrates (Fox and Vevers, 1960). Trace levels of copper are required in many enzyme systems and other biological catalysts (Frieden, 1968). Cytochrome oxidase, a copper-containing enzyme, is essential to oxidative metabolism in all animals.

When biological requirements are exceeded, copper becomes toxic to aquatic biota. However, toxic concentrations vary widely for different test conditions and animal species. Doudoroff and Katz (1953) concluded that exposure levels below 25 μg Cu/l were not acutely toxic to most common species of fish. Juvenile Atlantic salmon (*Salmo salar*) did not suffer mortality when exposed to 28 μg Cu/l for 168 hr (Sprague and Ramsay, 1965), but a concentration of 4 μg/l produced avoidance responses (Sprague, 1964). On the basis of laboratory bioassays, the no-effect level for copper appears to range from 5 to 15 μg/l, and this closely approaches mean ambient concentrations for numerous freshwater resources (Kopp and Kroner, 1967; EPA, 1976).

Effects of copper on aquatic life vary with certain physicochemical parameters of water. Generally, toxicity decreases with increases in alkalinity, pH, or hardness (Pickering and Henderson, 1966; Mount and Stephan, 1969; EPA, 1976). Such phenomena appear to be associated in part with variations in the chemical speciation of aqueous copper (Black et al., 1973; Culp, 1975). For example, the soluble forms of copper that tend to predominate at different pH ranges include Cu^{2+} up to pH 6, $CuCO_3$ from pH 6 to 9, $Cu(CO_3)_2^{2-}$ from pH 9 to 11, and $Cu(OH)_4^{2-}$ above pH 13. Though these results were obtained in simple closed systems, they indicate the complexity of the aquatic chemistry of copper (see Leckie and Davis, this volume, Part I). As this element complexes readily with many organic compounds, aquatic biota can tolerate higher concentrations of copper in water that contains appreciable levels of detritus and organic ligands (Black, 1974; Shaw and Brown, 1974; Chynoweth et al., 1976).

Numerous reports have shown copper to be more toxic to embryonic and juvenile stages than to adult animals. Therefore, the principal objective of this chapter is to examine the effects of copper on the development and reproduction of aquatic vertebrates.

2. TESTING METHODS

In recent investigations, embryo-larval bioassays were conducted on four species of fish and five amphibians. The fish included the largemouth bass (*Micropterus salmoides*), channel catfish (*Ictalurus punctatus*), goldfish (*Carassius auratus*), and rainbow trout (*Salmo gairdneri*). The amphibians selected for study were the northern leopard frog (*Rana pipiens*), southern gray treefrog (*Hyla chrysoscelis*), Fowler's toad (*Bufo fowleri*), narrow-mouthed toad (*Gastrophryne carolinensis*), and marbled salamander (*Ambystoma opacum*). Test animals were chosen for economic importance and for variations in ecological and geographic distribution. This selection also included species with different patterns of reproduction, involving a number of developmental variables that could respond differentially to copper exposure. As shown in Table 1, habitat requirements and reproductive characteristics varied substantially for the amphibian species.

Bioassays were performed using the static renewal method, in which test water and copper toxicant were changed at regular 12-hr intervals. Stock solutions were prepared daily at 10 to 100 mg Cu/l, using copper sulfate. Lower exposure concentrations were prepared by dilution, and analyses for copper were performed with a Perkin-Elmer atomic absorption spectrophotometer (model 503) equipped with a graphite furnace (Perkin-Elmer Corporation, 1973).

Eggs and larvae were maintained in 500-ml Pyrex test chambers, and exposure to copper was initiated after fertilization and continued through 4 days posthatching. Tests on the narrow-mouthed toad and goldfish were conducted at a water hardness of 197 mg $CaCO_3$/l. Hardness averaged 100 mg $CaCO_3$/l for other species, with an overall range of 93 to 105 mg/l. The pH varied from 7.2 to 7.8, and moderate aeration was used to maintain dissolved oxygen near saturation. Temperature was regulated at 12 to 13°C for trout and at 20 to 24°C in all other bioassays. Hatching times averaged 24 days for the trout, 6 days for the channel catfish, and 3 to 4 days for the remaining species. Survival frequencies were control adjusted, and log probit analysis (Daum, 1969) was used to determine LC_1 and LC_{50} values with 95% confidence limits. Grossly

Table 1. Ecological and Reproductive Characteristics of Amphibian Test Species[a]

Family and Species	Reproductive Characteristics: Days of Development[b]				Geographic Distribution		Predominnt Species Habitat
	Hatching	Metamorphosis	Egg Production	Breeding Site and Season	Family (Worldwide)	Species[c] (North America)	
Ambystomatidae *Ambystoma opacum*	4 (22°C)	300–600[1,2]	100–250 laid singly; female tends nest[1,2]	Dry ponds later filled by rain; eggs must be submerged to hatch[2] August–November[1-4]	Cold to subtropical regions of North America[3]	Eastern and south central U.S. except Florida[4] 30–42° lat. 70–95° long.	Terrestrial Moderately adapted within geographic range
Bufonidae *Bufo fowleri*	3 (22°C)	40–60[5]	4000–10,000 laid in strings on submerged vegetation[2,5]	Permanent and temporary water[4,5] April–August[5]	Cold to tropical regions on all land masses except Madagascar, New Guinea, Australia[6]	Eastern and south central U.S. except southeastern Coastal Plain[4,5] 30–45° lat. 70–95° long.	Terrestrial Very broadly adapted within geographic range
Hylidae *Hyla chrysoscelis*	3 (22°C)	45–65[5]	30–40 laid in packets on surface film[5]	Ponds and temporary water[5] March–July[4,5]	Cold to tropical regions of North America and tropical to warm temperate regions of Central and South America; cool	Southeastern U.S. except southern Florida[4,7] 30–37° lat. 75–97° long.	Arboreal Narrowly adapted within geographic range

					temperate to subtropical regions of Eurasia[6]		
Microhylidae *Gastrophryne carolinensis*	3 (22°C)	20–70[5]	10–90 laid in packets on surface film[5]	Ponds and temporary water[5] March–August[5]	Cool temperate to tropical regions of Asia and Indo-Malaya; warm temperate, subtropical, and tropical regions of North, Central, and South America[6]	Southeastern U.S.[4,5] 27–37° lat. 75–97° long.	Fossorial Narrowly adapted within geographic range
Ranidae *Rana pipiens*	4 (22°C)	60–90[2,5]	2000–4500 laid in mass on submerged vegetation[2,4,5]	Permanent water[5] March–May[2,5]	Cold to tropical regions on all land masses except New Guinea and Australia[6]	Southern Canada and north central and northeastern U.S.[4,5,8] 38–55° lat. 57–105° long.	Semiaquatic Broadly adapted within geographic range

[a] Data for reproductive characteristics and geographic distribution were taken from (1) Bishop (1943); (2) Rugh (1962); (3) Anderson (1967); (4) Conant (1975); (5) Wright and Wright (1949); (6) Vial (1973); (7) B. Monroe, personal communication, 1977; (8) Dunlap and Kruse (1976).

[b] Hatching times were determined under laboratory conditions. Time intervals given are for the full span of development through the completion of metamorphosis.

[c] Coordinates are given to designate approximate latitude and longitude limits for species distribution.

deformed survivors were counted as lethals when computing percent mortality. Control survival ranged from 83 to 98%, and minimum egg sample size was 35 for the marbled salamander and 100 for other species. Detailed test procedures, including acquisition and care of animals, have been described elsewhere (Birge and Black, 1977a; Birge, 1978; Birge et al., 1978).

3. TOXICITY OF COPPER TO DEVELOPMENTAL STAGES OF FISH AND AMPHIBIANS

3.1. Fish Embryo-Larval Bioassays

The effects of copper on developmental stages of fish are summarized in Table 2. An exposure concentration of 1 mg/l produced complete lethality of trout stages, and 6% mortality was observed at 1 μg Cu/l. By comparison, control-adjusted survival frequencies for other species ranged from 95 to 102% at an exposure level of 0.1 mg Cu/l, and complete mortality occurred at 25 to 50 mg total Cu/l.

Percent hatchability for trout eggs treated by static renewal procedures did not differ substantially from results obtained in the continuous-flow administration of copper by Sauter et al. (1976). They reported values for eggs of the brook trout (*Salvelinus fontinalis*) which were treated with copper at two levels of water hardness (37 and 187 mg $CaCO_3$/l). When results were adjusted to

Table 2. Toxicity of Copper to Embryo-Larval Stages of Fish

Concentration	Percent Survival at 4 Days Posthatching[a]			
(mg/l)	Trout	Catfish	Goldfish	Bass
0.001	94	99	—	—
0.005	81	98	—	—
0.010	85	100	—	—
0.025	75	99	101	—
0.050	58	99	101	—
0.100	61	95	102	100
0.250	39	92	—	—
0.500	15	89	93	99
1.000	0	83	84	97
5.000	—	71	50	74
10.000	—	46	21	25
25.000	—	15	0	0
50.000	—	0	0	0

[a] All survival frequencies were control adjusted. Values above 100% indicate higher survival rates for copper-treated animals than for corresponding controls. Grossly teratic larvae were counted as lethals.

their control baseline, hatching frequencies were about 33, 74, and 94% at exposure levels of 74, 49, and 21 μg Cu/l hard water. The best correlation was at an exposure range of 49 to 51 μg Cu/l, which reduced hatchability of brook trout eggs to approximately 5 and 74% in soft and hard water, respectively. The hatchability of rainbow trout eggs averaged 58% when copper was administered at 50 μg/l in static renewal bioassays which were conducted in water having a hardness of 100 mg $CaCO_3$/l (Table 2). In continuous-flow tests, McKim and Benoit (1974) reported hatchability frequencies of 65 to 99% for brook trout eggs exposed to 9.4 μg Cu/l, and 82 to 91% embryonic survival occurred at 6.1 μg/l. Control eggs were maintained in test water that contained background copper at 2.7 μg/l, and hatchability was 92%. These figures are not inconsistent with those given for rainbow trout (Table 2).

The poorest correlations between results for static renewal and continuous-flow test procedures occurred at higher copper concentrations. McKim et al. (1978) observed nearly complete lethality for rainbow trout eggs when copper was administered at 37 μg/l in a flow-through system. The more extended dose-response range observed in static tests with trout eggs was due in part to the precipitation of copper at the higher exposure levels. As noted above, test water and toxicant were changed at regular 12-hr intervals. Retention of total copper at the end of 12 hr, determined by direct analysis of test water, was nearly 100% for exposure levels of 0.01 and 0.02 mg/l and 94% at 0.1 mg/l. However, retention after 12 hr dropped to 56% for an exposure level of 0.75 mg Cu/l. Actual concentrations, averaged for three replicate measurements, were 0.748, 0.482, and 0.421 mg/l at 0, 6, and 12 hr, respectively. Results of static renewal and flow-through bioassays probably would have compared more closely at higher concentrations had exposure levels been based on soluble rather than total copper (Brungs et al., 1976).

To evaluate and compare further the toxic effects of copper on different species of fish, median lethal concentrations (LC_{50}) were determined by probit analysis (Daum, 1969). The threshold for copper toxicity was estimated using an LC_1 value, defined as the exposure concentration that produced 1% control-adjusted impairment. All lethal concentrations were calculated at hatching and 4 days posthatching, and included accumulative test responses incurred from the onset of copper exposure.

Copper LC_{50}'s at 4 days posthatching varied from 0.11 mg/l for trout to 5.20 to 6.62 mg/l for goldfish, bass, and catfish. On the basis of LC_{50} values, the rainbow trout was about 60 to 70 times more sensitive to copper than were other species (Table 3). Hatching times averaged 3, 4, 6, and 24 days for goldfish, bass, catfish, and trout, respectively. The longer egg exposure period probably contributed to the higher sensitivity of trout embryo-larval stages. However, as LC_{50} values did not differ significantly for the three remaining species, no direct correlation could be established between egg exposure time and species sensitivity to copper.

Table 3. Log Probit LC_1 and LC_{50} Values for Copper[a]

Animal Species	Exposure Time (days)	LC_{50} (mg/l)	95% Confidence Limits	LC_1 (μg/l)	95% Confidence Limits
Fish					
Trout[b]	24	0.12	0.10–0.15	4.1	1.9–7.1
	28	0.11	0.09–0.14	3.4	1.6–5.9
Catfish	6	7.56	5.24–10.20	191	44–446
	10	6.62	4.68–9.50	167	42–379
Goldfish	3	5.20	3.08–5.40	261	88–504
	7	5.20	4.13–6.41	299	101–571
Bass	4	6.97	6.11–7.92	1511	929–2065
	8	6.56	5.66–7.54	1592	893–2234
Amphibians					
Narrow-mouthed	3	0.05	0.04–0.06	0.6	0.3–1.1
toad	7	0.04	0.03–0.05	0.7	0.3–1.3
Southern gray	3	0.06	0.04–0.08	0.2	0.1–0.4
treefrog	7	0.04	0.02–0.08	0.3	0.0–1.2
Leopard frog	4	0.06	0.04–0.08	0.4	0.2–0.9
	8	0.05	0.03–0.08	0.5	0.1–1.5
Marbled	4	3.59	2.35–5.56	37.2	7.9–97.5
salamander	8	0.77	0.52–1.11	21.4	5.8–48.9
Fowler's toad	3	35.99	24.42–52.35	2415	411–5475
	7	26.96	19.13–37.52	1436	361–3065

[a] Survival frequencies were determined at hatching and 4 days posthatching. Values for the latter included accumulative responses for the entire exposure period. Regression analysis included survival values for a few exposure concentrations not universally administered to all test species and which for sake of simplicity were not shown in Tables 2 and 4.

[b] Values were calculated by the method of Finney (1971).

The copper LC_1's for trout embryo-larval stages were 4.1 and 3.4 μg/l at hatching and 4 days posthatching (Table 3). The difference was not significant, because of overlap of the 95% confidence limits. Taken 4 days after hatching, LC_1 values for catfish and goldfish were 167 and 299 μg Cu/l, but the largemouth bass was far more tolerant. These findings are in general agreement with those of McKim et al. (1978), who found embryo-larval stages of the rainbow trout to be more sensitive to copper than those of seven other species of freshwater fish. In addition, they reported that embryos of the smallmouth bass (*Micropterus dolomieui*) were the most tolerant to copper. When administered at 517 μg/l in a flow-through system, copper had no effect on smallmouth bass eggs. As seen in Table 2, copper at 500 μg/l also produced control-level survival in static renewal tests with the largemouth bass, although this concentration obviously was not far below the threshold for acute toxic effects.

Sauter et al. (1976) determined maximum acceptable toxicant concentrations (MATC) of 12 to 18 and 13 to 19 μg Cu/l for eggs and fry of the channel catfish. These ranges are about an order of magnitude lower than the LC_1 given in Table 3, but their results were based on considerably longer tests in which continuous-flow exposure was maintained for 60 days after hatching. The low MATC's were largely due to the high toxicity of copper to catfish fry. As reported by Sauter et al., egg hatchability was not significantly reduced by copper concentrations of 66 μg/l in hard water or 24 μg/l in soft water. These results were consistent with the static test data for hatchability given in Table 7. However, the probit values presented in Table 3 for catfish, goldfish, and bass undoubtedly were skewed upward somewhat by the precipitation of copper at the higher exposure levels.

The copper LC_1 value reported for the rainbow trout (Table 3) is close to the MATC's of 3 to 5 and 5 to 8 μg Cu/l determined in flow tests on eggs and fry of the brook trout (Sauter et al., 1976). In chronic reproductive studies with brook trout, McKim and Benoit (1974) reported the "no effect" level for copper at 9.4 μg/l. Based on these results for fish embryo-larval stages, it appeared that low concentrations of copper were as toxic in static renewal tests as in flow-through bioassays. As noted above, the lower sensitivity observed at higher exposure levels appeared to be due to increased problems with copper precipitation. To test this premise, static renewal bioassays were repeated for the catfish, and test water was analyzed for both total and soluble copper. Soluble copper was determined for water samples passed through a 0.45-μm Millipore filter, as described by Brungs et al. (1976). Mean retention of total copper at the end of the 12-hr renewal interval ranged from 94% at 0.1 mg/l to 83% at 5 mg/l, and decreased to 52% at 25 mg/l (Figure 1). The 12-hr retention of soluble copper ranged from 67 and 64% at exposure levels of 0.1 and 1.0 mg/l, respectively, to only 7% at 25 mg/l. When soluble copper at 12 hr was expressed as percent of initial total copper, retention dropped even more dramatically as exposure concentration was increased. For example, as seen in Figure 1, there was a decrease from 51% at 1 mg/l to only 18% at 5 mg/l, and these observations on soluble copper were consistent with data reported by Brungs et al. (1976) in studies with the fathead minnow. Brungs et al. suggested that acute toxicity was coincidentally related to the concentration of dissolved copper. Thus, the solubility properties of copper probably accounted for the reduced sensitivity observed at higher exposure concentrations in static renewal tests. In trout bioassays, as noted above, total copper after 12 hr was reduced to 56% for an exposure concentration of 0.75 mg/l. The higher values reported in Figure 1 presumably were due largely to the elevated temperature (26°C) of test water used for catfish eggs.

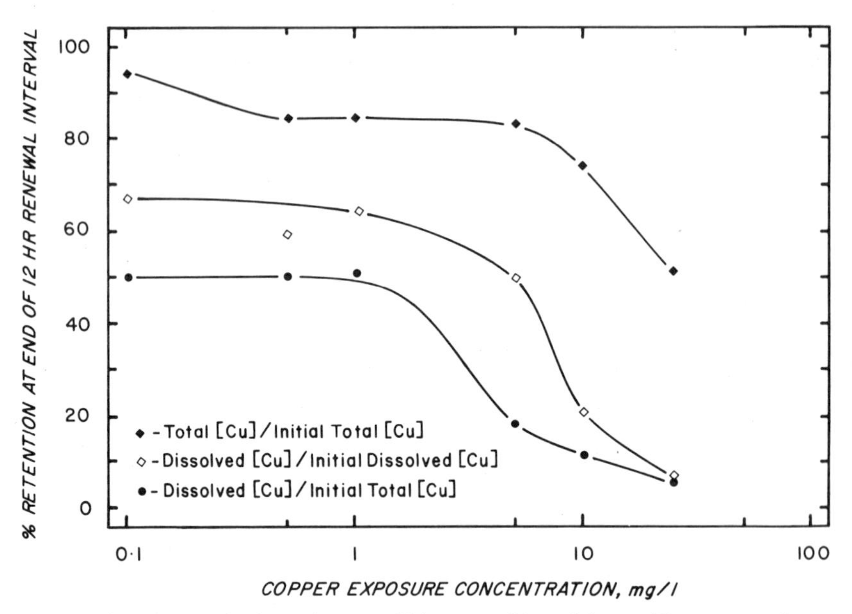

Figure 1. Copper retention in static renewal bioassays with catfish eggs. Test water and copper toxicant were changed every 12 hr. Copper was analyzed at the end of the renewal interval and expressed as percent of initial copper for exposure levels of 0.1 to 25 mg/l. Standard errors were based on 10 replicate assays and ranged from only 0.3 to 3.1, except for a value of 7.2 obtained for total copper at 10 and 25 mg/l.

3.2. Comparative Sensitivity of Developmental and Posthatched Stages

When static renewal procedures were used, copper was more toxic to fish eggs and embryos than to larvae or early fry. As seen in Table 3, LC_{50} and LC_1 values did not change appreciably when copper exposure was extended from hatching through 4 days posthatching. Similar results were obtained in static and continuous-flow bioassays with boron, mercury, and certain organic toxicants (Birge et al., 1974, 1979a, 1979b; Birge and Black, 1977a, 1977b). Working with the herring (*Clupea harengus*), Blaxter (1977) reported substantial reproductive impairment when eggs were treated at fertilization with 30 μg Cu/l, but an exposure concentration of 1000 μg/l was required to produce high mortality of newly hatched larvae. Steele et al. (1973) observed low survival and premature hatching of herring eggs at exposure levels above 10 μg Cu/l. Grande (1966), using static renewal procedures with trout and salmon, found zinc to be more toxic to eggs than to fry. Rainbow trout fry survived a concentration of 0.06 mg Cu/l, which produced complete mortality of eggs. In flow-through bioassays with brook trout, frequencies of mortality often were higher for eggs than for fry when exposure to copper was extended for 30 days beyond hatching (Sauter et al., 1976). In the same investigation, 7 to 13 μg Cu/l hard water (186 mg

$CaCO_3$/l) produced generally comparable mortality for catfish eggs and fry. Exposure levels of 19 to 66 μg Cu/l gave survival values of 23 to 47% for eggs and 0 to 14% for fry. In view of the difference in exposure time for eggs (6 to 8 days) and fry (30 days), these data do not appear to reflect disproportionately higher sensitivity for catfish fry.

When administered in soft water, copper often has been reported to be more toxic to fish larvae and fry than to eggs and embryos. At a water hardness of 36 mg $CaCO_3$/l, catfish eggs survived copper concentrations that proved highly toxic to fry (Sauter et al., 1976), and larval stages of brook trout and bluegill (*Lepomis macrochirus*) were shown to be more sensitive than embryos to copper in soft water (McKim and Benoit, 1971; Benoit, 1975). Hazel and Meith (1970) also reported that fry of the king salmon (*Oncorhynchus tshawytscha*) were more sensitive than eggs when copper was administered in water with a hardness of 44 mg $CaCO_3$/l, but this comparison was inconclusive, as the eggs were not exposed for the first 24 days of development. The embryopathic effects of copper are difficult to assess when exposure does not include the period of egg hardening and the full range of development. Early and intermediate embryonic stages exhibited high mortality rates in static tests with rainbow trout, and Blaxter (1977) noted far less reproductive impairment when copper treatment of herring eggs was delayed until 4 days after fertilization. Furthermore, the early post-hatched mortality observed by Hazel and Meith (1970) could have resulted in part from the effects of copper on late embryonic stages. In an extensive investigation, McKim et al. (1978) compared the effects of copper on embryonic, larval, and juvenile stages of eight species of fish. Tests were conducted in soft water (45 mg $CaCO_3$/l) and were initiated in most instances at early to late eyed stages. These results also indicated that larvae and early juveniles were more sensitive to copper than were eggs and embryos. However, when copper was administered for varying periods of time to eggs of the brown trout (*Salmo trutta*), with treatment continuing 60 days beyond hatching, reductions in larval standing crop increased with the duration of embryonic exposure. McKim et al. suggested that embryonic exposure probably produced subtle changes which subsequently affected survival and growth of larvae and early juveniles.

Comparing the tolerances of early life cycle stages is complicated by numerous factors. As noted by McKim et al. (1978), embryonic impairments may contribute to posthatched mortalities. This contention is further supported by the high incidence of copper-induced teratogenesis in fish, as discussed below. McKim and Benoit (1974) also showed that copper residues in brook trout eggs increased from 6.3 to 27.8 μg/g when treatment was maintained for 3 weeks at 9.4 μg/l. On the assumption that egg yolk accumulates significant concentrations of copper, larvae may receive increased exposure during final resorption and assimilation of the principal yolk mass. In addition, hatching time and other developmental variables may affect the maturity and tolerance of larvae and early juvenile stages. The egg:larval ratio of sensitivity to copper, based on percent mortality, may increase for species with longer egg hatching periods. Various characteristics of test water also may differentially affect copper toxicity to different developmental stages. There is increasing evidence that, for copper and certain other toxicants, mortality of larvae and fry is favored by soft water, while embryopathic effects may be greater in hard water (Birge and Black, 1977b; Birge et al., 1979a). Most studies that have shown higher copper tolerances for eggs and embryos have been conducted in water of low hardness, but the differential sensitivity between embryos and posthatched stages tends to decrease or reverse when hardness is increased. Abbreviated exposure of eggs (e.g., eyed stages) and extended treatment of larvae (e.g., 30 to 90 days) also have complicated comparisons on the relative sensitivities of different life cycle stages. Though the toxicological effects on development are not completely understood, it is clear that embryos, larvae, and early juvenile stages of fish are highly susceptible to copper and other trace contaminants (Birge et al., 1979b; McKim et al., 1978). As suggested by McKim et al. (1978) and Birge et al. (1979b), short-term bioassays with sensitive developmental stages should provide a rapid and economical means of quantifying the toxic effects of aquatic pollutants.

3.3. Effects of Copper on Amphibian Development

Test responses for the five amphibian species are summarized in Table 4. Copper was highly toxic to developmental stages of the narrow-mouthed toad, southern gray treefrog, and northern leopard frog. No survival of eggs and larvae occurred

Table 4. Toxicity of Copper to Embryo-Larval Stages of Amphibians

	Percent Survival at 4 Days Posthatching[a]				
Concentration (mg/l)	Narrow-Mouthed Toad	Southern Gray Treefrog	Leopard Frog	Marbled Salamander	Fowler's Toad
0.001	100	100	99	—	—
0.005	94	90	87	—	—
0.010	66	61	66	100	—
0.050	48	51	54	97	—
0.100	30	40	41	82	99
0.500	13	24	18	70	99
1.000	0	0	0	45	97
5.000	0	0	0	15	92
10.000	—	—	—	0	74
25.000	—	—	—	—	53
50.000	—	—	—	—	42
100.000	—	—	—	—	0

[a] All survival frequencies were control adjusted.

at 1 mg Cu/l, and mortality varied from 6 to 13% at 5 μg/l. Embryo-larval stages of the marbled salamander and Fowler's toad withstood higher exposures. The latter species suffered no appreciable impairment at 0.1 to 1.0 mg total Cu/l, and a concentration of 50 mg/l only reduced survival to 42%. Survival frequencies observed for the salamander were intermediate to those obtained for trout and bass (Figure 2). Median lethal concentrations for the narrow-mouthed toad, treefrog, and leopard frog were surprisingly low, ranging from 0.04 to 0.05 mg Cu/l when taken 4 days after hatching (Table 3). The LC_1's for these species varied from only 0.3 to 0.7 μg/l. The LC_{50}'s for the marbled salamander and Fowler's toad were 0.77 and 27 mg/l, respectively. Of the five amphibian species tested, the embryo-larval stages of three were more sensitive to copper than were trout embryos and alevins (Table 3). Such data raise great concern for the fact that little attention has been given to amphibians in establishing freshwater criteria for copper and other trace contaminants. The high sensitivity of eggs and embryos of many amphibian species may be contributing to the reduction of natural populations, as reported by Gibbs et al. (1971) and others (Anonymous, 1973). Amphibian embryos are far more susceptible to copper than are the adults. Mature leopard frogs reportedly can survive about 4 mg Cu/l for 30 days (Kaplan and Yoh, 1961), compared with an LC_1 of 0.5 μg/l for embryo-larval stages of this species (Table 3).

Amphibian embryos were substantially more sensitive to copper than were larval stages. When treatment was extended from hatching through 4 days

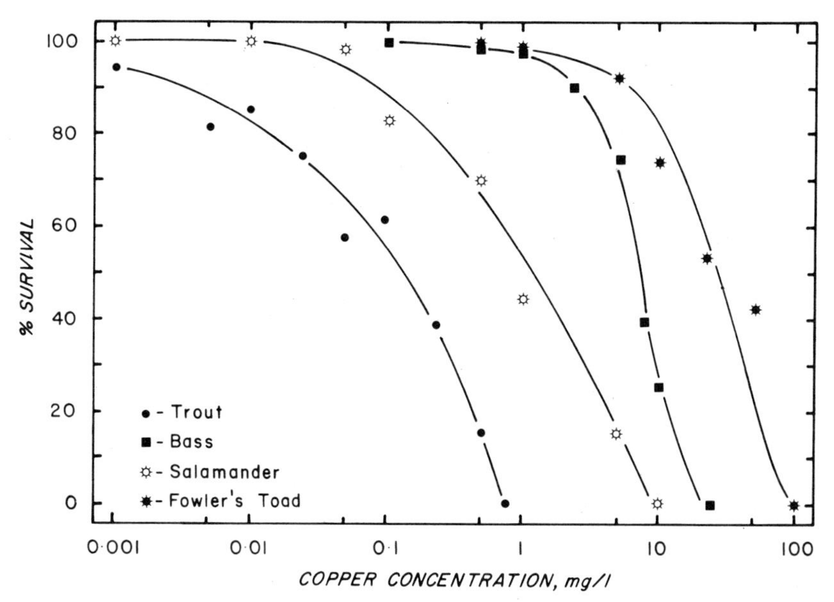

Figure 2. Effects of copper on embryo-larval stages of fish and amphibians. Eggs were treated from fertilization through 4 days posthatching. Exposure concentrations were based on milligrams of total copper per liter administered in static renewal bioassays.

posthatching, the only significant reduction in LC_{50} values occurred for the salamander. These results are at variance with certain tests in which copper salts reportedly produced mortality and growth inhibition of frog tadpoles at concentrations that were not lethal to eggs (Dilling and Healey, 1926; Lande and Guttman, 1973). In the study by Lande and Guttman (1973), egg sample size was about 120 for a test water volume of 1.5 liters, and the toxicant solution apparently was not replenished during treatment. However, after hatching the larvae were recultured using five animals per chamber, with renewal of water and toxicant at 2- to 3-day intervals. In view of these considerations, it is possible that biomass overloading diminished the effects of copper on the eggs. The poor retention of copper also indicated that test conditions were inadequate for the biomass volume. At a nominal concentration of 0.16 mg/l, the actual copper detected in the water was only 0.03 mg/l. In the study summarized above (Table 4), copper retention at 0.1 mg/l averaged 94%.

In an extensive investigation of pollution stress on amphibian species, attempts are being made to correlate sensitivity of embryo-larval stages with particular ecological and reproductive adaptations. Results for copper and certain other trace metals are now being analyzed, and initial findings indicate that more narrowly adapted species generally are more susceptible to aquatic pollutants. Geographic range, habitat, and breeding characteristics have been summarized in Table 1 for the five amphibian species. Though species tolerance did not necessarily correlate with the extent of geographic distribution, amphibians adapted to broader ecological niches were less affected by copper. This concept is best illustrated by Fowler's toad, which is a broadly adapted species with great tolerance to copper. By comparison, the narrow-mouthed toad and the southern gray treefrog are substantially more specialized and restricted in their ecological requirements and reproductive shelter. In a previous investigation (Birge, 1978), embryo-larval stages of the narrow-mouthed toad were found to be more sensitive than trout eggs and alevins to 18 of 22 inorganic toxicants. The ability of the different amphibian species to withstand the effects of trace contaminants appeared to increase with tolerance to natural environmental stresses. The only exception was the northern leopard frog, a rather broadly adapted species which showed high susceptibility to copper. However, in recent bioassays with mercury, the leopard frog exhibited sensitivity that was intermediate among eight amphibian species. The LC_{50} values at hatching were 0.001, 0.01, and 0.07 mg/l for the narrow-mouthed toad, northern leopard frog, and Fowler's toad, respectively (Westerman, 1977).

4. INTERACTIONS OF COPPER WITH OTHER METALS

Copper generally occurs in water resources in combination with various other aquatic contaminants, particularly other metals. Therefore it is important to determine whether additive, antagonistic, or synergistic interactions govern the combined effects of such copper-containing mixtures. Numerous investigations with adult fish have provided some insight into this problem (Doudoroff, 1952; Cairns and Scheier, 1968). A classic study by Doudoroff (1952) prompted attention to possible synergism between copper and zinc. Subsequently, Lloyd (1961) and Sprague and Ramsay (1965) found that copper and zinc generally exerted additive effects on salmonid species but interacted synergistically when higher concentrations were administered in soft water. Additive toxic effects on rainbow trout also were reported for mixtures of copper, zinc, and nickel, as well as for copper, zinc, and phenol (Brown and Dalton, 1970).

Though the reproductive process is particularly sensitive to copper and other aquatic contaminants, little is known about the combined effects of toxicants on developmental stages. Therefore investigations were undertaken to evaluate the embryopathic effects of copper-zinc and copper-mercury mixtures. Metals for each combination were mixed in equal proportions, and fish eggs were treated using the bioassay procedures described above. Theoretical additive values were calculated from the survival frequencies observed for individual metals, and chi-square analysis was used to determine the significance for antagonism or synergism. The results for copper-zinc are summarized in Table 5. The hatchability of catfish and goldfish eggs closely approached the frequencies for additive effects, except that synergism occurred at high concentrations. Copper-zinc also produced additive responses when bass eggs were treated at near-threshold levels, but synergism occurred when the exposure concentration reached 1 mg/l ($P < .001$). These results were consistent with findings for juvenile and adult salmonids (Lloyd, 1961; Sprague and Ramsay, 1965), in which

Table 5. Effects of Copper-Zinc on Fish Embryos

Concentration[a] (mg/l)	Percent Survival at Hatching[b]		
	Catfish	Goldfish	Bass
0.010	103 (100)	—	—
0.050	94 (100)	103 (102)	—
0.100	84 (83)	102 (101)	98 (99)
0.500	70 (67)	97 (97)	97 (99)
1.000	44 (46)	75 (78)	77 (97)
5.000	9 (36)	42 (51)	56 (65)
10.000	0 (25)	7 (21)	21 (31)
Probit LC_{50} values (mg/l)	0.70 (1.55)	2.63 (3.99)	4.23 (6.73)

[a] Concentrations were based on equal proportions of copper and zinc.

[b] Survival frequencies were control adjusted, and values calculated for additive effects are given parenthetically.

copper and zinc were additive at lower exposure levels but synergistic at higher concentrations.

Mixtures of copper and mercury were tested on four species of fish (Table 6). Responses at near-threshold concentrations were additive for catfish and goldfish eggs, but as the exposure level was increased, hatchability rapidly dropped below the levels expected for additive effects. Synergism became highly significant ($P < .001$) at 0.01 and 0.05 mg/l for catfish and goldfish, respectively. In the treatment of bass eggs, possible antagonism occurred at 0.010 to 0.025 mg Cu-Hg/l ($P < .02$). However, additive responses were observed at intermediate exposure levels, and synergism resulted at and above 1.0 mg/l ($P < .001$). The only clear instance of antagonism was observed for trout eggs exposed to 0.001 to 0.010 mg Cu-Hg/l ($P < .005$). Throughout this exposure range the hatchability of trout eggs consistently exceeded frequencies for additive effects, but synergism occurred at 0.05 mg Cu-Hg/l ($P < .001$). On the basis of LC_{50} values, mercury was approximately 25 times more toxic to trout eggs than was copper. However, the copper-mercury mixture was less toxic than copper at low concentrations, but equally as toxic as mercury at or above 50 μg/l (Figure 3). This unusual combination of antagonism at low exposure levels and synergism at high concentrations resulted in a copper-mercury LC_{50} value that did not differ appreciably from the one calculated for additive effects. This was in

Table 6. Effects of Copper-Mercury on Fish Embryos

Concentration[a] (mg/l)	Percent Survival at Hatching[b]			
	Trout	Catfish	Goldfish	Bass
0.001	99 (91)	99 (100)	—	—
0.002	96 (78)	—	—	—
0.005	94 (67)	103 (96)	—	—
0.007	82 (61)	—	—	—
0.010	78 (56)	84 (95)	102 (99)	99 (93)
0.025	51 (44)	47 (83)	101 (98)	100 (91)
0.050	0 (29)	20 (68)	85 (95)	93 (90)
0.075	—	0 (65)	74 (83)	93 (85)
0.100	—	0 (64)	52 (76)	82 (84)
0.250	—	0 (61)	42 (62)	73 (70)
0.500	—	0 (47)	30 (53)	51 (57)
0.750	—	0 (45)	26 (47)	41 (49)
1.000	—	0 (43)	0 (43)	16 (48)
2.500	—	—	—	0
Probit LC_{50} values (mg/l)	0.018 (0.015)	0.023 (0.42)	0.18 (0.59)	0.42 (0.87)

[a] Concentrations were based on equal proportions of copper and mercury.
[b] Survival frequencies were control adjusted, and values calculated for additive effects are given parenthetically.

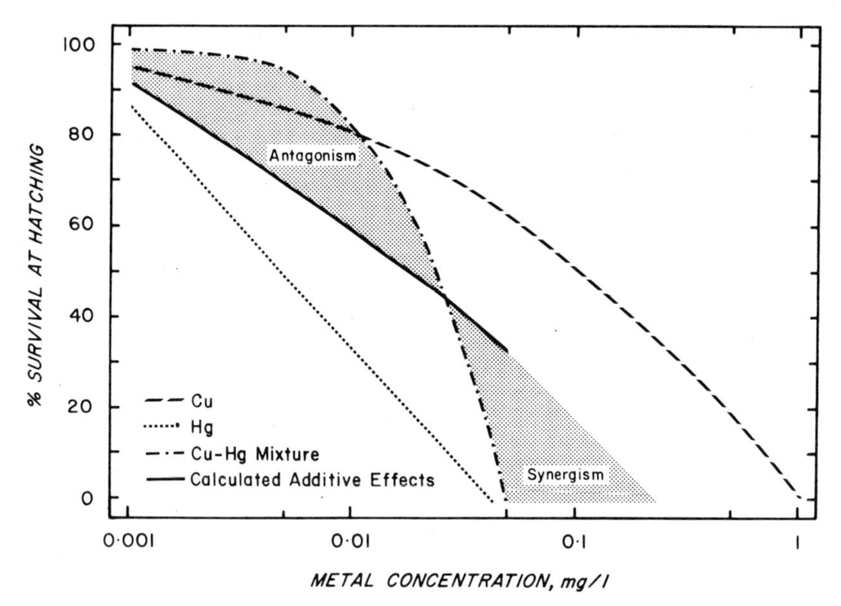

Figure 3. Effects of a 1:1 mixture of copper-mercury on embryo-larval stages of the rainbow trout. The mixture was less toxic than copper at low exposure levels but equally toxic as mercury when concentration was increased to 0.05 mg/l. When LC50's for individual metals were compared, mercury was approximately 25 times more toxic than copper. The curve for calculated additive effects served as a basis for assessing antagonism and synergism.

contrast to the results for all other copper-mercury and copper-zinc mixtures, in which probit LC_{50} values were reduced because of synergism at higher exposure concentrations (Tables 5 and 6).

In fish embryo-larval bioassays the interaction of copper with other metals was highly dependent on exposure concentration. Depending on the particular metal combination and animal test species, effects at low exposure levels were predominantly additive but occasionally antagonistic. At higher concentrations, usually above those found in most fresh waters, copper-containing mixtures consistently produced synergism. Mercury-selenium mixtures also were found to be synergistic to fish eggs (Huckabee and Griffith, 1974; Birge et al., 1979b).

5. TERATOGENIC EFFECTS OF COPPER

The debilitating effects of copper on development were not limited simply to embryonic and larval mortality. In the treatment of fish eggs, copper produced anomalous embryos, and significant numbers of teratic larvae were observed in newly hatched populations. Observations were limited to gross anatomical deformities considered likely to produce eventual mortality. Frequencies of egg hatchability included both normal and defective animals (Table 7), and per-

Table 7. Teratogenic Effects of Copper on Fish Embryos

Concentration	Percent Egg Hatchability[a]			
(mg/l)	Trout	Catfish	Goldfish	Bass
0.001	97 (4)	99	—	—
0.005	90 (10)	98	—	—
0.010	92 (7)	100	—	—
0.025	82 (8)	99	—	—
0.050	65 (11)	99	—	—
0.100	63 (4)	96	100	—
0.500	24 (31)	92 (1)	93	100
1.000	0	87 (2)	83	97
5.000	0	74 (3)	52 (3)	78 (2)
10.000	—	55 (8)	24 (8)	34 (13)
25.000	—	25 (16)	0	3 (25)
50.000	—	0	0	0

[a] Percentages of newly hatched survivors affected by gross teratogenic defects are given parenthetically.

centages of anomalous survivors were determined upon completion of hatching. The number of defective larvae generally increased with exposure level. Goldfish eggs were the least affected, and teratic bass larvae occurred only at higher exposure levels. Low frequencies of anomalous catfish alevins were observed at copper concentrations of 0.5 to 5.0 mg/l, but trout suffered significant teratogenic impairment at 1 to 5 μg/l. Except with the trout, embryonic mortality was a more sensitive test response than teratogenesis.

For species with higher copper tolerances (e.g., bass, goldfish), gross teratogenic impairment does not appear to be a principal limiting factor in larval survival. However, species such as the rainbow trout, which produce more sensitive developmental stages, may be affected as much by anomalous development as by egg mortality when copper is administered at near-threshold concentrations. In certain instances, teratogenesis appears to be the more sensitive response. Blaxter (1977) observed no reduction in the hatchability of herring eggs exposed to 30 μg Cu/l, but about 70% of the early larvae were deformed. Freshwater criteria for copper may not adequately protect reproduction in natural fish populations unless teratogenic defects are considered. Such copper-induced impairments may contribute to increased mortality of larval and juvenile fish. In future studies, attention should be given to more subtle embryonic anomalies which cannot be deduced on the basis of gross morphological observations.

Terata produced by the treatment of fish eggs with copper did not differ substantially from those previously described for boron and certain heavy metals (Birge and Black, 1977a, 1979b). The skeletal system was the most sensitive indicator of teratogenic responses, and the most common defects included moderate to acute lordosis, scoliosis, kyphosis, and rigid coiling of the vertebral

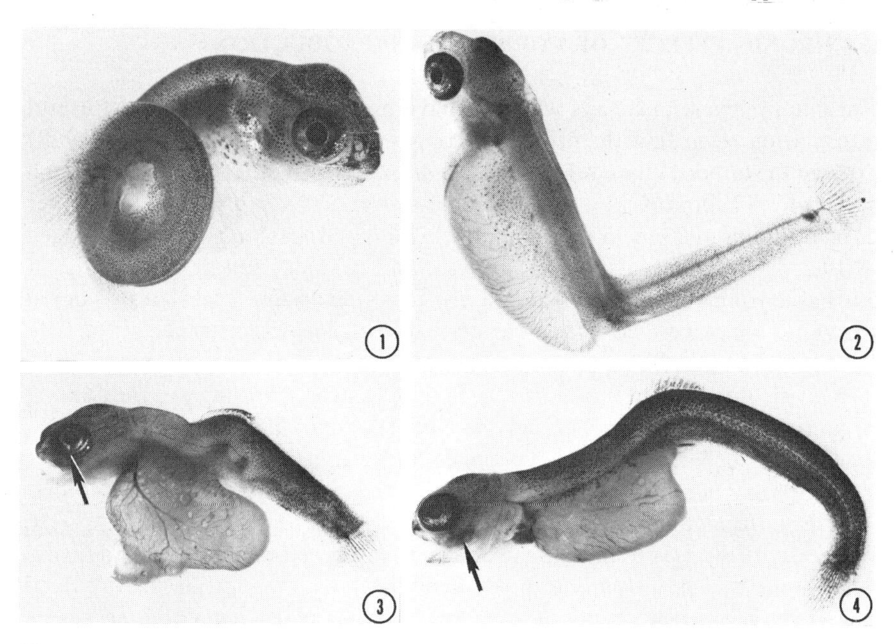

Figure 4. Teratogenic effects of copper on rainbow trout. Trout eggs were treated with copper or a 1:1 copper-mercury mixture from fertilization through hatching. Concentrations of copper as low as 1 μg/l produced gross anomalies. As shown in parts 1 to 4, teratogenic defects were most evident in the skeletal system, particularly the vertebral column. Photographs were taken 1 to 2 weeks subsequent to hatching at a magnification of 4 to 8×. 1, Acute inflexible coiling of the vertebral column produced by treatment with a copper-mercury mixture. 2, Acute lateral flexion (scoliosis) of the spinal column produced by a copper-mercury mixture. 3, Anomalous kyphotic spine, dwarfed trunk, and retinal coloboma. The last is discernible as a vertical cleft at the mid-inferior border of the eye (arrow). 4, Kyphosis of the spinal column and immobile, synarthrodic jaws (arrow).

column (Figure 4). Immobile or synarthrodic jaws were often observed, and there were less frequent occurrences of partial twinning, dwarfed bodies, defective fins, absent or reduced eyes, and anomalous yolk sacs. Scoliosis and other vertebral anomalies also have been described for frog tadpoles (Hardy, 1964; Underhill, 1966), and copper has been reported to be teratogenic to avian and mammalian embryos, producing defects in the skeleton and other organ systems (Ferm and Hanlon, 1974; Rest, 1976). Copper treatment affected chondrocytes and reduced the cartilage matrix in skeletal tissues of the chick embryo, and Rest (1976) suggested that such conditions possibly could contribute to osteoarthritis, cessation of osteogenesis, or osteoporosis. As vertebral differentiation and osteogenesis are particularly sensitive to copper and other trace contaminants, skeletal defects are especially useful criteria in screening exposed populations for terata. Copper also may produce genetic alterations resulting in such conditions as scoliosis and albinism (Underhill, 1966; Westerman and Birge, 1978).

6. CHRONIC EFFECTS OF COPPER ON REPRODUCTION

Chronic life cycle bioassays with fish have provided some of the most useful information regarding the effects of copper on reproduction. Mount (1968) exposed the fathead minnow (*Pimephales promelas*) to $CuSO_4$ in a flow-through system for 11 months, at a water hardness level of 198 mg $CaCO_3$/l. Copper concentrations of 33 to 95 μg/l completely inhibited spawning, but fish exposed to 14.5 μg/l exhibited survival, growth, and reproduction at rates indistinguishable from controls. At 33 μg/l, the lowest exposure level that prevented spawning, there were no effects on survival or physical appearance of the fish. When egg samples from control fish were treated at the higher concentrations, no reductions in hatchability occurred. In this instance, spawning was the limiting response, and Mount emphasized the fact that reproduction can be curtailed by copper concentrations which do not reduce egg hatchability or survival.

Mount and Stephan (1969) reported a similar investigation in which tests with the fathead minnow were conducted at a water hardness of 31 mg $CaCO_3$/l. Copper at 18.4 μg/l produced about 50% mortality, retarded growth and sexual development of survivors, and prevented spawning. In addition, larvae from unexposed (control) parents suffered complete mortality when treated at this concentration. Chronic exposure at 10.6 μg Cu/l produced no observable effects on adults and did not inhibit reproduction. Comparing these results with the earlier study by Mount (1968), it may be assumed that growth and reproduction of the fathead minnow are somewhat more affected when copper is administered in soft water.

Brungs et al. (1976) also studied the chronic effects of copper on reproduction of the fathead minnow. At a mean water hardness of 267 mg $CaCO_3$/l exposure levels of 64 to 67 μg Cu/l did not affect spawning but appeared to reduce egg production. A concentration of 120 μg Cu/l sharply curtailed both spawning and fecundity, but there was no significant reduction in the hatchability of the eggs produced. Spawning was totally blocked at an exposure level of 180 μg/l. Water for these chronic bioassays was taken from a natural stream which received effluent from a secondary waste treatment plant. The effluent outfall was 3 to 4 miles upstream from the study area, at which point recovery of the water source was sufficient to maintain a varied, natural biota. Higher hardness and other characteristics of the test water may have contributed to the reduced effects of copper on reproduction of the fathead minnow, as compared with earlier findings (Mount, 1968; Mount and Stephan, 1969). Depending on chronic test conditions, either egg production or spawning activity may be the more sensitive response to copper.

McKim and Benoit (1971) maintained brook trout under chronic exposure to copper for a period of 22 months. Copper at 32.5 μg/l decreased survival and growth of adult fish and substantially reduced egg fecundity and hatchability. Exposure at 17.4 μg Cu/l did not adversely affect adults or hatchability of eggs, but survival of alevins and juvenile stages was reduced by 12% at 3 months and complete mortality occurred after about 6 months. The no-effect level, considered safe for all life cycle stages, was established at 9.5 μg/l, a value confirmed in a subsequent study (McKim and Benoit, 1974). All tests were performed with soft water having a hardness of 45 mg $CaCO_3$/l.

A similar investigation was undertaken with the bluegill (Benoit, 1975). At 162 μg/l, copper in soft water (45 mg $CaCO_3$/l) reduced survival and growth of adults and precluded spawning. Egg hatchability and survival through 4 days posthatching were not affected at an exposure level of 77 μg/l, and no deformities were observed among the larval populations. However, larvae that hatched from control eggs suffered high mortality when treated for 90 days at 77 and 40 μg Cu/l, although no effects were observed at 21 μg/l. Benoit (1975) considered the sensitivity of larvae to be the best indicator of copper toxicity to the bluegill. Low concentrations of copper also have been shown to affect reproduction of numerous invertebrate species. Biesinger and Christensen (1972) reported 16% reproductive impairment of *Daphnia magna* treated for 3 weeks in soft water (45 mg $CaCO_3$/l) at 22 μg Cu/l, and larvae of various marine invertebrates (e.g., shrimp) were shown to be considerably more sensitive to copper than were the adults (Connor, 1972).

7. CONCLUSIONS

In the promulgation of protective standards for freshwater and marine aquatic life, it was concluded that exposure levels of 5 to 15 μg Cu/l produced no harmful effects for several aquatic species, and that such concentrations were very close to mean ambient values for fresh waters affected by copper (EPA, 1976). It appears from experimental data that growth and reproduction of most freshwater fish would be unobstructed by either short- or long-term exposure to copper at or below this concentration range. However, the limiting criterion for copper was set at "0.1 times the 96-hour LC_{50} as determined through nonaerated bioassays using a sensitive aquatic resident species" (EPA, 1976, p. 54). Using the adult fathead minnow as the test species and taking a 96-hr TL_m of 430 μg/l (Mount, 1968) gives a criterion for copper of 43 μg/l. Exposure levels of 5 to 10 μg/l produced appreciable frequencies of mortality and teratogenesis in static renewal tests with eggs of the rainbow trout and several species of amphibians. Even when calculations are based on data for 14-month-old brook trout (McKim and Benoit, 1971), the limit for copper would be 10 μg/l, and this would not appear to provide adequate protection for embryos and larvae of certain vertebrate species.

On the basis of these findings, there is some need to re-evaluate the freshwater standard for copper. As the perpetuation of natural species is dependent in substantial measure on reproduction, criteria for copper should be compatible with the reproductive process. Reductions of only 5 to 10% in reproductive potential may prove deleterious when coupled with the effects of natural stresses on developmental stages. Therefore it would be more appropriate to base the limiting criteria for copper on tolerances of embryos, larvae, or early juveniles of sensitive resident species. On the basis of 96-hr TL_{50} values of 18 to 20 $\mu g/l$ for juvenile chinook salmon and rainbow trout (EPA, 1976), the copper limit would be about 2 $\mu g/l$, and this is in close agreement with LC_1's obtained in static renewal bioassays for rainbow trout eggs exposed from fertilization through 4 days posthatching. The toxicity of copper to developmental stages of fish and amphibians may be affected by hardness, alkalinity, and other characteristics of the test water. Larval and juvenile fish usually are more susceptible to copper administered in soft water, but copper in hard water may be equally or more toxic to eggs and embryos than to posthatched stages.

The present criterion for copper also may not be adequate to protect certain species of amphibians. Particular attention should be drawn to the fact that the developmental stages of three of the five amphibian species tested exhibited sensitivity equal to or greater than that observed for embryos and early larvae of the rainbow trout. Taking into account data presented in Table 2 and that given by Sauter et al. (1976), practical limits for copper probably should be established within ranges of 2 to 5 $\mu g/l$ in soft or medium hard water and 5 to 8 $\mu g/l$ in hard water to protect aquatic habitat frequented by sensitive fish and amphibian species.

Copper at 1 to 5 $\mu g/l$ produced significant frequencies of teratic larvae in trout, and teratogenesis was a more sensitive test response than embryo-larval mortality in studies with the herring (Blaxter, 1977). However, for species with higher tolerances (e.g., bass, goldfish), anomalous larvae seldom were observed when copper was administered below median lethal concentrations.

The interaction of copper with other metals was highly dependent on exposure concentration. Copper-zinc and copper-mercury mixtures generally were additive but occasionally were antagonistic when fish eggs were treated at low concentrations, and synergism consistently was observed at higher exposure levels. Except for the channel catfish, synergism was not significant below copper-mercury concentrations of 50 $\mu g/l$.

ACKNOWLEDGMENTS

We are deeply grateful to Barbara Ramey, Albert Westerman, Jarvis Hudson, and Paul Francis for technical assistance. Research was supported from funds provided by the National Science Foundation (Grant AEN 74-00768 A01), Office of Water Resources Technology, Department of the Interior (Grant B-44-KY), and Kentucky Institute for Mining and Minerals Research (Grant 7576-EZ).

REFERENCES

Ahmad, M. U. (1973). "Strip Mining and Water Pollution," *Ground Water,* **11**(5), 37–41.

Anderson, J. D. (1967). "*Ambystoma opacum.*" In H. G. Dowling, Ed., *Catalogue of American Amphibians and Reptiles.* American Society of Ichthyologists and Herpetologists, New York, 1963–1970, pp. 46.1–46.2.

Anonymous (1973). "Where Have All the Frogs Gone?" (special article), *Mod. Med.,* **41,** 20–24.

Benoit, D. A. (1975). "Chronic Effects of Copper on Survival, Growth, and Reproduction of the Bluegill (*Lepomis macrochirus*)," *Trans. Am. Fish. Soc.,* **104,** 353–358.

Biesinger, K. E. and Christensen, G. M. (1972). "Effects of Various Metals on Survival, Growth, Reproduction, and Metabolism of *Daphnia magna*," *J. Fish. Res. Board Can.,* **29,** 1691–1700.

Birge, W. J. (1978). "Aquatic Toxicology of Trace Elements of Coal and Fly Ash." In J. H. Thorp and J. W. Gibbons, Eds., *Energy and Environmental Stress in Aquatic Systems.* DOE Symposium Series (CONF-771114), Washington, D.C., pp. 219–240.

Birge, W. J. and Black, J. A. (1977a). *Bioassay Protocol, a Continuous Flow System Using Fish and Amphibian Eggs for Bioassay Determinations on Embryonic Mortality and Teratogenesis.* U.S. Environmental Protection Agency, Office of Toxic Substances, 560/5-77-002, Washington, D.C., 59 pp.

Birge, W. J. and Black, J. A. (1977b). *Sensitivity of Vertebrate Embryos to Boron Compounds.* U.S. Environmental Protection Agency, Office of Toxic Substances, 560/1-76-008, Washington, D.C., 66 pp.

Birge, W. J., Westerman, A. G., and Roberts, O. W. (1974). "Lethal and Teratogenic Effects of Metallic Pollutants on Vertebrate Embryos." In *2nd Annual NSF-RANN Trace Contaminants Conference,* Asilomar, Calif. National Technical Information Service, Springfield, Va., pp. 316–320.

Birge, W. J., Hudson, J. E., Black, J. A., and Westerman, A. G. (1978). "Embryo-Larval Bioassays on Inorganic Coal Elements and *in situ* Biomonitoring of Coal-Waste Effluents." In D. E. Samuel, J. R. Stauffer, C. H. Hocutt, and W. T. Mason, Jr., Eds., *Surface Mining and Fish/Wildlife Needs in the Eastern United States. Proceedings of a Symposium.* Publication no. FWS/OBS-78/81, Office of Biological Sciences, Fish and Wildlife Service, U.S. Department of the Interior, pp. 97–104.

Birge, W. J., Black, J. A., Hudson, J. E., and Bruser, D. M. (1979a). "Embryo-Larval Toxicity Tests With Organic Compounds." In L. L. Marking and R. A. Kimerle, Eds., *Aquatic Toxicology.* Special Technical Publication 657, American Society for Testing and Materials, Philadelphia, pp. 131–147.

Birge, W. J., Black, J. A., and Westerman, A. G. (1979b). "Evaluation of Aquatic Pollutants Using Fish and Amphibian Eggs as Bioassay Organisms." In F. M. Peter, Ed., *Proceedings of the Symposium on Pathobiology of Environmental Pollutants: Animal Models and Wildlife as Monitors.* Institute of Laboratory Animal Resources, National Research Council-National Academy of Science, Washington, D.C. (in press).

Bishop, S. C. (1943). *Handbook of Salamanders: The Salamanders of the United States and Canada, and of Lower California.* Comstock Publ. Co., Ithaca, N.Y., 555 pp.

Black, J. A. (1974). "The Effect of Certain Organic Pollutants on Copper Toxicity to Fish (*Lebistes reticulatus*)." Ph.D. dissertation, University of Michigan, Ann Arbor, 129 pp.

Black, J. A., Roberts, R. F., Johnson, D. M., Minicucci, D. D., Mancy, K. H., and Allen, H. E. (1973). "The Significance of Physicochemical Variables in Aquatic Bioassays of Heavy Metals." In G. Glass, Ed., *Bioassay Techniques and Environmental Chemistry.* Ann Arbor Science Publ., Ann Arbor, Mich., pp. 259–275.

Blaxter, J. H. S. (1977). "The Effect of Copper on the Eggs and Larvae of Plaice and Herring," *J. Mar. Biol. Assoc. U.K.,* **57,** 849–858.

Bolton, N. E., Carter, J. A., Emery, J. F., Feldman, C., Fulkerson, W., Hulett, L. D., and Lyon, W. S. (1975). "Trace Element Mass Balance Around a Coal-Fired Steam Plant." In S. P. Babu, Ed., *Trace Elements in Fuel.* Symposium, Division of Fuel Chemistry at the 166th Meeting of the American Chemical Society, Adv. Chem. Series, Vol. 141, pp. 175–187.

Brown, V. M. and Dalton, R. A. (1970). "The Acute Lethal Toxicity to Rainbow Trout of Mixtures of Copper, Phenol, Zinc and Nickel," *J. Fish Biol.,* **2,** 211–216.

Brungs, W. A., Geckler, J. R., and Gast, M. (1976). "Acute and Chronic Toxicity of Copper to the Fathead Minnow in a Surface Water of Variable Quality," *Water Res.,* **10,** 37–43.

Cairns, J., Jr., and Scheier, A. (1968). "A Comparison of the Toxicity of Some Common Industrial Waste Components Tested Individually and Combined," *Prog. Fish Cult.,* **30**(1), 3–8.

Cherry, D. S., Guthrie, R. K., Rodgers, J. H., Jr., Cairns, J., Jr., and Dickson, K. L. (1976). "Responses of Mosquitofish (*Gambusia affinis*) to Ash Effluent and Thermal Stress," *Trans. Am. Fish. Soc.,* **105**(6), 686–694.

Chu, T. J., Nicholas, W. R., and Ruane, R. J. (1975). "Complete Reuse of Ash Pond Effluents in Fossil-Fueled Power Plants." Paper presented at 68th Annual Meeting of the American Institute of Chemical Engineers Symposium on Water Reuse in the Chemical Industry, Los Angeles, Calif., Nov. 17, 1975, 33 pp.

Chynoweth, D. P., Black, J. A., and Mancy, K. H. (1976). "Effects of Organic Pollutants on Copper Toxicity to Fish." In R. W. Andrew, P. V. Hodson, and D. E. Konasewich, Eds., *Toxicity to Biota of Metal Forms in Natural Water.* Great Lakes Research Advisory Board, pp. 145–157.

Conant, R. (1975). *A Field Guide to Reptiles and Amphibians of Eastern and Central North America,* 2nd ed. Houghton Mifflin, Boston, 429 pp.

Connor, P. M. (1972). "Acute Toxicity of Heavy Metals to Some Marine Larvae," *Mar. Pollut. Bull.,* **3,** 190–192.

Culp, B. R. (1975). "Aqueous Complexation of Copper with Sewage and Naturally Occurring Organics." Ph.D. dissertation, University of Michigan, Ann Arbor, 118 pp.

Daum, R. J. (1969). "A Revision of Two Computer Programs for Probit Analysis," *Bull. Entomol. Soc. Am.,* **16,** 10–15.

Dilling, W. J. and Healey, C. W. (1926). "Influence of Lead and the Metallic Ions of Copper, Zinc, Thorium, Beryllium and Thallium on the Germination of Frogs' Spawn and on the Growth of Tadpoles," *Ann. Appl. Biol.,* **13,** 177–188.

Doudoroff, P. (1952). "Some Recent Developments in the Study of Toxic Industrial Wastes." In *Proceedings of the 4th Annual Pacific Northwest Industrial Waste Conference,* State College, Pullman, Wash., pp. 21–25.

Doudoroff, P. and Katz, M. (1953). "Critical Review of Literature on the Toxicity of Industrial Wastes and Their Components to Fish. II: The Metals, as Salts," *Sewage Ind. Wastes,* **25,** 802–839.

Dunlap, D. G. and Kruse, K. C. (1976). "Frogs of the *Rana pipiens* Complex in the Northern and Central Plains States," *Southwest. Nat.,* **20**(4), 559–571.

Environmental Protection Agency (1976). *Quality Criteria for Water.* Washington, D.C., 256 pp.

Ferm, V. H. and Hanlon, D. P. (1974). "Toxicity of Copper Salts in Hamster Embryonic Development," *Biol. Reprod.,* **11,** 97–101.

Finney, D. J. (1971). *Probit Analysis,* 3rd ed. Cambridge Press, N.Y., 333 pp.

Fox, H. M. and Vevers, G. (1960). *The Nature of Animal Pigments.* Sidgwick and Jackson, London.

Frieden, E. (1968). "The Biochemistry of Copper," *Sci. Am.,* **218**(5), 102–114.

Gibbs, E. R., Nace, G. W., and Emmons, M. B. (1971). "The Live Frog Is Almost Dead," *BioScience,* **21**(20), 1027–1034.

Grande, M. (1966). "Effect of Copper and Zinc on Salmonid Fishes." In O. Jaag and H. Liebmann, Eds., *Advances in Water Pollution Research, Proceedings of the 3rd International Conference,* Munich, Vol. 1. Water Pollution Control Federation, Washington, D.C., pp. 97–111.

Hardy, J. D., Jr. (1964). "The Spontaneous Occurrence of Scoliosis in Tadpoles of the Leopard Frog, *Rana pipiens*," *Chesapeake Sci.,* **5,** 101–102.

Hazel, C. R. and Meith, S. J. (1970). "Bioassay of King Salmon Eggs and Sac Fry in Copper Solutions," *Calif. Fish Game,* **56**(2), 121–124.

Hildebrand, S. G., Cushman, R. M., and Carter, J. A. (1976). "The Potential Toxicity and Bioaccumulation in Aquatic Systems of Trace Elements Present in Aqueous Coal Conversion Effluents." In D. D. Hemphill, Ed., *Trace Substances in Environmental Health*—X. University of Missouri, Columbia, pp. 305–313.

Huckabee, J. W. and Griffith, N. A. (1974). "Toxicity of Mercury and Selenium to the Eggs of Carp (*Cyprinus carpio*)," *Trans. Am. Fish. Soc.,* **103,** 822–825.

Kaplan, H. M. and Licht, L. (1955). "Evaluation of Chemicals Used in Control and Treatment of Disease in Fish and Frogs Caused by *Pseudomonas hydrophila*," *Am. J. Vet. Res.,* **59,** 342–344.

Kaplan, H. M. and Yoh, L. (1961). "Toxicity of Copper for Frogs," *Herpetologica,* **17,** 131–135.

Kopp, J. F. and Kroner, R. C. (1967). *Trace Metals in Waters of the United States, Oct.*

1, 1962, to Sept. 30, 1967. Federal Water Pollution Control Administration, U.S. Department of the Interior, Cincinnati, Ohio, 206 pp.

Lande, S. P. and Guttman, S. I. (1973). "The Effects of Copper Sulfate on the Growth and Mortality Rate of *Rana pipiens* Tadpoles," *Herpetologica,* **29,** 22–27.

Lloyd, R. (1961). "The Toxicity of Mixtures of Zinc and Copper Sulphates to Rainbow Trout (*Salmo gairdneri* Richardson)," *Ann. Appl. Biol.,* **49,** 535–538.

McDermott, G. N., Moore, W. A., Post, M. A., and Ettinger, M. B. (1963). "Copper and Anaerobic Sludge Digestion," *J. Water Pollut. Control Fed.,* **35,** 655–662.

McKim, J. M. and Benoit, D. A. (1971). "Effects of Long-Term Exposure to Copper on Survival, Growth, and Reproduction of Brook Trout (*Salvelinus fontinalis*)," *J. Fish. Res. Board Can.,* **28,** 655–662.

McKim, J. M. and Benoit, D. A. (1974). "Duration of Toxicity Tests for Establishing "No Effect" Concentrations for Copper with Brook Trout (*Salvelinus fontinalis*)," *J. Fish. Res. Board Can.,* **31,** 449–452.

McKim, J. M., Eaton, J. G., and Holcombe, G. W. (1978). "Metal Toxicity to Embryos and Larvae of Eight Species of Freshwater Fish. II: Copper," *Bull. Environ. Contam. Toxicol.,* **19,** 608–616.

Mount, D. I. (1968). "Chronic Toxicity of Copper to Fathead Minnows (*Pimephales promelas* Rafinesque)," *Water Res.,* **2,** 215–223.

Mount, D. I. and Stephan, C. E. (1969). "Chronic Toxicity of Copper to the Fathead Minnow (*Pimephales promelas*) in Soft Water," *J. Fish. Res. Board Can.,* **26,** 2449–2457.

Perkin-Elmer Corporation. (1973). *Analytical Methods for Atomic Absorption Spectrophotometry.* Perkin-Elmer Corporation, Norwalk, Conn.

Pickering, Q. H. and Henderson, C. (1966). "The Acute Toxicity of Some Heavy Metals to Different Species of Warmwater Fishes," *Air Water Pollut. Int. J.,* **10,** 453–463.

Rest, J. R. (1976). "The Histological Effects of Copper and Zinc on Chick Embryo Skeletal Tissues in Organ Culture," *Br. J. Nutr.,* **36,** 243–253.

Ruch, R. R., Gluskoter, H. J., and Shimp, N. F. (1974). "Distribution of Trace Elements in Coal." In F. A. Ayer, Ed., *Symposium Proceedings: Environmental Aspects of Fuel Conversion Technology.* U.S. Environmental Protection Agency, Office of Research and Development, EPA-650/2-74-118, Washington, D.C., pp. 49–53.

Rugh, R. (1962). *Experimental Embryology,* 3rd ed. Burgess, Minneapolis, Minn., 501 pp.

Sauter, S., Buxton, K. S., Macek, K. J., and Petrocelli, S. R. (1976). *Effects of Exposure to Heavy Metals on Selected Freshwater Fish. Toxicity of Copper, Cadmium, Chromium and Lead to Eggs and Fry of Seven Fish Species.* U.S. Environmental Protection Agency, EPA-600/3-76-105, Duluth, Minn., 75 pp.

Shaw, T. L. and Brown, V. M. (1974). "The Toxicity of Some Forms of Copper to Rainbow Trout," *Water Res.,* **8,** 377–382.

Sheibley, D. W. (1975). "Trace Elements by Instrumental Neutron Activation Analysis for Pollution Monitoring." In S. P. Babu, Ed., *Trace Elements in Fuel.* Symposium,

Division of Fuel Chemistry at the 166th Meeting of the American Chemical Society, Adv. Chem. Series, Vol. 141, pp. 98–117.

Sprague, J. B. (1964). "Avoidance of Copper-Zinc Solutions by Young Salmon in the Laboratory," *J. Water Pollut. Control Fed.,* **36**(8), 990–1004.

Sprague, J. B. and Ramsay, B. A. (1965). "Lethal Levels of Mixed Copper-Zinc Solutions for Juvenile Salmon," *J. Fish. Res. Board Can.,* **22**(2), 425–432.

Steele, J. H., McIntyre, A. D., Johnston, R., Baxter, I. G., Topping, G., and Dooley, H. D. (1973). "Pollution Studies in the Clyde Sea Area," *Mar. Pollut. Bull.,* **4,** 153–157.

Underhill, D. K. (1966). "An Incidence of Spontaneous Caudal Scoliosis in Tadpoles of *Rana pipiens* Schreber," *Copeia,* **1966**(3), 582–583.

Vial, J. L. (1973). *Evolutionary Biology of the Anurans.* University of Missouri Press, Columbia, 470 pp.

Westerman, A. G. (1977). "Lethal and Teratogenic Effects of Inorganic Mercury and Cadmium on Embryonic Development of Anurans." M.S. thesis, University of Kentucky, Lexington, 76 pp.

Westerman, A. G. and Birge, W. J. (1978). "Accelerated Rate of Albinism in Channel Catfish Exposed to Metals," *Prog. Fish Cult.,* **40**(4), 143–146.

Wright, A. H. and Wright, A. A. (1949). *Handbook of Frogs and Toads of the United States and Canada,* 3rd ed. Cornell University Press, Ithaca, N.Y., 640 pp.

13

USE OF COPPER AS A MOLLUSCICIDE

Thomas C. Cheng

Institute for Pathobiology, Center for Health Sciences, Lehigh University, Bethlehem, Pennsylvania

1. INTRODUCTION

The necessity of developing molluscicides became apparent when parasitologists discovered that many of the helminthic diseases of human beings and domesticated animals involve direct or indirect transmission by molluscs. The need for molluscicides to be employed in disease control has increased in urgency as attempts to develop chemotherapy and vaccines have not been totally successful, especially when viewed from the standpoints of undesirable side effects, reinfection, and the necessity of widespread deployment.

The first reports of the use of molluscicides for the control of human schistosomiasis, a parasitic disease caused by the trematodes *Schistosoma mansoni, S. haematobium,* and *S. japonicum,* appear to be those of Fujinami and Narabayashi (1913a, 1913b) and Narabayashi (1915), who used lime, and that of Miyagama (1913), who employed calcium cyanamide. The use of molluscicides against the snail *Oncomelania hupensis,* the vector for the oriental type of schistosomiasis, in Japan has been reviewed by Komiya (1961). Apparently Narabayashi (1915) was the first to record the molluscicidal properties of copper sulfate; however, he considered the use of this inorganic copper salt to be impracticable because of its toxicity to other animals and plants, especially rice. In spite of Narabayashi's pessimism, copper sulfate became a popular molluscicide.

Chandler (1920) reported that copper sulfate is effective in killing the molluscan intermediate host of the liver fluke, *Fasciola hepatica,* and other snails, and as a result of his publication, copper sulfate has been in essentially continuous use as a molluscicide. However, only relatively recently have the pathophysiology and mechanisms responsible for the molluscicidal properties of the copper been studied. This fact has not deterred the mass application of copper sulfate as a lethal agent against the molluscan intermediate hosts of schistosomes, especially in the Sudan and Egypt (Sharaf and El Nagar, 1955; El-Ghindy, 1957a, 1957b). A bibliography of papers published between 1852 and 1962 dealing with the use of copper sulfate and other molluscicides has been contributed by Warren and Newill (1967).

Copper sulfate, like all other major molluscicides (e.g., niclosamide, sodium pentachlorophenate, *N*-tritylmorpholine) has certain disadvantages (Ritchie, 1973). In the case of copper sulfate, in addition to total or partial inactivation in natural waters due to adsorption by soil and organic materials, it is ineffective at alkaline pH's and is toxic to other nontarget organisms, especially young fish and certain aquatic vegetation (Birge and Black, this volume). On the other hand, the cost of copper sulfate is sufficiently low so that most developing nations where domestic animal and human schistosomiasis and other snail-borne diseases occur can afford to use it as a molluscicide on a large scale. Also, since copper is mined in many areas of the world, such as Gambia, where snail-transmitted diseases, especially schistosomiasis, are serious health problems, such nations do not have to spend funds to import molluscicides.

The use of copper-containing molluscicides has by no means been limited to Africa. Copper sulfate has received attention by a number of workers in Brazil (Pinto and Penido, 1948; Chaia et al., 1956; Chaia and Paulini, 1957; Camey and Paulini, 1962). Jansen (1943), also in Brazil, tested the efficacy of $CuSO_4$ mixed with lime, and Perlawagora-Szumlewicz (1955) in Brazil tested the use of copper sulfate simultaneously with pentachlorophenate in the hope that the resulting CuPCP precipitate would have a residual effect.

In addition, Pitchford et al. (1960) reported success with $CuSO_4$ in controlling the snail hosts of human-infecting schistosomes in eastern Transvaal, South Africa, and El-Ghindy (1957a, 1957b) reported on the use of $CuSO_4$ as a molluscicide in Egypt. The numerous published accounts of the use of $CuSO_4$ in Egypt have been reviewed by Ayad (1961). It is probably in the Sudan that $CuSO_4$ has been most extensively employed to control snails (Sharaf and El Nagar, 1955).

2. COPPER-CONTAINING MOLLUSCICIDES

This section reviews what is known about several copper-containing compounds other than $CuSO_4$ that have been tested in the field. For a more detailed discussion of this topic, the review of Paulini (1974) is recommended.

2.1. Copper Pentachlorophenate (CuPCP)

This compound was first tested by Dobrovolny and Dobbin (1955), who used both a wettable and a nonwettable powder manufactured by the Monsanto Chemical Co. This molluscicide was applied to static water ponds or ditches either as a dust mixed with inert material, using a duster, or as a suspension in water, using a compression sprayer. They reported >90% reduction of the snail *Biomphalaria glabrata* in 80% of the trials when concentrations of 5 ppm or more were used.

Puma and de Zagustin (1969) tested CuPCP in Venezuela against *B. glabrata.* The product that they used was prepared at the test sites by mixing equal volumes of a 10% $CuSO_4$ solution with a 20% sodium pentachlorophenate solution. The $CuSO_4$ solution also included 1 g/l of sodium citrate to prevent precipitation by bicarbonates. These investigators reported that their preparation was lethal to the snails at a minimum concentration of 1.5 ppm with 24-hr exposure. Furthermore, different field conditions required varying concentrations. In streams with sandy bottoms and clear water, 6 ppm Cu as CuPCP was applied for 18 hr with effective results, while in streams with muddy bottoms and turbid water a concentration of 6 ppm applied for 24 hr was successful. In relatively clear waters the effective range may extend an average of 5000 m from the point of application, whereas in very turbid waters the effective distance is <1500 m.

When $CuSO_4$ is mixed with sodium pentachlorophenate in a 1:1 (w/w) ratio, the resulting "colloidal" suspension is less influenced by turbidity. This formulation has been used with success at a concentration of 3 ppm for 48 hr in both clear and turbid waters (Puma and de Zagustin, 1969).

Because of the satisfactory use of CuPCP in Venezuela, it was the molluscicide of choice against *B. glabrata* from 1953 to 1968.

2.2. Copper Carbonate

Barbosa et al. (1956) tested the efficacy of copper carbonate against planorbid snails in northeastern Brazil. Small streams that ranged from 27 to 370 m in length and from 0.4 to 1.7 m in width were treated with concentrations ranging from 30 to 430 g/m^2. The results were excellent in five streams where complete elimination of snails was achieved for 12 months with doses of 40 to 80 g/m^2. In other streams the results were good, with the snails eliminated for 2 to 6 months with doses of 80 to 430 g/m^2. Finally, in three of the test streams the results were poor, that is, the snails were eliminated for only 1 month when copper carbonate was applied at concentrations of 30 to 60 g/m^2.

One of the undesirable side effects of copper carbonate is that it also kills fish. Barbosa et al. (1956) reported that fish disappeared from their test streams for 2 to 3 months, although they eventually reappeared. Also, like $CuSO_4$, $CuCO_3$ settles to the bottom rapidly.

2.3. Copper-Tartaric Acid

Puma and de Zagustin (1969) reported from Venezuela that when tartaric acid was added to $CuSO_4$ in small quantities, with the intent of dissolving the basic carbonate precipitate, it prevented the precipitation of copper in alkaline water.

This desirable feature was achieved most effectively when 1 part tartaric acid was added to 5 parts $CuSO_4$. It was also reported that a concentration of 50 ppm applied over a 6-hr period eliminated the snail *B. glabrata* for a distance of 1500 m in clear, rapid streams, but was less effective in turbid streams with muddy bottoms. Thus it is apparent that mud and organic matter absorb copper even in the presence of tartaric acid.

2.4. Copper Arsenite-Acetate (Paris green)

Castro (1954) was the first to report the molluscicidal property of Paris green. Subsequently, Barbosa et al. (1956) reported that this compound, applied at a concentration of 3 g/m^2 or more, killed all of the exposed *B. glabrata*.

Paulini (1974) tested Paris green in the field and found that an aqueous suspension sprayed over small irrigation canals at a concentration of 1 g/m^2 produced variable results. In one test area 100% reduction of snails lasted for 2 weeks, but in another area <50% reduction was achieved after 7 days.

2.5. Copper(I) Oxide

This compound was reported by Deschiens et al. (1961) to be molluscicidal. Subsequently, Paulini et al. (1963) demonstrated that Cu(I) oxide at a minimum concentration of 0.7 ppm Cu was lethal to all snails after 72 hr of exposure and after 24 hr of exposure at 3.5 ppm. However, when the laboratory bioassays were conducted in the presence of mud collected from natural habitats, only 10 to 20% of *B. glabrata* was killed after 24-hr exposure to a concentration of 3.5 ppm Cu. Continuous exposure for 3 to 4 days to mud covered with Cu(I) oxide was necessary to kill the snails. Paulini et al. also demonstrated experimentally that mud adsorbs this compound.

2.6. Copper(I) Chloride

Paulini (1974) stated that this copper compound was tested in his laboratory in Brazil during 1965. He reported that Cu(I) chloride, under the trade name of Procida 22-07, contains 62% copper. It was dissolved in a saturated NaCl solution, and subsequently diluted to working concentrations with tap water. A 24-hr exposure, followed by a 7-day recovery period, resulted in a LC_{50} value of 0.4 ppm and a LC_{90} value of 1.2 ppm for adult *B. glabrata*. Furthermore, the LC_{50} and LC_{90} values for 1-day-old eggs of *B. glabrata* were 6.0 and 10 ppm, respectively.

Chu et al. (1968) tested Cu(I) chloride in both the laboratory and the field against *Bulinus truncatus* in Iran. They reported that this compound showed molluscicidal activity at 5 ppm with 24 hr of exposure. The toxicity of the compound declined slowly during the following 90 days.

When Chu et al. (1968) tested Cu(I) chloride in the presence of mud (100 g/l water), its molluscicidal activity was lost rapidly. Specifically, after 11 days of contact with mud, 25 ppm of the compound did not kill more than 50% of the snails that had been exposed for 72 hr. Furthermore, 100 ppm of this compound was biologically inactive 50 days postexposure to mud. The water incorporated in these tests had a pH of 7.4 to 7.9, and the total solid content was 350 to 420 ppm.

2.7. Copper(II) Acetylacetonate

Dobrovolny and Dobbin (1955) tested this solution emulsified with Tween 80. They reported that, when applied at concentrations of 5 and 10 ppm, it was not effective as a molluscicide against *Biomphalaria glabrata* at the lower concentration; however, at the higher one, 95 to 100% reduction of snails was obtained in the field.

2.8. Copper Dimethyldithiocarbamate

This commercial fungicide was tested by Paulini and Camey (1965). They reported that at 7 ppm Cu this compound was effective at the LC_{90} level against mature *B. glabrata* with 24 hr of exposure. When tested in the field in static water, the same concentration gave 100% reduction of snails. Furthermore, when applied twice to canals with intermittent flow within a 2-week period, complete elimination of the snails was achieved.

2.9. Copper Ricinoleate

Gonçalves and Soares (1955) tested the effectiveness of copper ricinoleate. They prepared the compound by mixing a concentrated soap solution (30% w/w of ricinus oil) with copper sulfate solution (17% w/w), separating the copper soap by decantation, and extracting the soap with hot alcohol. The alcoholic solution contained most of the copper ricinoleate since the other copper soaps are insoluble.

Gonçalves and Soares stated that a relatively stable emulsion can be prepared by mixing the alcoholic solution with water. The active component is reported to be CuO.

Unfortunately, the molluscicidal property of copper ricinoleate reported by

Gonçalves and Soares cannot be appropriately compared with the results of others because of the bioassay method that they employed. Specifically, they used retraction of the snail into its shell as the criterion for "effectiveness." They found that 1 ppm or less of copper ricinoleate was required to elicit this response in 75% of the exposed snails, while parallel tests involving NaPCP caused this response in only 50% of the snails.

Gonçalves and Soares also tested copper ricinoleate in the field. The suspension at a calculated concentration of 1 ppm was applied with a compression sprayer to a pond; however, the results were inconclusive.

2.10. Copper Rosinate

Paulini (1974) reported his results of testing copper rosinate in the field. The product used was a mixture of equal volumes of copper sulfate (16.5%) and rosin soap (20%) in "a large volume of water" just before application. This preparation was distributed into small irrigation canals in Brazil with a watering can at a rate of 1 g/m^2. As a result of monitoring *B. glabrata* weekly, it was found that 100% reduction in the snail population was achieved for 10 weeks in one region; however, there was only <50% reduction in another within 1 week after application.

3. APPLICATION METHODS

Because of differences in physicochemical characteristics in the aquatic habitats where undesirable molluscs are found, a variety of application methods have been devised. This topic has been reviewed in part by Amin (1974); however, for the sake of completeness the essential features of each method are described below.

3.1. Drip-Feed Method

With this method the molluscicide is applied to the headwaters of any flowing water (e.g., small rivers, streams, or irrigation canals) and is carried by the current to penetrate the main and subsidiary tributaries. The dispensing apparatuses are placed at narrow or turbulent spots along the stream to ensure complete mixing of the molluscicide with the water.

Whether or not to employ the drip-feed method depends on a number of factors, such as formulation of the molluscicide, ambient conditions, irrigation practices, and water management methods. For example, there is no advantage to using this method to deliver a molluscicide at the point of entry of a canal if the water is stagnant. Also, there is no advantage in drip-feeding a chemical that will sink or become inactivated by adsorption onto silt.

Amin (1972) has reported that in the Gezira Scheme, Sudan, drip-feeding a copper sulfate solution from plastic dispensers resulted in very poor downstream carriage. The reason was the rapid precipitation of copper and its continuous uptake by mud and vegetation. On the other hand, Webbe (1963) reported the successful reduction of *Biomphalaria sudanica tanganyicensis* in the Mirona River in Africa by drip-feeding the wettable formulation of Baylucide at 1 ppm for 8 hr from one application point. Similar satisfactory results were obtained by Fenwick (1970) and others with this application method involving other non-copper-containing molluscicides.

3.2. Ground Spraying

For relatively small bodies of stagnant or slowly flowing water, spraying the molluscicide from the ground is preferred. Hand sprayers are sufficient for small ponds and streams; however, for larger bodies of water more sophisticated sprayers are necessary. For example, in the case of stagnant and shallow water covered with considerable vegetation, such as paddies, swamps, and seepage areas, the use of high-pressure pumps is necessary. Furthermore, if long distances have to be covered, tractor-mounted sprayers may be required.

3.3. Aerial Spraying

If vast areas have to be covered, aerial spraying can be employed. First tried on a lakeshore in Rhodesia by Barnish and Shiff (1970), this method was subsequently used by Degremont et al. (1972) to spray rice fields and large drains in the Malagasy Republic, by Barnish and Sturrock (1973) to spray a swamp in St. Lucia, and by Amin and Fenwick (1973) on large stagnant and slow-flowing courses in the Sudan. Although Barnish and Shiff (1970), Degremont et al. (1972), and Barnish and Sturrock (1973) achieved only limited success, Amin and Fenwick (1973) reported a very high snail kill. They utilized a flying speed of 160 km/hr and a spraying height of 2 to 3 m above the water surface. It is noted that they reported the cost of spraying 350 km of canals with *N*-tritylmorpholine to be only 5% of the cost of the molluscicide.

3.4. Gravity Liquid Feeder

Because of its simplicity and low cost, the liquid feeder has been used widely in dispensing molluscicides. This apparatus, which has been modified by various people, is essentially a container with an opening controlled by a valve. There are two disadvantages to this type of apparatus: (*a*) to maintain a constant flow of molluscicide, the valve must be continually adjusted; and (*b*) the orifice commonly becomes blocked because of crystalization of the molluscicide. These disadvantages can be overcome, however, by modifying the feeder (Shiff, 1961; Webbe, 1963; World Health Organization, 1965).

3.5. Dipping Sacks

In 1956 copper sulfate was applied to irrigation canals, ponds, and other types of habitats in the Sudan by employing small burlap sacks containing this chemical. In this method the sacks, each attached to a pole, are dipped in the water at spatial intervals from along the banks. The major disadvantage is the difficulty of covering every meter of very long watercourses. Also, because the water is often flowing, there is continuous dilution of the molluscicide and the desired concentration is not easily attained.

3.6. Chemical Barriers

Burlap bags containing soluble solid molluscicides, such as $CuSO_4$, are commonly employed at main intersections of flowing water to serve as chemical barriers against the passage of snails that transmit schistosomes. For example, El Nagar (1958) in the Sudan used bags of $CuSO_4$ which gave a concentration of 0.125 ppm in the immediate vicinity to stop snails from the main irrigation canals from entering tributaries. Also, Teesdale et al. (1961), working in Kenya, reported that such a barrier, which gave an average of 0.25 ppm $CuSO_4$, was effective in preventing the passage of snails; however, when the concentration in the immediate area was 0.125 ppm, the results were intermediate.

3.7. Floating Formulations

The use of floating formulations in mollusc control may be useful in large bodies of water (e.g., lakes and drains). The molluscicide usually is in liquid form. Strufe et al. (1965) used corn cobs saturated with a commercial molluscicide (Baylucide) to treat the upper levels of deep, stagnant waters.

The major disadvantage of floating formulations is that, if there is a wind, the floats drift away from the intended target areas.

3.8. Slow-Release Formulations

The origin and history of what are commonly referred to as "slow-release" or "controlled-release" formulations relative to molluscicides have been reviewed by Cardarelli (1974). This important advancement in molluscicide delivery technology consists of continuous release of the chemical from an elastomeric matrix for an extended period of time. Rubber is now the most commonly employed matrix, and a variety of molluscicides have been tested (Cardarelli, 1974). Although this method has many advantages, it is not applicable to all environments. For example, preliminary tests with $CuSO_4$ embedded in rubber pellets in the relatively fast-flowing mountain streams of St. Lucia have indicated that it does not achieve a high degree of effectiveness in killing *B. glabrata*.

For a definitive review of slow-release technology, see Cardarelli (1974).

4. UPTAKE OF COPPER AND ITS LETHAL MECHANISM

In spite of the widespread use of copper compounds, especially $CuSO_4$, as molluscicides in recent years, with the development of a more acute awareness of environmental problems there has been concern that the continuous application of copper sulfate may result in irreversible damage to the environment. Consequently, it was deemed important to conduct research aimed at taking advantage of the lethal property of the cupric ion on snails and developing alternative copper compounds that are effective molluscicides but are specific for molluscs, especially the species that serve as intermediate hosts for animal- and human-infecting parasites. However, it was recognized that, to achieve the stated objective, a fuller understanding of several basic phenomena associated with the uptake and toxicity of copper had to be achieved. Specifically, the following questions had to be answered:

1. How is copper taken up from the aqueous environment by molluscs?
2. Are there target cells and/or tissues that are most vulnerable to the toxic cupric ion, and does the death of the mollusc result from the effect of copper on such cells and/or tissues?
3. What is the specific mode of action of the cupric ion that causes lethality?
4. In what stereochemical form must copper be in order to possess lethal properties?
5. Can the toxicities of candidate copper-containing molluscicides be tested in a more efficient, reliable, and quantitative manner than the conventional LC_{50} method?

These are but some of the basic questions that needed to be answered before a more rational approach could be taken to develop new copper-containing molluscicides that include the desirable properties, that is, are stable under a variety of ambient conditions, are readily taken up but only by the target species of molluscs, are safe to handle, and are economically feasible to produce on the competitive market.

A series of studies has been carried out at this institute and elsewhere to answer the questions posed above. Below is a review of some of the available answers.

4.1. Uptake of Copper

By employing ^{67}Cu in the form of $CuCl_2$ as a tracer, Cheng and Sullivan (1974) demonstrated that, when specimens of *Biomphalaria glabrata* were dissected into three regions: the head (head-foot region), the mid, and the digestive gland regions (Figure 1), the uptake of ^{67}Cu, as determined by scintillation counting, was significantly greater in the head-foot region (Figure 2) on a per gram of wet weight basis. The mid region took up radioactive copper at approximately the same rate as the soft tissues of the entire snail, and the digestive gland region took up the least amount of ^{67}Cu (Figure 2). These results appear to conflict with those of Yager and Harry (1964), who reported that the highest uptake of another radioactive heavy metal, cadmium, was by the digestive gland of *B. glabrata*. It is uncertain, however, because of the procedure used by Yager and Harry, whether cadmium was actually concentrated in the digestive gland. This needs to be verified by employing a more suitable technique. The findings of Cheng and Sullivan (1974), however, are in agreement with those of Ryder and Bowen (1977a). These investigators demonstrated that in the slug *Agriolimax reticulatus* the uptake of copper in the form of $CuSO_4$ occurs through the epithelial lining of the foot. Moreover, by employing potassium ferrocyanide to precipitate $CuSO_4$ to form copper ferrocyanide, followed by transmission

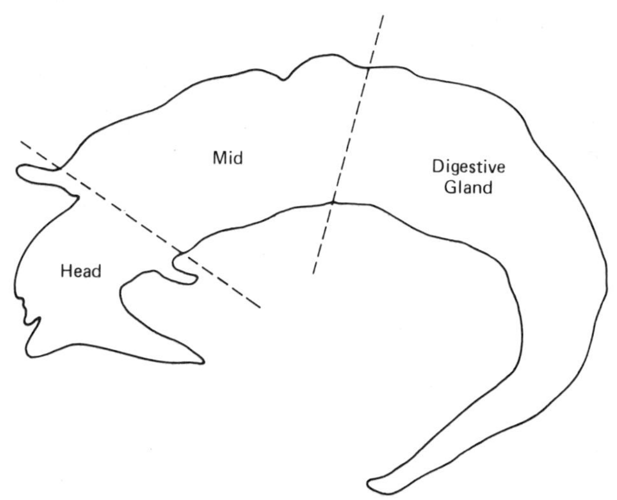

Figure 1. Schematic drawing showing dissection of *Biomphalaria glabrata* into head (head-foot), mid, and digestive gland regions. After Cheng and Sullivan (1974), with permission of Academic Press.

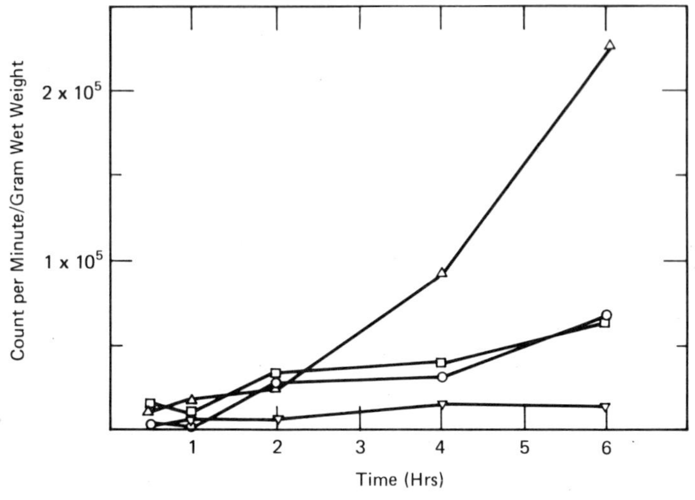

Figure 2. Graph showing uptake of ^{67}Cu by head (△), mid (○), and digestive gland, (▽) segments and by the entire snail (□), *B. glabrata,* on a per gram wet weight basis. After Cheng and Sullivan (1974), with permission of Academic Press.

electron microscopy, Ryder and Bowen showed that the precipitate was localized in the zonula adherens and septate junctions between adjacent epithelial cells. This indicates that the uptake of copper occurs by diffusion between the cells lining the surface of the foot of *A. reticulatus.* However, it is of importance to note that Ryder and Bowen found that copper ferrocyanide was also present in some of the pinocytotic vesicles of the epithelial cells. Thus it is being emphasized that there is some uptake of copper into the epithelial lining cells of the foot.

By employing autoradiography, Cheng and Sullivan (1974) demonstrated that ^{67}Cu is accumulated in two regions on the surface of *B. glabrata:* on the epithelium covering the head-foot, and on the epithelium covering the rectal ridge. Furthermore, the accumulation of ^{67}Cu on the head-foot was greater in snails that had been exposed to the radioactive copper for 6, 8, and 10 hr than in snails exposed for 0.5, 2, and 4 hr. The copper deposits are embedded in a thick layer of mucus that coats the head-foot. This mucus is secreted in response to contact with toxic substances.

It has been stated that the rectal ridge is one of the two areas of *B. glabrata* at which copper accumulates. Pan (1958) showed that this ridge is composed histologically of a single layer of ciliated epithelium covering a core of heavy, loose connective tissue. Cheng and Sullivan (1974) postulated that the rectal ridge, which runs along the length of the snail, may well be an organ that serves in gaseous exchange since water is propelled by ciliary action across its surfaces. Sullivan et al. (1974), as a result of studying the rectal ridge by transmission and electron microscopy, concluded that the ultrastructure of the epithelium covering this ridge characterizes this tissue as a "transporting epithelium" (Berridge and Oschman, 1972). Specifically, the presence of microvilli and

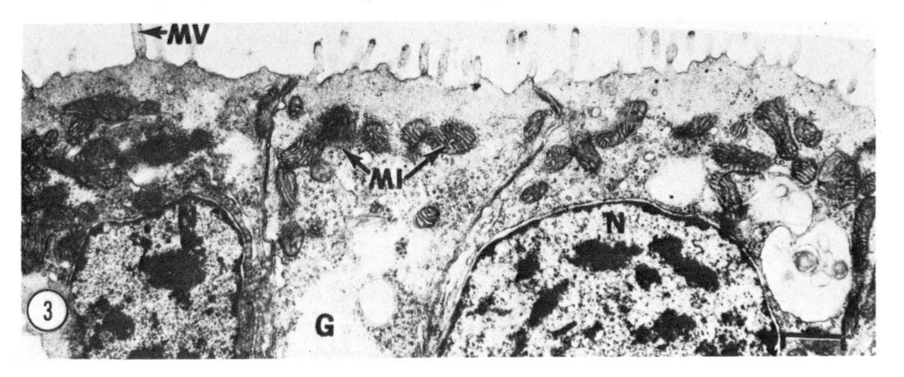

Figure 3. Transmission electron micrograph of microvillar surface and apical mitochondria in rectal ridge epithelial cells of *B. glabrata*. G, α-glycogen; MI, mitochondria; MV, microvilli, N, nucleus, Bar = 1 μm. After Sullivan et al. (1974), with permission of Springer-Verlag.

mitochondria at the apical pole (Figure 3), the abundance of ciliated cells, and the intimacy of the surface epithelium with the underlying loose, vascular connective tissue all suggest a transporting function.

Also, Sullivan et al. (1974) pointed out that, since *B. glabrata* is a freshwater species, the uptake and retention of ions is one of the mechanisms by which the osmotic gradient is maintained between the hemolymph of the snail and the aqueous environment. Therefore it is most likely that the transporting epithelium of the rectal ridge takes up ions from water swept into the mantle cavity by ciliary currents. Oxygen uptake also occurs across this epithelium by simple diffusion. The passage of metals, such as copper, as Ryder and Bowen (1977a) have demonstrated, occurs primarily between the epithelial cells. This confirms experimentally the hypothesis of Berridge and Oschman (1972) that the type of cell junction present in transporting epithelium is not a barrier to diffusion.

It needs to be re-emphasized that Ryder and Bowen (1977a, 1977b) demonstrated that epithelial cells of the foot of molluscs are capable of endocytosis. As is explained at a later point, this mode of uptake is intimately associated with the lethal mechanism of copper.

4.2. Target Tissue/System Hypothesis

The question that has been raised is: Is there a specific internal tissue or system in molluscs that is especially vulnerable to copper? This is an alternative to the hypothesis that the lethal action of copper is an attack on the surface epithelium (Sullivan and Cheng, 1975, 1976).

The initial approach taken to test the validity of the internal tissue/system hypothesis was to determine the possible poisoning effect of copper on some critical metabolic enzyme. Sullivan (1975) selected succinic dehydrogenase

(SDH) as the enzyme on which to test. By following the SDH assay method of King (1963), he was able to demonstrate that copper effects 100% inhibition of the mitochondrial membrane-bound SDH reaction at threshold concentrations of 75 to 187 μM (4.76 to 11.90 ppm).

In spite of the results obtained by Sullivan (1975), it was decided that an additional study should be performed to test the internal target tissue hypothesis. Consequently, Sullivan and Cheng (1976) carried out a study wherein various concentrations of copper in the form of $CuSO_4$ (100, 500, 750, 875, 1000, and 2500 ppm Cu) were injected into the hemocoel of *B. glabrata,* and the mortality of this gastropod was subsequently monitored. Also, the concentrations of copper in the hemolymph of injected snails were calculated, and additional specimens of *B. glabrata* were incubated in these concentrations, that is, the snails were placed in water containing concentrations of copper in the form of $CuSO_4$ equal to those in the hemolymph of injected snails. The results of this study revealed a greater mortality among snails that had been exposed to copper externally than among those internally exposed (Figure 4). Thus it was concluded that, in spite of the inhibition of Cu^{2+} on SDH and undoubtedly on other critical enzymes, copper is more lethal to *B. glabrata* externally than internally. This supports the hypothesis that the lethal action of copper on *B. glabrata,* and most probably on other species of freshwater gastropods, is due to damage inflicted on the surface epithelia rather than on some internal tissue/system.

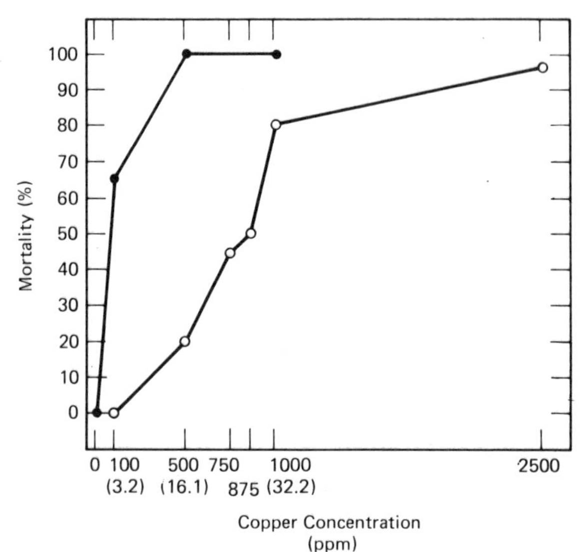

Figure 4. Graph showing mortality of *B. glabrata* that had been injected with or incubated in solutions of copper as $CuSo_4$. Concentrations in parentheses refer to copper solutions in which snails were incubated and which are equivalent to the concentrations injected, appearing immediately above (●) incubation, or (O) injection; $n = 20$. After Sullivan and Cheng, (1976), with permission of Academic Press.

4.3. Specific Lethal Mechanism

Having demonstrated that the lethal mechanism of copper is directed primarily at the outer epithelial surfaces (e.g., those covering the head-foot and the rectal ridge), studies were carried out to ascertain what cytopathologic alterations attributable to copper occur at the subcellular level.

Sullivan and Cheng (1975) conducted a cytopathologic study at the electron microscope level, selecting the ciliated epithelial cells covering the rectal ridge of *B. glabrata* as the model. The snails of the two experimental series were exposed to deionized water containing 60 ppm Cu as $CuSO_4$ for 12 hr and to water containing 0.06 ppm Cu as $CuSO_4$ for 60 hr, respectively. The rationale for selecting these two concentrations of copper and the associated exposure times was that molluscicides are usually applied in the field in one of two ways: either as a "slug dose," whereby a high initial concentration of molluscicide is attained, or by a controlled-release system (reviewed earlier), where a low-level application is maintained over an extended period of time. It is known from laboratory tests that $CuSO_4$ at a concentration of 0.06 ppm Cu^{2+} is lethal to *B. glabrata* (Malek and Cheng, 1974).

In specimens of *B. glabrata* that have not been exposed to copper, a single layer of microvilli-bearing epithelial cells covers the rectal ridge surface in alternating series of projections and furrows (Figure 5). The apical cytoplasm of these cells is packed with mitochondria. Ciliated cells and mucus-secreting goblet cells are distributed throughout this epithelium. The digitiform bases of the epithelial cells interdigitate with constrictions of the underlying basal lamina.

Mediad to the basal lamina are bands of myofibers, which are not organized as a continuous sheet. Below these myofibers is the matrix of the rectal ridge. It is composed of loose vascular connective tissue in which are embedded pigment cells, amoebocyte-like cells, scattered myofibers, and calcium spherites. Membrane-bound vesicles, some enclosing glycogen, and cytoplasmic strands also occur in the matrix.

The fine structure of the rectal ridge epithelium of snails that have been exposed to 60 ppm Cu^{2+} as $CuSO_4$ for 12 hr is markedly different from that of the controls (Figure 6). The corrugated appearance of the surface is not as prominent, although cilia are present, and the cytoplasmic organelles appear normal. Also, the number of digitiform projections at the bases of the epithelial cells is greatly reduced; they are totally absent in some cells. Moreover, the basal lamina is not present, although fibrous remnants are evident. Finally, the loose vascular connective tissue appears "empty," that is, large spaces occur in the matrix. The various types of cells embedded in the connective tissue matrix, however, appear normal.

In snails that have been exposed to 0.06 ppm Cu as $CuSO_4$ for 60 hr, distinct pathologic features also occur. Specifically, in addition to those already described

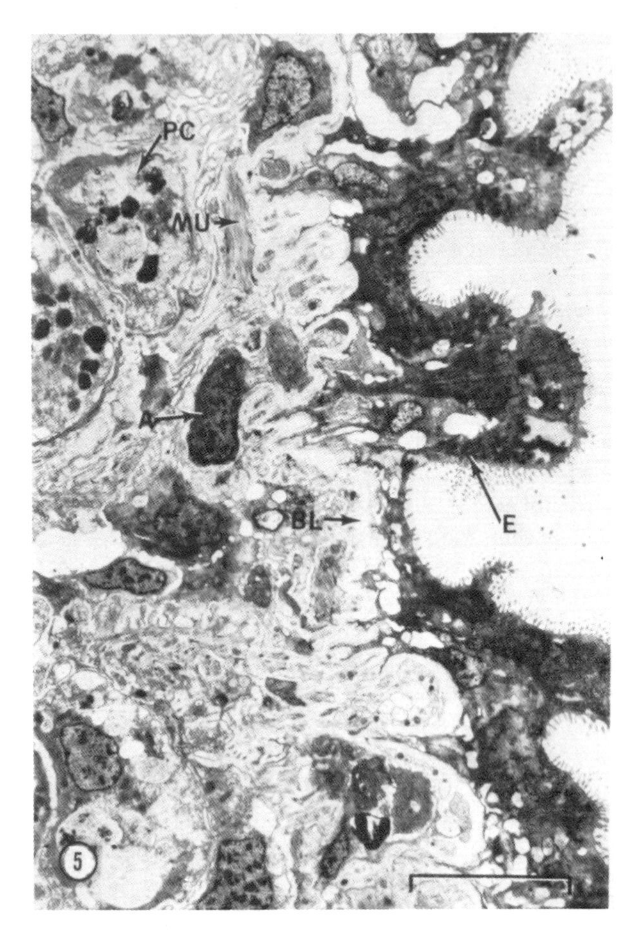

Figure 5. Transmission electron micrograph of epithelium of rectal ridge and underlying connective tissue of *B. glabrata*. Note digitiform projections at bases of epithelial cells. A, amoebocyte-like cell; BL, basal lamina; E, epithelial cells; MU, muscle; PC, pigment cell. Bar = 10 μm. After Sullivan et al. (1974), with permission of Springer-Verlag.

for the rectal ridge of snails exposed to 60 ppm Cu^{2+} for 12 hr, certain cells of the epithelium are highly vacuolated (Figure 7). Also, other epithelial cells have unusually smooth surfaces and are devoid of microvilli and cilia (Figure 8). Numerous vesicles occur in the matrix of the rectal ridge (Figure 9). Some of these resemble cisternae of rough endoplasmic reticulum normally found in pigment cells (Figure 10). Some of these vesicles contain a crystalline substance, which represents fragments of microtubule-like structures that occur in pigment cells. Other vesicles contain an electron-dense substance and closely resemble the pigment granules of pigment cells (Figure 11). Finally, degenerating mitochondria are free in the matrix (Figure 12). The inner and outer membranes

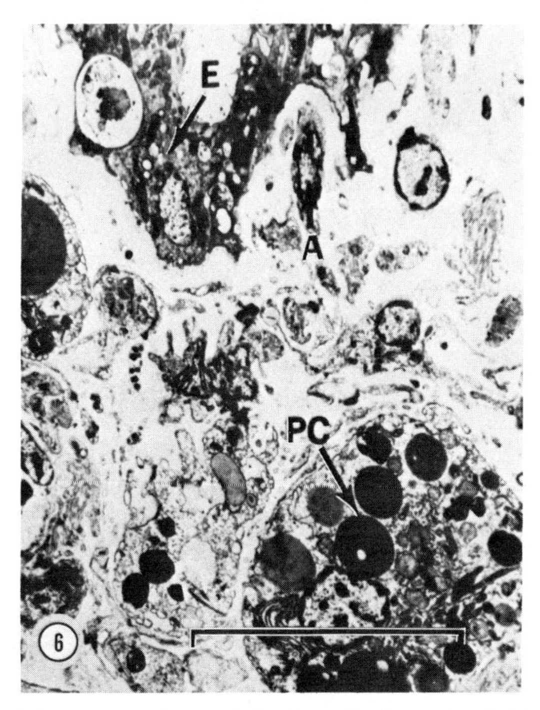

Figure 6. Transmission electron micrograph showing epithelium and underlying connective tissue from *B. glabrata* that had been exposed to 60 ppm Cu as $CuSO_4$ for 12 hr. A, amoebocyte-like cell; E, epithelial cell; PC, pigment cell. Bar = 20 μm. After Sullivan and Cheng (1975), with permission of The New York Academy of Sciences.

of these mitochondria are separated by a wider space than is normally observed. Furthermore, many of the mitochondria, such as the one depicted in Figure 12, are abnormally distended. It is noted that normal-appearing pigment cells and hemocytes also occur in the connective tissue of these snails.

It is apparent from the above that the epithelial cells of the rectal ridge of *B. glabrata* that has been exposed to lethal doses of copper become greatly distended. Specifically, the loss of the corrugated appearance of the epithelium, the loss of digitiform projections from the bases of the cells, and the disintegration of the underlying basal lamina imply that a swelling of the tissues mediad to the epithelium has occurred, which, in turn, exerts a stretching force on that epithelium and the basal lamina. Furthermore, the distended, empty appearance of the loose vascular connective tissue of the rectal ridge suggests an influx of water into this tissue. Thus a failure in the osmoregulatory physiology of the snail, resulting in the accumulation of water in the tissues, is indicated.

A secondary effect of the influx and accumulation of water in the tissues of snails that have been exposed to copper is the lysis of pigment cells, the vesicular remnants of which are consequently found free in the matrical intercellular spaces. Although the function of the pigment cells remains uncertain, Sminia

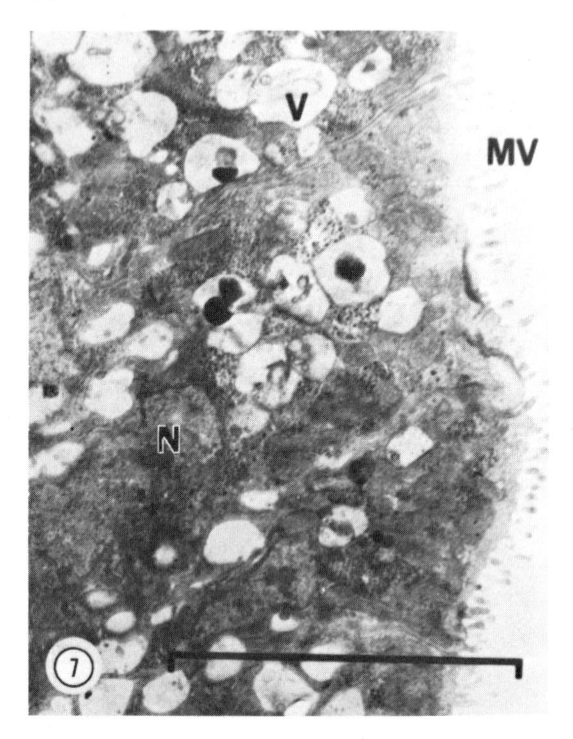

Figure 7. Transmission electron micrograph showing vacuolated epithelial cells of rectal ridge from *B. glabrata* that had been exposed to 0.06 ppm Cu as $CuSO_4$ for 60 hr. MV, microvilli; N, nucleus; V, vacuole. After Sullivan and Cheng (1975), with permission of The New York Academy of Sciences.

et al. (1972) attributed a hemoglobin-producing role to them. If such is indeed the case, their destruction should result in a reduction in the amount of hemoglobin in snails that have been exposed to a lethal dose of copper. However, Cheng (1975) demonstrated that this does not occur. Thus it has been concluded that the lethal mechanism of copper does not involve a pathologic reduction in the hemoglobin level.

Although the precise toxic effect of copper on the surface epithelium of *B. glabrata* which, in turn, brings about the pathologic alterations described above remains uncertain, three possibilities have been proposed by Sullivan and Cheng (1975):

1. Copper elicits the secretion of mucus by goblet cells embedded in the epithelial surface. This mucus may have a suffocating effect by preventing exchange of ions and gases across the epithelium. Indeed, Cheng and Sullivan (1973a) demonstrated that a copper compound, Cu(II)-bis-*N,N*-dihydroxyethylglycine [$Cu(DEG)_2$], at concentrations of 2.5, 12.5, and 20.0 ppm Cu, inhibits the respiration of *B. glabrata* and is also lethal. The LC_{50} of this compound is 3.140 ppm for 2 hr of exposure and 0.0123 ppm for 24 hr of exposure.

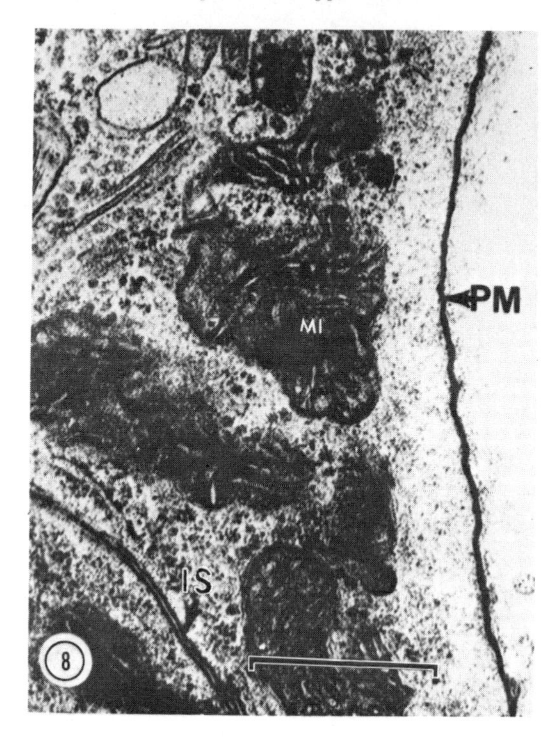

Figure 8. Transmission electron micrograph of apical plasma membrane of epithelial cell of rectal ridge from *B. glabrata* that had been exposed to 0.06 ppm Cu as $CuSO_4$ for 60 hr. IS, intercellular space; MI, mitochondrian; PM, plasma membrane. Bar = 1 μm. After Sullivan and Cheng (1975), with permission of The New York Academy of Sciences.

2. Copper may become bound to hydrophilic regions of the external plasma membrane of the epithelial cells and, as a result, disrupt the permeability of the membrane by altering its biochemical and biophysical properties.
3. Copper entering surface epithelial cells by pinocytosis, as has been demonstrated by Ryder and Bowen (1977a), may disrupt their normal functions. Ryder and Bowen (1977b) reported that the epithelial cells of *Agriolimax reticulatus* endocytose both peroxidase and ferritin, which ultimately become incorporated into the lysosomal system of the cells.

If indeed the primary lethal effect of copper is disruption of the transporting surface epithelium of snails, then, as has been recommended by Ryder and Bowen (1977a), it would be interesting to ascertain whether the lethal property of Cu^{2+} can be enhanced if it is coupled to a pinocytosis inducer. Ryder and Bowen (1977c) have demonstrated that the uptake of lanthanum is enhanced when coupled to such pinocytosis inducers as ribonuclease and peroxidase.

Before proceeding to the next topic, it should be restated that, although snails can take up copper by diffusion between the surface epithelial cells, as Sullivan and Cheng (1976) have shown, copper circulated in hemolymph to various tissues

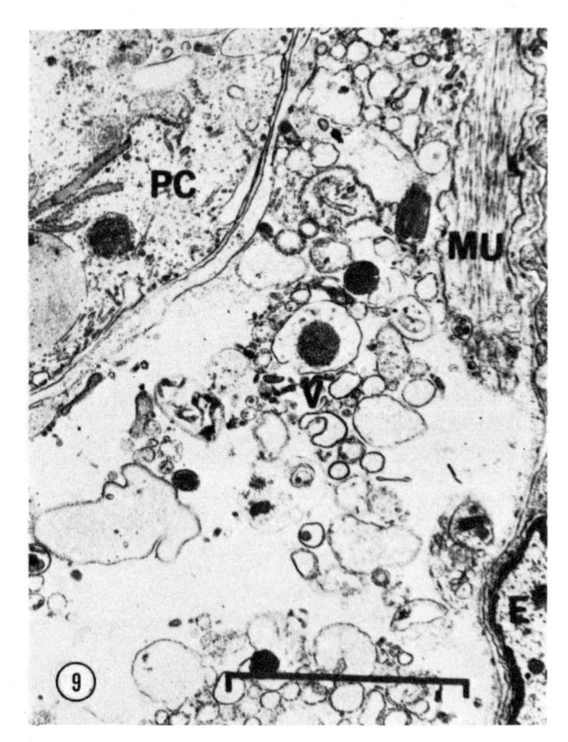

Figure 9. Transmission electron micrograph of loose vascular connective tissue of rectal ridge from *B. glabrata* that had been exposed to 0.06 ppm Cu as $CuSO_4$ for 60 hr. Note occurrence of numerous vesicles. E, epithelial cell; MU, muscle; PC, pigment cell; V, vesicles. Bar = 5 μm. After Sullivan and Cheng (1975), with permission of The New York Academy of Sciences.

is not as lethal as external copper adhering to and taken up by epithelial cells. It is noted that, although the studies carried out by Sullivan and this author have been directed primarily at the rectal ridge of *B. glabrata,* essentially identical results have been obtained with the surface epithelia of the head-foot (Cheng, unpublished).

4.4. Effect of Molluscicide Stereochemistry

To date, a total of 26 copper compounds has been tested at this laboratory for their possible molluscicidal properties (Cheng and Sullivan, 1974). These compounds are listed in Table 1. As a result of finding only some of these to be lethal, a hypothesis was developed in consultation with Dr. Earl H. Hess of the Lancaster Laboratories, Inc., Lancaster, Pennsylvania. Specifically, it was postulated that it is the copper ion that is lethal; therefore any compound in which Cu^{2+} is sterically hindered is less effective as a molluscicide.

The hypothesis stated above was tested by Cheng and Sullivan (1973a). The

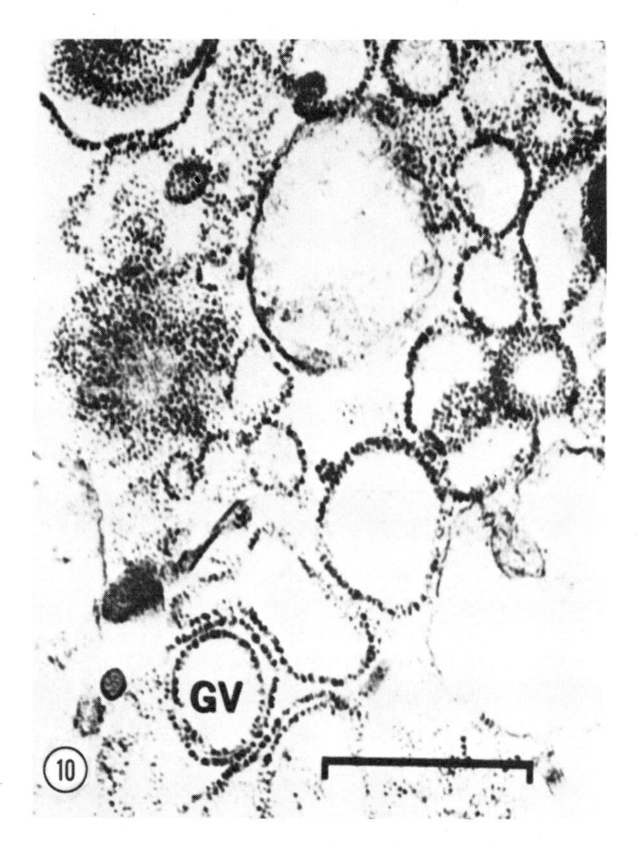

Figure 10. Transmission electron micrograph of granular vesicles in intercellular spaces of loose vascular connective tissue of rectal ridge from *B. glabrata* that had been exposed to 0.06 ppm Cu as $CuSO_4$ for 60 hr. GV, granular vesicles. Bar = 1 μm. After Sullivan and Cheng (1975), with permission of The New York Academy of Sciences.

two copper compounds selected for testing were copper(II)-ethylenediamine-*N,N,N′,N′*-tetraacetic acid (Cu-EDTA) and copper(II)-bis-*N,N*-dihydroxyethylglycine [$Cu(DEG)_2$]. These compounds are water-soluble chelates, and the concentrations at which both compounds were tested were 2.5, 12.5, and 20.0 ppm ionic copper. Two bioassay methods were employed: respirometry and LC_{50} mortality measurements. It was shown that Cu-EDTA does not inhibit the respiration of *B. glabrata,* nor is it toxic to this snail, but that $Cu(DEG)_2$ significantly depresses oxygen utilization as well as being lethal. An analysis of the properties and stereochemistry of these compounds gave some insight into the mechanism underlying this difference. It is known that DEG forms a 1:1 tridentate complex with copper when dissociated in solution and forms a square planar configuration (Chaberek et al., 1953; Martell et al., 1957; Frost et al., 1957). When in the dissociated form (Figure 13), the copper moeity, although stereochemically encapsulated, is bonded in part to H_2O; and since this bonding has a low stability (i.e., the H_2O can be displaced by ligands with greater elec-

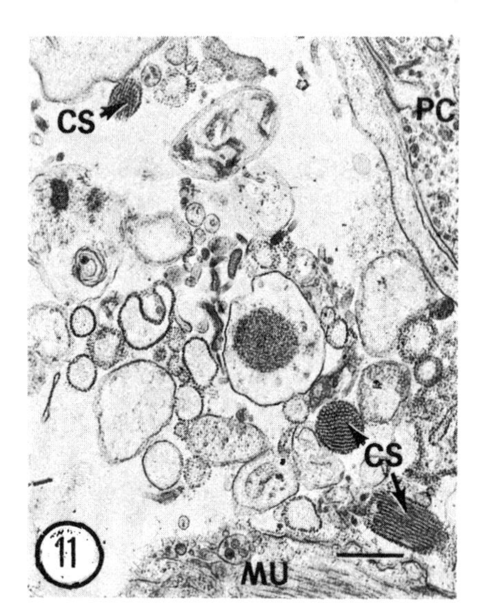

Figure 11. Transmission electron micrograph showing loose vascular connective tissue of rectal ridge from *B. glabrata* that had been exposed to 0.06 ppm Cu as $CuSO_4$ for 60 hr. Notice crystal-like substance in three vesicles. CS, crystal-like substance; MU, muscle; PC, pigment cell. Bar = 1 μm. After Sullivan and Cheng (1975), with permission of The New York Academy of Sciences.

tronegative charges), the copper is in essence biologically exposed. Thus it is not surprising that $Cu(DEG)_2$ is capable of inhibiting respiration and effecting the death of snails, as dose $CuSO_4$.

On the other hand, the copper moiety of Cu-EDTA is also stereochemically encapsulated; however, it differs from that of $Cu(DEG)_2$ in that it is not aquated except at one position (Figure 14). Consequently, it is considerably less exposed biologically. It is known that EDTA forms a 1:1 quinque-dentate complex with copper when dissociated in solution. A proposed three-dimensional model of Cu-EDTA, based on X-ray diffraction studies on Ni-EDTA by Smith and Hoard (1958), is presented (Figure 15) to illustrate the nearly complete encapsulation of the copper moeity. It would thus appear that the chemical basis for the lethal property of $Cu(DEG)_2$ is the fact that its copper moeity is biologically exposed, while the nonlethal property of Cu-EDTA is due to the nearly total encapsulation of its copper moeity. In other words, it seems that the toxicity of copper compounds to gastropods, especially *B. glabrata,* is inversely correlated with the steric hindrance of the chelated copper. It is interest that copper ethylenediamine-*N*-hydroxyethyl-*N,N′,N′*-triacetic acid (Cu-HEEDTA), which differs from the biologically inert Cu-EDTA only in the substitution of a hydroxyl for an acetic group and hence has one less negative charge, displays a lethal effect that is intermediate in magnitude (Cheng and Sullivan, 1974).

Further studies on several copper compounds have substantiated our hy-

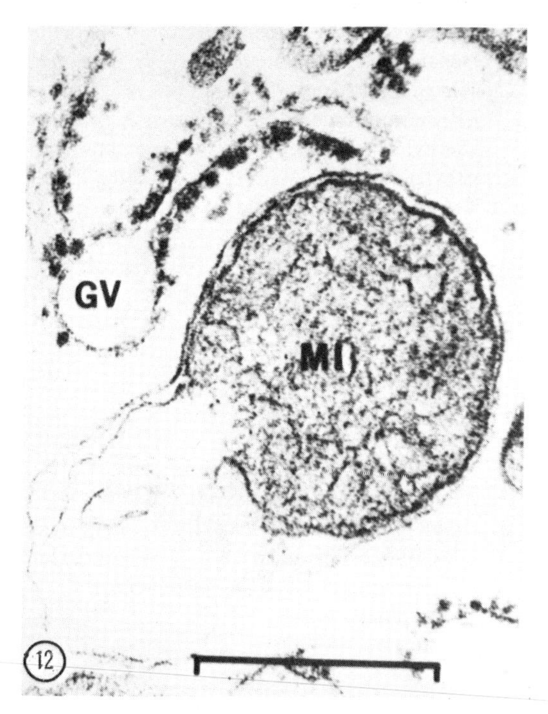

Figure 12. Transmission electron micrograph of extracellular degenerating mitochondrion in loose vascular connective tissue of rectal ridge of *B. glabrata* that had been exposed to 0.06 ppm Cu as $CuSO_4$ for 60 hr. GV, granular vesicle, MI, degenerating mitochondrion. Bar = 0.5 μm. After Sullivan and Cheng (1975), with permission of The New York Academy of Sciences.

pothesis concerning the relationship between the stereochemistry of a molecule and its molluscicidal property (Table 2). Therefore it is now known that the development of new copper compounds as molluscicides should take into consideration two important features: (*a*) the copper moeity should be sufficiently exposed to allow for biological activity, and (*b*) the copper should be sufficiently encapsulated to prevent its being adsorbed onto molecules, colloids, and ions in the aquatic environment.

4.5. Assaying Molluscicides

With the development of new candidate molluscicides, the question will undoubtedly continue to be raised of whether there is a more sensitive as well as more quantitative method for assaying the efficacy of such compounds.

The World Health Organization (WHO) (1971) recommended that potential molluscicides be subjected to three levels of laboratory screening: (*a*) a preliminary screening during which the potential toxicity of the candidate mol-

Table 1. Copper Compounds Tested to Date against *Biomphalaria glabrata*

CuDequest 2041:	Ethylenediamine-*N,N,N′,N′*-tetramethylenephosphonic acid
CuDequest 2051:	Hexamethylenediamine-*N,N,N′,N′*-tetramethylenephosphonic acid
Cu-HEEDTA:	Ethylenediamine-*N*-hydroxyethyl-*N,N′,N′*-triacetic acid
Cu-DTPMPA:	Diethylenetriaminepentamethylenephosphonic acid
Cu-EDTE:	Ethylenediamine-*N,N,N′,N′*-tetraethan-2-ol
Cu-TEPA:	Tetraethylenepentamine
Cu-TETA:	Triethylenetetramine
Cu-DETPA:	Diethylenetriamine-*N,N,N,N′,N″*-pentaacetic acid
Cu-*trans*-(14)-diene:	5,7,7,12,14,14,-Hexamethyl-1,4,8,11-tetraazacyclodeca-4,11-diene copper(II) perchlorate
Cu-(13)-1-ene:	11,13,13-trimethyl-1,4,7,10-tetraazacyclotrideca10-ene
Cu-TAAB:	Tetrabenzotetraazacyclohexadecine copper(II) nitrate
Cu-EBTA:	Ethylene-bis(oxyethylenedinitrilo)-*N,N,N′,N′*-tetracetic acid
Cu-TETHA:	Triethylenetetramine-*N,N,N′N″,N″*-hexaacetic acid
Cutrine:	Copper(II)-bistriethanolamine
Cu-DEG:	*N,N*-Dihydroxyethylglycine
Cu-EDTA:	Ethylenediamine-*N,N,N′,N′*-tetraacetic acid
Cu-CDTA:	1,2-Cyclohexanedinitro-*N,N,N′,N′*-tetraacetic acid
Cu-EDDA:	Ethylenediamine-*N,N′*-diacetic acid
Cu-EDG:	*N*-Ethanoldiglycine
Cu-NTA:	Nitrilotriacetic acid
Cu-histidine	
Cu-mucate	
Cu-saccharate	
Cu-cyanuric acid:	2,4,6-Trihydroxy-S-triazine
$CuNH_4$ alginate	
CuGlucoquest Ac	

Table 2. Lethal Properties and Stereochemistry of Selected Copper Compounds

Compound	Stereochemistry	Lethal Effect
CuDequest 2041	Encapsulated	None
CuDequest 2051	Square planar	Yes
Cu-HEEDTA	Encapsulated (incomplete)	Slight
Cu-DTPMPA	Encapsulated	None
Cu-EDTE	Square planar	Yes
Cu-EBTA	Square planar	Yes
$Cu(DEG)_2$	Square planar	Yes
Cu-EDTA	Encapsulated	None
Cu-CDTA	Encapsulated	None
Cu-EDDA	Square planar	None
Cu-EDG	Encapsulated (incomplete)	Slight

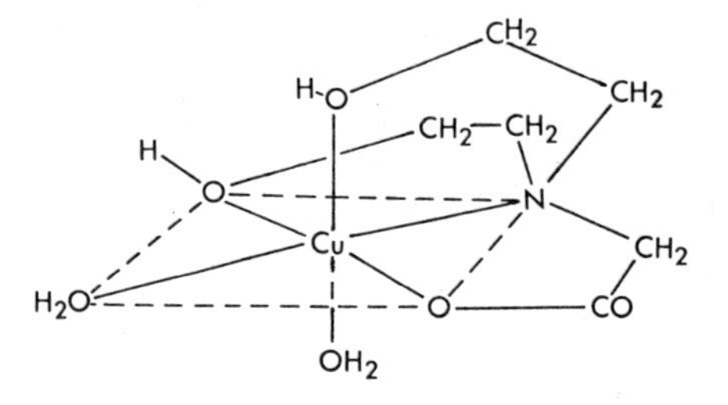

Figure 13. Structural formula of Cu-DEG.

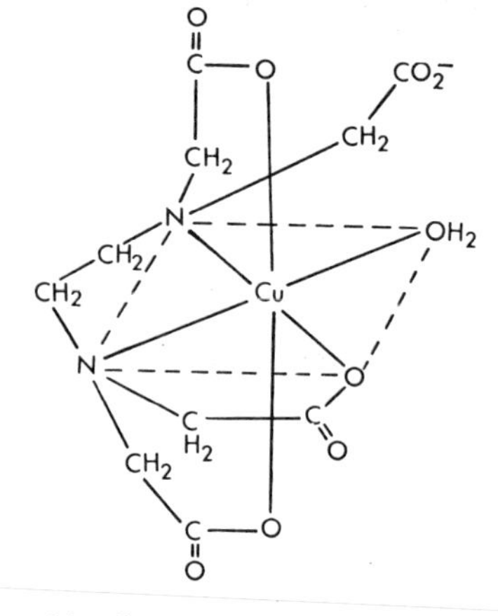

Figure 14. Structural formula of Cu-EDTA.

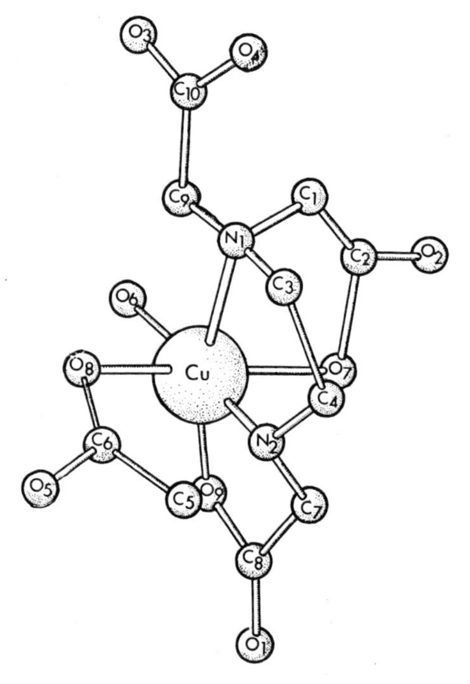

Figure 15. A proposed three-dimensional model of Cu-EDTA.

luscicide is tested; (*b*) a definitive screening during which the LC_{50} and LC_{90} values are determined; and (*c*) comprehensive laboratory evaluating during which the molluscicidal activity of the compound is tested against a number of variables (e.g., sunlight, temperature, and pH) before field testing. These levels represent a lengthy process which could be abbreviated.

Cheng and Sullivan (1973b) showed that concentrations of copper that produce the so-called distress syndrome in *B. glabrata* (Harry and Aldrich, 1963) also effect a measurable decrease in heart rate. A snail manifests the distress syndrome when it is exposed to concentrations of metallic ions below those causing complete retraction into its shell. The gastropod's cephalopedal mass remains partially or fully extended from the shell aperture, and since the snail is not able to attach its foot to the substratum, crawling attempts are ineffectual. Harry and Aldrich also reported that the tentacles of snails portraying the distress syndrome become swollen, cells are sloughed in the region of the tentacular bases, sand grains normally retained in the stomach are defecated, and the heart rate slows down. The last observation is of particular interest since the heart rate response not only lends itself to quantification but also may provide a clue as to the toxic action of inorganic ions on molluscs.

Cheng and Sullivan (1973b) demonstrated that, when the so-called albino strain of *B. glabrata* (Newton, 1955) is exposed to lower concentrations of copper

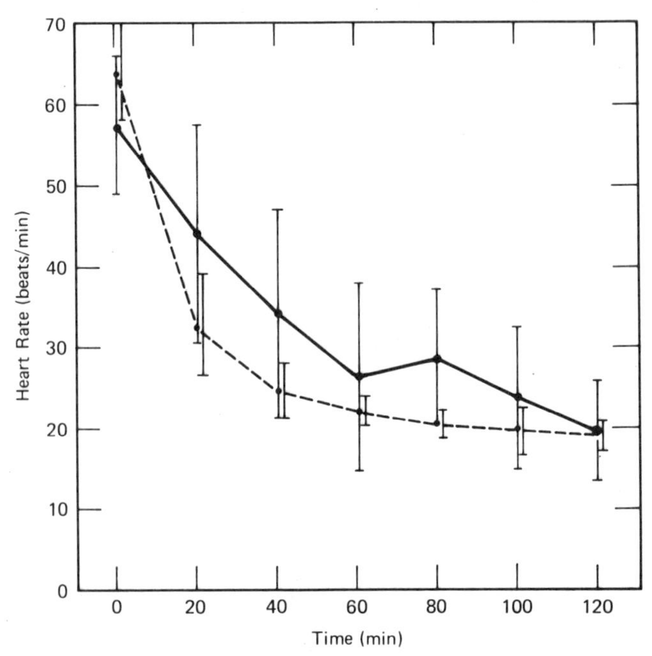

Figure 16. Graph showing mean heart rates of *B. glabrata* exposed to 1.0 ppm (- - -) and 0.1 ppm (—) Cu as $CuSO_4$ as a function of time. Verticle lines represent standard deviations. After Cheng and Sullivan, (1973b), with permission of Scientechnica.

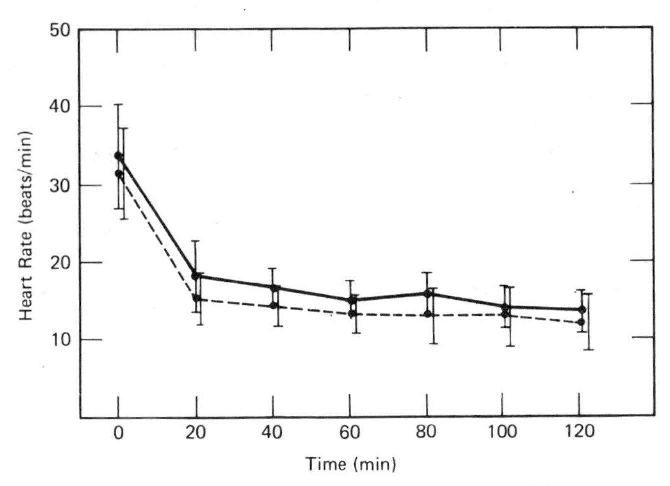

Figure 17. Graph showing mean heart rates of *B. glabrata* exposed to 5.0 ppm (---) and 2.5 ppm (—) Cu as $CuSO_4$ as a function of time. Verticle lines represent standard deviations. After Cheng and Sullivan (1973b), with permission of Scientechnica.

(i.e., 1 ppm and below), the reduction in heart rate is proportional to the concentration of copper as $CuSO_4$ to which the snail is exposed (Figure 16). At higher concentrations (i.e., 2.5, 5.0, 12.5, and 20.0 ppm Cu) the effect on heart rate is maximal, regardless of the concentration (Figures 17 and 18). Thus the albino strain of *B. glabrata,* through the shell of which the heart is readily visible, is a useful organism for quantitatively assaying low concentrations of candidate molluscicides. In fact, this method has been submitted to WHO and has been widely distributed by that organization (Cheng and Sullivan, 1973c).

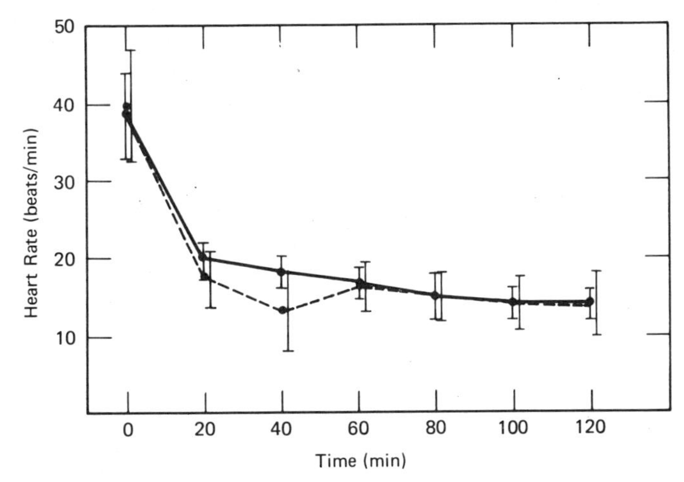

Figure 18. Graph showing mean heart rates of *B. glabrata* exposed to 20.0 ppm (---) and 12.5 ppm (—) Cu as $CuSO_4$ as a function of time. Verticle lines represent standard deviations. After Cheng and Sullivan, (1973b), with permission of Scientechnica.

In addition to quantifying heart rates, Cheng and Sullivan (1973b) demonstrated that quantitative determinations of oxygen consumption are also useful for assaying candidate molluscicides. Von Brand et al. (1949) were the first to show that molluscicides, including $CuSO_4$, significantly reduce the oxygen consumption of *B. glabrata*. The simple respirometric technique employed in this method of bioassaying candidate molluscicides can be found in Cheng and Sullivan (1973a) and Malek and Cheng (1974). This technique too has been widely disseminated by WHO (Cheng and Sullivan, 1973c).

Both the heart rate and respirometric methods are preferred over the conventional LC_{50} and LC_{90} method for the following reasons: (*a*) the evaluation of a compound can be completed within hours; (*b*) these tests are able to distinguish between different molecular configurations of the same toxic ion; (*c*) the apparatuses required are simple, involving a dissection microscope, a water bath, and a respirometer; and (*d*) the methods are extremely sensitive and accurate. Cheng and Sullivan (1973a, 1973b) demonstrated that all of the copper-containing compounds tested that depressed heart rate and respiration and effected a distress response in the same concentrations as does $CuSO_4$ had molluscicidal properties approaching those of $CuSO_4$, and, conversely, compounds without these effects proved to be nontoxic in economically feasible concentrations.

5. CONCLUSIONS

It is apparent from the above discussion that additional research on the use of copper as a molluscicide is warranted. Specifically, it now appears that the formulation of a copper compound so as to enhance its being pinocytosed by the transport epithelium on the surfaces on gastropods is the next logical step. Furthermore, the same compound should not be inactivated by ambient fluctuating factors such as pH, temperature, and quantity of organic matter. Also, more efficient technology should be developed to deliver the molluscicide to difficult environments. These and other aspects of future research with copper-containing molluscicides have been presented in detail by Cheng (1974). It is noted, however, that because of the large quantity of the substance that will be needed to control undesirable snails, especially those that transmit schistosomiasis, in entire subcontinents, the ideal molluscicide must be commercially competitive.

ACKNOWLEDGMENT

The original data included in this contribution are the results of research sponsored by Grant INCRA-193 from the International Copper Research Association, Inc.

REFERENCES

Amin, M. A. (1972). "The Control of Schistosomiasis in the Gezira, Suda. The Use of Copper Sulfate and Mechanical Barriers," *Sudan. Med. J.*, **10,** 75–82.

Amin, M. A. (1974). "Methods of Application of Molluscicides." In T. C. Cheng, Ed., *Molluscicides in Schistosomiasis Control.* Academic Press, New York, pp. 67–76.

Amin, M. A. and Fenwick, A. (1973). "Snail Control in the Gezira Irrigated Area of the Sudan by Aerial Application of *N*-Tritylmorphine." In *Proceedings of the 9th International Congress on Tropical Medicine and Malaria,* Athens, Greece.

Ayad, N. (1961). "The Use of Molluscicides in Egypt," *Bull. WHO,* **25,** 712–721.

Barbosa, F. S., J. Carneiro Filho, J. G. De Moraes, and Carneiro, E. (1956). "Altividade Moluscocida dos Sais Insolúveis do Cobre," *Publ. Avuls. Inst. Aggeu. Magalhães,* **5,** 7–20.

Barnish, G. and Shiff, C. J. (1970). "Aerial Application of the Molluscicide Frescon at Lake McIlwaine," *Rhodesia Agric. J.*, **67,** 2.

Barnish, G. and Sturrock, R. F. (1973). "Aerial Application of a Molluscicide to a Marsh," *Trans. R. Soc. Trop. Med. Hyg.*, **67,** 610–611.

Berridge, M. J. and Oschman, J. L. (1972). *Transporting Epithelia.* Academic Press, New York.

Camey, T. and Paulini, E. (1962). "Observations on the Effect of Copper Sulphate on the Egg Laying of *Taphius glabratus*," *Rev. Brasil. Biol.*, **22,** 47–53.

Cardarelli, N. F. (1974). "Slow Release Molluscicides and Related Materials," In T. C. Cheng, Ed., *Molluscicides in Schistosomiasis Control.* Academic Press, New York, pp. 177–240.

Castro, G. M. O. (1954). "Verde de Paris como Planorbicida," *Rev. Brasil. Med.*, **11,** 166–168.

Chaberek, S. Jr., R. C. Courtney, and Martell, A. E. (1953). "Stability of Metal Chelates. V: *N,N*-Dihydroxyethylglycine," *J. Am. Chem. Soc.*, **75,** 2185–2190.

Chaia, G. and Paulini, E. (1957). "Effects of Sodium Pentachlorophenate and of Copper Sulphate on the Eggs and Miracidia of *Schistosoma mansoni*," *Rev. Brasil. Malar.*, **9,** 511–514.

Chaia, G., Paulini, E., and Quieroz, A. B. (1956). "The Effects of Sodium Pentachlorophenate and Copper Sulphate on the Eggs of *Australorbis glabratus* (Say)," *Rev. Brasil. Malar.*, **8,** 605–612.

Chandler, A. C. (1920). "Control of Fluke Diseases by Destruction of the Intermediate Host," *J. Agric. Res.*, **20,** 193–208.

Cheng, T. C. (1975). "Does Copper Cause Anemia in *Biomphalaria glabrata?" J. Invert. Pathol.*, **26,** 421–422.

Cheng, T. C. and Sullivan, J. T. (1973a). "Comparative Study of the Effects of Two Copper Compounds on the Respiration and Survival of *Biomphalaria glabrata* (Mollusca: Pulmonata)," *Comp. Gen. Pharmacol.*, **4,** 315–320.

Cheng, T. C. and Sullivan, J. T. (1973b). "The Effect of Copper on the Heart-Rate of

Biomphalaria glabrata (Mollusca: Pulmonata)," *Comp. Gen. Pharmacol.*, **4**, 37–41.

Cheng, T. C. and Sullivan, J. T. (1973c). "A New Method for the Preliminary Testing of Chemical Molluscicides," *WHO/SCHISTO/73.27*, pp. 1–10, Washington, D.C.

Cheng, T. C. and Sullivan, J. T. (1974). "Mode of Entry, Action, and Toxicity of Copper Molluscicides." In T. C. Cheng, Ed., *Molluscicides in Schistosomiasis Control*. Academic Press, New York, pp. 89–153.

Chu, K. Y., Massoud, J., and Arfaa, F. (1968). "Comparative Studies of the Molluscicidal Effect of Cuprous Chloride and Copper Sulphate in Iran," *Bull. WHO*, **39**, 320–326.

Degremont, A. A., Geigy, R., and Perret, P. (1972). "Preliminary Results of the Project for Controlling and Preventing Schistosomiasis in the Lower Mangoky (Malagasy Republic)," *Acta Trop.*, **29**, 138–174.

Deschiens, R., Le Coroller, Y., and Pastac, S. (1961). "Utilization du Prototoxyde de Cuivre comme Molluscicide Selectif en Prophylaxie des Bilharzioses," *C. R. Acad. Sci.*, **252**, 4221–4222.

Dobrovolny, C. and Dobbin, I. E., Jr. (1955). "Initial Field Trials with Some Molluscicides in Brazil," *Rev. Brasil. Malar. Doen. Trop. Publ. Avul.* 1, Julho, pp. 1–19.

El-Ghindy, M. S. (1957a). "Laboratory Studies on the Effect of Copper Sulphate on *Biomphalaria boissyi*, the Snail Vector of *Schistosomiasis mansoni* in Egypt," *J. Egypt. Med. Assoc.*, **40**, 45–53.

El-Ghindy, M. S. (1957b). "Field Studies on the Action of Copper Sulphate as a Molluscicide for the Control of the Snail Vectors of Schistosomiasis in Egypt," *J. Egypt. Med. Assoc.*, **40**, 111–121.

El Nagar, H. (1958). "Control of Schistosomiasis in the Gezira, Sudan," *J. Trop. Med. Hyg.*, **61**, 231–235.

Fenwick, A. (1970). "The Development of Snail Control Methods on an Irrigated Sugar Estate in Northern Tanzania," *Bull. WHO*, **42**, 589.

Frost, A. E., Chaberek, S., and Bicknell, N. J. (1957). "The Preparation and Chelating Properties of *N,N*-Bis(2-hydroxypropyl)glycine," *J. Am. Chem. Soc.*, **79**, 2755–2758.

Fujinami, K. and Narabayashi, H. (1913a). "Prevention of Schistosomiasis, Especially by Mixing Lime in Infective Water," *Chugai Iji Shimp.*, **34**, 649–657.

Fujinami, K. and Narabayashi, H. (1913b). "The Preventive Measure for Japanese Schistosomiasis, Particularly the Technique for Mixing Lime," *Hiroshima Eishei Iji Geppo*, **174**, 179–190.

Gonçalves, B. N. and Soares, R. R. L. (1955). "Altividade Moluscocida do Recinoleato de Cobre," *Mem. Inst. Oswaldo Cruz*, **53**, 397–409.

Harry, H. W. and Aldrich, D. V. (1963). "The Distress Syndrome in *Taphius glabratus* (Say) as a Reaction to Toxic Concentrations of Inorganic Ions," *Malacologia*, **1**, 283–289.

Jansen, G. (1943). "Observations on the Combat of Human Schistosomiasis in the Municipality of Catende, Pernumbuco, Brazil; Index of Infection in *Australorbis* and the Use of Slaked Lime and Copper Sulphate in the Combat of Molluscs," *Mem. Inst. Oswaldo Cruz,* **39,** 335–347.

King, T. E. (1963). "Reconstitution of Respiratory Chain Enzyme Systems. XI: Use of Artificial Electron Acceptors in the Assay of Succinate-Dehydrogenase Enzyme," *J. Biol. Chem.,* **238,** 4032.

Komiya, Y. (1961). "Study and Application of Molluscicides in Japan," *Bull. WHO,* **25,** 573–579.

Malek, E. A. and Cheng, T. C. (1974). *Medical and Economical Malacology.* Academic Press, New York.

Martell, A. E., Chaberek, S., Jr., Courtney, R. C., Westerback, S., and Hyytianinen, H. (1957). "Hydrolytic Tendencies of Metal Chelate Compounds. I: Cu(II) Chelates," *J. Am. Chem. Soc.,* **79,** 3036–3041.

Miyagama, Y. (1913). [Control of Snails]. *Iji Shimbun,* **890,** 1; **891,** 5 (in Japanese).

Narabayashi, H. (1915). [Prophylaxis of *Schistosomiasis japonica,* Especially on the Destroying of the Intermediate Host and Lining of the Infected Water]. *Chugai Iji Shimpo,* **855,** 1381–1416 (in Japanese).

Newton, W. L. (1955). "The Establishment of a Strain of *Australorbis glabratus* Which Combines Albinism and a High Susceptibility to Infection with *Schistosoma mansoni,*" *J. Parasitol.,* **41,** 526–528.

Pan, C. T. (1958). "The General Histology and Topographic Microanatomy of *Australorbis glabratus,*" *Bull. Mus. Comp. Zool. Harvard,* **119,** 237–299.

Paulini, E. (1974). "Copper Molluscicides: Research and Goals." In T. C. Cheng, Ed., *Molluscicides in Schistosomiasis Control.* Academic Press, New York, pp. 155–170.

Paulini, E. and Camey, T. (1965). "Ensaios de Laboratorio e de Campo com o "Planticuivre" como Moluscicida," *Rev. Brasil. Malar. Doen. Trop.,* **17,** 49–53.

Paulini, E., Camey, T., and Pereira, J. P. (1963). "Ensaios de Laboratorio com um Novo Moluscicida (Sal de Chevreul)," *Rev. Brasil. Malar. Doen. Trop.,* **15,** 41–46.

Perlawagora-Szumlewicz, A. (1955). "Laboratory Experiments on the Residual Planorbicidal Effect of Sodium Pentachlorophenate and Copper Sulphate," *Rev. Brasil. Malar. Doen. Trop. Publ. Avul.* 2, Julho, pp. 1–19.

Pinto, D. B. and Penido, H. M. (1948). "Note on the Planorbicidal Effect of a Copper Sulphate-Tartaric Acid Complex," *Rev. Serv. Saude. Publ.,* **2,** 509–514.

Pitchford, R. J., Meyling, A. H., Brummer, J. J., Duroit, J. F., and Vorster, S. V. (1960). "An Assessment of Copper Sulphate as a Molluscicide in the Eastern Transvaal Lowveld," *Cent. Afr. J. Med.,* **6,** 97–108.

Puma, R. and de Zagustin, T. (1969). "Mollusquicidas en Uso en la Campaña de Lucha Contra la Bilharziasis." Second meeting of the Caribbean Commission on Bilharziasis Research, Maracay, Venezuela.

Ritchie, L. S. (1973). "Chemical Control of Snails." In N. Ansari, Ed., *Epidemiology and Control of Schistosomiasis.* S. Karger, Basel, Switzerland, pp. 458–532.

Ryder, T. A. and Bowen, I. D. (1977a). "The Slug Foot as a Site of Uptake of Copper Molluscicide," *J. Invert. Pathol.,* **30,** 381–386.

Ryder, T. A. and Bowen, I. D. (1977b). "Endocytosis and Aspects of Autophagy in the

Foot Epithelium of the Slug *Agriolimax reticulatus* (Müller)," *Cell Tiss. Res.*, **181,** 129–142.

Ryder, T. A. and Bowen, I. D. (1977c). "Studies on Transmembrane and Paracellular Phenomena in the Foot of the Slug *Agriolimax reticulatus* (Mü)," *Cell Tiss. Res.*, **183,** 143–152.

Sharaf, E. D. and El Nagar, H. (1955). "Control of Snails by Copper Sulphate in the Canals of the Gezira Irrigated Area of the Sudan," *J. Trop. Med. Hyg.*, **58,** 260–263.

Shiff, C. J. (1961). "Trials with a New Molluscicide, Bayer 73, in S. Rhodesia," *Bull. WHO,* **25,** 533–542.

Sminia, T., Boer, H. H., and Niemantsverdriet, A. (1972). "Haemoglobin Producing Cells in Freshwater Snails," *Z. Zellforsch.*, **135,** 563–658.

Smith, G. S. and Hoard, J. L. (1958). "The Structure of Dihydrogen Ethylenediaminetetraacetatoaquonickel," *J. Am. Chem. Soc.*, **81,** 556–561.

Strufe, R., Dazo, B. C., and Dawod, I. K. (1965). "Field and Laboratory Trials with Bayluscide in Bilharziasis Control Pilot Project Egypt 49," *Bayer,* **18,** 110–122.

Sullivan, J. T. (1975). "Mechanisms of Copper Toxicity to *Biomphalaria glabrata* (Mollusca: Pulmonata)." Ph.D. dissertation, Lehigh University, Bethlehem, Pa.

Sullivan, J. T. and Cheng, T. C. (1975). "Heavy Metal Toxicity to *Biomphalaria glabrata* (Mollusca: Pulmonata)," *Ann. N.Y. Acad. Sci.*, **266,** 437–444.

Sullivan, J. T. and Cheng, T. C. (1976). "Comparative Mortality Studies on *Biomphalaria glabrata* (Mollusca: Pulmonata) Exposed to Copper Internally and Externally," *J. Invert. Pathol.*, **28,** 255–257.

Sullivan, J. T., Rodrick, G. E., and Cheng, T. C. (1974). "A Transmission and Scanning Electron Microscopical Study of the Rectal Ridge of *Biomphalaria glabrata* (Mollusca: Pulmonata)," *Cell Tiss. Res.*, **154,** 29–38.

Teesdale, C., Hadman, D. F., and Nguriathi, J. N. (1961). "The Use of Continuous Low Dosage of Copper Sulphate as a Molluscicide on an Irrigation Scheme in Kenya," *Bull. WHO,* **25,** 563–571.

Von Brand, T., Mehlman, B., and Nolan, M. O. (1949). "Influence of Some Potential Molluscicides on the Oxygen Consumption of *Australorbis glabratus*," *J. Parasitol.*, **35,** 475–481.

Warren, K. S. and Newill, V. A., Eds. (1967). *Schistosomiasis.* Year Book Medical Publ., Chicago.

Webbe, G. (1963). "The Application of a Molluscicide to the Mironga River in Mwanza, Tanganyika," *Bayer,* **16,** 244–252.

World Health Organization. (1965). *Snail Control in the Prevention of Bilharziasis.* WHO Monogr. Series 50.

World Health Organization (1971). "Meeting of Directors of Collaborating Laboratories on Molluscicide Testing and Evaluation, 1970," *WHO/SCHISTO/71.6,* Washington, D.C.

Yager, C. M. and Harry, H. W. (1964). "The Uptake of Radioactive Zinc, Cadmium, and Copper by the Freshwater Snail, *Taphius glabratus*," *Malacologia,* **1,** 339–353.

14

INTERACTIONS OF COPPER WITH OTHER TRACE ELEMENTS

M. Kirchgessner

F. J. Schwarz

E. Grassmann

H. Steinhart

Institut für Ernährungsphysiologie der Technischen Universität München, Freising-Weihenstephan, Germany

1. INTRODUCTION

Copper may interact inside the organism with the essential trace elements Fe, Zn, Mo, Mn, Ni, and Se and also with the nonessential elements Ag, Cd, Hg, and Pb. Figure 1 shows in a simplified scheme the most important of these interactions. Influences that are mediated by the involvement of another element (multielement interactions) are not specifically depicted. The arrows indicate whether copper is affected by another element or exerts an effect on another element. These interactions may be either positive (+) or negative (−). Positive interactions require the presence of one or several other trace elements for the normal metabolic efficacy of the trace element in question, whereas negative interactions imply that the metabolic role of this element is inhibited by the relative excess of another or several other trace elements (Davies, 1974). The terms "synergistic" and "antagonistic" are also used to describe interrelationships between trace elements. It must, however, be specified whether synergistic or antagonistic interactions refer to metabolic function or to storage. Interactions are considered direct only if competitive reactions occur at the binding sites of the ligands. All other interactions are indirect. Davies (1974) introduced the terms "competitive" and "noncompetitive" for these situations. Interactions

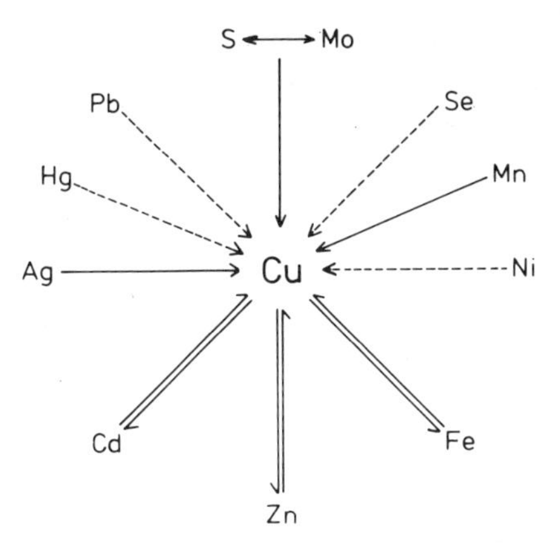

Figure 1. Interactions of copper with other trace elements. The arrows indicate the unidirectional or bidirectional (mutual) interactions. Solid lines indicate primary interactions; dotted lines, interactions which result from effects on the general metabolism or from multielement interrelationships.

may take effect both during digestion and absorption and during intermediary metabolism.

Direct interactions of copper with other trace elements occur when copper is displaced from its complexes by other trace elements or when copper displaces other trace elements from their complexes. These interactions may be explained by the thermodynamic and kinetic stabilities of the particular complexes. The stability of a complex depends mainly upon the electron configuration of the trace elements and on the kind of ligands, but the radius and charge of the metal ions and the reaction milieu are also important. Table 1 shows the thermodynamic stabilities of various complexes. Since Cu^{2+} ions obviously form very stable complexes with proteins as well as with other ligands, their displacement from complexes is likely to occur only at a high excess of other trace elements. Conversely, however, Cu^{2+} ions may displace other trace elements from their complexes. Because of similarities in the electron configuration and ion radius, the interaction between zinc and copper may be expected to be particularly intensive. This is also evident from the similar stability constants of Cu^{2+} and Zn^{2+} in complexes with proteins as ligands. Thus it is possible to substitute zinc for Cu^{2+} ions in several metalloenzymes, which may, however, often lose their physiological activity with this exchange. Also, interactions occur between copper and zinc during absorption, whereby competing reactions at the intestinal metal-binding sites may be involved. High zinc doses inhibit the absorption of copper; conversely, high copper doses affect the absorption of zinc (Van Campen and Scaife, 1967).

The interrelationships between copper and calcium may serve as an example of an indirect interaction mediated by the influence of the physiological environment. Elevated calcium contents in the diets of ruminants lowered copper absorption (Kirchgessner, 1959). In the case of pigs (Kirchgessner et al., 1960)

Table 1. Stability of Metallocarboxypeptidase and Chelates of Known Structure[a]

Metal	Log k Metallocarboxy-peptidase	Log k_1	
		Glycine	Ethylendiamine
Mn^{2+}	5.6	3.4	2.7
Ni^{2+}	8.2	6.2	7.7
Zn^{2+}	10.5	5.2	5.7
Cu^{2+}	10.6	8.6	10.7

[a] Vallee and Coleman (1964).

and chicks (Hampel and Kirchgessner, 1970; Grassmann et al., 1971), however, no such dependence of copper absorption on increased dietary calcium contents could be established. These findings may be interpreted by the hypothesis that in ruminants even minor calcium supplementation increases the pH value of the ruminal contents so much that the copper is precipitated as hydrated oxide and hence rendered less absorbable, whereas in the monogastric animal the milieu remains sufficiently acid to prevent copper from being precipitated.

Interactions of copper with other trace elements also occur in connection with their function of stabilizing the structure of proteins and, specifically, of enzymes. Thus pepsin was either activated or inhibited *in vitro,* depending upon the amount of copper added (Beyer et al., 1976; Kirchgessner et al., 1976). Also, three modes of interactions of copper with other trace elements were found with regard to their effect on *in vitro* pepsin activity (Steinhart et al., 1976): an additive, a competing, and an indifferent effect. An additive effect is involved in the interaction between Cu^{2+} and Fe^{2+} ions; this means that the separate effects of the Cu^{2+} and Fe^{2+} are additive when they are applied in combination. In the case of the simultaneous addition of Cu^{2+} and Ni^{2+}, the effect was competitive. The turnover-rate curves for the Cu^{2+}-Ni^{2+} combination took an intermediary course between those obtained with the addition of either Cu^{2+} alone or Ni^{2+} alone. In contrast, the simultaneous addition of Zn^{2+} ions to a medium with Cu^{2+} ions had no influence on the peptic activity obtained with the sole addition of Cu^{2+} ions. The autocatalytic hydrolysis of pepsin in solution was evidently slowed down by metal complex formation, involving either the same (competing or indifferent effect) or different (additive effect) binding sites.

Although the aforementioned interactions of copper with other trace elements may, for the most part, be explained by applying chemical laws, this is often not possible with physiological interactions in metabolism. In these instances the various findings from different studies must, puzzlelike, be pieced together. One of the most thoroughly studied examples represents the *in vivo* interaction of copper and iron.

2. COPPER-IRON INTERACTIONS

Interactions between copper and iron have been known since the finding of Hart et al. (1928) that milk-anemic rats could be cured only if, along with iron, copper was also supplied. Deficient copper supply affects iron utilization, so that hypochromic microcytic anemia occurs, which cannot morphologically be distinguished from uncomplicated iron-deficiency anemia (Lahey et al., 1952). In this connection, other species-related forms of anemia were also noted and discussed (Matrone, 1960). However, the hematological parameters were also observed to be altered and microcytic and hemolytic anemias to occur when high copper doses were applied (Underwood, 1977).

Because of the particular importance of the relationship between copper and iron in human and animal nutrition, many studies concentrated on the interaction between these two trace elements, but primarily on the influence of copper on iron metabolism. Only in recent times have studies been concerned also with the effect of iron on copper metabolism.

2.1. Influence of Copper on Iron Metabolism

Iron Retention

The influence of deficient copper nutrition on iron storage has been known since the investigations by Elvehjem and Sherman (1932). Copper deficiency leads to an elevated accumulation of iron in the liver and, in part, also in other organs, especially the spleen. This response has been demonstrated in rats (Bunn and Matrone, 1966; Sourkes et al., 1968; Symes et al., 1969; Marston et al., 1971; Thompson and Evans, 1977a), sheep (Marston, 1952), and cattle (Chapman and Kidder, 1964). The time course of the iron accumulation in the liver of rats during progressive copper depletion is illustrated in Figure 2 (Grassmann and Kirchgessner, 1973a). The iron content of the liver steeply increased after about 15 days and approached an upper limit after 46 days. At this time the mean hepatic iron concentration of the copper-deficient animals was about twice as high as that of copper-supplemented controls. Considerably higher concentrations were reported for sheep by Marston (1952). In contrast to these reports, no increase in the iron concentration could be found in the liver of copper-deficient pigs, compared with adequately supplied animals (Gubler et al., 1952; Lee et al., 1968a). The iron concentrations of the spleen and intestinal wall, however, were elevated in copper-deficient pigs (Lee et al., 1968a).

The increased storage of iron in the liver is not restricted to nutritional copper-deficiency. It has been reported that the hepatic iron concentration markedly increases also in zinc-induced copper deficiency (Bunn and Matrone, 1966). This response, however, depends upon the dosage of zinc, insofar as at very high

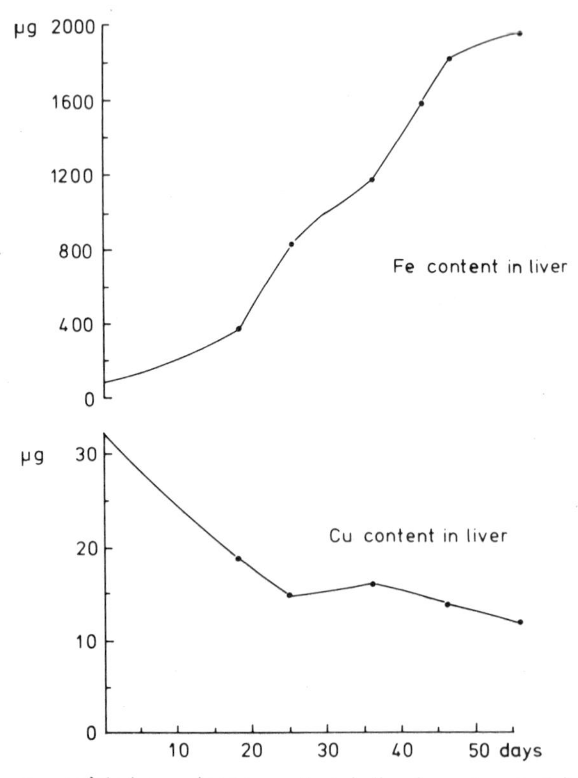

Figure 2. Time course of the iron and copper contents in liver in response to deficient copper nutrition. Grassmann and Kirchgessner (1973a).

zinc intakes the iron storage in the liver may again be greatly reduced (Magee and Matrone, 1960; Lee and Matrone, 1969).

The iron accumulation resulting either from dietary copper depletion or from zinc-induced copper deficiency normalizes again with dietary copper supplementation (Ritchie et al., 1963; Marston et al., 1971). This decline of the elevated iron stores may, however, require a longer period of time. In studies with rats the increased iron depots were found to be removed only partially within 14 days after supplementing the diet with 3.2 mg Cu/kg as copper sulfate (Grassmann and Kirchgessner, 1973a). The rate of this iron mobilization may be highly dose dependent. Thus Evans and Abraham (1973) noted a 30% reduction of the liver iron content as early as 2 days after dietary repletion with 35 ppm Cu and a decline to the control level or even below that within 5 days. Also, as noted by Marston et al. (1971), there seem to be differences in the dynamics of iron exchange between different organs.

In addition to the effect of copper deficiency, the significance of very high copper doses on iron accumulation was also a subject for study. Various authors reported the iron concentration of the liver to be reduced with increasing dietary copper contents or even after copper injection (Cassidy and Eva, 1958; Ritchie et al., 1963; Ruszczyc and Gacek, 1963; Marston et al., 1971). Kinnamon (1966) found no iron response to 200 mg Cu/kg of diet. On the other hand, Kirchgessner and Weser (1963) noted in balance trials that growing pigs supplemented with 125 and 250 mg Cu/kg of diets retained significantly more iron than control animals.

Iron Metalloproteins

Despite the increased iron storage in various organs during copper deficiency, the hematological parameters such as hemoglobin concentration, hematocrit, and erythrocyte count are greatly reduced. This distinct effect was observed both in rats (Sourkes et al., 1968; Evans and Abraham, 1973; Grassmann and Kirchgessner, 1973a) and (especially) in swine (Lahey et al., 1952; Gubler et al., 1957). Differences in the extent of reduction of the hemoglobin concentration must be viewed in relation not only to the copper nutrition but also to the simultaneous iron supply, as is evident from studies with rats (Sourkes et al., 1968; Grassmann and Kirchgessner, 1973a, 1973b).

There is little information regarding the influence of copper nutrition on iron proteins other than hemoglobin. Whereas the myoglobin concentration of muscle greatly decreases in response to mere iron deficiency, it may remain unaffected during copper deficiency, according to studies by Gubler et al. (1957).

As to the cytochromes, Cohen and Elvehjem (1934) had observed a slight decrease in these enzymes in the heart and liver tissue of rats supplied inadequately with iron and copper. The cytochrome contents could be increased only by the simultaneous administration of these two trace elements. According to all findings, cytochromes respond much more sensitively to deficient iron supply than to copper deficiency (Gubler et al., 1957; Matrone, 1960). The cytochrome oxidase, however, which also contains copper, is an exception among the cytochromes. In all tissues studied, it was affected primarily by inadequate copper supply (Schultze, 1939, 1941; Gallagher, 1957; Matrone, 1960). This holds true also for zinc-induced copper deficiency (Duncan et al., 1953; Van Reen, 1953; Grant-Frost and Underwood, 1958; Magee and Matrone, 1960).

Changes in the catalase activity during copper deficiency result from the diminished metabolic efficiency of iron utilization associated with the increased iron fixation. After 7 weeks of copper depletion, the catalase activity of blood was lower in depleted rats than in copper-supplemented controls (Grassmann and Kirchgessner, 1973a). The fact that its response was delayed in comparison with the hemoglobin concentration explains why, in other trials of shorter duration (Grassmann and Kirchgessner, 1973b), the catalase was not affected. A similar effect was observed during zinc-induced copper deficiency (Van Reen, 1953). Although Schultze and Kuiken (1941) also found greatly reduced catalase activities in blood, liver, and kidneys of rats, no definite relationship can be deduced from their study because the dietary supply of both iron and copper was lowered. For pigs, Lahey et al. (1952) observed reduced catalase activity in erythrocytes. Later the same research group (Gubler et al., 1957) also reported a significant reduction of the catalase activity in the liver.

According to studies by Theorell et al. (1951) and Miller (1958), biosynthesis of the heme portion of both the catalase and the hemoglobin follows the same route. Nevertheless, the depletion- and repletion-response curves of these two iron metalloproteins do not take strictly parallel courses, as was noted as early as 1941 by Schultze and Kuiken. Matrone (1960), in reviewing the literature, assumed that the catalase activity must be influenced by additional factors and, consequently, that its use for studying copper-iron interactions is of limited value. In support of this assumption are findings of the effect of dietary protein content and quality on the activity of this enzyme (Kirchgessner et al., 1977; Grassmann et al., 1978).

On the Mode of Interaction

The processes occurring during absorption (Gubler et al., 1952; Lahey et al., 1952; Bush et al., 1956), intermediary metabolism (Lee et al., 1968a), and erythropoesis (Lahey et al., 1952; Matrone, 1960; Lee et al., 1968b, Dunlap et al., 1974) have been considered as possible sites for the influence of copper on iron metabolism. With the finding that ceruloplasmin acts as ferroxidase (Osaki et al., 1966; Osaki and Johnson, 1969), the effect of copper on iron utilization could unequivocally be attributed to problems of mobilization (Lee et al., 1969; Ragan et al., 1969; Frieden, 1970; Roeser et al., 1970). Ceruloplasmin oxidizes divalent iron, which is first liberated in this form from the depot tissue, to the trivalent state and thereby makes possible its transfer to transferrin (Reinhold, 1975). For a detailed account of iron mobilization see the reviews by Frieden in this volume and elsewhere (Frieden, 1973). Since ceruloplasmin rapidly responds to changes in copper supply (Starcher et al., 1964; Gomez-Garcia and Matrone, 1967; Milne and Weswig, 1968; Kirchgessner and Grassmann, 1970; see also Figure 3 of Grassmann and Kirchgessner, 1973a), it is understandable that the deposition and mobilization of iron in the liver and other depot organs are affected with similar rapidity. This effect on iron mobilization also explains the differences in the concentrations of plasma iron and iron metalloproteins and, hence, the copper-iron interactions during deficient copper nutrition.

The copper-iron interaction in intermediary metabolism may also influence iron absorption. During copper deficiency the iron content of the intestinal mucosa is elevated (Lee et al., 1968a). The response to marked increases in the iron concentration of the intestinal wall is lowered iron absorption (Forth and Rummel, 1973). Accordingly, it has been observed that copper deficiency reduces iron absorption (Chase et al., 1952; Matrone, 1960; Lee et al., 1968a; Schwarz and Kirchgessner, 1974b).

Altered iron concentration in the liver as the result of high copper supply cannot be explained by the action of the ferroxidase. Here displacement reactions

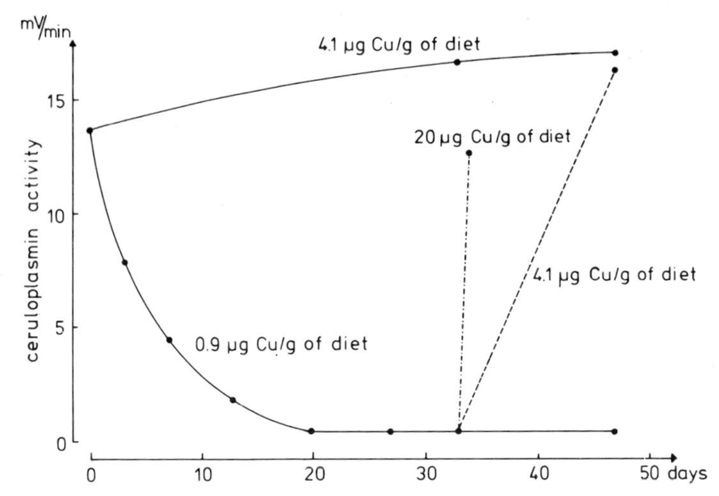

Figure 3. Ceruloplasmin activity of rats in relation to copper supply. Grassmann and Kirchgessner (1973a).

occurring either intraluminally or in intermediary metabolism must be considered on the basis of the complex chemistry of these two trace elements and findings from *in vitro* studies (Kirchgessner and Weser, 1963).

2.2. Influence of Iron on Copper Metabolism

The copper-iron interaction is determined not only by the influence of copper on iron metabolism, but also by the effect of deficient or excessive iron supply on copper metabolism. Early evidence for an effect of deficient iron nutrition on copper utilization stems from studies with pigs by Lahey et al. (1952); the animals showed a severalfold increase in the copper contents of their livers in spite of a normal supply of this element. With rats also, the liver copper contents were found to be significantly increased after iron depletion for more than 8 weeks (Sourkes et al., 1968; Symes et al., 1969). In other studies (Grassmann and Kirchgessner, 1973b, 1973c; Grassmann, 1976) this copper accumulation was already observed after 3 weeks of deficient iron supply. In several studies (Gubler et al., 1957; Grassmann, 1977; Thompson and Evans, 1977b), however, this effect of iron supply on copper retention, particularly in the liver, was not detected. These discrepant results may be related to the particular level of iron supply (Grassmann, 1977), the duration of the experiments, and the growth rate of the animals.

Corresponding to the different findings on the copper storage of several organs during iron deficiency, there are differences regarding the effect of iron supply on the activity of various copper metalloproteins. Thus a markedly reduced ceruloplasmin activity was found in the liver despite an elevated copper control after about 3 weeks of iron depletion (Grassmann and Kirchgessner, 1973b, 1973c). This might be interpreted as a reduced metabolic efficiency of copper utilization during deficient iron nutrition, analogous to the impaired iron mobilization resulting from copper deficiency, and possibly leading to a reduced synthesis of ceruloplasmin. Also, the decreased activities of the cytochrome oxidase (Dagg et al., 1966) and the monoamine oxidase of hepatic mitochondria (Symes et al., 1969) encountered during iron deficiency could be explained in this way. Certainly, an overall lower metabolic rate during iron deficiency may be of importance, because hemoglobin levels below 6 to 7 have been associated with significant growth retardation in various species (Kirchgessner et al., 1971; Grassmann and Kirchgessner, 1973c, 1973d; Kirchgessner and Weigand, 1975). Therefore reduced enzyme activities might be attributed to a slower rate of synthesis of body substance (Kirchgessner et al., 1977; Grassmann et al., 1978). For example, ceruloplasmin is formed only during functioning protein synthesis (Gomez-Garcia and Matrone, 1967; Gregoriadis and Sourkes, 1968; Evans et al., 1969).

A direct influence of iron on copper absorption was considered unlikely by Matrone (1960). This view is supported by subsequent studies which have found no evidence for this mode of interaction (Van Campen, 1971). Accordingly, the intestinal transport of copper, in contrast to iron, is not stimulated by iron deficiency (Schwarz and Kirchgessner, 1973, 1974b).

There are few studies on the effect of high iron supply on copper storage and distribution in the liver. Dick (1954a) reported that a daily intake of 5.2 g Fe per animal markedly reduced copper levels. Since he used iron sulfate, he could not rule out the possibility that the anion, in the presence of low molybdenum contents, was responsible for this effect. High iron doses also led to copper depletion in the liver of cattle (Standish et al., 1969). Studies (Suttle and Mills, 1966a) showing that increasing the iron supply could alleviate the toxic effects of excessive copper doses may be interpreted similarly.

3. COPPER-ZINC INTERACTIONS

The interaction between copper and zinc may be considered to be mutually antagonistic, that is, copper affects zinc metabolism and vice versa. The adverse effect of excessive nutritional doses of one element may be repressed by increasing the supply of the other element. Of decisive importance for the effect of these two elements on health and productive functions is the particular ratio in which they are present in the diet. In this respect the analytically defined content is not as relevant as the quantities available for absorption, because factors affecting absorption, such as chemical binding of the element in the gastrointestinal tract, and other dietary ligands, minerals, or trace elements may either lower or increase the incidence of interactions. Therefore the numerous studies on the copper-zinc interaction must be discussed with reference to the particular ratio of absorbable copper and zinc.

3.1. Effects of Varying the Copper Supply on Zinc Metabolism

The mutual relationship of copper and zinc in metabolism may, for one, be disturbed by either deficient or excessive copper nutrition even if the zinc supply remains adequate. Consequently, when the copper supply is low, zinc will be present at a relatively high level compared to the copper. In this case the zinc status is little affected. Usually, a slight increase of the zinc retention in the liver or bone is indicated in copper-deficient animals as compared to controls given a normal copper supply (Alfaro and Heaton, 1973; Gipp et al., 1973). Also, in a recent study (Schwarz and Kirchgessner, 1979a), the zinc contents in bone, liver, and serum of copper-deficient rats were found to be increased by about 5 to 10%.

When pharmacologically high copper doses ranging from 200 to 250 mg/kg are administered to growing pigs as growth promotants (see Braude, 1967), the copper intake assumes a manifold excess over the zinc supply provided by practical-type rations. Controversial findings are described in the literature regarding the effects on grossly apparent symptoms of the zinc status such as parakeratosis of pigs. Copper supplements of 125 to 300 mg/kg to diets with suboptimum zinc and/or high calcium or phytate levels mitigated or completely reversed parakeratotic skin lesions and were almost as effective as zinc supplements (Hoefer et al., 1960; Wallace et al., 1960; Ritchie et al., 1961, 1963; De Goey et al., 1971). Similarly, additional copper in soybean protein diets for chicks seems to improve zinc availability slightly (Nielsen et al., 1966). Dialysis experiments provide evidence that copper ions may displace other metals (e.g., zinc), when bound with lower affinity, from their complexes, so that improved conditions for zinc absorption could result (Kirchgessner and Weser, 1963). The increased zinc retention in the liver, observed in almost all studies using moderate to very high copper supplementation (Kirchgessner and Weser, 1963; Ritchie et al., 1963; Suttle and Mills, 1966a, 1966b; Gipp et al., 1973), would support this hypothesis of an improved zinc absorption. On the other hand, O'Hara et al. (1960) and Suttle and Mills (1966a, 1966b) reported intensified clinical zinc-deficiency symptoms following subtoxic copper doses, despite elevated liver zinc concentrations. The opposite response of the copper and zinc contents in serum to widening of the dietary Cu:Zn ratio (Suttle and Mills, 1966b; Gipp et al., 1973; Murthy et al., 1974) also supports these observations. Consequently, interactions between copper and zinc due to high copper doses may possibly occur both in the lumen of the digestive tract and in the intermediary metabolism; whichever site prevails may depend upon dietary composition.

Using highly zinc-deficient diets with 2 mg Zn/kg and from 12 to 200 mg

Cu/kg to obtain varying Cu:Zn ratios, Pallauf and Kirchgessner (1974) studied the possibility that copper could assume the functions of zinc during extreme zinc deficiency, since such an interrelation between zinc and copper was indicated as possible from *in vitro* studies on enzyme activation (O'Dell and Campbell, 1971). The effects on the development of the live weights and zinc status of the zinc-deficient rats were, however, inconsistent, so that no substitutive effect of copper for zinc is evident from the data presently available.

3.2. Effects of Varying the Zinc Supply on Copper Metabolism

Analogously to the studies with varying copper supply (Section 3.1), imbalances are encountered when the zinc nutrition is altered while the copper supply is kept adequate. In zinc deficiency the result is an excess of copper, and in excessive zinc administration the relationship changes to the disadvantage of copper.

The effects of zinc depletion on copper metabolism need to be differentiated. Whereas various authors (Prasad et al., 1967; Reinhold et al., 1967; Cox et al., 1969; Chu et al., 1970; Schwarz and Kirchgessner, 1974a) found no differences in the copper contents of various organs and tissues between depleted and control animals, others (Moses and Parker, 1964; Prasad et al., 1969; Petering et al., 1971; Burch et al., 1975; Magee et al., 1975; Roth and Kirchgessner, 1977) reported increased copper concentrations in liver and bone along with reductions in zinc concentrations (see Table 2; Roth and Kirchgessner, 1977). These two trace elements also showed opposing responses regarding the level of their excretion in the milk of highly zinc-depleted dairy cows (Kirchgessner et al., 1978). The occurrence of these unequivocal interactions may, however, be related, in part, to differences in the feeding practices (pair-fed or *ad libitum*) or the dietary copper content. Also, the changes in the element concentrations in favor of copper

Table 2. Changes in the Copper and Zinc Contents of Liver and Bone during Zinc Deficiency[a]

	Zinc Deficiency (1.2 mg Zn/kg + 5 mg Cu/kg in the diet)	Control (Pair-Fed) (96 mg Zn/kg + 5 mg Cu/kg in the diet)
Liver (μg/g dry matter)		
Zinc	98 ± 4	100 ± 4
Copper	26 ± 4	22 ± 2
Bone (μg/g dry matter)		
Zinc	52 ± 6	232 ± 14
Copper	18 ± 4	10 ± 3

[a] Roth and Kirchgessner (1977).

are not consistent between various organs. Thus in kidneys and heart the copper content significantly decreases during zinc-deficient nutrition (Prasad et al., 1967, 1969; Burch et al., 1975). Similarly the serum and blood copper levels, analogously to the zinc levels, tend to be lower (Petering et al., 1971; Schwarz and Kirchgessner, 1974a; Burch et al., 1975).

On the other hand, copper falls to a relative minimum when the zinc supply is excessively increased while the copper supply is kept at a constant level. The effects on the copper status therefore correspond to an induced copper deficiency and are manifested by a reduction of the copper concentrations in liver, heart, and serum, and in the activities of copper-metalloenzymes such as ceruloplasmin and cytochrome oxidase (Sutton and Nelson, 1937; Smith and Larson, 1946; Duncan et al., 1953; Van Reen, 1953; Grant-Frost and Underwood, 1958; Cox and Harris, 1960; Magee and Matrone, 1960; Schlicker and Cox, 1968; Lee and Matrone, 1969; Whanger and Weswig, 1970; Campbell and Mills, 1974; cf. also Lantzsch, 1973). However, the effective dose level varies over a very wide range from about 100 to more than 1000 mg Zn/kg dietary dry matter, because various dietary constituents may influence the efficiency of zinc absorption and, thereby, the effects on copper metabolism. Also, species-related differences may be significant. The interactions become particularly distinct when, in addition, the dietary copper content is low (Hill et al., 1963). The changes imposed on the copper status may be reversed by raising the copper supply in order to narrow the Zn:Cu ratio.

In the same way, increasing the zinc supply during excessive copper intake may counteract copper toxicosis. Thus the additional feeding of zinc to pigs given pharmacologically high copper doses lowered the copper retention of the liver (Ritchie et al., 1963; Suttle and Mills, 1966a, 1966b). Likewise, the activity of the serum aspartate aminotransferase decreased (Suttle and Mills, 1966a, 1966b). Similar relationships apply to the ruminant, especially the sheep (Dick, 1954a; Bremner et al., 1976), in which relatively small increments of the copper dose may suffice to exert subtoxic effects. Therefore, in taking advantage of these interactions, the zinc supply may be raised as a protective measure against copper toxicity (Bremner et al., 1976).

3.3. On the Mode of the Copper-Zinc Interactions

Although data on changes in the concentrations of copper and zinc in various organs and tissues or on differences in total retention in response to altered relations of the dietary copper and zinc supply may often give the first indication of interactions, for the most part they show only net results. Moreover, the majority of studies provide data on just a few selected organs or on total retention without supplemental studies concerning distribution. The basic physiological events explaining the copper-zinc interactions may be sought particularly in the processes of absorption, excretion, or possible binding of these trace elements in the various compartments.

Interactions during Absorption and Excretion of Copper and Zinc

Van Campen and co-worker (Van Campen, 1969; Van Campen and Scaife, 1967) established that copper and zinc inhibited the intestinal absorption of each other. The simultaneous intraluminal application of a high excess of zinc significantly reduced copper absorption; conversely, a high copper dose reduced zinc absorption (Table 3). In contrast, intraperitoneal administration of the respective elements did not affect absorption. Likewise, the absorption of copper given intraluminally to rats on empty stomachs, after 5 weeks of extremely high zinc nutrition, was not influenced by the zinc supply status (Magee and Matrone, 1960). According to Evans et al. (1974), depression of zinc absorption in the presence of excessive copper is observed only in rats offered an adequate zinc supply, since in zinc-deficient rats zinc absorption could not be inhibited by copper. It has further been shown by several studies (Schwarz and Kirchgessner, 1973, 1974a, 1974b) that the absorption of both zinc and copper, administered separately, is improved by zinc deficiency (Table 4). Whereas increased absorption of zinc has been found to also result from copper deficiency, stimulated zinc absorption has hitherto not been established (Schwarz and Kirchgessner, 1974b; 1979b).

Table 3. Influence of High Zinc or High Copper Doses on the Absorption of Copper and Zinc[a]

	Copper Absorption (% of ^{64}Cu dose in sampled tissues)	Zinc Absorption (% of ^{65}Zn dose in sampled tissues)
Control	22	
+ Zinc[b] (intraduodenally)	11	
+ Zinc[b] (intraperitoneally, 2 hr in advance)	20	
Control		19
+ Copper[c] (intraduodenally)		8
+ Copper[c] (interperitoneally, 2 hr in advance)		23

[a] Van Campen and Scaife (1967) and Van Campen (1969).
[b] 1000 μg Zn.
[c] 200 μg Cu.

Table 4. Interactions in the Intestinal Transfer of Copper and Zinc during Either Zinc or Copper Deficiency (ng/intestinal sac·60 min)[a]

	Intestinal Copper Transfer	Intestinal Zinc Transfer
Trial 1		
Copper deficiency (0.5 mg Cu/kg dry matter)	29 ± 7	22 ± 5
Control (4 mg Cu/kg dry matter)	9 ± 2	19 ± 4
Trial 2		
Zinc deficiency (2 mg Zn/kg dry matter)	77 ± 23	68 ± 11
Control (70 mg Zn/kg dry matter)	17 ± 4	17 ± 2

[a] Schwarz and Kirchgessner (1974b).

The interactions described, which result in lower copper concentrations in various organs (because of high zinc supply) or in markedly elevated copper retention (because of deficient zinc supply) might be attributed primarily to the absorptive process. On the other hand, one could also assume that the negative influence of high copper contents on zinc absorption entails making additional zinc available for absorption from poorly soluble dietary zinc because of possible displacement by copper from complex compounds in the intestinal contents. Thus, under these circumstances, one actually might have expected an increase in zinc absorption.

A satisfactory explanation for the mechanisms underlying the copper-zinc interactions at the site of absorption cannot, at present, be offered. The experimental findings on intestinal zinc binding after copper is supplied *in vitro* or to fasted rats differ considerably: they show either no effect (Oberleas et al., 1966; Pearson et al., 1966; Evans et al., 1974), a lower zinc binding when the Cu:Zn ratio is extremely wide, or even an increased binding (Van Campen and Scaife, 1967). Therefore it appears that the actual cellular transport process is influenced. In this connection a mutual displacement in the mucosal copper- and zinc-binding proteins isolated by various authors (Starcher, 1969; Evans et al., 1970) or in the metallothionein (Richards and Cousins, 1976; Cousins, 1978) might be considered.

A mutual influence of copper and zinc on their excretion may hardly be important in comparison to the effect on their absorption. Thus nutritional or parenteral copper loading does not affect endogenous zinc excretion (Cotzias et al., 1962; Kinnamon and Bunce, 1965). Similarly, copper deficiency has no influence on endogenous zinc excretion (Schwarz and Kirchgessner, 1979b). A high zinc supply status, however, causes a significant increase in urinary copper excretion (Duncan et al., 1953; Magee and Matrone, 1960). Nevertheless, this urinary loss remains small compared with the fecal excretion of copper.

Interactions of Copper and Zinc in Intermediary Metabolism

Additional insight into the interactions of copper and zinc may be gained from their subcellular distribution (see Owen et al., 1977) and their binding to amino acids, peptides, or proteins in intermediary metabolism (Mills, 1974). Thus various copper- and zinc-binding proteins were found by fractionation of the soluble fractions from liver and kidney homogenates. Several studies, especially those of Bremner and co-workers (Bremner and Marshall, 1974a; Bremner, 1976; Bremner and Davies, 1976; Bremner et al., 1977), have shown that copper and zinc appear, for the most part, in association with proteins of three major fractions characterized by molecular weights of >65,000 to 75,000, about 35,000, and 8000 to 12,000. The copper and zinc contents of the 35,000-molecular-weight fraction are relatively independent of the particular copper and zinc supply status (Bremner and Marshall, 1974a). In contrast, the relative copper and zinc levels in the low molecular weight fraction (around 10,000) and in the high molecular weight fraction (around 70,000) change in relation to the nutritional supply of these traces elements (Bremner and Marshall, 1974a, 1974b).

The low molecular weight fraction (8000 to 12,000) holds the copper- and zinc-thioneins (Bremner and Marshall, 1974b; Bremner and Davies, 1975; Bremner and Young, 1976a, 1976b; Richards and Cousins, 1976; Rupp and Weser, 1978), which may possibly play a key role in the metal interactions. A detailed description of the copper-thionein and its possible functions is given in the chapter by Weser et al. (this volume). Interactions between copper and zinc in thioneins are likely in view of the mutual displacement reactions at the binding sites (Bremner and Marshall, 1974a, 1974b) and also the findings that zinc stimulates the biosynthesis of thioneins (Bremner and Davies, 1975; Richards and Cousins, 1976), increases their stability, and reduces their degradation (Bremner, 1976; Bremner and Davies, 1976). During zinc deficiency the fraction containing these copper- and zinc-binding proteins with molecular weights in the range of 8000 to 12,000 is relatively small, while the binding, especially of copper, by the high molecular weight fraction is markedly enforced (Bremner and Marshall, 1974a, 1974b). Similar effects are indicated when the zinc supply is adequate and the copper supply exceedingly high (Bremner and Marshall, 1974a, 1974b). However, certain species related differences are apparent between the sheep and calf, on the one hand, and the pig and rat, on the other (Bremner, 1976; Bremner and Davies, 1976).

Although the possible functions of the copper- and zinc-thioneins must still be discussed with some reservation, and their importance has been observed in relation to the energy transfer and efficient use of redox potentials (Prinz and Weser, 1975), thioneins may also play a role in the storage and detoxification of metals. This function could be related to the increased copper retention in response to nutritional zinc depletion or to relative zinc deficiency induced by excessive copper supply. In this respect it is of interest that thioneins have a very short half-life (Bremner, 1976; Bremner and Davies, 1976; Cousins, 1978), which could mediate rapid availability of the metals in intermediary metabolism (Bremner and Davies, 1975).

4. INTERACTIONS BETWEEN COPPER, MOLYBDENUM, AND SULFUR

Ferguson et al. (1938) reported, for the first time, the incidence of a disease in cattle grazing pastures on molybdenum-rich soils in England. The symptoms of this disease, called teart, were diarrhea and debilitation. The molybdenum contents in the forage (dry basis) ranged from 20 to 100 mg/kg instead of the 3 to 5 mg/kg in normal forage. Spraying the pastures or watering the animals with $CuSO_4$ solutions cured or prevented this syndrome (Ferguson et al., 1943), which was attributed to interactions between molybdenum and copper. Evidently, a copper deficiency induced by an excess of molybdenum was involved (Dick and Bull, 1945), because the Cu:Mo ratio in the feed should generally be near 2:1 to 4:1 (Miltmore and Mason, 1971; Alloway, 1973). Similar syndromes in cattle were also found in the United States (Britton and Coss, 1946), New Zealand (Cunningham, 1950), Canada (Cunningham et al., 1953), and Sweden (Hallgren et al., 1954). In studies with sheep, Dick (1953a, 1953b, 1954b) established that, in addition, the sulfate content of the diet is of crucial importance in respect to the interaction between copper and molybdenum.

Since there are evidently species differences in this multielement interaction, the experimental findings for ruminants and monogastrics will first be described separately.

4.1. Interactions between Copper, Molybdenum, and Sulfur in Ruminants

The effects of molybdenum and sulfur on copper metabolism must especially be viewed with respect to the ratios between copper, molybdenum, and sulfur, because different results are obtained depending upon the particular level of dietary supply of these elements. The change in liver copper content in response to variation in the molybdenum supply, while keeping the dietary sulfate content constant, is shown in Figure 4, while Figure 5 illustrates the relationship when the sulfate supply is varied and the molybdenum content kept constant (Dick, 1954a). The copper accumulation in the liver decreases both with increasing molybdenum intake and with increasing sulfate intake. The interrelationship between these three factors is made clear once more in Table 5 by data from Goodrich and Tillman (1966) on growth and copper storage. Molybdenum and

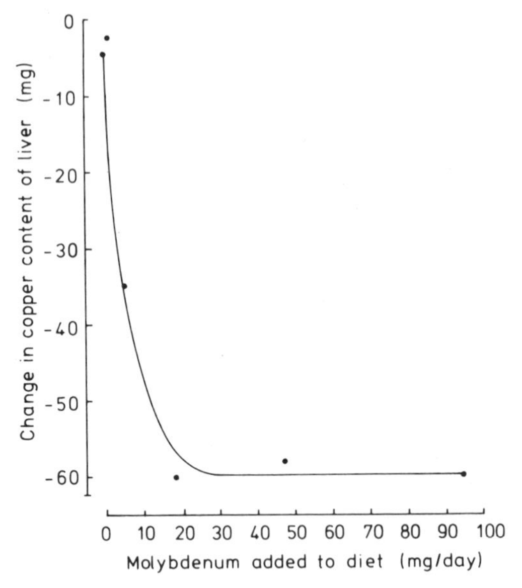

Figure 4. Relationship between added molybdenum to the diet and amount of copper in the liver at constant sulfur contents in the diet. Dick (1954a).

SO_4 supplementation reduce the daily gain and the copper retention in the liver. Matching responses of the liver copper storage to molybdenum and SO_4 supplements were also found by Wynne and McClymont (1956).

In many studies, however, supplementation of molybdenum and sulfur, either alone or in combination, did not affect copper status (Allcroft and Lewis, 1956; Butler and Barlow, 1963; Butler et al., 1964; Hogan et al., 1966). Whether or not interactions are observed depends greatly upon the basal content of molybdenum and sulfur and upon their ratio. Whereas increasing dietary molybdenum levels (from 2.4 to 9.2 mg/kg) did not influence the liver and plasma copper contents when the diet contained 0.03% sulfate, such an increase greatly reduced the contents of copper when the diet contained 0.55% sulfate (Mylrea, 1958). Similar results for the liver were also observed by Dick (1954a, 1954b), Marcilese et al. (1969), Pierson and Aanes (1958), and Suttle and Field (1968a).

From the experimental results it may be concluded that high molybdenum and sulfate intakes impair the metabolic efficiency of copper utilization and therefore induce copper deficiency. From studies by Marcilese et al. (1969) with labeled copper, it is evident that molybdenum and sulfate supplementation decreases the incorporation of copper into the liver and the synthesis of ceruloplasmin. Accordingly, high copper supplements are again required to counteract these effects of molybdenum and sulfate. Mills et al. (1963) reported that $CuSO_4$ injection improved the growth of calves grazing high-molybdenum pastures, and similar results were described by Poole (1970) for cattle. These relationships

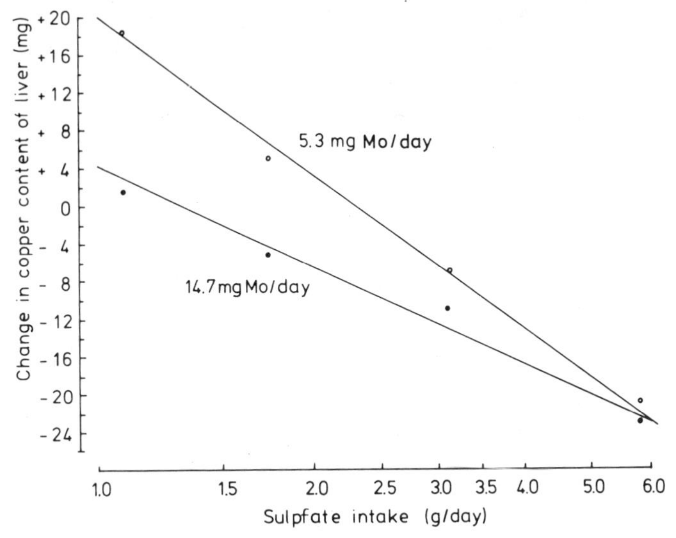

Figure 5. Relationship between change in liver copper and logarithm of daily inorganic sulfate intake at constant molybdenum contents in the diet. Dick (1954a).

are also evident from studies with lambs in which a copper-deficiency syndrome (neonatal ataxis, swayback) was induced by giving their dams molybdenum and sulfate supplements during pregnancy (Mills and Fell, 1960; Fell et al., 1965; Suttle and Field, 1968b).

The interaction between copper and molybdenum is influenced not only by

Table 5. Effects of Sulfur, Molybdenum, and Copper Levels on Sheep[a]

	Sulfur Content (%)							
	0.10				0.40			
	Molybdenum Content (mg/kg)							
	2		8		2		8	
	Copper Content (mg/kg)							
Parameter	10	40	10	40	10	40	10	40
Daily gains (g)	103	99	88	78	50	40	95	61
Hemoglobin (g/100 ml blood)	13.1	13.5	12.4	13.0	12.2	12.9	13.2	12.6
Plasma copper (μg/100 ml)	111	112	104	107	100	105	111	116
Liver copper (μg/g dry matter)	154	268	118	179	117	120	67	114

[a] Goodrich and Tillman (1966).

sulfate but by other sulfur compounds as well. Dick (1956a) observed with sheep that the adverse effects of molybdenum and sulfate on copper utilization may be reversed by high dietary contents of manganese, which presumably blocks the limiting action of molybdenum and sulfate. Mylrea (1958), however, could find no such effect of high manganese intakes on steers.

4.2. Interactions between Copper, Molybdenum, and Sulfur in Monogastrics

In most of the studies on the interaction of copper with molybdenum and sulfur in monogastric animals, much higher dietary molybdenum contents were applied than in studies with ruminants. These high supplements (800 to 1000 mg Mo/kg) caused marked growth depression in rats (Gray and Ellis, 1950; Kulwich et al., 1951, 1953; Gray and Daniel, 1954; Halverson et al., 1960), which also developed anemia and diarrhea. During deficient copper nutrition these symptoms also occurred with lower molybdenum supplements (20 to 140 mg/kg) (Jeter and Davis, 1950). Copper supplementation relieved the anemia and diarrhea (Halverson et al., 1960) and partially corrected the growth depression (Gray and Daniel, 1954). Also, in studies with pigs given a very high molybdenum supply (1500 mg/kg), pronounced growth depressions could be established which were again alleviated by adding 0.4% sulfate to the diet (Standish et al., 1975). Feeding diets with relatively low molybdenum (20 to 50 mg) and sulfate (0.1 to 0.4%) contents did not depress growth (Gipp et al., 1967; Hays and Kline, 1969; Kline et al., 1971; Dale et al., 1973). However, the copper supply was adequate or high in these studies.

In several studies copper retention was increased by molybdenum or sulfate supplementation (Kulwich et al., 1951; Compere et al., 1965; Dale et al., 1973; Standish et al., 1975). This reduction in fecal copper excretion cannot be assumed, however, to result solely from a change in the true copper absorption, as the endogenous excretion could also be affected. Further studies with swine found no influence, not even a reduction in liver copper contents, in response to either molybdenum or sulfate or molybdenum plus sulfate supplementation (Gipp et al., 1967; Kline et al., 1971; Dale et al., 1973; Standish et al., 1975). These different findings are essentially related to the copper and sulfate contents and, in addition, to other sulfur compounds in the diet. Thus sulfide markedly lowers copper availability (Mills, 1960; Hampel and Kirchgessner, 1970; Marston et al., 1971; Kline et al., 1973).

The metabolic efficiency of copper also is altered by interaction with molybdenum and sulfur. By this mode of action the differences in ^{64}Cu clearance and ceruloplasmin activity found, for example, by Dale et al. (1973) and Standish et al. (1975) may be interpreted. Dowdy and Matrone (1968a, 1968b) and Dowdy et al. (1969) attributed the diminished ceruloplasmin activity to the formation of a nonavailable copper-molybdenum complex, which was, however, added to the diet as $2CuMoO_4 \cdot (OH)_2$ (lindgrenit).

4.3. On the Mode of the Interactions

The effects of molybdenum and sulfur depend upon the copper status of the animals and the ratio of these three nutrients in the diet. Thus, in rats with suboptimum copper intake, a low molybdenum supplement (10 mg/kg) was sufficient to induce copper deficiency. The symptoms that were aggravated by the simultaneous addition of sulfate could be prevented by additional copper. With adequate copper intake, high molybdenum supplements were needed to bring about copper deficiency (Gray and Daniel, 1964). In addition, other factors are also of importance, for example, the pH value, the presence of other trace elements (e.g., manganese), and the oxidation state of the sulfur.

Influence of the Sulfur Sources on the Copper-Molybdenum-Sulfur Interactions

Ruminants, in contrast to monogastrics, are able to reduce sulfate to sulfide to a major extent in the rumen (Huisingh and Matrone, 1972) because of the action of *Desulfovibrio* bacteria, which were isolated for the first time by Huisingh et al. (1974) from the ruminal contents of sheep. The addition of copper to washed cultures of these microorganisms from the ovine rumen depressed the reduction of sulfate to sulfide only slightly; molybdenum addition, however, did so very effectively even at low concentrations. When copper and molybdenum were added together to the culture medium, the effect of the latter element was diminished. This is apparently due to the formation of a "Cu-Mo complex" in which molybdenum is less effective (Dowdy et al., 1969). According to Huisingh et al. (1974), molybdenum inhibits irreversibly a major enzyme involved in sulfur reduction, namely, ATP sulfurylase.

Sulfide formed in the rumen may be either absorbed and detoxified in the liver, or incorporated into sulfur-amino acids, or precipitated as a heavy metal sulfide. In the presence of copper, insoluble CuS may be formed in the rumen (Mills, 1960), so that copper becomes less absorbable. Since molybdenum inhibits the reduction of sulfate to sulfide, molybdenum supplements should lead to an increase in the amount of copper available for absorption. The experimental results by Gomez-Garcia (1972) may be interpreted in this way. He showed that molybdenum mitigated the copper-deficiency symptoms if sulfate was the sole sulfur source and urea the sole nitrogen source and thus no sulfur-containing amino acids were present in the feed.

Mills (1960) and Hartmans and Bosman (1970) reported, however, that the sulfide contents in the rumen increased if the protein-holding ration contained sulfate and molybdenum. Huisingh et al. (1975) attributed these results to the fact that sulfide is also liberated from sulfur-containing amino acids in the rumen. While molybdenum inhibits the production of sulfide from sulfate, it activates sulfide liberation from sulfur-containing amino acids, so that the total sulfide production in the rumen depends upon the ratio of sulfate and sulfur from organic sources in the diet. These concepts are also discussed by Gawthorne and

Nader (1976), who assumed, in addition, that the absorption of sulfide from the rumen is decreased by molybdenum, since they found that the sulfide contents in the ruminal fluid rose overall, while the *de novo* synthesis of sulfide was significantly lowered by molybdenum supplementation.

These results may explain some of the differences between ruminants and monogastrics. In contrast to the findings by Huisingh and Matrone (1972, 1976), Suttle (1975) observed that sulfur from methionine and cysteine and sulfur from sulfate play similar roles in copper-molybdenum-sulfur interactions. All sulfur sources were equally effective in their adverse influence on copper metabolism in the presence of molybdenum. However, sulfide production depended upon the molybdenum concentration (Suttle, 1975). The molybdenum supplementation of 4 mg/kg used in these studies is fairly low. Moreover, Huisingh and Matrone employed diets in which they could accurately control the nitrogen and sulfur contents by the use of urea, sulfur-amino acids, and sulfate.

Possible Influence of Copper and Molybdenum on Sulfide Oxidase

The sulfide produced in the rumen is, in part, very rapidly absorbed through the rumen wall, oxidized to sulfate in the liver, and excreted in this form in the urine (Huisingh and Matrone, 1976). The activity of the sulfide oxidase involved in the oxidation of sulfide to sulfate appears to depend upon the content of available copper in the liver (Siegel and Monty, 1961). Whereas copper increases sulfide oxidase activity, molybdenum lowers it (Van Reen, 1954). Reduced sulfide oxidase activity leads to accumulation of sulfide in the liver and hence to precipitation of copper as insoluble nonavailable CuS (Halverson et al., 1960; Siegel and Monty, 1961; Spais et al., 1968). This influence is determined by the amount of absorbable molybdenum, since the concentration of this element controls the activity of the sulfide oxidase in the liver.

Dick (1953b, 1956a, 1956b) postulated that sulfate changes the absorption of molybdenum. Later this was clearly demonstrated with *in vitro* studies by Cardin and Mason (1975) and by Huisingh and Matrone (1976). Thus, in the presence of a large amount of sulfate, less molybdate is absorbed, and vice versa. On the average, more molybdate is absorbed by ruminants, since less sulfate generally reaches the intestinal tract in ruminants than in monogastrics because of the reduction of sulfate to sulfide by the rumen bacteria.

Production of Copper-Molybdenum Complexes

As an additional possibility for explaining the copper-molybdenum interaction, Davis (1958) discussed the production of a poorly available copper-molybdenum complex. Dowdy and Matrone (1968a, 1968b) conducted studies on this aspect (see Section 4.2). A compound that they referred to simply as "Cu-Mo complex" is apparently absorbed, but the copper seems to be hardly available from this complex. Electron paramagnetic resonance (EPR) studies by Huisingh and Matrone (1976) provided evidence that this complex is not stable in serum. Under the conditions in the rumen, however, it appeared to be

Table 6. Effects of Molybdate, Sulfide, or Thiomolybdate on Fate of Dietary ^{64}Cu in Rats[a]

Dietary Supplement		Control	Molybdate	Molybdate + Sulfide	Sulfide	Thiomolybdate
^{64}Cu recovered from carcass (% of dose)		34 ± 1.6	36 ± 2.7	34 ± 6.4	27 ± 6.4	4 ± 1.1
Percent of carcass ^{64}Cu in	blood	6 ± 0.7	9 ± 0.9	11 ± 0.9	8 ± 0.6	68 ± 0.5
	liver	33 ± 2	33 ± 2	31 ± 2	33 ± 2	10 ± 1
	kidney	13 ± 0.1	13 ± 0.1	12 ± 1.4	14 ± 1.8	2 ± 0.5

[a] Mills et al. (1978).

stable. On the basis of the EPR studies it may be assumed that the "Cu-Mo complex" is synthesized in the rumen if molybdenum is present.

In newer reports the assumption is favored that, in addition to or instead of the "Cu-Mo complex" described, thiomolybdate complexes play a major role. The coordinating ligand is postulated to have the general formula $(MoO_xS_{4-x})^{2-}$ (Dick et al., 1975). Mills et al. (1978) found that thiomolybdate is generated *in vitro* by rumen contents when combined with ammonium molybdate and sulfur sources under anaerobic conditions. An unequivocal proof of the *in vivo* synthesis of this ligand had hitherto not been demonstrated. It is evident from absorption studies with ^{64}Cu in rats that the addition of synthetic thiomolybdate significantly depresses copper absorption (see Table 6), whereas the addition of molybdate, molybdate plus sulfide, or sulfide does not affect copper absorption (Mills et al., 1978).

The absorbed thiomolybdate is rapidly bound to albumin. There are signs that thiomolybdate is also bound to various other plasma proteins in both ruminant and mongastric animals. These thiomolybdate-protein complexes exhibit an increased reactivity toward copper ions. Under these experimental conditions, copper is obviously bound very tightly to plasma, since it is hardly liberated by TCA precipitation of the proteins (Bremner and Young, 1978; Mills et al., 1978).

These studies showed that rats given 2 mg Mo/kg diet or more as thiomolybdate for longer periods of time developed symptoms that were similar to those of the copper-molybdenum-sulfur interactions in ruminants fed sulfur and high molybdenum supplements. The symptoms included increased plasma copper levels, reduced ceruloplasmin activity, and lowered liver copper contents.

4.4. Differences in the Copper-Molybdenum-Sulfur Interactions between Ruminants and Monogastrics

The different effects of the copper-molybdenum-sulfur interactions in ruminants and in monogastrics may be associated with the different digestive systems. The ruminant digestive system provides the basis for three processes which are responsible for the graver incidence of interactions in these animals. Thus first, the microbial activity in the forestomachs involves greater production of sulfide by the reduction of sulfate as well as from the degradation of sulfur-containing amino acids. Sulfide reacts with copper to form poorly soluble and hardly absorbable copper sulfide. The second process is the passing of less sulfate to the intestine of ruminants that occurs with monogastrics. Sulfate inhibits the absorption of molybdate, and hence less molybdenum may reach the liver of monogastrics. Third, molybdenum inhibits sulfide oxidase, so that more sulfide may accumulate in the livers of ruminants than in those of monogastrics. Since sulfide supposedly also reacts with copper in the liver to form sulfide, the amount of available copper for metabolism in the ruminant body is reduced.

Sulfide is also required for the synthesis of the copper-thiomolybdate complex. Since the sulfide supply is higher in ruminants than in monogastrics, a greater portion of the copper may be present as the poorly utilizable copper-thiomolybdate complex. This leads to an aggravation of the copper deficiency in ruminants. The synthesis of copper-thiomolybdate, and of copper-molybdenum complexes in general, may quantitatively contribute most to the differences between monogastrics and ruminants.

5. INTERACTIONS BETWEEN COPPER AND MANGANESE, NICKEL, AND SELENIUM

A few publications report on interactions between copper and the essential trace elements manganese, nickel, and selenium. The increased manganese retention in pigs supplied with pharmacologically high copper doses (Kirchgessner and Weser, 1963) may be due to intraluminal displacement reactions from complex compounds and hence to an improved absorption condition for manganese. On the other hand, severe manganese deficiency was associated with markedly reduced concentrations of copper (and manganese as well) in liver, kidneys, spleen, and pancreas (Kirchgessner and Heiseke, 1977). At the same time the iron concentration increased in the liver, while it decreased in the bones (Heiseke and Kirchgessner, 1978). Moreover, the hemoglobin and hematocrit values were lower in manganese-deficient animals than in controls given a normal supply of this element (Heiseke and Kirchgessner, 1977). This response of iron and of the blood parameters was also observed during nutritional copper deficiency. Therefore the copper-manganese interaction might be considered to be primary.

Symptoms of iron anemia also occurred with excessive manganese administration (Gubler et al., 1954; Hartman et al., 1955; Matrone et al., 1959). Furthermore, Vohra and Heil (1969) observed a reduction in manganese storage in the liver of turkeys given zinc and copper doses that exceeded requirements. This finding can be interpreted to indicate an interaction with zinc, since zinc and manganese respond in the same direction (Roth and Kirchgessner, 1977; Heiseke and Kirchgessner, 1978).

Schnegg and Kirchgessner (1975a, 1976a) could also show that severe nickel deficiency causes a significant reduction in the growth of rats, in the nickel contents, and, overall, in the iron, copper, and zinc contents of various organs. Yet there also were changes in the concentrations of trace elements. Thus the copper concentrations in the livers of nickel-deficient male rats, compared with controls on a normal supply, were markedly increased in two experiments. However, all nickel-depleted rats were distinctly anemic (Schnegg and Kirchgessner, 1975b, 1976b). Therefore the interaction between nickel and copper could be a secondary effect, especially since increased copper concentrations in the livers of iron-anemic animals have repeatedly been reported (Lahey et al., 1952; Sourkes et al., 1968; Grassmann and Kirchgessner, 1973a, 1973b). Whether the improved growth, the increased ceruloplasmin activity, and the higher hemoglobin and hematocrit values found when copper-deficient rats were given a supplement of 20 mg Ni/kg diet (Spears et al., 1977) could be due to improved iron metabolism needs further investigation.

Godwin et al. (1978) noted a markedly increased sensitivity of selenium-deficient rats to copper. This enforced copper toxicity was attributed, however, to intensification of the damage to the erythrocyte membrane, since the first consequence of deficient selenium intake is increased fragility of the erythrocyte membrane. Therefore this interaction cannot be considered as being directly between these two trace elements but rather should be regarded as an aggravation of the deficiency symptoms.

6. INTERACTIONS BETWEEN COPPER AND CADMIUM, SILVER, MERCURY, AND LEAD

6.1. Interactions between Copper and Cadmium

Effect of Cadmium on Copper Utilization

Interactions between copper and cadmium were observed in fowl, ruminants, and monogastrics. Thus Hill et al. (1963) reported that cadmium greatly reduces copper utilization in baby chicks. The mortality of the chicks dramatically increased with high dietary cadmium contents. Bunn and Matrone (1966) found similar effects in rats. In these animals and also in goats (Henning et al., 1974) the copper concentrations in the liver and other tissues decreased. Whanger and Weswig (1970) observed, in addition, a significant reduction in ceruloplasmin activity.

The interaction between cadmium and copper depends not only upon the level of the cadmium supply but, of course, also upon the level of the copper supply. Campbell et al. (1978) pointed out that copper-cadmium interactions are particularly pronounced during suboptimum copper intake. Thus the plasma copper content of rats fell by about 50% in response to the addition of about 6 mg Cd/kg diet when the dietary copper contents were low, whereas there was no cadmium effect when the copper intake was adequate. Since cadmium strongly interacts with zinc, and copper metabolism is closely associated with zinc nutrition (see Section 3.2), the influence of cadmium on copper utilization must also be seen in relation to the dietary zinc content (Campbell and Mills, 1974; Petering, 1974; Campbell et al., 1978). It is evident from the data of Table 7 that 6.1 mg Cd/kg diet reduces the activity of ceruloplasmin to the same extent as 1000 mg Zn/kg diet. Thus cadmium has an effect on copper utilization comparable to that of zinc (see Sections 3.2 and 3.3); however, much smaller levels of dietary cadmium are sufficient to exert the same effect.

On the Mode of Interaction

From studies by Hill et al. (1963), Van Campen (1966), and Campbell et al. (1978) it is evident that interactions may occur during absorption. Campbell et al. (1978) found that 17.6 mg Cd/kg diet reduced copper absorption by about 45% in rats given a suboptimum copper supply. Since, on the one hand, cadmium in low concentrations uncouples oxidative phosphorylation (Jacobs et al., 1956), and, on the other hand, the transport of trace elements across the intestinal wall is energy dependent (Crampton et al., 1965; Hutagalung et al., 1977), this defect, among others, could very well be responsible for the reduced absorption (Van Campen, 1966).

These interactions during absorption occur only if both elements are administered simultaneously (Campbell et al., 1978). Therefore displacement reactions at the site of the intestinal binding or transport must also be considered. Here the intestinal metallothioneins could be of major importance (Evans et al., 1970). The occurrence of metallothioneins in the liver and kidneys (see Weser

Table 7. Effects of Dietary Supplements of Cadmium and Zinc on Plasma Ceruloplasm in Activity (U/l) in Rats Maintained for 9 Weeks on Diets Providing 2.6 mg Cu/kg Dry Matter[a]

	Cd (mg/kg)			
Zn (mg/kg)	0.16	1.5	6.1	18
30	43.5	18.2	9.3	6.5
300	27.4	17.0	5.2	4.9
1000	9.5	5.0	5.1	3.7

[a] Campbell and Mills (1974).

et al., this volume) and the binding of copper and cadmium to these proteins could, however, also bring about an interaction between these two elements at the site of intermediary metabolism.

6.2. Interactions between Copper and Silver

According to Whanger and Weswig (1970), silver is the strongest copper antagonist. Hill et al. (1964) showed, for the first time, that 100 and 200 mg Ag/kg diet aggravated the copper-deficiency symptoms in chicks receiving a suboptimum copper supply. Effects induced by high dietary silver contents (900 mg/kg) could partially be reversed by supplements of 50 mg Cu/kg in the case of turkey poults (Jensen et al., 1974) and chicks (Peterson and Jensen, 1975).

According to studies by Van Campen (1966), the absorption of copper is not influenced by silver. Therefore the interactions are to be sought especially in metabolism. The strong affinity of silver to SH groups makes it likely that thionein especially (Evans et al., 1970) and/or other SH-containing proteins are involved.

6.3. Interactions between Copper and Mercury

In contrast to the copper-silver interactions, there may be no direct interaction between copper and mercury. In studies by Hill et al. (1964) the growth of copper-deficient chicks was not affected by a mercury load, whereas that of chicks given 25 mg Cu/kg diet was severely depressed. There was no change in ceruloplasmin activity due to the administration of 400 mg Hg/kg diet (Whanger and Weswig, 1970). Likewise, Van Campen (1966) could not detect an influence of mercury on copper absorption. Rather, the effect of mercury on copper metabolism may come about because of a general damaging of the organism.

6.4. Interactions between Copper and Lead

Despite normal copper and molybdenum intakes, lambs from mothers grazing high-lead pastures contracted neonatal ataxia (Alloway, 1969). The incidence of this disease was attributed by Alloway (1969) to a lead-induced copper deficiency. In rats, high lead doses (0.5%) combined with 2.5 mg Cu/kg diet caused a reduction in plasma copper contents and ceruloplasmin activity. Lowering the dietary copper content to 0.5 mg/kg also brought about severe reduction of the hematocrit and the hemoglobin (Klauder et al., 1973). In addition, these authors found that the reduced growth, the lead retention in liver and kidneys, and the lead contents of the erythrocytes were inversely proportional to the dietary copper contents. In reviewing these findings, Klauder et al. (1973) concluded that copper lowers the toxicity of lead and, conversely, that lead induces copper-deficiency symptoms.

REFERENCES

Alfaro, B. and Heaton, F. W. (1973). "Relationships between Copper, Zinc and Iron in the Plasma, Soft Tissue and Skeleton of the Rat during Cu Deficiency," *Br. J. Nutr.,* **29,** 73–85.

Allcroft, R. and Lewis, G. (1956). "Relationship of Copper, Molybdenum and Inorganic Sulphate Content of Feeding Stuffs to the Occurrence of Copper Deficiency in Sheep and Cattle," *Landbouwkd. Tijdschr.,* **68,** 711–722.

Alloway, B. J. (1969). "The Soils and Vegetation of Areas Affected by Mining for Non-ferrous Metalliferrous Ores, with Special Reference to Cadmium, Copper, Lead and Zinc." Ph.D. thesis, University of Wales, Aberystwyth, Chapter 13, pp. 629–683.

Alloway, B. J. (1973). "Copper and Molybdenum in Swayback Pastures," *J. Agric. Sci.,* **80,** 521–524.

Beyer, M. G., Kirchgessner, M., and Steinhart, H. (1976). "Kinetische Untersuchungen zur Hemmung von Pepsin durch Spurenelemente," *Landwirtsch. Forsch.,* **29,** 53–58.

Braude, R. (1967). "Copper as a Stimulant in Pig Feeding," *World Rev. Anim. Prod.,* **3,** 69–82.

Bremner, I. (1976). "The Relationship between the Zinc Status of Pigs and the Occurrence of Copper- and Zinc-Binding Proteins in Liver," *Br. J. Nutr.,* **35,** 245–252.

Bremner, I. and Davies, N. T. (1975). "The Induction of Metallothionein in Rat Liver by Zinc Injection and Restriction of Food Intake," *Biochem. J.,* **149,** 733–738.

Bremner, I. and Davies, N. T. (1976). "Studies on the Appearance of a Hepatic Copper-Binding Protein in Normal and Zinc-Deficient Rats," *Br. J. Nutr.,* **36,** 101–112.

Bremner, I. and Marshall, R. B. (1974a) "Hepatic Copper- and Zinc-Binding Proteins in Ruminants. 1: Distribution of Cu and Zn among Soluble Proteins of Livers of Varying Cu and Zn Content," *Br. J. Nutr.,* **32,** 283–291.

Bremner, I. and Marshall, R. B. (1974b). "Hepatic Copper- and Zinc-Binding Proteins in Ruminants. 2: Relationship between Cu and Zn Concentrations and the Occurrence of a Metallothionein-like Fraction," *Br. J. Nutr.,* **32,** 293–300.

Bremner, I. and Young, B. W. (1976a). "Isolation of (Copper, Zinc)-Thioneins from Pig Liver," *Biochem. J.,* **155,** 631–635.

Bremner, I. and Young, B. W. (1976b). "Isolation of (Copper,Zinc)-Thioneins from the Livers of Copper-Injected Rats," *Biochem. J.,* **157,** 517–520.

Bremner, I. and Young, B. W. (1978). "Effects of Dietary Molybdenum and Sulphur on the Distribution of Copper in Plasma and Kidneys of Sheep," *Br. J. Nutr.,* **39,** 325–336.

Bremner, I., Young, B. W., and Mills, C. F. (1976). "Protective Effect of Zinc Supplementation against Copper Toxicosis in Sheep," *Br. J. Nutr.*, **36,** 551–561.

Bremner, I., Williams, R. B., and Young, B. W. (1977). "Distribution of Copper and Zinc in the Liver of the Developing Sheep Foetus," *Br. J. Nutr.*, **38,** 87–92.

Britton, J. W. and Goss, H. (1946). "Chronic Molybdenum Poisoning in Cattle," *J. Am. Vet. Med. Assoc.*, **108,** 176–178.

Bunn, C. and Matrone, G. (1966). "*In vivo* Interactions of Cadmium, Copper, Zinc and Iron in the Mouse and Rat," *J. Nutr.*, **90,** 395–399.

Burch, R. E., Williams, R. V., Hahn, H. K. J., Jetton, M. M., and Sullivan, J. F. (1975). "Serum and Tissue Enzyme Activity and Trace Element Content in Response to Zinc Deficiency in the Pig," *Clin. Chem.*, **21,** 568–577.

Bush, J. A., Jensen, W. N., Athens, J. W., Ashenbrucker, H., Cartwright, G. E., and Wintrobe, M. M., (1956). "Studies on Copper Metabolism. XIX: The Kinetics of Iron Metabolism and Erythrocyte Life-Span in Copper-Deficient Swine," *J. Exp. Med.*, **103,** 701–712.

Butler, E. J. and Barlow, R. M. (1963). "Copper Deficiency in Relation to Swayback in Sheep. I: Effect of Molybdate and Sulfate Supplements during Pregnancy," *J. Comp. Pathol.*, **73,** 208–213.

Butler, E. J., Barlow, R. M., and Smith, B. S. W. (1964). "Copper Deficiency in Relation to Swayback in Sheep. II: Effect of Dosing Young Lambs with Molybdate and Sulfate," *J. Comp. Pathol.*, **74,** 419–426.

Campbell, J. K. and Mills, C. F. (1974). "Effect of Dietary Cadmium and Zinc on Rats Maintained on Diets Low in Copper," *Proc. Nutr. Soc.*, **33,** 15A–17A.

Campbell, J. K., Davies, N. T., and Mills, C. F. (1978). "Interactions of Cadmium, Copper and Zinc in Animals Chronically Exposed to Low Levels of Dietary Cadmium." In M. Kirchgessner, Ed., *Trace Element Metabolism in Man and Animals*—3. ATW Weihenstephan, Germany, pp. 553–556.

Cardin, C. J. and Mason, J., (1975). "Sulphate Transport by Rat Ileum. Effect of Molybdate and Other Anions," *Biochem. Biophys. Acta*, **394,** 46–54.

Cassidy, J. and Eva, J. K. (1958). "Relationship between the Copper and Iron Concentrations in Pigs," *Proc. Nutr. Soc.*, **17,** XXXI.

Chapman, H. L., Jr., and Kidder, R. W. (1964). *Copper and Cobalt for Beef Cattle.* Fla. Agric. Exp. Sta. Bull. 674.

Chase, M. S., Gubler, C. J., Cartwright, G. E., and Wintrobe, M. M. (1952). "Studies on Copper Metabolism. IV: The Influence of Copper on the Absorption of Iron," *J. Biol. Chem.*, **199,** 757–763.

Chu, R. C., Schlicker, S. A., and Cox, D. H. (1970). "A Zinc-Biotin Interrelationship in the Rat," *Nutr. Rep. Int.*, **1,** 11–18.

Cohen, E. and Elvehjem, C. A. (1934). "The Relation of Iron and Copper to the Cytochrome and Oxidase Content of Animal Tissues," *J. Biol. Chem.*, **107,** 97–105.

Compere, R., Burney, A., Francois, E., and Vanuytrecht, S. (1965). "Use of Cu^{64} in the Study of Metabolism of Cu in the Presence of Mo and Inorganic Sulfate," *Bul. Inst. Agronom. Gembloux*, **33,** 202–216.

Cotzias, G. C., Borg, D. C., and Selleck, B. (1962). "Specificity of Zinc Pathway through the Body: Turnover of Zn-65 in the Mouse," *Am. J. Physiol.*, **202,** 359–363.

Cousins, R. J. (1978). "Synthesis and Degradation of Hepatic Zinc-Thionein and Its Role in Zinc Metabolism." In M. Kirchgessner, Ed., *Trace Element Metabolism in Man and Animals*—3. ATW, Weihenstephan, Germany, pp. 57–63.

Cox, D. H. and Harris, D. L. (1960). "Effect of Excess Dietary Zinc on Iron and Copper in the Rat," *J. Nutr.,* **70,** 514–520.

Cox, D. H., Chu, R. C., and Schlicker, S. A. (1969). "Zinc Deficiency in the Maternal Rat during Gestation, and Zinc, Iron, Copper, and Calcium Content and Enzyme Activity in Maternal and Fetal Tissues," *J. Nutr.,* **98,** 449–458.

Crampton, R. F., Matthews, D. M., and Poisner, R. (1965). "Observations on the Mechanism of Absorption of Copper by the Small Intestine," *J. Physiol.,* **178,** 111–126.

Cunningham, I. J. (1950). "Copper and Molybdenum in Relation to the Diseases of Cattle and Sheep in New Zealand." In W. D. Mc Elroy and B. Glass, Eds., *Copper Metabolism.* John Hopkins Press, Baltimore, Md., pp. 246–273.

Cunningham, I. J., Brown, J. M., and Edie, A. E. (1953). "Molybdenum Poisoning of Cattle in the Swan River Valley of Manitoba," *Can. J. Agric. Sci.,* **33,** 254–260.

Dagg, J. H., Jackson, J. M., Curry, B., and Goldberg, A. (1966). "Cytochrome Oxidase in Latent Iron Deficiency (Sideropenia)," *Br. J. Haematol.,* **12,** 331–333.

Dale, S. E., Ewan, R. C., Speer, V. C., and Zimmerman, D. R. (1973). "Copper, Molybdenum and Sulfate Interaction in Young Swine," *J. Anim. Sci.,* **37,** 913–917.

Davies, N. T. (1974). "Recent Studies of Antagonistic Interactions in the Aetology of Trace Element Deficiency and Excess," *Proc. Nutr. Soc.,* **33,** 293–298.

Davis, G. K. (1958). "Mechanism of Trace Mineral Function," *Soil Sci.,* **85,** 59–62.

De Goey, L. W., Wahlstrom, R. C., and Emerick, R. J. (1971). "Studies of High Level Copper Supplementation to Rations for Growing Pigs," *J. Anim. Sci.,* **33,** 52–57.

Dick, A. T. (1953a). "The Effect of Inorganic Sulphate on the Excretion of Molybdenum in the Sheep," *Austr. Vet. J.,* **29,** 18–26.

Dick, A. T. (1953b). "The Control of Copper Storage in the Liver of Sheep by Inorganic Sulfate and Molybdenum," *Austr. Vet. J.,* **29,** 233–239.

Dick, A. T. (1954a). "Studies on the Assimilation and Storage of Copper in Crossbred Sheep," *Austr. J. Agric. Res.,* **5,** 511–544.

Dick, A. T. (1954b). "Preliminary Observations on the Effect of High Intakes of Molybdenum and Inorganic Sulfate on Blood Copper and on Fleece Character in Crossbred Sheep," *Austr. Vet. J.,* **30,** 196–202.

Dick, A. T. (1956a). "Molybdenum and Copper Relationship in Animal Nutrition." In W. D. McElroy and R. Glass, Eds., *Inorganic Nitrogen Metabolism.* John Hopkins Press, Baltimore, Md., pp. 445–473.

Dick, A. T. (1956b). "Molybdenum in Animal Nutrition," *Soil Sci.,* **81,** 229–236.

Dick, A. T. and Bull, L. B. (1945). "Some Preliminary Observations on the Effect of Molybdenum on Copper Metabolism in Herbivorous Animals," *Austr. Vet. J.,* **21,** 70–72.

Dick, A. T., Dewey, D. W., and Gawthorne, J. M. (1975). "Thiomolybdates and the Copper-Molybdenum-Sulfur Interaction in Ruminant Nutrition," *J. Agric. Sci.,* **85,** 567–568.

Dowdy, R. P. and Matrone, G. (1968a). "Copper-Molybdenum Interaction in Sheep and Chicks," *J. Nutr.,* **95,** 191–196.

Dowdy, R. P. and Matrone, G. (1968b). "A Copper-Molybdenum Complex: Its Effects and Movement in the Piglet and Sheep," *J. Nutr.,* **95,** 197–201.

Dowdy, R. P., Kunz, G. A., and Sauberlich, H. E. (1969). "Effect of a Copper-Molybdenum Compound upon Copper Metabolism in the Rat," *J. Nutr.,* **99,** 491–496.

Duncan, G. D., Gray, L. F., and Daniel, L. J. (1953). "Effect of Zinc on Cytochrome Oxidase Activity," *Proc. Soc. Exp. Biol. Med.,* **83,** 625–627.

Dunlap, W. M., James, G. W., and Hume, D. M. (1974). "Anemia and Neutropenia Caused by Copper Deficiency," *Ann. Intern. Med.,* **80,** 470–476.

Elvehjem, C. A. and Sherman, W. C. (1932). "The Action of Copper in Iron Metabolism," *J. Biol. Chem.,* **98,** 309–319.

Evans, G. W., Myron, D. R., and Wiederanders, R. E. (1969). "Effect of Protein Synthesis Inhibitors on Plasma Ceruloplasmin in the Rat," *Am. J. Physiol.,* **216,** 340–342.

Evans, G. W., Majors, P. F., and Cornatzer, W. E. (1970). "Mechanism for Cadmium and Zinc Antagonism of Copper Metabolism," *Biochem. Biophys. Res. Commun.,* **40,** 1142–1148.

Evans, G. W., Grace, C. I., and Hahn, C. (1974). "The Effect of Copper and Cadmium on ^{65}Zn Absorption in Zinc-Deficient and Zinc-Supplemented Rats," *Bioinorg. Chem.,* **3,** 115–120.

Evans, J. L. and Abraham, P. A. (1973). "Anemia, Iron Storage and Ceruloplasmin in Copper Nutrition in the Growing Rat," *J. Nutr.,* **103,** 196–201.

Fell, B. F., Mills, C. F., and Boyne, R. (1965). "Cytochrome-Oxidase Deficiency in the Motor Neurones of Copper-Deficient Lambs: A Histochemical Study," *Res. Vet. Sci.,* **6,** 170–177.

Ferguson, W. S., Lewis, A. H., and Watson, S. J. (1938). "Action of Molybdenum in Nutrition of Milking Cattle," *Nature,* **141,** 553.

Ferguson, W. S., Lewis, A. H., and Watson, S. J. (1943). "The Teart Pastures of Somerset. I: The Cause and Cure of Teartness," *J. Agric. Sci.,* **33,** 44–51.

Forth, W. and Rummel, W. (1973). "Iron Absorption," *Physiol. Rev.,* **53,** 724–792.

Frieden, E. (1970). "Ceruloplasmin, a Link between Copper and Iron Metabolism," *Nutr. Rev.,* **28,** 87–91.

Frieden, E. (1973). "The Ferrous to Ferric Cycles in Iron Metabolism," *Nutr. Rev.,* **31,** 41–44.

Gallagher, C. H. (1957). "The Pathology and Biochemistry of Copper Deficiency," *Austr. Vet. J.,* **33,** 311–317.

Gawthorne, J. M. and Nader, C. J. (1976). "The Effect of Molybdenum on the Conversion of Sulphate to Sulphide and Microbial-Protein-Sulphur in the Rumen of Sheep," *Br. J. Nutr.,* **35,** 11–23.

Gipp, W. F., Pond, W. G., and Smith, S. E. (1967). "Effects of Level of Dietary Copper, Molybdenum, Sulfate and Zinc on Bodyweight Gain, Hemoglobin and Liver Copper Storage of Growing Pigs," *J. Anim. Sci.,* **26,** 727–730.

Gipp, W. F., Pond, W. G., Tasker, J., Van Campen, D., Krook, L., and Visek, W. J.

(1973). "Influence of Level of Dietary Copper on Weight Gain, Hematology, and Liver Copper and Iron Storage of Young Pigs," *J. Nutr.*, **103,** 713–719.

Godwin, K. O., Partick, E. J., and Fuss, C. N. (1978). "Adverse Effects of Copper, and to a Lesser Extent Iron, When Administered to Selenium-Deficient Rats." In M. Kirchgessner, Ed., *Trace Element Metabolism in Man and Animals*—3. ATW Weihenstephan, Germany pp. 185–187.

Gomez-Garcia, G. G. (1972). "Investigation of the Biological Interactions of Copper, Molybdate and Sulfate." Ph.D. thesis, North Carolina State University.

Gomez-Garcia, G. G. and Matrone, G. (1967). "Copper Metabolism in the Early Postnatal Period of the Piglet," *J. Nutr.*, **92,** 237–244.

Goodrich, R. D. and Tillman, A. D. (1966). "Copper, Sulfate and Molybdenum Interrelationships in Sheep," *J. Nutr.*, **90,** 76–80.

Grant-Frost, D. R. and Underwood, E. J. (1958). "Zinc Toxicity in the Rat and Its Interrelationship with Copper," *Austr. J. Exp. Biol. Med. Sci.*, **36,** 339–345.

Grassmann, E. (1976). "Zur Verwertung verschiedener Eisenverbindungen bei der Ratte," *Zbl. Vet. Med.*, **A23,** 292–306.

Grassmann, E. (1977). "Zur Verwertung suboptimaler Zulagen verschiedener Eisenverbindungen durch die Ratte," *Zbl. Vet. Med.*, **A24,** 817–826.

Grassmann, E. and Kirchgessner, M. (1973a). "Zur Eisenverwertung bei unterschiedlicher Kupferversorgung," *Arch. Tierernähr.*, **23,** 261–271.

Grassmann, E. and Kirchgessner, M. (1973b). "Zur Fe- und Cu-Verfügbarkeit im Stoffwechsel bei unterschiedlicher Fe- and Cu-Versorgung," *Z. Tierphysiol., Tierernähr. Futtermittelkde.*, **31,** 113–120.

Grassmann, E. and Kirchgessner, M. (1973c). "Katalase-, Coeruloplasmin-Aktivität und Hämoglobingehalt im Blut wachsender Ratten nach De- und Repletion an Eisen," *Z. Tierphysiol., Tierernähr. Futtermittelkd.*, **31,** 38–44.

Grassmann, E. and Kirchgessner, M. (1973d). "Katalaseaktivität des Blutes von Saugferkeln und Mastkälbern bei mangelnder Eisenversorgung," *Zbl. Vet. Med.*, **A20,** 481–486.

Grassmann, E., Kirchgessner, M., Hampel, G., and Roth-Maier, D. A. (1971). "Untersuchungen zum Kupferstoffwechsel wachsender Küken," *Arch. Geflügelkd.*, **35,** 67–71.

Grassmann, E., von Krziwanek, S., and Kirchgessner, M. (1978). "Zum Einfluss verschiedener Proteingehalte in der Diät auf die Aktivität einiger Metalloenzyme," *Arch. Tierernähr.*, **28,** 451–458.

Gray, L. F. and Daniel, L. J. (1954). "Some Effects of Excess Molybdenum on the Nutrition of the Rat," *J. Nutr.*, **53,** 43–51.

Gray, L. F. and Daniel, L. J. (1964). "Effect of the Copper Status of the Rat on the Copper-Molybdenum-Sulfate Interaction," *J. Nutr.*, **84,** 31–37.

Gray, L. F. and Ellis, G. H. (1950). "Some Interrelationships of Copper, Molybdenum, Zinc, and Lead in the Nutrition of the Rat," *J. Nutr.*, **40,** 441–452.

Gregoriadis, G. and Sourkes, T. L. (1968). "Role of Protein in Removal of Copper from Liver," *Nature*, **218,** 290–291.

Gubler, C. J., Lahey, M. E., Chase, M. S., Cartwright, G. E., and Wintrobe, M. M.

(1952). "Studies on Copper Metabolism. III: The Metabolism of Iron in Copper-Deficient Swine," *Blood,* **7,** 1075–1092.

Gubler, C. J., Lahey, M. E., Chase, M. S., Cartwright, G. E., and Wintrobe, M. M. (1954). "Studies on Copper Metabolism. 12: Influence of Manganese on Metabolism of Copper," *Proc. Soc. Exp. Biol. Med.,* **86,** 223–227.

Gubler, C. J., Cartwright, G. E., and Wintrobe, M. M. (1957). "Studies on Copper Metabolism. XX: Enzyme Activities and Iron Metabolism in Copper and Iron Deficiencies," *J. Biol. Chem.,* **224,** 533–546.

Hallgren, W., Karlsson, K., and Weamby, G. (1954). "Molybden Forgiftuing ("Molybdenosis") hos Notkreatur i Sverige," *Nord. Vet. Med.,* **6,** 469–480.

Halverson, A. W., Phifer, J. H., and Monty, K. J. (1960). "A Mechanism for the Copper-Molybdenum Interrelationship," *J. Nutr.,* **71,** 95–100.

Hampel, G. and Kirchgessner, M. (1970). "Zum Einfluss von Calcium-, Sulfid- und Phytinzulagen auf die Kupferverwertung bei Mastküken," *Arch. Geflügelkd.,* **34,** 181–184.

Hart, E. B., Steenbock, H., Waddell, J., and Elvehjem, C. A. (1928). "Iron in Nutrition. VII: Copper as a Supplement to Iron for Hemoglobin Building in the Rat," *J. Biol. Chem.,* **77,** 797–812.

Hartman, R. H., Matrone, G., and Wise, G. H. (1955). "Effect of High Dietary Manganese on Hemoglobin Formation," *J. Nutr.,* **57,** 429–440.

Hartmans, J. and Bosman, M. S. M. (1970). "Differences in the Copper Status of Grazing and Housed Cattle and Their Biochemical Background." In C. F. Mills, Ed., *Trace Element Metabolism in Animals.* E. and S. Livingstone, London, pp. 362–365.

Hays, V. W. and Kline, R. D. (1969). "Copper-Molybdenum-Sulfate Interrelationships in Growing Pigs," *Feedstuffs,* **41**(44), 18.

Heiseke, D. and Kirchgessner, M. (1977). "Experimenteller Mn-Mangel bei Ratten durch Frühabsetzen," *Z. Tierphysiol., Tierernähr. Futtermittelkd.,* **39,** 197–203.

Heiseke, D. and Kirchgessner, M. (1978). "Eisen- und Zinkgehalte in verschiedenen Organen der Ratte bei Mangan-Mangel," *Zbl. Vet. Med.,* **A, 25,** 307–311.

Henning, A., Anke, M., Groppel, B., and Lüdke, H. (1974). "Secondary Copper Deficiency in Ruminants." In W. G. Hoekstra, J. W. Suttie, H. E. Ganther, and W. Mertz, Eds., *Trace Element Metabolism in Animals*—2. University Park Press, Baltimore, Md., pp. 726–729.

Hill, C. H., Matrone, G., Payne, W. L., and Barber, C. W. (1963). "*In vivo* Interactions of Cadmium with Copper, Zinc and Iron," *J. Nutr.,* **80,** 227–235.

Hill, C. H., Starcher, B., and Matrone, G. (1964). "Mercury and Silver Interrelationships of Cadmium with Copper, Zinc, and Iron," *J. Nutr.,* **80,** 227–235.

Hoefer, J. A., Miller, E. R., Ullrey, D. E., Ritchie, H. D., and Luecke, R. W. (1960). "Interrelationships between Calcium, Zinc, Iron and Copper in Swine Feeding," *J. Anim. Sci.,* **19,** 249–259.

Hogan, K. G., Ris, D. R., and Hutchinson, A. J. (1966). "An Attempt to Produce Copper Deficiency in Sheep by Dosing Molybdate and Sulfate," *N. Z. J. Argic. Res.,* **9,** 691–698.

Huisingh, J. and Matrone, G. (1972). "Copper-Molybdenum Interactions with Sulfate-Reducing System in Rumen Microorganisms," *Proc. Soc. Exp. Biol. Med.*, **139,** 518–521.

Huisingh, J. and Matrone G. (1976). "Interactions of Molybdenum in Animal Nutrition." Chapter 12 in W. R. Chappell and K. K. Petersen, Eds., *Molybdenum in the Environment,* Vol. 1. Marcel Dekker, New York.

Huisingh, J., McNeill, J. J., and Matrone, G. (1974). "Sulfate Reduction by a *Desulfovibrio* Species Isolated from Sheep Rumen," *Appl. Microsc.*, **28,** 489–497.

Huisingh, J., Milholland, D. C., and Matrone, G. (1975). "Effects of Molybdate on Sulfide Production from Methionine and Sulfate by Ruminal Microorganisms of Sheep," *J. Nutr.*, **105,** 1199–1205.

Hutagalung, R. I., Schwarz, F. J., and Kirchgessner, M. (1977). "Intestinale Aufnahme und Durchtritt von Zink nach Zulage von 2,4-Dinitrophenol und Ouabain," *Arch. Tierernähr.*, **27,** 347–357.

Jacobs, E. E., Jacob, M., Sanadi, D. R., and Bradley, L. B. (1956). "Uncoupling of Oxidative Phosphorylation by Cadmium Ion," *J. Biol. Chem.*, **223,** 147–156.

Jensen, L. S., Peterson, R. P., and Falen, L. (1974). "Inducement of Enlarged Hearts and Muscular Dystrophy in Turkey Poults with Dietary Silver," *Poult. Sci.*, **53,** 57–64.

Jeter, M. A. and Davis, G K. (1950). "Influence of Varying Levels of Molybdenum upon Growth and Hemoglobin of Rats," *J. Anim. Sci.*, **9,** 660.

Kinnamon, K. E. (1966). "The Role of Iron in the Copper-Zinc Interrelationship in the Rat," *J. Nutr.*, **90,** 315–322.

Kinnamon, K. E. and Bunce, G. E. (1965). "Effects of Copper, Molybdenum, and Zinc on Zinc-65 Tissue Distribution and Excretion in the Rat," *J. Nutr.*, **86,** 225–230.

Kirchgessner, M. (1959). "Wechselbeziehungen zwischen Spurenelementen in Futtermitteln und tierischen Substanzen sowie Abhängigkeitsverhältnisse zwischen einzelnen Spurenelementen bei der Retention. VI: Mitteilung: Abhängigkeitsverhältnisse zwischen verschiedenen Spurenelementen bei der Retention," *Z. Tierphysiol., Tierernähr. Futtermittelkd.*, **14,** 278–283.

Kirchgessner, M. and Grassmann, E. (1970). "Untersuchungen zur Kupferverfügbarkeit mit Coeruloplasmin als Testenzym," *Z. Tierphysiol., Tierernähr. Futtermittelkd.*, **26,** 340–344.

Kirchgessner, M. and Heiseke, D. (1977). "Wechselbeziehungen von Mangan zu Kupfer in verschiedenen Organen der Ratte bei ausreichender und mangelnder Mn-Versorgung," *Z. Tierphysiol., Tierernähr. Futtermittelkd.*, **39,** 272–276.

Kirchgessner, M. and Weigand, E. (1975). "Zur Wirksamkeit von einmalig oral zu verabreichenden Eisen-Depot-Tabletten bei der Eisenmangelanämie von Saugferkeln," *Z. Tierphysiol., Tierernähr. Futtermittelkd.*, **34,** 205–211.

Kirchgessner, M. and Weser, U. (1963). "Spurenelemente im Stoffwechsel wachsender Schweine bei unterschiedlichen Kupferzulagen," *Z. Tierphysiol., Tierernähr. Futtermittelkd.*, **18,** 181–186.

Kirchgessner, M., Munz, W., and Oelschläger, W. (1960). "Der Einfluss einer $CaCO_3$-Zulage auf die Retention von Mengen- und Spurenelementen bei wachsenden Schweinen," *Arch. Tierernähr.*, **10,** 1–19.

Kirchgessner, M., Grassmann, E., Krippl, J., and Müller, H. L. (1971). "Zum Einfluss einmaliger und kontinuierlicher Fe-Zulagen in der Kälbermast auf ernährungsphysiologische Wirksamkeit und Fleischfarbe," *Züchtungskunde,* **43,** 336–345.

Kirchgessner, M., Beyer, M. G. and Steinhart, H. (1976). "Activation of Pepsin (EC 3.4.4.1) by Heavy-Metal Ions Including a Contribution to the Mode of Action of Copper Sulphate in Pig Nutrition," *Br. J. Nutr.,* **36,** 15–22.

Kirchgessner, M., von Krziwanek, S., and Grassmann, E. (1977). "Zum Einfluss der Proteinqualität in der Diät auf die Aktivitäten einiger Metallo-Enzyme," *Z. Tierphysiol., Tierernähr. Futtermittelkd.,* **38,** 273–282.

Kirchgessner, M., Schwarz, F. J., Roth, H.-P., and Schwarz, W. A. (1978). "Wechselwirkungen zwischen den Spurenelementen Zink, Kupfer und Eisen nach Zinkdepletion und -repletion von Milchkühen," *Arch. Tierernähr.,* **28,** 723–733.

Klauder, D. S., Murthy L., and Petering, H. G. (1973). "Effect of Dietary Intake of Lead Acetate on Copper Metabolism in Male Rats." In D. D. Hemphill, Ed., *Trace Substances in Environmental Health,* Vol. 6, University of Missouri, Columbia, pp. 131–135.

Kline, R. D., Hays, V. W., and Cromwell, G. L. (1971). "Effects of Copper, Molybdenum and Sulfate on Performance, Hematology and Copper Stores of Pigs and Lambs," *J. Anim. Sci.,* **33,** 771–779.

Kline, R. D., Corzo, M. A., Hays, V. W., and Cromwell, G. L. (1973). "Related Effects of Copper, Molybdenum and Sulfide on Performance, Hematology and Copper Stores of Growing Pigs," *J. Anim. Sci.,* **37,** 936–941.

Kulwich, R., Hansard, S. L., Comar, C. L., and Davis, G. K. (1951). "The Effect of Molybdenum on the Metabolism of Copper in Swine and Rats," *J. Anim. Sci.,* **10,** 1052.

Kulwich, R., Hansard, S. L., Comar, C. L., and Davis, G. K. (1953). "Copper, Molybdenum and Zinc Interrelationships in Rats and Swine," *Proc. Soc. Exp. Biol. Med.,* **84,** 487–491.

Lahey, M. E., Gubler, C. J., Chase, M. S., Cartwright, G. E., and Wintrobe, M. M. (1952). "Studies on Copper Metabolism. II: Hematologic Manifestations of Copper-Deficiency in Swine," *Blood,* **7,** 1053–1074.

Lantzsch, H.-J. (1973). "Alimentär bedingte Zink-Toxizität," *Übers. Tierernähr.,* **1,** 57–88.

Lee, D., Jr., and Matrone, G. (1969). "Iron and Copper Effects on Serum Ceruloplasminactivity of Rat with Zinc-Induced Copper Deficiency," *Proc. Soc. Exp. Biol. Med.,* **130,** 1190–1194.

Lee, G. R., Nacht, S., Lukens, J. N., and Cartwright, G. E. (1968a). "Iron Metabolism in Copper-Deficient Swine," *J. Clin. Invest.,* **47,** 2058–2069.

Lee, G. R., Cartwright, G. E., and Wintrobe, M. M. (1968b). "Heme biosynthesis in Copper-Deficient Swine," *Proc. Soc. Exp. Biol. Med.,* **127,** 977–981.

Lee, G. R., Ragan, H. A., Nacht, S., and Cartwright, G. E. (1969). "The Effect of Ceruloplasmin on Plasma Iron in Copper-Deficient Swine," *Clin. Res.,* **17,** 152–159.

Magee, A. C. and Matrone, G. (1960). "Studies on Growth, Copper Metabolism and Iron Metabolism of Rats Fed High Levels of Zinc," *J. Nutr.,* **72,** 233–242.

Magee, A. C., Jackson, M. Y., and Wade, W. (1975). "Interrelationships between Copper, Iron and Zinc in Young Rats Fed Zinc-Deficient Diets," *Fed. Proc.*, **34,** 906.

Marcilese, N. A., Ammerman, C. B., Valsecchi, R. M., Dunavant, B. G., and Davis, G. K. (1969). "Effect of Dietary Molybdenum and Sulfate upon Copper Metabolism in Sheep," *J. Nutr.*, **99,** 177–183.

Marston, H. R. (1952). "Cobalt, Copper and Molybdenum in the Nutrition of Animals and Plants," *Physiol. Res.*, **32,** 66–121.

Marston, H. R., Allen, S. H., and Swaby, S. L. (1971). "Iron Metabolism in Copper-Deficient Rats," *Br. J. Nutr.*, **25,** 15–30.

Matrone, G. (1960). "Interrelationships of Iron and Copper in the Nutrition and Metabolism of Animals," *Fed. Proc.*, **19,** 659–665.

Matrone, G., Hartman, R. H., and Clawson, A. J. (1959). "Studies of a Manganese-Iron Antagonism in the Nutrition of Rabbits and Baby Pigs," *J. Nutr.*, **67,** 309–317.

Miller, H. (1958). "The Relationship between Catalase and Hemoglobin in Human Blood," *Biochem. J.*, **68,** 275–282.

Mills, C. F. (1960). "Comparative Studies of Copper, Molybdenum and Sulphur Metabolism in the Ruminant and the Rat," *Proc. Nutr. Soc.*, **19,** 162–169.

Mills, C. F. (1974). "Trace Element Interactions: Effect of Dietary Composition on the Development of Imbalance and Toxicity." In W. G. Hoekstra, J. W. Suttie, H. E. Ganther, and W. Mertz, Eds., *Trace Element Metabolism in Animals*—2. University Park Press, Baltimore, Md., pp. 79–90.

Mills, C. F. and Fell, B. F. (1960). "Demyelination in Lambs Born of Ewes Maintained on High Intakes of Sulphate and Molybdate," *Nature*, **185,** 20–22.

Mills, C. F., Williams, R. B., and Poole, D. B. R. (1963). "Tissue Cytochrome-Oxidase in Copper-Deficient Cattle," *Biochem. J.*, **87,** 10–11.

Mills, C. F., Bremner, I., El-Gallad, T. T., Dalgarno, A. C., and Young, B. W. (1978). "Mechanism of the Molybdenum-Sulphur Antagonism of Copper Utilisation by Ruminants." In M. Kirchgessner, Ed., *Trace Element Metabolism in Man and Animals*—3. ATW, Weihenstephan, Germany, pp. 150–158.

Milne, D. B. and Weswig, P. H. (1968). "Effect of Supplementary Copper on Blood and Liver Copper-Containing Fractions in Rats," *J. Nutr.*, **95,** 429–433.

Miltmore, J. E. and Mason, J. L. (1971). "Copper to Molybdenum Ratio and Copper Concentration in Ruminant Feeds," *Can. J. Anim. Sci.*, **51,** 193–200.

Moses, H. A. and Parker, H. E. (1964). "Influence of Dietary Zinc and Age on the Mineral Content of Rat Tissues," *Fed. Proc.*, **23,** 132.

Murthy, L., Klevay, L. M., and Petering, H. G. (1974). "Interrelationships of Zinc and Copper Nutriture in the Rat," *J. Nutr.*, **104,** 1458–1465.

Mylrea, P. J. (1958). "Copper-Molybdenum-Sulfate-Manganese Interaction and the Copper Status of Cattle," *Austr. J. Agric. Res.*, **9,** 373–384.

Nielsen, F. H., Sunde, M. L., and Hoekstra, W. G. (1966). "Effect of Dietary Amino Acid Source on the Zinc-Deficiency Syndrome in the Chick," *J. Nutr.*, **89,** 24–34.

Oberleas, D., Muhrer, M. E., and O'Dell, B. L. (1966). "Dietary Metal-Complexing Agents and Zinc Availability in the Rat," *J. Nutr.,* **90,** 56–62.

O'Dell, B. L. and Campbell, B. J. (1971). "Trace Elements Metabolism and Metabolic Function." In M. Florkin and E. H. Stotz, Eds., *Metabolism of Vitamins and Trace Elements,* Vol. 21. Elsevier, Amsterdam, pp. 203–217.

O'Hara, P. J., Newman, A. P., and Jackson, R. (1960). "Parakeratosis and Copper Poisoning in Pigs Fed a Copper Supplement," *Austr. Vet. J.,* **36,** 225–229.

Osaki, S. and Johnson, D. A. (1969). "Mobilization of Liver Iron by Ferroxidase (Ceruloplasmin)," *J. Biol. Chem.,* **244,** 5757–5758.

Osaki, S., Johnson, D. A., and Frieden, E. (1966). "The Possible Significance of the Ferrous Oxidase Activity of Ceruloplasmin in Normal Human Serum," *J. Biol. Chem.,* **241,** 2746–2751.

Owen, C. A., Dickson, E. R., Goldstein, N. P., Baggenstoss, A. W., and McCall, J. T. (1977). "Hepatic Subcellular Distribution of Copper in Primary Biliary Cirrhosis," *Mayo Clin. Proc.,* **52,** 73–80.

Pallauf, J. and Kirchgessner, M. (1974). "Effekt verschiedener Mn- bzw. Cu-Zulagen bei mangelnder Zinkversorgung," *Zbl. Vet. Med.,* **A21,** 562–571.

Pearson, W. N., Schwink, T., and Reich, M. (1966). "*In vitro* Studies of Zinc Absorption in the Rat." In A. S. Prasad, Ed., *Zinc Metabolism.* Thomas, Springfield, Ill., pp. 239–249.

Petering, H. G. (1974). "The Effect of Cadmium and Lead on Copper and Zinc Metabolism." In W. G. Hoekstra, J. W. Suttie, H. E. Ganther, and W. Mertz, Eds., *Trace Element Metabolism in Animals*—2. University Press, Baltimore, Md., pp. 311–325.

Petering, H. G., Johnson, M. A., and Horwitz, J. P. (1971). "Studies of Zinc Metabolism in the Rat," *Arch. Environ. Health,* **23,** 93–101.

Peterson, R. P. and Jensen, L. S. (1975). "Interrelationships of Dietary Silver with Copper in the Chick," *Poult. Sci.,* **54,** 771–775.

Pierson, R. E. and Aanes, W. A. (1958). "Treatment of Chronic Copper Poisoning in Sheep," *J. Am. Vet. Med. Assoc.,* **133,** 307–310.

Poole, D. B. R. (1970). "Cytochrome-Oxidase in Induced Hypocuprosis." In C. F. Mills, Ed., *Trace Element Metabolism in Animals.* Livingstone, Edinburgh, pp. 465–469.

Prasad, A. S., Oberleas, D., Wolf, P., and Horwitz, J. P. (1967). "Studies on Zinc Deficiency: Changes in Trace Elements and Enzyme Activities in Tissues of Zinc-Deficient Rats," *J. Clin. Invest.,* **46,** 549–557.

Prasad, A. S., Oberleas, D., Wolf, P., Horwitz, J. P., Miller, E. R., and Luecke, R. W. (1969). "Changes in Trace Elements and Enzyme Activities in Tissues of Zinc-Deficient Pigs," *Am. J. Clin. Nutr.,* **22,** 628–637.

Prinz, R. and Weser, U. (1975). "A Naturally Occurring Cu-Thionein in *Saccharomyces cerevisiae,*" *Hoppe-Seylers Z. Physiol. Chem.,* **356,** 767–776.

Ragan, H. A., Nacht, S., Lee, G. R., Bishop, C. R., and Cartwright, G. E. (1969). "Effect of Ceruloplasmin on Plasma Iron in Copper-Deficient Swine," *Am. J. Physiol.,* **217,** 1320–1323.

Reinhold, J. G. (1975). "Trace Element—a Selective Survey," *Clin. Chem.*, **21,** 476–500.

Reinhold, J. G., Kfoury, G. A., and Thomas, T. A. (1967). "Zinc, Copper and Iron Concentrations in Hair and Other Tissues: Effects of Low Zinc and Low Protein Intakes in Rats," *J. Nutr.*, **92,** 173–182.

Richards, M. P. and Cousins, R. J. (1976). "Metallothionein and Its Relationship to the Metabolism of Dietary Zinc in Rats," *J. Nutr.*, **106,** 1591–1599.

Ritchie, H. D., Miller, E. R., Luecke, R. W., Ullrey, D. E., and Hoefer, J. A. (1961). "Copper and Zinc Interrelationships in Swine Feeding," *J. Anim. Sci.*, **20,** 950.

Ritchie, H. D., Luecke, R. W., Baltzer, B. V., Miller, E. R., Ullrey, D. E., and Hoefer, J. A. (1963). "Copper and Zinc Interrelationships in the Pig," *J. Nutr.*, **79,** 117–123.

Roeser, H. P., Lee, G. R., Nacht, S., and Cartwright, G. E. (1970). "The Role of Ceruloplasmin in Iron Metabolism," *J. Clin. Invest.*, **49,** 2408–2417.

Roth, H.-P. and Kirchgessner, M. (1977). "Zum Gehalt von Zink, Kupfer, Eisen, Mangan und Calcium in Knochen und Lebern von an Zink depletierter und repletierter Ratten," *Zbl. Vet. Med.*, **A24,** 177–188.

Rupp, H. and Weser, U. (1978). "Circular Dichroism of Metallothioneins," *Biochim. Biophys. Acta,* **533,** 209–217.

Ruszczyc, Z. and Gacek, K. (1963). "The Concentration of Copper and Iron in Tissues and Organs of Fattening Pigs Receiving High-Copper Supplements in the Ration," *Rocz. Nauk Roln.*, **B81,** 563–568.

Schlicker, S. and Cox, D. H. (1968). "Maternal Dietary Zinc, and Development and Zinc, Iron, and Copper Content of the Rat Fetus," *J. Nutr.*, **95,** 287–294.

Schnegg, A. and Kirchgessner, M. (1975a). "Zur Essentialität von Nickel für das tierische Wachstum," *Z. Tierphysiol., Tierernähr. Futtermittelkd.*, **36,** 63–74.

Schnegg, A. and Kirchgessner, M. (1975b). "Veränderungen des Hämoglobingehaltes, der Erythrocytenzahl und des Hämatokrits bei Nickelmangel," *Nutr. Metabol.*, **19,** 268–278.

Schnegg, A. and Kirchgessner, M. (1976a). "Zur Interaktion von Nickel mit Eisen, Kupfer und Zink," *Arch. Tierernähr.*, **26,** 543–549.

Schnegg, A. and Kirchgessner, M. (1976b). "Zur Absorption und Verfügbarkeit von Eisen bei Nickelmangel," *Int. J. Vitam. Nutr. Res.*, **46,** 96–99.

Schultze, M. O. (1939). "The Effect of Deficiencies in Copper and Iron on the Cytochrome Oxidase of Rat Tissues," *J. Biol. Chem.*, **129,** 729–737.

Schultze, M. O. (1941). "The Relation of Copper to Cytochrome Oxidase and Hematopoietic Activity of the Bone Marrow of Rats," *J. Biol. Chem.*, **138,** 219–224.

Schultze, M. O. and Kuiken, K. A. (1941). "The Effect of Deficiencies in Copper and Iron on the Catalase Activity of Rat Tissues," *J. Biol. Chem.*, **137,** 727–734.

Schwarz, F. J. and Kirchgessner, M. (1973). "Intestinale Cu-Absorption *in vitro* nach Fe- oder Zn-Depletion," *Z. Tierphysiol., Tierernähr. Futtermittelkd.*, **31,** 91–98.

Schwarz, F. J. and Kirchgessner, M. (1974a). "Absorption von Zink-65 und Kupfer-64 im Zinkmangel," *Int. J. Vitam. Nutr. Res.*, **44,** 258–266.

Schwarz, F. J. and Kirchgessner, M. (1974b). "Wechselwirkungen bei der intestinalen Absorption von ^{64}Cu, ^{65}Zn und ^{59}Fe nach Cu-, Zn- oder Fe-Depletion," *Int. J. Vitam. Nutr. Res.*, **44,** 116–126.

Schwarz, F. J., and Kirchgessner, M. (1979a), "Kupfer-, Zink-, Eisen- und Mangankonzentrationen im Serum, in Knochen und der Leber nach Kupferdepletion," *Zbl. Vet. Med.*, **A.** (in press).

Schwarz, F. J. and Kirchgessner, M. (1979b). "Veränderungen der Kupfer- und Zink-Absorption und -Exkretion bei mangelnder Kupferversorgung," *Z. Tierphysiol.*, Tierernähr. Futtermittelkd. (in press).

Siegel, L. M. and Monty, K. J. (1961). "A Mechanism for the Copper-Molybdenum Interrelationships. II: Response of Liver Sulfide Oxidase Activity to Nutritional Factors," *J. Nutr.*, **74,** 167–170.

Smith, S. E. and Larson, E. J. (1946). "Zinc Toxicity in Rats: Antagonistic Effects of Copper and Liver," *J. Biol. Chem.*, **163,** 29–38.

Sourkes, T. L., Lloyd, K., and Birnbaum, H. (1968). "Inverse Relationship of Hepatic Copper and Iron Concentrations in Rat Fed Deficient Diets," *Can. J. Biochem.*, **46,** 267–271.

Spais, A. G., Lazaridis, T. K., and Agiannidis, A. K. (1968). "Studies on Sulphur Metabolism in Sheep in Association with Copper Deficiency," *Res. Vet. Sci.*, **9,** 337–344.

Spears, J. W., Hatfield, E. E., and Forbes, R. M. (1977). "Nickel as a Substitute for Copper and Zinc," *Fed. Proc.*, **36,** 1106.

Standish, J. F., Ammerman, C. B., Simpson, C. F., Neal, F. C., and Palmer, A. Z. (1969). "Influence of Graded Levels of Dietary Iron as Ferrous Sulfate on Performance and Tissue Mineral Composition of Steers," *J. Anim. Sci.*, **29,** 496–503.

Standish, J. F., Ammerman, C. B., Wallace, H. D., and Gombs, G. E. (1975). "Effect of High Dietary Molybdenum and Sulfate on Plasma Copper Clearance and Tissue Minerals in Growing Swine," *J. Anim. Sci.*, **40,** 509–513.

Starcher, B. C. (1969). "Studies on the Mechanism of Copper Absorption in the Chick," *J. Nutr.*, **97,** 321–326.

Starcher, B. C., Matrone, G., and Hill, C. H. (1964). "Specific Activity of Ceruloplasmin in Sheep, Pigs, and Chicks," *Fed. Proc.*, **23,** I, 133.

Steinhart, H., Kirchgessner, M., and Beyer, M. G. (1976). "Interaktionen zwischen einigen Übergangselementen in ihrem Einfluss auf die Pepsinaktivität," *Arch. Tierernähr.*, **26,** 629–635.

Suttle, N. F. (1975). "The Role of Organic Sulphur in the Copper-Molybdenum-S Interrelationship in Ruminant Nutrition," *Br. J. Nutr.*, **34,** 411–420.

Suttle, N. F. and Mills, C. F. (1966a). "Studies on the Toxicity of Copper to Pigs. 1: Effects of Oral Supplements to Zinc and Iron Salts at the Development of Copper Toxicosis," *Br. J. Nutr.*, **20,** 135–148.

Suttle, N. F. and Mills, C. F. (1966b). "Studies on the Toxicity of Copper to Pigs. 2: Effect of Protein Source and Other Dietary Components on the Response to High and Moderate Intakes of Copper," *Br. J. Nutr.*, **20,** 149–161.

Suttle, N. F. and Field, A. C. (1968a). "Effect of Intake of Copper, Molybdenum and Sulphate on Copper Metabolism in Sheep. 1: Clinical Condition and Distribution of Copper in Blood of the Pregnant Ewe," *J. Comp. Pathol.*, **78,** 351–362.

Suttle, N. F. and Field, A. C. (1968b). "Effect of Intake of Copper, Molybdenum and Sulphate on Copper Metabolism in Sheep. 2: Copper Status of the Newborn Lambs," *J. Comp. Pathol.*, **78**, 363–370.

Sutton, W. R. and Nelson, V. E. (1937). "Studies on Zinc," *Proc. Soc. Exp. Biol. Med.*, **36**, 211–213.

Symes, A. L., Sourkes, T. L., Youdim, M. B. H., Gregoriadis, G., and Birnbaum, H. (1969). "Decreased Monoamine Oxidase Activity in Liver of Iron-Deficient Rats," *Can. J. Biochem.*, **47**, 999–1002.

Theorell, H., Beznak, M., Bonnichsen, R., Paul, K. G., and Akeson, A. (1951). "On the Distribution of Injected Radioactive Iron in Guinea Pigs and Its Rate of Appearance in Some Hemoproteins and Ferrities," *Acta Chem. Scand.*, **5**, 445–475.

Thompson, H. J. and Evans, J. L. (1977b). "Influence of Level of Dietary Iron on the Distribution of Copper and Iron in the Rat," *Nutr. Rep. Int.*, **15**, 279–286.

Thompson, H. J. and Evans, J. L. (1977a). "Influence of Dietary Copper on Assimilation and Distribution of Copper and Iron into Metabolic Compartments," *Nutr. Rep. Int.*, **15**, 287–293.

Underwood, E. J. (1977). *Trace Elements in Human and Animal Nutrition.* Academic Press, New York.

Vallee, B. L. and Coleman, J. E. (1964). "Metal Coordination and Enzyme Action. In M. Florkin and E. H. Stotz, Eds., *Comprehensive Biochemistry,* Vol. 12. Elsevier, Amsterdam, pp. 165–235.

Van Campen, D. R. (1966). "Effects of Zinc, Cadmium, Silver and Mercury on the Absorption and Distribution of Copper-64 in Rats," *J. Nutr.*, **88**, 125–130.

Van Campen, D. R. (1969). "Copper Interference with the Intestinal Absorption of Zinc-65 by Rats," *J. Nutr.*, **97**, 104–108.

Van Campen, D. R. (1971). "Absorption of Copper from the Gastrointestinal Tract." In S. C. Skoryna and D. Waldron-Edward, Eds., *Intestinal Absorption of Metal Ions, Trace Elements and Radionuclides.* Pergamon Press, Oxford, pp. 211–227.

Van Campen, D. R. and Scaife, P. U. (1967). "Zinc Interference with Copper Absorption in Rats," *J. Nutr.*, **91**, 473–476.

Van Reen, R. (1953). "Effects of Excessive Dietary Zinc in the Rat and the Interrelationship with Copper," *Arch. Biochem. Biophys.*, **46**, 337–344.

Van Reen, R. (1954). "The Influence of Excessive Dietary Molybdenum on Rat Liver Enzymes," *Arch. Biochem. Biophys.*, **53**, 77–84.

Vohra, P. and Heil, J. R. (1969). "Dietary Interactions between Zn, Mn and Cu for Turkey Poults," *Poult. Sci.*, **48**, 1686–1691.

Wallace, H. D., McCall, J. T., Bass, B., and Combs, G. E. (1960). "High Level Copper for Growing-Finishing Swine," *J. Anim. Sci.*, **19**, 1153–1163.

Whanger, P. D. and Weswig, P. H. (1970). "Effect of Some Copper Antagonists on Induction of Ceruloplasmin in the Rat," *J. Nutr.*, **100**, 341–348.

Wynne, K. N. and McClymont, G. L. (1956). "Copper-Molybdenum-Sulphate Interaction in Induction of Ovine Hypocupraemia and Hypocuprosis," *Austr. J. Agric. Res.*, **7**, 45–56.

INDEX

ENERGY UTILIZATION AND ENVIRONMENTAL HEALTH
Richard A. Wadden, Editor

METHODOLOGICAL APPROACHES TO DERIVING ENVIRONMENTAL AND OCCUPATIONAL HEALTH STANDARDS
Edward J. Calabrese

FOOD, CLIMATE AND MAN
Margaret R. Biswas and Asit K. Biswas, Editors

CHEMICAL CONCEPTS IN POLLUTANT BEHAVIOR
Ian J. Tinsley

RESOURCE RECOVERY AND RECYCLING
A. F. M. Barton

QUANTITATIVE TOXICOLOGY
V. A. Filov, A. A. Golubev, E. I. Liublina, and N. A. Tolokontsev

BREEDING PLANTS RESISTANT TO INSECTS
Fowden G. Maxwell and Peter Jennings, Editors

ATMOSPHERIC MOTION AND AIR POLLUTION
Richard A. Dobbins

INDUSTRIAL POLLUTION CONTROL—VOLUME I: AGRO-INDUSTRIES
E. Joe Middlebrooks

COPPER IN THE ENVIRONMENT, Parts I and II
Jerome O. Nriagu, Editor